AIR POLLUTION

THIRD EDITION

VOLUME VI

Supplement to Air Pollutants, Their Transformation, Transport, and Effects

Edited by

Arthur C. Stern
Department of Environmental Sciences and Engineering
School of Public Health
University of North Carolina at Chapel Hill
Chapel Hill, North Carolina

1986

ACADEMIC PRESS, INC.
Harcourt Brace Jovanovich, Publishers
Orlando San Diego New York Austin
Boston London Sydney Tokyo Toronto

ACADEMIC PRESS, INC.
Orlando, Florida 32887

United Kingdom Edition published by
ACADEMIC PRESS INC. (LONDON) LTD.
24–28 Oval Road, London NW1 7DX

Library of Congress Cataloging in Publication Data
(Revised for v. 6-8)

Stern, Arthur Cecil.
Air pollution.

(Environmental sciences)
Includes bibliographical references and indexes.
Contents: – [etc.] – v. 6. Supplement to Air pollutants, their transformation, transport, and effects – v. 7. Supplement Measurements, monitoring, surveillance, and engineering control – v. 8. Supplement to managing air quality.
1. Air–Pollution–Collected works. 2. Air pollution. I. Title. II. Series.
TD883.S83 1976 363.7'392 76-8256
ISBN 0–12–666606–7 (v. 6)

PRINTED IN THE UNITED STATES OF AMERICA

86 87 88 89 9 8 7 6 5 4 3 2 1

To Benjamin Samuel, Daniel Arthur, Diana Jocelyn,
Lara Helen, and Sophie Katherine

Contents

Contributors

Numbers in parentheses indicate the pages on which the authors' contributions begin.

Norbert S. Baer (145), Conservation Center, Institute of Fine Arts, New York University, New York, New York 10021

Donald L. Fox (61), School of Public Health, Department of Environmental Sciences and Engineering, University of North Carolina at Chapel Hill, Chapel Hill, North Carolina 27514

John R. Goldsmith (391), Ben Gurion University of the Negev, Beer Sheva, Israel

Magda Havas (351), Institute of Environmental Studies, University of Toronto, Toronto, Ontario, Canada M5S 1A4

Allen S. Heagle (247), United States Department of Agriculture/Agricultural Research Service, North Carolina State University, Raleigh, North Carolina 27606

Walter W. Heck (247), United States Department of Agriculture, Agricultural Research Service, Botany Department, North Carolina State University, Raleigh, North Carolina 27606

Elmer Robinson (145), Mauna Loa Observatory, National Oceanic and Atmospheric Administration, Hilo, Hawaii 96720

David S. Shriner (247), Oak Ridge National Laboratory, Oak Ridge, Tennessee 37831

D. Bruce Turner (95), United States Environmental Protection Agency, Research Triangle Park, North Carolina 27709

Paul Urone (1), Environmental Engineering Department, University of Florida, Gainesville, Florida 32611

John E. Yocom (145), TRC Environmental Consultants, Inc., East Hartford, Connecticut 06108

Preface

There have been a great number of significant developments in the science, technology, and public policy of air pollution and its control in the decade since 1975. Many aspects of this problem that were of minor concern are now of major concern. These include acidic deposition, asbestos, carbon dioxide, indoor air pollution, lead, long-range transport, emissions from nuclear accidents, nonionizing radiation, stratospheric ozone, toxic substances, visibility, and risk assessment and management.

These aspects need to be addressed and the material in the first five volumes of the 1976–1977 third edition brought up to date to maintain the viability of this treatise.

Since the material in the five volumes of the third edition and the three volumes of the second edition is still basic and valid, we rejected the option of publishing a fourth edition in favor of these supplement volumes. In publishing this supplement (which we consider to be a part of the third edition), we have therefore presumed that its users will have access to either the first five volumes of the third edition or the three volumes of the second edition, if they need to tie material in the supplement to the background of these earlier volumes.

Our instructions to contributors were to not repeat in this supplement material (text, tables, figures, or references) already in the published volumes, and to limit their presentations to material and issues that have appeared or developed since the late 1970s, but also to include significant references of prior dates that did not appear in the first five volumes. This should not greatly disservice those who hold the second, but not the third, edition because the combination of the second edition and this supplement makes a viable whole.

As noted in the preface to the earlier volumes, this treatise is intended for professionals in the sciences, engineering, meteorology, biology, medicine, law and public administration; and it is assumed that the reader has an adequate background in his or her profession.

The first five volumes of the third edition have 15 parts. We originally intended to organize this supplement into 15 chapters, each to update its corresponding part. As this supplement developed, however, it became apparent that some of these chapters would either be quite long or would have to be split into more than one chapter, or would bring together in one chapter subject matter better presented as separate chapters. There was no part or chapter in the first five volumes on effects of air pollution on the aquatic environment. Such a new chapter has been added to this supplement. Chapter 7 of Volume V was the only chapter not updated (because of its irrelevance).

Thus this supplement has 21 chapters. Because of their length, it has been necessary to print them in three volumes with Chapters 1 through 7 in Volume VI *(Air Pollutants, Their Transformation, Transport, and Effects);* Chapters 1 through 9 in Volume VII *(Measurement, Monitoring, Surveillance, and Engineering Control of Air Pollution);* and Chapters 1 through 5 in Volume VIII *(Management of Air Quality)*. Each volume has its own subject index. The correspondence between the chapters of the first five volumes of the third edition, the three volumes of the second edition, and the three volumes and 21 chapters of this supplement is shown below.

The draftman for this book was Peter Bedick, University of North Carolina at Chapel Hill. I wish to thank my secretary, Delores Plummer, for her assistance as well.

Arthur C. Stern

Supplement		*Third edition*		*Second edition*	
Volume	*Chapter*	*Volume*	*Chapter*	*Volume*	*Chapter*
VI	1	I	1–5	I	1–5 (Vol. II, Ch. 25)
VI	2	I	6–7	I	6
VI	3	I	8–12	I	7–10 (Vol. II, Ch. 24)
VI	4	II	1–3	I	11;15
VI	5	II	4	I	12
VI	6	II	5	I	13 (Vol. III, Ch. 49)
VI	7	II	6–7	I	14
VII	1	II	8	III	54
VII	2	III	1–8	II	16–23;27
VII	3	III	9–13	II	26;31
VII	4	III	14–17	II	28–30
VII	5	IV	1	III	41
VII	6	IV	2–4	III	42–44;46
VII	7	IV	5–9	III	45–48
VII	8	IV	10–14	III	32–33
VII	9	IV	15–21	III	34–40
VIII	1	V	1–6	III	50;52
VIII	2	V	8	III	50
VIII	3	V	9	III	50;52
VIII	4	V	10	III	53
VIII	5	V	11–13	III	51

Contents of Other Supplement Volumes[1]

VOLUME VII SUPPLEMENT TO MEASUREMENTS, MONITORING, SURVEILLANCE, AND ENGINEERING CONTROL

[1] Contents of Volumes I–V are in those volumes.

VOLUME VIII SUPPLEMENT TO MANAGEMENT OF AIR QUALITY

1

The Pollutants

Paul Urone

Environmental Engineering Department
University of Florida
Gainesville, Florida

I. Classification and Extent of Air Pollution Problems

The problems discussed in Vol. I, Chapter 1 (1976) are still with us. However, a group of new problems have emerged over the past decade. They are, in alphabetical order, acidic deposition (acid rain), asbestos, carbon dioxide, indoor pollution, lead, long-range transport, nuclear accidents, nonionizing radiation, risk assessment and management, stratospheric ozone, toxic substances, and visibility.

A. Acidic Deposition

Deposition is called acidic when its hydrogen ion content, measured as pH, indicates an acidity greater than that which would result from a simple equilibrium with atmospheric carbon dioxide. The pH scale indicates the negative logarithm of the hydrogen ion (H^+) concentration of a solution: the smaller the number, the greater the acidity (Fig. 1) *(1)*. Atmospheric carbon dioxide dissolves in raindrops to form carbonic acid equilibrating at a pH of 5.6. However, other acidic and alkaline substances in air are absorbed and contribute to a resultant impact of both wet and dry deposition upon water bodies and surface soils (Fig. 2) *(2)*. Acidity problems develop because there are an excess of acidic gases and particles in the atmosphere and insufficient neutralizing alkaline substances in the air and the lakes, streams, and soils on which deposition occurs. This effect has been most evident in Scandinavia, eastern Canada, and northeastern United States. The

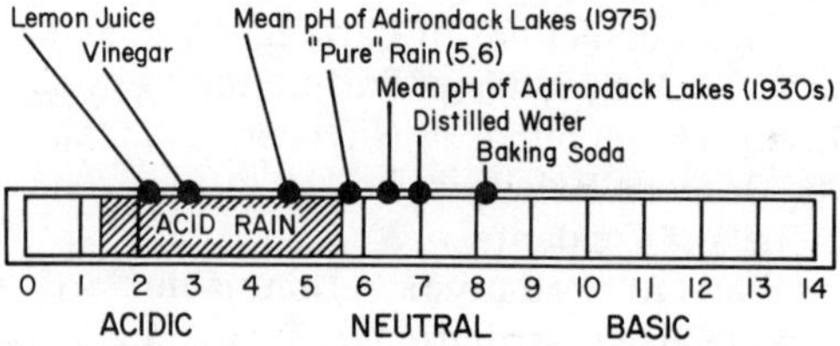

Figure 1. The pH scale as a measure of hydrogen ion concentration. The pH of common substances is shown with various values along the scale *(1)*.

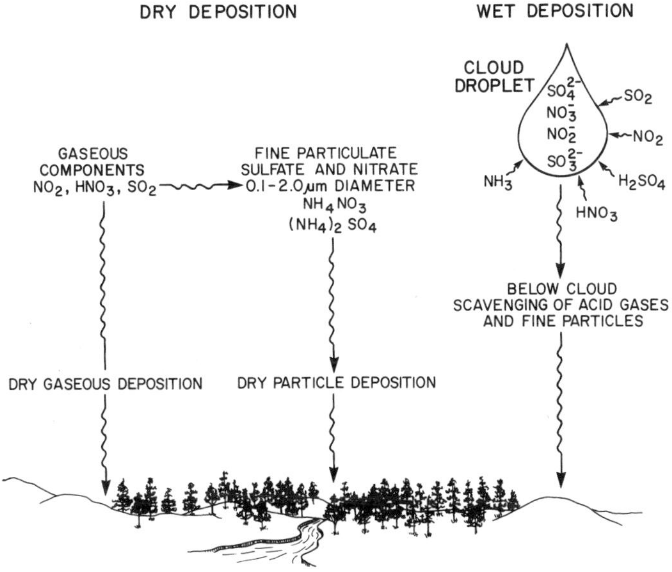

Figure 2. Atmospheric processes involved in acidic deposition. The two principal deposition pathways are dry deposition (nonrain events) and wet deposition (rain events) [From Stern *et al.* *(2)*. Reproduced with permission].

adverse effects of excess acidity include loss of aquatic life in poorly buffered lakes and damage to impacted forests.

Cowling *(3)* has written a history of acidic precipitation, showing recognition of the problem in the 19th and earlier centuries. Particularly concerted investigations have been conducted in the decade before this chapter was written (1985) *(3–23)*. There are sincere disagreements on the extent of the problem, the exact mechanisms operating, and the kinds and amounts of remedial actions that need to be taken. Many studies are being conducted to assess the problem more exactly *(6–10)*. Table I, for example, shows that rain in unpolluted maritime areas has pH values ranging from 3.3 to 7.2 *(23)*.

It has been reported that in the eastern United States sulfur dioxide accounts for roughly 65% and nitrogen oxides 30% of the acidity in rain (*4;* see also II.A.1, this chapter, and Chapters 2 and 3). Actions directed toward sulfur dioxide control alone may shift atmospheric acidic impact patterns in unexpected and unknown directions. Figure 3 *(24)* shows pH isopleths in the United States. In September 1983, an Acid Rain Peer Review Panel reported to the White House (President's) Office of Science and Technology Policy *(25)*. Realizing that it

Table I The pH of Precipitation in Some Remote (or Unpolluted) Maritime Areas [Reproduced from Sequeira *(23)*]

Location	Observation period	Collection mode	pH				Number of obs.
			Min.	*Max.*	*Median*	*Mean*[a]	
Cape Matatula, Am Samoa	1977–1979	Wet, monthly	4.5	6.0	5.3	3.6	36
Cape Grim, Tasmania[b]	1977	Wet, monthly	5.8	7.2	6.3	6.1	12
West Coast,[c] India (4 Stations)	1975	Bulk, monthly	5.8	6.6	6.3		13
Mauna Loa, Hawaii	1973–1979	Wet, monthly	3.3	6.7	5.0	4.9	47
Hilo, Hawaii[d]	1975–1978	Wet, daily or 2/week	3.7	5.4	4.6	~4.6	Several hundred

[a] Precipitation volume-weighted.
[b] Yet to be published in NOAA/WMO/EPA report.
[c] Unpolluted areas receiving heavy rainfall, 500–1000 cm during 4–5 monsoon months; wet soils during the season.
[d] J. M. Miller, personal communication.

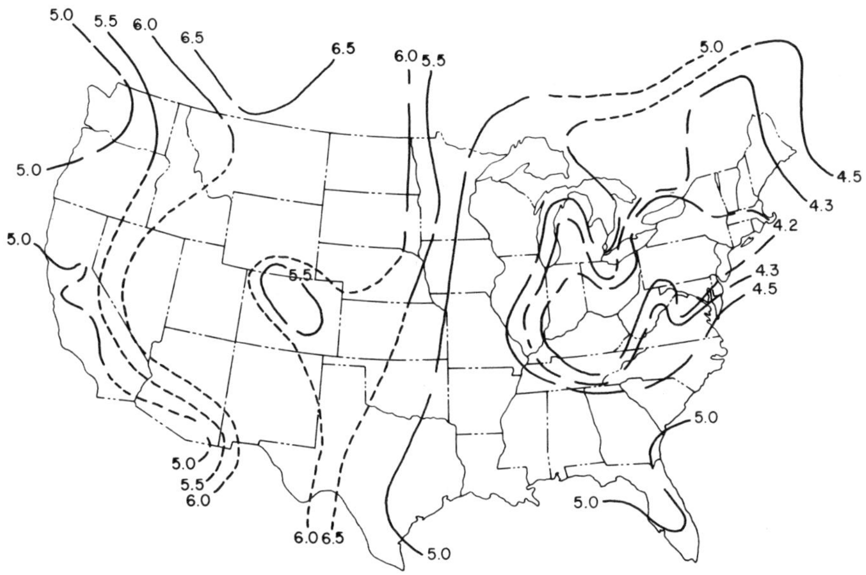

Figure 3. Continental scale of the air pollution problem. Average pH isopleths of precipitation as determined from laboratory analyses of precipitation samples, weighted by the reported quantity of precipitation (1976–1979). (From J. Wisniewski and E. L. Keitz *Water, Air, Soil Pollution* **19**, 327 [1983]. Reproduced with permission.)

would take years to determine exact relationships, the panel advised immediate action to curtail acid rain damage, particularly in the northeastern part of the United States.

B. Asbestos

Asbestos is an important, naturally occurring fibrous mineral widely used for its unique thermal and electrical insulating properties. It occurs in a number of forms, as shown in the following diagram *(26)*.

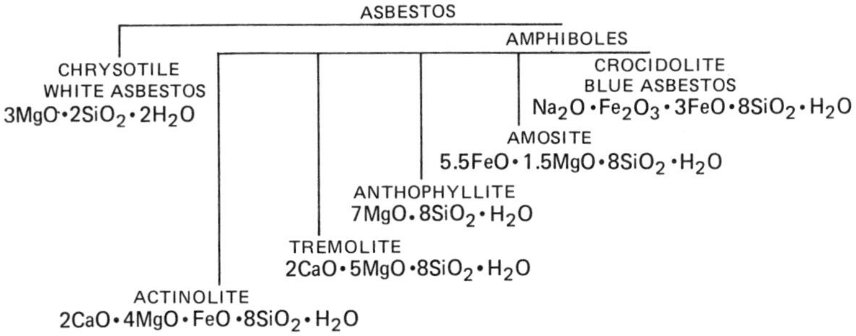

Each form has its own special properties and uses. In general, the bundles of fibers found in natural asbestos ores are separated and woven into textiles or mixed with other materials for insulation, flexibility, and tensile strength. Asbestos, particularly under dry, abrasive conditions, easily fragments to give dusts consisting of fibers ranging from inches in length to microscopic sizes. The microscopic fibers tend to remain suspended in air and enter into the respiratory and digestive systems of exposed persons. Many of the small fibers (fibriles) are not removed by natural protective processes for the rest of an individual's life. The fibers (>5 μm in length) initiate fibrogenic reactions resulting in asbestosis and asbestos-related cancers. Increased incidences of occupational lung, stomach, bowel, and larynx cancers and mesotheliomas of the pleura and peritoneum have been linked to asbestos exposure (see Chapter 7). Latent periods of 20 to 30 years have been hypothesized *(26)*.

Asbestos fibers are found in the general environment and in the lungs of most adults in urban areas of the Western world *(27)*. Table II summarizes the levels of asbestos found in the urban areas of the United States. An average lifetime exposure of 3 ng/m^3 for urban areas and 0.1 ng/m^3 for rural areas has been estimated. Calculated lifetime risks of 100 mesotheliomas and 2 lung cancers per million population have been projected but not proved from occupational data. The 50 to 1 discrepancy in the projected ratios also cannot be explained at this time *(28)*.

Table II Concentrations of Asbestos in Urban Areas of the United States as Determined in Two Studies [Reproduced from Enterline *(28)*]

Asbestos concentration ng/m^3	*Mt. Sinai number of samples*	*Battelle Institute number of samples*
<1	61	27
1–2	58	33
2–5	45	42
5–10	12	22
10–20	8	1
20–50	1	2
50–100	2	0
Mean	3.58	3.62
Median	1.57	2.29

The United States Environmental Protection Agency has listed asbestos as one of seven officially designated hazardous air pollutants. Emission standards for asbestos product manufacturers and processers were set at "no visible emissions." Official methods for demolition, renovation, disposal, and other potential asbestos-emitting activities were also specified *(29)*. No ambient air quality standards for asbestos have been set. A great deal of activity has been centered around potential exposures from removing asbestos-containing insulation, ceilings, and ductwork in school rooms, dormitories, offices, and other public buildings *(29a, 29b)*.

C. Carbon Dioxide

The concentration of carbon dioxide in the atmosphere continues its rise. From an estimated 290 ppm in 1880, the concentration has risen to approximately 340 ppm, a 12-ppm rise in the 1970–1980 decade alone. Continuously recorded measurements from 1958 at Mauna Loa, Hawaii, give indisputable evidence of the rising curve (Fig. 4). The combustion of fossil fuels for transportation, industry, commercial activities, and residential heating are the principal contributors to the buildup of carbon dioxide in the air. Forest fires and deforestation also contribute.

Approximately half of worldwide emissions account for the rise in carbon dioxide concentrations in air. The other half apparently enters into the natural carbon reservoirs and flux cycles, which involve very large amounts of carbon (Fig. 5). Although the amount of atmospheric carbon dioxide is small compared to the world reservoirs, the long-

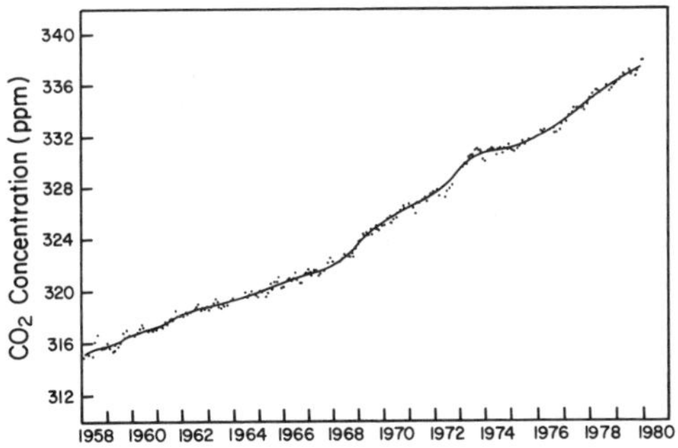

Figure 4. Mauna Loa monthly averages of atmospheric CO_2 concentrations, with seasonal effect removed *(30)*.

range cumulative effects of increasing atmospheric concentrations are predicted to bring about major changes and disruptions in the world's climate *(30, 31)*. Predictions vary considerably, depending on the assumptions (or scenerios) used in the calculations. One large factor involves the prediction of the amount of fossil fuels that will be used in the future. Conservative projections of little to no growth in fossil fuel usage still indicate a potential for doubling the carbon dioxide concentration in the air by the end of the next century. Growth rates of 4 percent per year would give exponential increases: doubling the carbon dioxide concentration by 2030 (Fig. 6).

The principal concern over the worldwide increase in carbon dioxide is what is popularly called the "greenhouse" effect. Somewhat like a greenhouse, carbon dioxide in air is transparent to the sunlight which warms the earth. At the same time it absorbs to some extent the heat-balancing infrared rays emitted by the earth. Over a long period of time, as this energy is retained, the earth will gradually warm up if all else remains constant. A reasonably predicted doubling of the carbon dioxide concentration by 2050 may cause an increase of 3 ± 1.5°C in the world's average temperature. The increase will concentrate more in the higher northern and southern latitudes. As a consequence, high percentages of the polar ice caps will melt, and sea levels will rise an estimated 5 m, flooding many coastal cities and flatlands *(31)*. Worldwide weather patterns will change, affecting forest types and the areas they cover, desert areas and plains regions, plant growth patterns and types, food production, sea composition, and sea life.

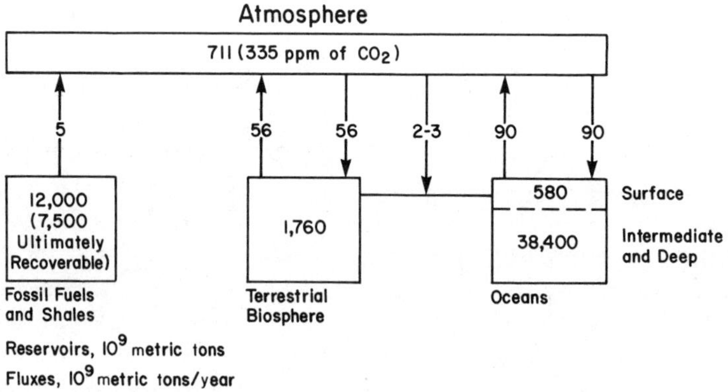

Figure 5. Exchangeable carbon reservoirs and fluxes *(30)*.

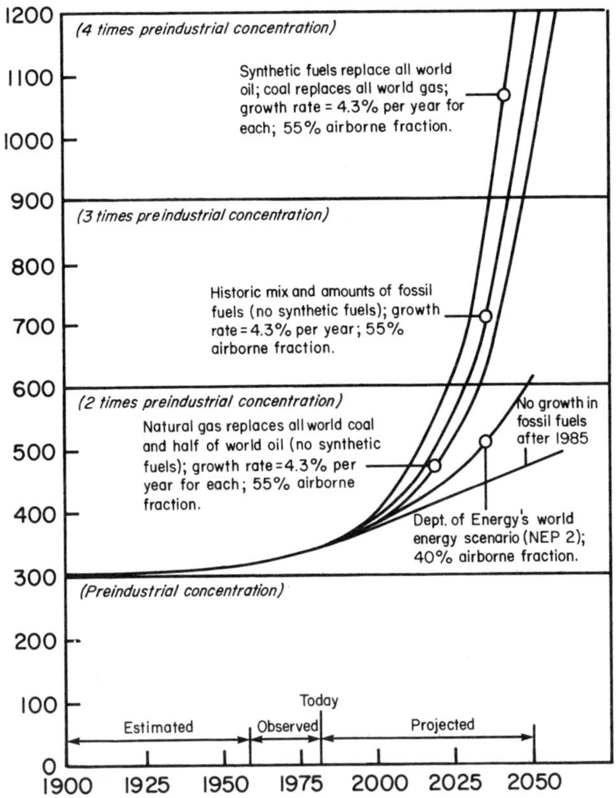

Figure 6. Carbon dioxide concentrations implied by various energy scenarios. Synthetic fuels derived from coal are assumed to release 3.4×10^{15} g of carbon in CO_2 per 100 quads of energy. Airborne fraction is the percentage of emitted CO_2 that remains in the atmosphere (U.S. Department of Energy) *(32)*.

Either pessimistic or optimistic projections involve many known and unknown factors. These include photosynthetic enhancement; air–ocean and air–land equilibria patterns; oceanic chemical-, micro-, and macrobiological sinks; terrestrial sinks; variations in solar luminosity and atmospheric reflectivity from increased or decreased cloud coverage, volcanic eruptions, and particulate pollution. In addition, there are the short-term and long-term variations in temperature, including the large variations between the glacial periods (Fig. 7) *(30–35)*. At best, it will be some 10 years or more before a definite trend can be detected *(34)*. In the meanwhile large-scale international programs to develop nonfossil energy sources are being urged.

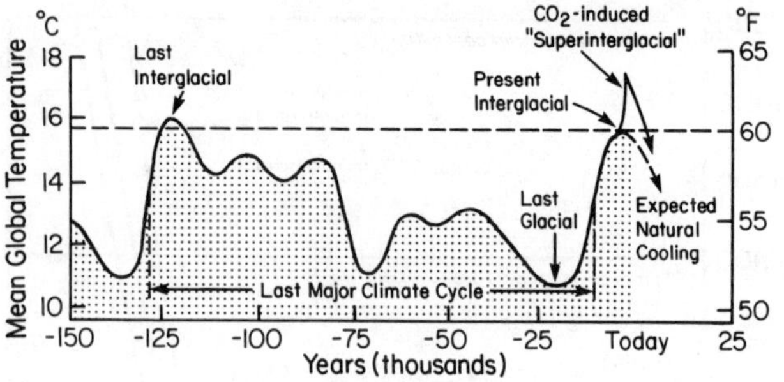

Figure 7. Carbon dioxide potential effect on global scale climate *(30)*.

D. Indoor Air Pollution

The harmful effects of air contaminants were historically first recognized in the work place, where very high doses were encountered, and later found in the general atmosphere. There is now widespread and growing interest in the causes and effects of indoor pollutants (*36–46;* see also Chapters 4 and 7). The earliest studies involved comparisons of outdoor pollution concentrations with indoor concentrations, particularly with respect to carbon monoxide, nitrogen oxides, sulfur dioxide, and particulate matter. The scope of studies rapidly broadened into areas of air contamination peculiar to home and office activities: cooking, heating, smoking, spraying insecticides, radon, and TV and microwave radiation, as well as pet- and plant-associated pollutants. Table III *(37–42)* lists the major indoor air pollutants. Not mentioned are such other indoor environmental stresses as noise, vibration, heat, cold, and indoor lighting.

Nitrogen oxides (NO_x) are common indoor air pollutants in homes with gas stoves, wood-burning fireplaces, and gas and kerosene space heaters. Several researchers have indicated increased respiratory difficulties for children exposed to indoor nitrogen oxides (see Ch. 6). Spengler and others have conducted extensive studies of indoor and outdoor levels of nitrogen oxides *(38–42b)*. Measurements made during a 1-year study of 137 homes in Portage, Wisconsin, showed mean nitrogen dioxide concentrations of 8.4, 65.5, and 65.6 $\mu g/m^3$ in kitchens with electric, natural gas, and liquid petroleum (LP) gas cooking stoves, respectively. Outdoor concentrations near the same types of kitchens were 12.8, 15.8, and 11.8 $\mu g/m^3$ *(39)*. In general, homes with electric cooking stoves had nitrogen oxide concentrations lower than the outside air. Homes using gas cooking in more polluted cities such as St. Louis, Missouri, and Watertown, Massachusetts, had

Table III Major Indoor Air Pollutants *(37–42)*

Contaminant	*Typical Sources*
Aerosols	Spray products
Allergens	House dust, animal dander, insect parts, mildew
Ammonia	Human and animal respiration and metabolites, ammonia-containing cleaners
Asbestos	Pipe, duct and ceiling insulation, roofing, firewalls, insulation board
Carbon dioxide	Unvented combustion, human and animal respiration
Carbon monoxide	Gas, wood, coal and oil stoves and furnaces, cigarettes, fireplaces, unvented space heaters, combustion engines
Chlorinated organic compounds	Some spot and paint removers, pet sprays, liquid drain cleaners
Chlordane	Pesticide used for professional termite control
Chlorine	Laundry bleaches, pool chlorinators
Dinitrobenzene	Furniture, shoe polishes
Copolymer resins	Hair sprays, paint sprays
Formaldehyde	Urea–formaldehyde foam insulation, veneer furniture, plywood, particle board, paneling, ceiling tiles, draperies, carpets, human activities
Insecticides	Ant, roach, flea, moth, housefly sprays
Lead	Automobile exhausts, some old paints
Mercury	Fungicides, some paints, thermometer breakage
Nitrogen oxides	(See Carbon monoxide)
Organic substances	Paints, solvents, wood preservatives, spot removers
Ozone	Electronic air cleaners, photocopiers, some electric motors
Pollens	Plants, trees, weeds, grass
Radon	Soil, stone, concrete, bricks, tiles, gypsum board
Respirable particles	Smoking, cooking, gas stoves, dust, dusting, aerosol sprays
Sodium hydroxide (Lye)	Oven cleaning sprays
Spores	Fungi, molds
Sulfur dioxide	Combustion of coal, oil, diesel fuel, kerosene space heaters
Tobacco smoke	Indoor smoking, exposed clothing and household articles
Water vapor	Human, plant, and animal activities; showers; washing

annual indoor to outdoor ratios of 47 : 37 and 52 : 49 $\mu g/m^3$, respectively *(40)*. Peak concentrations during cooking periods ranged from 500 to 1000 $\mu g/m^3$. For comparison, the United States' national ambient air quality standard is 100 $\mu g/m^3$ annual arithmetic average.

Carbon monoxide is by far the most acutely dangerous pollutant in the home. It is a colorless, odorless gas produced by incomplete combustion of natural gas, oil, coal, wood, tobacco, and so on. Defective, unvented, or poorly vented heating systems can produce carbon monoxide levels that cause headaches, hazy vision, mental confusion, nausea, and death. Kitchen and house concentrations commonly build up to subacute levels when ventilation rates significantly fall below one change of air per hour. Cooking and smoking also contribute to carbon monoxide levels in the home *(43)*.

Formaldehyde is released from urea–formaldehyde insulation and various building and furniture materials. Mobile homes in particular may contain formaldehyde-emitting substances *(44, 46a,c)*.

Aerosol sprays are used to dispense room fresheners, insecticides, hair sprays, odorants, oven cleaners, clothing prewash treatments, and furniture and shoe polishes. In all cases, the propellant and a gas, liquid, and/or a solid aerosol are dispersed at rather high concentrations. Proper ventilation, in addition to following the manufacturer's instructions, is imperative *(45)*.

Sick buildings have been described *(41, 46b)*. Many of them are the result of energy conservation measures whereby air movement into and out of a building is limited without proper allowance for the removal or dilution of indoor air contaminants. Many buildings will contain, at varying times, excessive amounts of the types of pollutants indicated in Table III. Some sick buildings are the result of poor or improper maintenance of filters or air-cleaning systems from which diseases such as legionnaires' disease can be spread *(41, 42)*.

Radon is a gaseous radioactive element that enters homes and buildings from the soil, building materials, and tap water. It is a daughter of radium-226, a trace element in most rocks and soils. Most of the radon found in buildings diffuses from the soil into the building through joints and cracks in the foundations and floors. In a few areas, drinking water from aquifers in radium-bearing strata becomes a major contributor to indoor radon *(46)*. Highest concentrations are generally found in the basements and lower floors of buildings (Table IV) *(47)*. The ventilation rate for buildings is a large factor in determining the concentration level *(42a, 48)*. It is concluded that even in well-ventilated homes ($>$1 air change per hour), high soil and water radon levels can lead to radon concentrations in excess of health guidelines *(46, 49)*.

Tobacco smoke is a universal source of particulate matter, organic combustion gases, and aerosols in homes, office buildings, bars, nightclubs, restaurants, and public transportation systems. Not only does the tobacco smoke affect the smoker, but it also affects the nonsmoker, who is considered to be a "passive" smoker (see Fig. 20).

Table IV Average Indoor Radon and Daughter Values [Reproduced from Gesell *(47)*, by permission of the Health Physics Society]

Location and description	*Average radon concentration pCi/l*
Sweden, four towns, 300 dwellings	
Wood houses	0.4
Brick house	1.1
Aerated concrete houses	3.1
Gävle, Sweden	
63 dwellings	4.9
Innsbruck, Austria	
12 mostly brick structures	1.2
Great Britain	
87 dwellings	0.57[a]
Hungary	
409 brick dwellings	1.56
247 panel dwellings	3.90
356 block dwellings	4.60
170 adobe dwellings	1.70
Canada	
70 basements	0.42
Norway	
42 wood structures	1.3
42 concrete structures	2.0
36 brick structures	1.0
United States	
18 residence basements	1.7[b]
18 residence first floors	0.83[b]
9 residence second floors	0.77[b]
29 Florida, background	0.0028[c]
29 Grand Junction, background	0.0072[c]

[a] Assuming one air change per hour.
[b] Geometric mean, New York, New Jersey.
[c] Working level (WL) geometric mean.

E. Lead (See also Chapter 7)

The United States National Ambient Air Quality Standard (NAAQS) for lead, promulgated in October 1978, is 1.5 $\mu g/m^3$ maximum arithmetic average per calendar quarter *(50)*. Automobiles account for 80% of atmospheric emissions, and the nonferrous and battery industries account for most of the rest. Since implementation of lead gasoline additive restrictions, the lead consumption for gasoline in the United States dropped from 170,000 tons/year in 1977 to an estimated 55,000 tons/year in 1981 *(51)*. Figure 8 shows the maximum quarterly ambient air average concentrations of lead for the 1975–1981 period. All values fall below the NAAQS. The nearly 60% drop in ambient air levels reflects the change caused by the use of unleaded gasoline *(51a)*. Improvements in ambient air levels are reflected in reduced blood lead levels reported by the United States National Center for Health Statistics. An average of 15.8 mg/dl lead in blood in 1976 dropped to 10.0 mg/dl in the last 6 months of 1980 *(52)*.

Continuing international studies of ambient air, soil, and blood lead levels show global interest in assessing and resolving exposure problems. Among these studies are reviews by the World Health Organization *(53)*, Chamberlain [Table V; *(54)*] and Millar and Cooney *(55)*. Historical data were also obtained from deep layer snow deposits *(56)*.

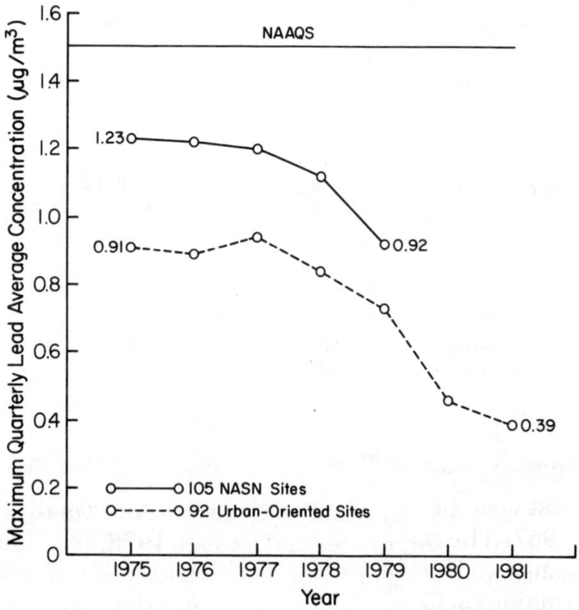

Figure 8. National trend in maximum quarterly average lead levels, 1975–1981 *(51)*.

Table V Blood Lead in Children and Mothers (Number of Subjects in Parentheses) [Reprinted with permission from Chamberlain *(54)*. Copyright 1983, Pergamon Press, Ltd.]

Area	*Age of children*	*PbB (μg/dl) Child*	*Mother*	*Ratio*
Derbyshire, U.K.				
Soil lead < 1000 ppm	2	20.7 (29)	14.1 (29)	1.5
1000–10,000 ppm	2	23.8 (43)	18.7 (43)	1.3
>10,000 ppm	2	29.0 (10)	14.8 (10)	2.0
All	2	23.3 (82)	16.6 (82)	1.4
London, U.K.	<5	33 (144)	28 (107)	1.2
Boston, Mass.	2–6	26.8 (168)	18.7 (249)	1.4
Clifton, Nr Manchester, U.K. leadworkers' families	<5	33.1 (192)	23.6 (147)	1.4
Others	<5	27.0 (273)	21.9 (193)	1.2
North Wales copper pipes	<16	19.2 (19)	13.9 (19)	1.4
Lead pipes	<16	41.6 (38)	39.5 (38)	1.1
Himalayas	3–12	3.5	2.9	1.2

F. *Long-Range Transport*

Long-range transport is discussed in Chapter 3.

G. *Nuclear Accidents*

Prior to 1979 there had been only two major nuclear reactor accidents. The first was the Windscale, England, military reactor accident of October 1957. The second was the Idaho Falls, United States, excursion of January 1961. Both are discussed in Vol. I, Chapter 5.

The third major accident occurred in the United States at the Three Mile Island, Pennsylvania, nuclear electric power generating plant in

March 1979. Large quantities of gaseous, radioactive xenon-133 were released. Average individual exposures were estimated as ranging from 80 millirem (mR) at 1 mi to zero at 50 mi. A considerable amount of radioactivity was held in the containment chamber and in the cooling and processing waters and was disposed of under controlled conditions *(57, 58)*. This accident catalyzed widespread concerns over safety in the use of nuclear power that covered the range from the design, construction, and operation of the reactors to the training and supervision of the operators. The costs of such precautions and the continuing problem of radioactive waste disposal have slowed the planning of new nuclear reactor power plants, particularly in the United States *(59)*.

H. Nonionizing Radiation

There is a growing realization that radiation other than ionizing radiation may be harmful to humans and animals. The introduction and widespread use of microwave ovens in the home has catalyzed awareness that serious physical damage can come from careless or accidental exposures to strong radio and microwave sources. Powerful radio, television, and radar broadcasting and tracking stations

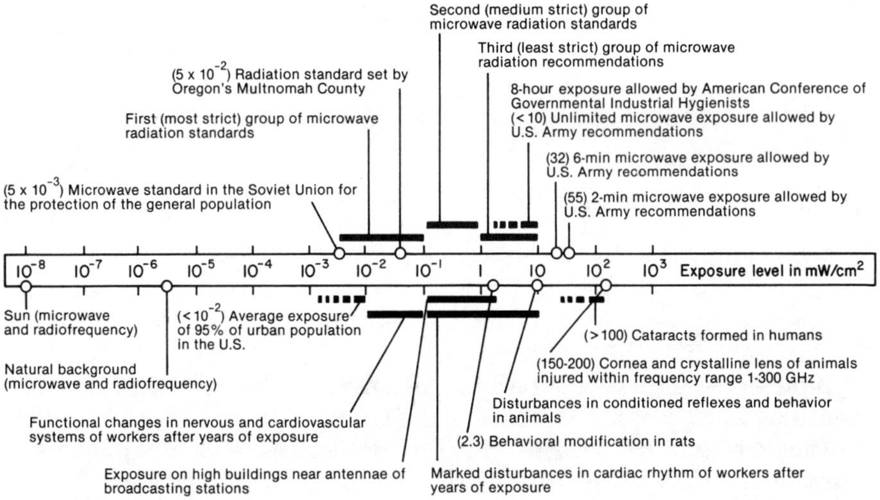

Figure 9. Standards and guidelines for exposure to microwave radiation (Reprinted with permission from B. Hileman, *Environmental Science and Technology* **16** 442A [1982]. Copyright 1982 American Chemical Society).

Table VI Population Exposures to Radiofrequency Radiation *(61)*

Power density (mW/cm^2)		Cumulative percentage of U.S. population exposed to radiofrequency radiation greater than level specified
10.	(U.S. occupational standard)	—
1.0	(U.S. standard for microwave oven at 5-cm distance)	—
0.01	(USSR occupational standard)	—
0.001	(USSR public standard)	0.5
0.0005		1.0
0.0002		2.5
0.0001		5.0
0.00005		8.0
0.00001		31.0
0.000005		51.0

also emit radiation which can cause physiological disfunctions. Figure 9 illustrates the wide range of microwave radiation as well as its physiological effects *(60)*. Table VI shows the estimated cumulative population exposure to radiofrequency radiation, as well as the United States and the USSR occupational and public standards *(61)*. Power densities at various locations near broadcasting towers varied from 0.003 to 0.18 mW/cm^2, whereas that from a microwave oven can be 1.00 mW/cm^2 *(61)*. Very serious burns have resulted from accidental exposures or defective microwave ovens and heating systems.

I. Risk Assessment and Management (See also Chapter 7 and Vol. VIII, Chapter 1)

Two important areas of specialization in air pollution control are (1) risk assessment, the process of evaluating the potential harmful effects air pollutants may have, and (2) risk management, the process of developing source emission and ambient air quality standards after risk assessments have been made.

The importance of these two areas has resulted from the scientific complexities and subjective decisions encountered in promulgating reasonable emission and ambient air quality standards. Great difficulties, for example, have been met in developing specific risk factors for the designated toxic substances (mercury, arsenic, benzene), pesticides, herbicides, and other known or suspected carcinogens. Extrapolations of animal or biological test data are often debatable and sometimes contradictory. Occupational exposure data are difficult to interpret with respect to the effects of intermittent versus continuous exposures and healthy versus nonhealthy or senitive human populations. Epidemiological studies of general population air pollution exposures are difficult to design, costly, very time-consuming, and consequently not easily used *(62–68b)*.

A National Academy of Sciences panel formed to study the complexities of risk assessment recommended four basic steps to study, classify, and quantify pertinent data from hazardous substance studies more systematically *(64)*.

1. Hazard Identification. Studies of human exposure (epidemiology), animal test, and/or *in vitro* biological test data to determine a degree of toxicity for suspect substances.

2. Dose–Response Assessment. Evaluation of health impacts from potential ambient air concentration and time-of-duration ranges.

3. Exposure Assessment. Calculation of potential pollutant exposure levels.

4. Risk Characterization. Provision of numerical estimates of the incidence of toxic impacts per unit of exposed populations.

Risk management decisions, once risk has been assessed with some degree of confidence, must also proceed through a tortuous path of extrapolations to acceptable risk factors with consideration of threshold and nonthreshold levels, safety margins, and cost–benefit analyses *(62, 65–66a)*.

J. Stratospheric Ozone

Ozone in the upper atmosphere provides a protective layer that absorbs the energetic 2900- to 3200-Å ultraviolet rays (UV-B rays) coming from the sun, which can cause cancer, break down biological matter, and pose a serious threat to life on earth. The total amount of ozone in a vertical column averages about 8×10^{18} molecules/cm^2: the equivalent of a 3-mm layer of ozone gas at standard temperature and pressure. The amount varies over 30% with latitude and season, and as much as 100% with weather variations. Total ozone amounts are negatively correlated with pressure and positively correlated with temperature in the lower stratosphere *(69)*.

The importance of the stratospheric ozone, coupled with its low concentration and its chemical and photochemical reactivity, has raised serious concerns that anthropogenic pollutant emissions would reduce the protective capacity of the ozone layer. First concerns surfaced when scientists realized that nitrogen oxides in the exhaust gases of high-altitude supersonic air flights could eventually react with and reduce the effectiveness of the ozone layer *(70)*. This was closely followed by the realization that highly stable gaseous chlorofluorocarbon compounds (CFCs) eventually would diffuse to the stratosphere and cause similar ozone depletion problems *(71)*.

Highly stable freon-11 and 12 (CCl_3F and CCl_2F_2) are universally used as refrigerants and as propellants in aerosol spray cans. Because of its stability, virtually every gram of chlorofluorohydrocarbon manufactured will be released into the atmosphere, either from spray cans or from leaking, worn-out, or discarded refrigeration units. On a worldwide basis spray cans use about half of the volatile CFCs. The United States' ban on the use of freons as aerosol propellants can reduce world emissions by only 20% *(72)*. Predictions of the depletion of ozone have been made [Fig. 10 *(73)*]. However, because of the complexity of the photochemical reactions, speculative rates of emissions of CFCs, impact of nitrogen oxides, uncertainties in the uses of nitrogenous fertilizers, possible beneficial effect of carbon dioxide, and natural variability of ozone concentrations, reliable predictions of the depletion rates are extremely difficult *(72)*.

An ad hoc group of scientists evaluating the problem for the United Nations Environment Program (UNEP) issued a summary statement which reiterated the potential seriousness of the problem and recommended continued indepth studies with advanced statistical and re-

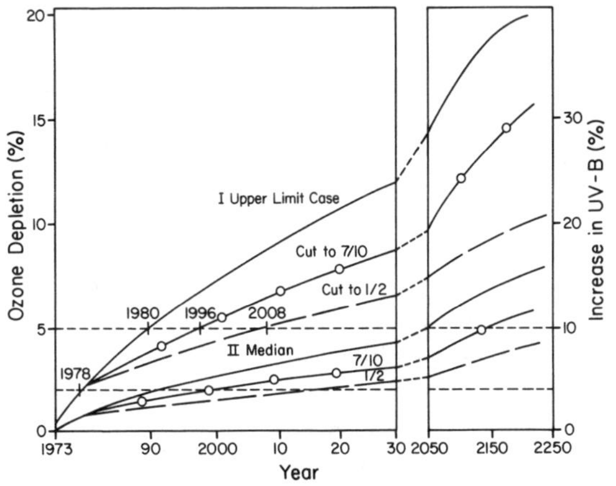

Figure 10. Model ozone depletion. The time scales for ozone depletions predicted by atmospheric photochemical models under various scenarios are depicted schematically (Reprinted with permission from B. Boville, "The Ozone Layer," United Nations Environment Programme, Proceedings, A. K. Biswas, ed., Copyright 1977, Pergamon Press Ltd.)

search techniques *(72–75a)*. Recent more refined studies predict lower depletion rates, but the uncertainty is still large *(75)*. Caution is advised, because once the effects of ozone depletion became confirmed, it would take a minimum of 10 years to correct it *(70, 72, 76)*.

K. Toxic Substances

Section 112 of the United States Clean Air Act of 1970 requires that the Environmental Protection Agency (EPA) develop standards to control the emission of hazardous pollutants which may reasonably be anticipated to result in increased human mortality or serious illness. The task has proved to be a formidable one. A major problem is that no straightforward set of chemical or physical properties or clear-cut definitions can be developed to identify toxic substances which may be hazardous to human health at air pollution concentrations. As of the date of this writing, seven pollutants have been promulgated as hazardous—arsenic, asbestos, benzene, beryllium, mercury, vinyl chloride, and radionuclides—but emissions standards have been promulgated for only four of these: asbestos, beryllium, mercury, and vinyl chloride *(77,* Part 61).

Operating through its own personnel, various subcontractors, and a Science Advisory Board of expert specialists, the EPA developed a list of 37 potentially toxic air pollutants [Table VII *(78)*]. A supplementary list of 184 substances has also been developed *(79)*. Screening these substances for appropriate priority action requires, in general, four difficult steps *(80, 81):* (1) identification, (2) dose–response assessment, (3) exposure assessment, and (4) risk characterization. Steps (3) and (4) are particularly complex when applied to air pollutants under a "reasonably" expected harmful air pollution effect requirement *(66a)*.

A problem in understanding the impact of toxic substances in the environment is that analytical methods have become extremely sensitive. It is not uncommon to be able to measure substances at parts per trillion concentrations or mass amounts less than picograms (10^{-12} g). Therefore, many substances can be detected in ordinary as well as possibly hazardous industrial emissions *(82–89)*. It is no longer a question of whether these substances are present in the air; more important is the amount present in relation to harmful levels, e.g., dioxins *(90)*, organic emissions from industrial and chemical waste sites *(87)*, and potential carcinogens in soot *(89)* and in vehicular exhaust *(91–92a)*.

There have been dramatic increases in the use of fungicides, herbi-

Table VII United States Environmental Protection Agency Hazardous Substance Study List of 37 *(78)*

Acetaldehyde	Formaldehyde
Acrolein	Hexachlorocyclopentadiene
Acrylonitrile	Maleic anhydride
Allyl chloride	Manganese
Benzyl chloride	Methyl chloroform (1,1,1-trichloroethane)
Beryllium	Methylene chloride (dichloromethane)
Cadmium	Nickel
Carbon tetrachloride	Nitrobenzene
Chlorobenzene	Nitrosomorpholine
Chloroform	Perchloroethylene
Chloroprene	Phenol
Coke oven emissions	Phosgene
o-, *m*-, *p*-Cresol	Polychlorinated biphenyls
p-Dichlorobenzene	Propylene oxide
Dimethyl nitrosamine	Toluene
Dioxin	Trichloroethylene
Epichlorohydrin	Vinylidene chloride
Ethylene dichloride	*o*-, *m*-, *p*-Xylene
Ethylene oxide	

cides, and insecticides throughout the world. The extensive use of such powerful chemicals is of wide concern because of potentially detrimental and possibly irreversible environmental effects. In the United States there has been a tenfold increase in the number of acres treated with herbicides since 1950, and the total production of herbicides, insecticides, and fungicides in 1980 was estimated to be 1.47 billion lb of active ingredients (Fig. 11) *(93)*.

Atmospheric impacts result from spraying or dusting applications, either from direct contact or downwind contamination *(94)*; manufacturing process emissions *(95)*; leakage from waste dump sites *(87)*; and household, office, or work space insect control spraying *(41, 93)*. Several notorious incidents include the Vietnam War exposures to contaminated Agent Orange; the Seveso, Italy, release of dioxin; the Love Canal waste dump contamination of homes; and the Times Beach, Missouri, dioxin contamination *(80, 90, 95, 95a)*.

Many toxic air pollutants are considered to be carcinogenic, mutagenic, or teratogenic. Table VII lists a number of the more strongly suspected carcinogens. At least 1000 additional chemicals have been on various lists as potential carcinogenic agents. A number of substances have been definitely associated with human cancer, particularly in certain occupations and trades: ionizing radiation, quartz,

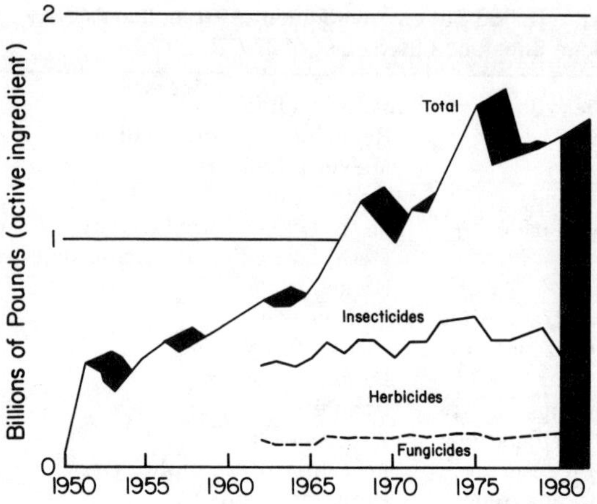

Figure 11. Synthetic organic pesticide production in the United States, 1950–1980. (Reprinted with permission from Bottrell and Smith *(93)*, Copyright 1982. American Chemical Society).

asbestos, beryllium, chromates, benzene, and polynuclear aromatics. Tobacco smoking has been clearly associated with excess lung cancer deaths. Many additional substances have produced tumors in laboratory test animals or have given positive cell pattern changes in biological tests (Ames test, *96*). Some 85% of all cancer cases are attributed to "environmental" factors, which in this case includes food and lifestyle patterns. Second- and third-generation descendents of United States immigrants show cancer susceptibility patterns similar to those in the United States rather than those of their ancestor countries. Five to 30% (at most) of cancers are estimated to be caused by environmental chemicals, including atmospheric pollutants *(80, 81)*.

L. Visibility

Visibility is discussed in Chapter 4.

II. The Primary Air Pollutants — Gaseous

Table VIII summarizes the global sources, concentrations, and important scavenging processes of the more important atmospheric trace gases *(2)*. Although, in many cases, the total mass of natural emissions surpasses pollutant emissions on a worldwide basis, they are usually widely dispersed. Pollutant emissions, on the other hand, are generally emitted from point sources or limited area sources. As a result, local ambient air quality levels may rise to environmentally undesirable and even health-endangering levels.

A. Gaseous Compounds of Carbon

1. The Hydrocarbons

Early concerns with hydrocarbons in air involved their role as precursors for photochemical oxidant formation. Motor vehicle exhaust, solvent evaporation, and related petroleum processing emissions were the primary targets of control efforts. The United States' national ambient air quality standard was based on measurements of nonmethane hydrocarbons (NMHC) to be taken 6 to 9 A.M. when rush hour traffic was at its highest. Los Angeles, with its serious photochemical oxidant problems, led in developing and enforcing control strategies for reducing hydrocarbon emissions. In addition to federally mandated catalytic exhaust devices, strong measures were undertaken to prevent evaporative losses from the automobile and from refining, storage, and transfer of petroleum products.

Figure 12 *(32, 51, 98–100)* shows the historical and projected emissions of nonmethane hydrocarbons in the United States. A leveling off with a slight drop in the 1970–1981 emissions is shown *(51)*. This came about despite an estimated 37% increase in vehicular miles traveled. Catalytic exhaust devices, evaporative loss prevention, and more efficient automobile engines were important factors in controlling hydrocarbon emission levels. Industrial process emissions have continued to grow, but at a more moderate rate. Continued legislative and economic pressures are expected to reduce emissions significantly in future decades *(32, 61)*.

Ambient air concentrations at first were measured and reported as nonmethane hydrocarbons (NMHC). Methane is relatively inert and a major natural trace component in air (1.5 ppm; 4-year half-life) *(97a)* and is not considered important in photochemical smog formation. The term *hydrocarbon* in itself has a chemically inert connotation and does not adequately describe the broad classes of gaseous organic compounds emitted into the atmosphere. Consequently, a more appropriate term, *volatile organic compounds* (VOCs), has come into common use.

Table IX lists average C_2 to C_{10} hydrocarbon concentrations mea-

Table VIII Summary of Sources, Concentrations, and Important Scavenging Processes of Atmospheric Trace Gases (Elmer Robinson, Mauna Loa Observatory, NOAH, Hilo, Hawaii *(2)*

Contaminant	*Major pollutant sources*	*Important natural sources*	*Estimated annual emissions Tg/yr*[a]		*Atmospheric background concentrations*	*Estimated atmospheric residence time*	*Removal reactions and sinks*	*Remarks*
			Pollutants	*Natural*				
SO_2	Fossil fuel combustion	Volcanoes, reactions of biogenic S emissions	212[b]	20[c]	About 0.1 ppb[d]	1–4 days	Oxidation to sulfate by photochemical reactions or in liquid droplets	High reaction rates in summer due to photochemical processes
H_2S and organic sulfides[e]	Chemical processes, sewage treatment	Volcanoes, biogenic processes in soil and water	3 (As sulfur)	84[f] (As sulfur)	H_2S: 0.05–0.1 ppb COS: 0.5 ppb[g] CS_2: 0.05 ppb[g]	H_2S: 1–2 days COS: 1–2 years (6)	Oxidation to SO_2 and SO_4	Atmospheric data are very incomplete; COS residence time may be 20 years[c]
CO	Auto exhaust general combustion	Forest fires, photochemical reactions	700[h]	2100[h]	0.1–0.2 ppm (N. Hemisphere) 0.04–0.06 ppm (S. Hemisphere)	1–3 months	Photochemical reactions with CH_4 and OH	No long-term changes in the atmosphere have been detected
NO, NO_2	Combustion	Biogenic processes in soil, lightning	75[i] (As NO_2)	180[j] (As NO_2)	About 0.1 ppb[k]	2–5 days	Oxidation to nitrate	Natural processes are mostly estimated; background concentrations are in doubt but may be as low as 0.01 ppb
NH_3	Waste treatment, combustion	Biogenic processes in soil	6[l]	260[j]	About 10 ppm[j]	1–7 days	Reaction with SO_2 to form $(NH_4)_2SO_4$ aerosol	Atmospheric measurements are very sparse
N_2O	Small amounts from combustion	Biogenic processes in soil	3[m]	340[n]	300 ppb	20–100 years	Photochemical in stratosphere	Some estimates place natural source at 100 Tg or less[m]
CH_4	Combustion, natural gas leakage	Biogenic processees in soil and water	160[o]	1050[o]	1.5 ppm	8 years[p]	Reaction with OH to form CO	Pollution source includes 60 Tg/year from biomass burning
Isoprene and terpenes	None	Biogenic plant emissions	None	830[h]	0	1–2 hours	Photochemical reactions with OH and O_3	Not found in ambient atmosphere away from

Total non-CH_4 hydrocarbons	Combustion	Biogenic processes in soil and vegetation	40[r]	2×10^4[q]	0–1 $\mu g/m^3$ for C_2's	Hours to a few days	Photochemical reactions with NO and O_3	Concentration given for C_2's in rural atmosphere
CO_2	Combustion	Biological processes	22,000[q]	10^6[n]	345 ppm (1981)	2–4 years	Biogenic processes, photosynthesis, absorption in oceans	Forest destruction and changes in earth's biomass may add 20–30 × 10^3 Tg CO_2/year to atmosphere[s]
CH_3Cl	Combustion	Oceanic biological processes	2[t]	4–6[q,t]	600 ppt[t,u]	1–2 years[t]	Stratospheric reactions	Photochemical reactions in stratosphere may impact on O_3 layer
HCl, Cl_2	Combustion, Cl manufacturing	Atmospheric reactions of NaCl, volcanoes	4[v]	100–200[x]	About 0.5 ppb[v]	About 1 week	Precipitation	Volcanoes may release 10–20 Tg Cl/year[w]

[a] Tg/yr = 10^{12} g/year or 10^6 metric tons/year.
[b] Based on 1978 global fuel usage and estimated sulfur contents.
[c] Major reference is R. D. Cadle, *Rev. Geophys Space Phys.* **18,** 746–752 (1980.)
[d] P. J. Maroulis, A. L. Torres, A. B. Goldberg, and A. R. Bandy, *JGR, J. Geophys Res.* **85,** 7345–7349 (1980).
[e] Includes COS, CS_2, $(CH_3)_2S_2$, CH_3, SH, etc.
[f] Adapted from D. F. Adams, S. O. Farwell, E. Robinson, and M. R. Pack, "Biogenic Sulfur Emissions in the SURE Region," EPRI Rep. No. EA-1516. Electric Power Research Institute, Palo Alto, California, 1980 (final report by Washington State University).
[g] A. L. Torres, P. J. Maroulis, A. B. Goldberg, and A. R. Bandy, *JGR, J. Geophys Res.* **85,** 7357–7360 (1980).
[h] P. R. Zimmerman, R. B. Chatfield, J. Fishman, P. J. Crutzen, and P. L. Hanst, *Geophys. Res Lett.* **5,** 679–682 (1978).
[i] Based on 1978 global combustion estimates.
[j] I. E. Galbally, *Tellus* **27,** 67–70 (1975).
[k] Approximate value combining values given in several references.
[l] R. Soderlund and B. H. Svensson, *Ecol. Bull.* 1976, 23–73. CA **86,** 57,995 h.
[m] 1978 fuel usage figures applied to the following references: R. F. Weiss and H. Craig, *Geophys Res. Lett.* **3,** 751–753 (1976); D. Pierotti and R. A. Rasmussen, *ibid.*, pp. 265–267.
[n] E. Robinson and R. C. Robbins, *in* "Air Pollution Control" (W. Strauss, ed.), Part 2, pp. 1–93. Wiley, New York, 1978.
[o] J. C. Sheppard, H. Westberg, J. F. Hopper, and K. Ganesan, *JGR, J. Geophys Res.* **87,** 1305–1312 (1982).
[p] L. E. Heidt, J. P. Krasnec, R. A. Leub, W. H. Pollock, B. E. Henry, and P. J. Crutzen, *JGR, J. Geophys Res.* **85,** 7329–7336 (1980).
[q] R. E. Graedel, *JGR, J. Geophys Res.* **84,** 273–286 (1979).
[r] Reference *n* tabulation updated to approximate 1978 emissions.
[s] G. M. Woodwell, R. H. Whittaker, W. A. Reiners, G. E. Likens, C. C. Delwiche, and D. B. Botkin, *Science* **199,** 141–146 (1978).
[t] R. A. Rasmussen, L. E. Rasmussen, M. A. K. Khalil, and R. W. Dalluge, *JGR, J. Geophys Res.* **85,** 7350–7356 (1980).
[u] E. Robinson, R. A. Rasmussen, J. Krasnec, D. Pierotti, and M. Jakubovic, *Atmos. Environ.* **11,** 213–215 (1977).
[v] J. A. Ryan and N. R. Mukherjee, *Rev. Geophys. Space Phys.* **13,** 650–658 (1975).
[w] R. D. Cadle, *Rev. Geophys. Space Phys.* **18,** 746–752 (1980).
[x] Based on estimated reaction of NaCl to form Cl_2.

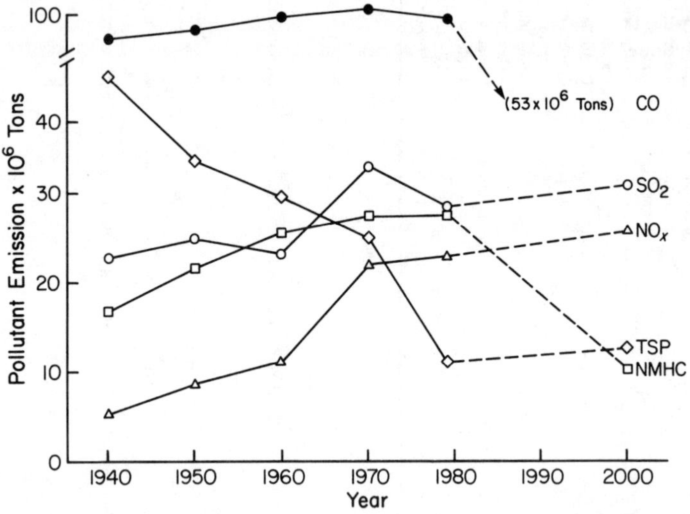

Figure 12. Historical and projected emissions of the principal pollutants in the United States *(32, 51, 98–100).*

sured in Sydney, Australia: quite typical of most NMHCs *(101, 102).* Table X gives the atmospheric sources, sinks, and background levels of 18 halogenated hydrocarbons, benzene, and formaldehyde in seven representative United States cities *(83, 84).* Concentrations in the cities studied were 15 to 30 times greater than background levels but still in the low parts per billion range. Table XI gives the concentrations of C_6 to C_9 hydrocarbons in selected worldwide cities *(85).* Organocontaminants caused by the 1980 Mount St. Helens volcanic eruption have been identified *(103).*

2. The Oxides of Carbon

A decreasing trend of carbon monoxide emissions has been observed for the 1970–1980 decade, and a large (40%) decrease has been projected to the year 2000 (Fig. 12). A principal factor in the observed and projected decrease is the economic and sometimes mandated incentive to burn fossil fuels as efficiently as possible. Catalytic afterburners required for new automobiles in the United States have very significantly reduced emissions of carbon monoxide.

Figure 13 *(51)* shows the annual average of the second highest 8-hour nonoverlapping concentrations of carbon monoxide at 37 United States National Air Monitorings Stations (NAMS) and for 224

Table IX Average Hydrocarbon Concentrations and Composition in Sydney, N.S.W., Air [Reprinted from Nelson and Quigley *(101)*. Copyright 1982 American Chemical Society.]

Hydrocarbon[a]	*Av concn,*[b] *ppbv*		*Av concn,*[b] *ppbv*
Ethane	7.5	2,3-Dimethylbutane	0.9
Ethylene	12.5	Methylcyclopentane	1.2
Acetylene	10.1	Cyclohexane	0.9
Propane	5.9	Benzene	2.6
Propylene	7.4	Heptane	0.7
Methylacetylene	0.5	2-Methylhexane	1.2
Butane	7.5	3-Methylhexane	0.8
Isobutane	4.7	2,4-Dimethylpentane	0.7
1-Butene	1.0	Methylcyclohexane	0.6
Isobutene	1.4	1,3-Dimethylcyclopentanes	0.2
trans-2-Butene	1.1	Toluene	8.9
cis-2-Butene	1.0	Octane	0.4
Pentane	5.0	2,2,4-trimethylpentane	1.2
Isopentane	9.0	Other C_8 alkanes	2.0
1-Pentene	0.4	Ethylbenzene	1.3
trans-2-Pentene	0.6	*m,p*-Xylenes	3.9
cis-2-pentene	0.7	*o*-Xylene	1.5
2-methyl-1-butene	0.5	Nonane	0.4
2-methyl-2-butene	1.3	Propylbenzene	0.4
Cyclopentane	0.8	1,2,4-Trimethylbenzene	1.3
Hexane	2.1	1,3,5-Trimethylbenzene	0.5
2-Methylpentane	2.6	*m,p*-Ethyltoluenes	1.1
3-Methylpentane	1.6	*o*-Ethyltoluene	0.4
2,2-Dimethylbutane	0.5	Decane	0.5

[a] [NMHC]—0.55 ppm C.
[b] Based on 140 samples at Eastern Suburbs Hospital (ESH), Goat Island (GI), and Rozelle Hospital (RH), September 1979 to June 1980.

total reporting stations in the United States. (The highest concentration is allowed to compensate for exceptional or other mitigating circumstances.) The NAMS data in general are collected in urban areas having greater vehicular traffic densities. The averages are compared to the National Ambient Air Quality Standard (NAAQS) of 9 parts per million (ppm). The data show a 25% decrease in ambient air concentrations for the 1975–1981 period.

Figure 14 *(51)* shows the annual average of the number of second highest 8-hour measurements that exceeded the 9 ppm NAAQS (called *exceedances*) for the same sets of sampling stations. Figure 14 shows

Table X Sources, Sinks, Background Levels, and Toxic Effects of Chemicals of Interest [Reprinted with permission from Singh *et al.* (83). Copyright 1982 American Chemical Society.]

Chemical	*Major source*[a]	*Dominant removal mechanism*[b]	*Daily loss,*[c] *%*	*Surface level background concentrations*[d] *ppt*	*ng/m*^{3}	*Toxicity*[e]
Methyl chloride	N(O), MM	HO	0.4	650	1340	BM
Methyl bromide	N(O), MM	HO	0.4	20	78	BM
Methyl iodide	N(O)	$h\nu$(T)	12.2	2	12	BM, SC
Dichloromethane	MM	HO	1.3	50	173	BM
Chloroform	MM	HO	0.9	20	97	BM, SC
Carbon tetrachloride	MM	$h\nu$(S)	~0.0	135	848	NBM, SC
1,2-Dichloroethane	MM	HO	1.9	40	168	BM, SC
1,2-Dibromoethane	MM	HO	2.2	2	15	BM, SC
1,1,1-Trichloroethane	MM	HO	<0.1	180	981	weak BM
1,1,2-Trichloroethane	MM	HO	2.8			NBM, SC
1,1,2,2-Tetrachloroethane	MM	HO	<0.1			BM, SC
1,2-Dichloropropane	MM	HO	10.2			BM
1,1-Dichloroethylene	MM	HO	29.2			BM, SC
Trichloroethylene	MM	HO	17.2	15	80	BM, SC
Tetrachloroethylene	MM	HO	1.5	50	337	SC
3-Chloro-1-propene	MM	HO	91.1			SC
Hexachloro-1,3-butadiene	MM	HO				
α-Chlorotoluene	MM	HO	22.8			BM, SC
Benzene	MM	HO	11.4			SC
Formaldehyde	N, MM	HO, $h\nu$(T)	88.2	400	490	BM, SC

[a] N, natural; O, oceanic; MM, man-made.
[b] HO, hydroxyl radical; $h\nu$, photolysis; T, troposphere; S, stratosphere.
[c] Within the boundary layer (12 sunlit hours); calculation based on estimated daytime (12 hours) average HO abundance of 2×10^6 molecules/cm^3 and mean temperature of 300 K.
[d] At 40° N.
[e] BM, bacterial mutagen (positive Ames test); NBM, not bacterial mutagen (negative Ames test); SC, suspect carcinogen. Cities Monitored: Houston, St. Louis, Denver, Riverside (Calif.), New York (Staten Island), Pittsburgh, Chicago.

a decrease of 84% in the number of exceedances. The discrepency between the 25% drop in ambient air concentrations and the 84% drop in exceedances is due to vehicular impact on the sampling stations. Peak height concentrations caused by rush hour traffic have apparently been reduced by a greater amount than the decrease in carbon monoxide emissions.

Carbon dioxide has been discussed in Section I.C.

Table XI **Average Concentrations (ppb, v) of Aromatic C_6–C_9 Hydrocarbons in Some Urban Atmospheres [Reprinted with permission from Isidorov *et al.* (85). Copyright 1983, Pergamon Press, Ltd.]**

Hydrocarbon	*Los Angeles (1968)*	*Houston (1974)*	*Zürich (1971)*	*The Hague (1975)*		*West Berlin (1978)*		*Johannesburg (1977)*	*Heston*[a]	*Luton*[b]	*Leningrad (1983)*
				Prinsengracht	*Mauritskade*	*Dahlem*	*Steglitz*		*(1979)*		
Benzene	15	1.3–15.0	54	4.6	13.1	2.2	14.6	3.4	91.5	122.5	29.8
Toulene	37	0.3–9.7	39	13.1	22.7	2.8	24.8	10.3	43.8	56.4	27.0
Ethylbenzene	6	3.1–4.5	8.7	1.2	5.0	0.7	5.7	1.9	18.5	28.7	4.2
p-Xylene	16	2.1–3.4	9	} 5.1	} 19.6	} 1.3	} 11.2	} 4.5			4.9
m-Xylene	16	5.4–7.8	21						} 26.0	} 34.4	12.7
o-Xylene	8	3.0–4.8	9.3	1.3	4.9	0.5	5.1	1.6			6.5
Cumene	3	—	—	—	—			—	—	—	1.7
n-Propylbenzene	2	1.5–4.0	1.5	—	—			0.4	1.1	3.3	1.8
m-, *p*-Ethyltoluene	8	—	9.0	2.4[c]	9.8[c]	} 1.9	} 12.5	1.6	—	—	3.4
1,3,5-Trimethylbenzene	3	—	2.1	1.2	1.9			0.6	—	—	1.5
1,2,4-Trimethylbenzene	9	—	9	4.4	10.1			1.6	—	—	3.4

[a] Near London Heathrow airport.
[b] City–county 50 km to the northwest of London with a population of about 155 thousand.
[c] Only a *p*-isomer.

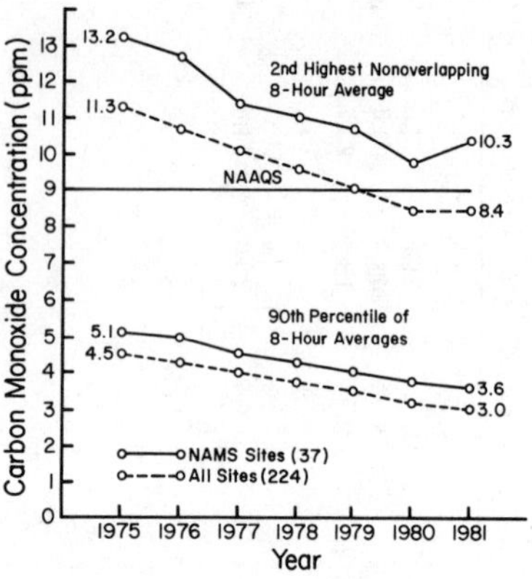

Figure 13. National trend in carbon monoxide levels. Comparing National Air Monitoring System (NAMS) with all sites and the second highest nonoverlapping 8-hour average with the 90th percentile of 8-hour averages, 1975–1981 *(51)*.

B. Gaseous Compounds of Sulfur

1. The Sulfur Oxides

Sulfur dioxide emissions in the United States have shown a drop for the 1970–1980 period (Fig. 12). Maximum urban sulfur dioxide levels are given in Table XII *(97)*. Efforts to reduce sulfur dioxide emissions because of their acid rain impact could, if legislated, reduce emissions

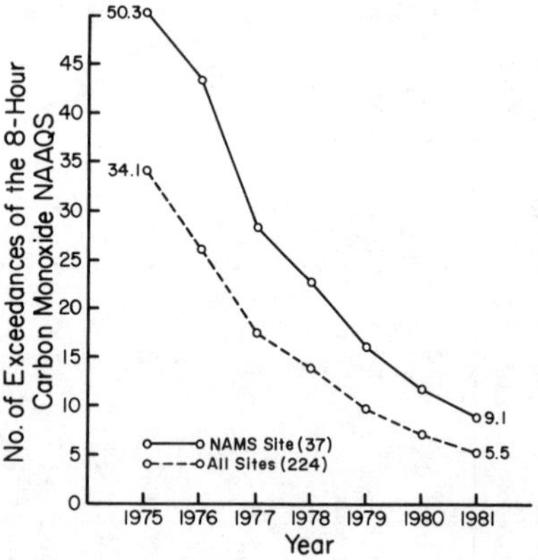

Figure 14. National trend in the composite average of the estimated number of exceedances of the 8-hour carbon monoxide National Ambient Air Quality Standard (NAAQS) at both National Air Monitoring System (NAMS) and all sites, 1975–1981 *(51)*.

Table XII Maximum Urban Sulfur Dioxide Levels in ppm, from 842 United States Sites, 1979 and 1980 by Averaging Time *(97)*

	Averaging time, hours			*Annual*
Percentile	*1*	*3*	*24*	—
99th—	0.78	0.60	0.22	0.03
95th—	0.46	0.35	0.14	0.02
50th (median)	0.16	0.12	0.06	0.01

by 6 to 10 million tons of the 31 million tons projected to the year 2000. However, efforts to reduce sulfur oxide emissions through the use of lower sulfur content fuels and sulfur oxide reduction techniques mandated by federal new source performance standards *(77)* have not overcome the effects of increases in per capita energy consumption *(32, 61)*. Although nuclear power was expected to furnish part of the new energy demand, its growth has been slowed considerably as a result of economic and public nuclear safety concerns.

Low sulfur content fuels are not readily accessible, and desulfurization of fuel oil and coal or their combustion gases is costly. Except for pressures to reduce acid rain impacts, public motivation to support costly control measures has not been so strong as the motivation to control particulate matter emissions.

Since 1963 sulfur dioxide concentrations in urban areas have dropped significantly (Fig. 15) because of the shift to natural gas and lower sulfur fuel oils for domestic heating and for smaller industrial power and heating plants. Major electric power plants are usually located in the outskirts of populated areas, and ground-level concentrations in populated areas tend to be low. Sulfur dioxide is one of the pollutants (the other is total suspended particulate matter [TSP]) measured by the World Health Organization–United Nations Environmental Program Global Environmental Monitoring System (GEMS) [Fig. 16; Table XIII *(104–104c)*].

2. Reduced Sulfur Compounds

a. Hydrogen Sulfide. The first systematic study of hydrogen sulfide was published in 1777 by Scheele, who referred to it as a stinkende (stinking or fetid) gas. Since that time there have been many studies concerned with its chemical properties, formation, uses, health effects, and odor impacts *(105, 106)*. No new significant

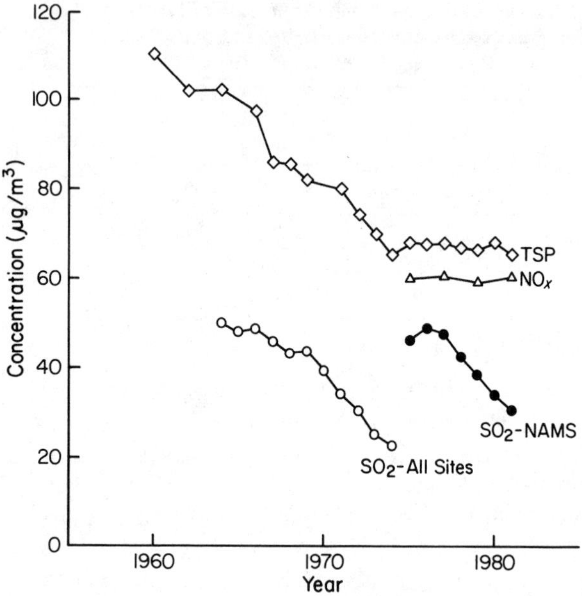

Figure 15. Ambient air concentrations for total suspended particular matter (TSP), nitrogen oxides (NO_x), and sulfur dioxide (SO_2). NAMS = National Air Monitoring System. *(32, 51, 98–100).*

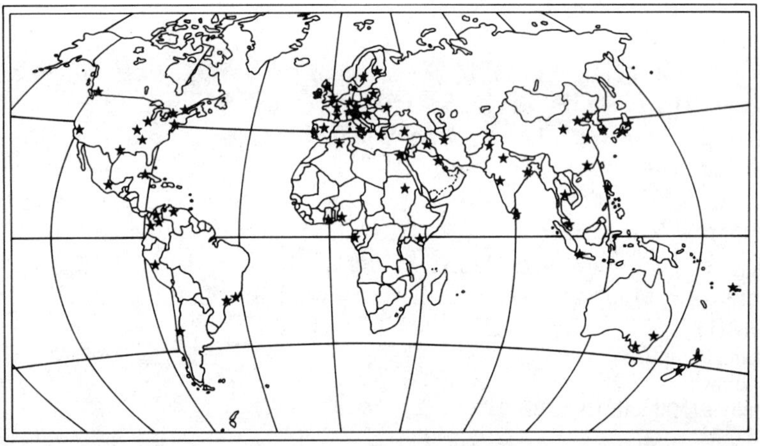

Figure 16. Monitoring locations, global environmental monitoring system (GEMS), World Health Organization—United Nations Environmental Program, 1979–1980. (From "Air Quality in Selected Urban Areas: 1977–1978," WHO offset Publication No. 76, Geneva, 1983. Reproduced with permission.)

Table XIII **Air Quality in Selected Urban Areas. Global Environmental Monitoring System, 1979–1980. [Reprinted from World Health Organization *(104)*.]**

Country (city)	*Particulate matter 24-hour geom. mean μg/m³*	*Sulfur dioxide 24-hour arith. mean μg/m³*
Australia (Melbourne)	68.8	—
(Sydney)	110.6	61.1
Belgium (Brussels)	15.5	76.3
Brazil (Sao Paolo)	71.5	134.6
Canada (Montreal)	77.9	131 (98%le)[c]
(Toronto)	90.9	79 (98%le)
(Vancouver)	71.9	52 (98%le)
Chile (Santiago)	77.8	58.5
Colombia (Bogota)	105.8	11.4
(Cali)	12.3	6 (98%le)
Cuba (Havana)	60.4	47.5
Denmark (Copenhagen)	37.2	30.1
Egypt (Cairo)	81.1	70 (98%le)
Finland (Helsinki)	101.5	134 (98%le)
Federal Republic of Germany (Frankfurt)	22.7	73.8
France (Gourdon)	43.3	95.3
(Toulouse)	19.5	35 (98%le)
Greece (Athens)	212.0	49.1
Hong Kong	85.6	70.0
India (Bombay)	188.5	48.4
(Calcutta)	423.6	48.0
(Delhi)	450.9	50.4
Indonesia (Jakarta)	260.1	—
Iran (Teheran)	357.9	159.7
Iraq (Baghdad)	503.8	18.9
Ireland (Dublin)	25.2	57.0
Israel (Tel Aviv)	—	52.6
Italy (Milan)	—	242.5
Japan (Osaka)	51.3	41.3
(Tokyo)	132.6	58.9
Kuwait (Kuwait)	521.7	1.6
Malaysia (Kuala Lumpur)	167.1	22.1
Netherlands (Amsterdan)	—	98 (98%le)
New Zealand (Auckland)	4.0	25.2
(Christchurch)	11.7	37.0
Pakistan (Lahore)	640.1	40.0
Peru (Lima)	28.9	38 (98%le)
Philippines (Manila)	109.2	89.7
Poland (Warsaw)	31.7	47.3
(Worclaw)	50.7	45.0
Portugal (Lisbon)	—	45.6
Romania (Bucharest)	140.2	16.8
(Craiova)	91.6	5.2

Table XIII (*Continued*)

Country (city)	*Particulate matter 24-hour geom. mean* $\mu g/m^3$	*Sulfur dioxide 24-hour arith. mean* $\mu g/m^3$
Spain (Madrid)	140.4	138.9
Sweden (Stockholm)	—	69.7
Thailand (Bangkok)	221.7	10.6
United Kingdom (Glasgow)	34.2	103.8
(London)	35.9	84.1
United States (Chicago)	118.5	47.4[b]
(Los Angeles)	150.8[b]	41.4[b]
(New York City)	59.5	76.2
Yugoslavia (Zagreb)	123.6	94.5

[a] Annual averages of selected sampling sites. The United States Primary Ambient Air Quality Standard for particulate matter is 75 $\mu g/m^3$ annual geometric mean, and for sulfur dioxide it is 80 $\mu g/m^3$ annual arithmetic mean.
[b] 1977–1978 data.
[c] 98%le: 98% of all measurements are below or at the indicated figure.

anthropogenic emission sources have been found, and worldwide emissions continue to be estimated at approximately 3 Tg/year (Table VIII). Natural emissions, which are of the order of 84 Tg/year, have elicited a considerable number of studies as to their nature and their emission rates throughout the biosphere *(107–111)*. Atmospheric background levels of hydrogen sulfide average from 0.05 to 0.1 ppb but vary considerably, depending upon climatic and terrestrial conditions. Odor threshold concentrations for hydrogen sulfide are of the order of 5 to 100 ppb. Olfactory paralysis occurs at 150 to 250 ppm, subacute poisoning occurs at between 100 and 1000 ppm, and acute poisoning and death occur at above 1000 ppm. The occurrence of olfactory paralysis makes accidental releases of high concentrations very dangerous *(105)*.

Anthropogenic pollution emission sources include petroleum refining, gas plants, sewage treatment plants, coke ovens, Kraft paper pulp plants, and waste disposal sites. There are at present no United States national emissions standards, but a number of states have adopted their own emission regulations to control, principally, malodor impacts. The state emission standards for hydrogen sulfide range from 0.6 grain/100 ft^3 (10 ppm) for California and New Mexico to 100 grains/100 ft^3 (1670 ppm) for Ohio and Michigan. Many countries of the world have emission standards ranging from 3.3 ppm (any process, Australia) to 1000 ppm (coke ovens, Federal Republic of Ger-

Table XIV Biogenic Emissions of Sulfur Compounds. Emission Rate gS/m²/yr. [Reprinted from Aneja *et al.* (107).]

		H_2S		DMS		DMDS		CH_3SH		CS_2		COS	
Investigator	*Source*	*Avg.*	*Max.*	*Avg.*	*Max.*	*Avg.*	*Max.*	*Avg.*	*Max.*	*Avg.*	*Max.*	*Avg.*	*Max.*
Aneja (1975) Hill, Aneja, Felder (1977)	Salt Marsh, N.Y.	0.55	41.5		3.84		0.81		1.92				
Maroulis, Bandy (1977)	Atlantic Ocean, Va.			0.006									
Aneja *et al.* (1979)	Salt marsh, NC	0.5	100	0.66	2.5						0.2		0.03
Aneja *et al.*	Fresh water marsh, N.C.	0.6	1.27										
Adams *et al.* (1979)	Salt marsh, N.C.	72	381	0.093		0.004			6.56		1.13		6.36
Adams *et al.* (1979)	Inland soils U.S.	0.001		0.001		0.001				0.001		0.002	
Jaeschke *et al.* (1978)	Swamps and tidal flats	0.044											
Hansen *et al.* (1978)	Sediments of shallow coastal area Denmark	~19	~2000										
Delmas *et al.* (1979)	Soils of humid equatorial forests Ivory Coast (near a salt marsh)	0.07	2.6										
Delmas *et al.* (1979)	Soils of temperate regions, France	0.044	0.24										

many). A number of the nations have adopted a 5.0-ppm emission standard *(105)*.

Ambient air quality standards have also not been set by the United States government. State standards vary from 0.003 ppm for New Mexico to 0.1 ppm for Pennsylvania. A number of the states specify a standard of 0.03 ppm. Countries other than the United States have ambient air quality standards varying from 0.003 in Spain to 0.3 ppm in Finland. Eastern European nations use a 0.005-ppm standard, and many other nations have standards ranging from 0.03 to 0.1 ppm hydrogen sulfide in ambient air *(105)*.

b. Biogenic Reduced Sulfur Compounds. Although hydrogen sulfide is the dominant reduced sulfur gas released in nature, significant amounts of dimethyl sulfide (CH_3SCH_3), dimethyl disulfide (CH_3SSCH_3), methyl mercaptan (CH_3SH), carbon disulfide (CS_2), and carbonyl sulfide (COS) have been identified and measured [Table XIV *(107)*]. Because the atmospheric residence time for carbonyl sulfide is 1 to 2 years, whereas that of hydrogen sulfide is 1 to 2 days, its average atmospheric background concentration is approximately five times that of hydrogen sulfide (0.5 ppb and 0.05–0.1 ppb, respectively; Table VIII). A considerable amount of continuing research to assess the sources and strengths of these emissions more accurately and to learn of their roles in the atmospheric sulfur cycles *(105–111)* is being conducted.

C. Gaseous Compounds of Nitrogen

1. The Oxides of Nitrogen

Nitrogen oxides (NO_x) pollution data are generally expressed in units of equivalent nitrogen dioxide (NO_2). Most of the nitrogen oxide emissions from combustion processes are in the form of nitric oxide (NO), which gradually oxidizes to nitrogen dioxide, a brown irritating gas. Nitrogen dioxide is photochemically active and is oxidized to nitric acid or its equivalent in the form of various nitrate compounds (see Chapter 2).

Emissions of nitrogen oxides have increased approximately 400% since 1940. Projections to the year 2000 show a continuing, but slower rate of increase, amounting to an increment of 10–15% of the 1980 emissions (Fig. 12). The 1975–1981 ambient air measurement data reflect the leveling off of the emission data over this time period (Fig. 15). The ambient air concentration levels are to a large extent well below the United States National Ambient Air Quality Standard of 100 $\mu g/m^3$ annual arithmetic mean. They are, however, important in that they play a complex and important role in the production of photochemical oxidants and acidic substances, which are of environmental concern.

On a strict weight-to-weight basis the acidic impact of nitrogen dioxide should be 70% that of sulfur dioxide since the molecular equivalent weight ratios are 46 to 32 (64 ÷ 2), respectively. However, since the emission tonnage for NO_2 is less than that for SO_2, the overall contribution of nitrogen oxides to atmospheric acidity should be in the order of 60% of sulfur dioxide contribution. In many cases the contribution of nitrogen oxides to the acid rain problem is not fully considered. The United States Environmental Protection Agency (USEPA), for example, attributes an amount that is less than half the sulfuric acid acidity for the impact of the nitric acid on the acidity of rain water *(4)*. The rainout, washout, dry deposition phenomena operate differently for SO_2 and NO_x. Sulfur dioxide is very soluble in rain water of pH 3 or greater, whereas NO and NO_2 are not very soluble. Emission patterns for SO_2 and NO_x also differ. A larger proportion of SO_2 is released from power plant smoke stacks, whereas a large portion of the NO_x comes from area sources such as automobile exhaust, commercial and residential heating, and trash burning *(112)*.

2. Ammonia

Ammonia is an important trace gas in the atmosphere. It plays an important role in the nitrogen cycle involving plant and animal life processes. Under certain circumstances it becomes a hazardous pollutant. Accidental releases from pipeline, railroad, or transport carrier and industrial storage vessels have caused death and serious injury *(113)*. On a worldwide basis, anthropogenic emissions to the atmosphere are relatively insignificant: 6 Tg/yr compared to 260 Tg/yr natural emissions (1 Tg = 10^6 metric tons) (Table VIII).

Major industrial atmospheric emission sources include ammonia manufacturers, petroleum refineries, diammonium and nitrate fertilizer plants, large cattle feed lots, Solvay process plants, and by-product and beehive coke ovens. In the United States these sources account for an estimated 130,000 tons of emissions compared to a 14.9 million tons/year ammonia production rate. World production has been projected to 84 million tons/year by 1985. Other anthropogenic atmospheric pollution sources include municipal sewage and waste incineration plants, direct application of ammonia as an agricultural fertilizer, fossil fuel combustion, and domestic and industrial ammonia cleansers *(113)*.

Ambient air ammonia concentrations vary according to nearness to urban areas or major pollutant sources. Background continental concentrations have been measured in the range of 4 to 7 ppb (2.8 to 4.9 $\mu g/m^3$). Urban concentrations are generally four to five times rural concentrations. Amounts varying from 50 to 400 ppb were measured in Cagliari, Italy, and as much as 300 ppb was present downwind from two pharmaceutical plants near Tokyo, Japan. Ground temperatures affect animal and plant matter decomposition rates and thus are a factor in ambient air levels *(114, 114a)*. Because of increased fossil fuel consumption during winter months, higher ambient air concentrations are found in urban areas. Ammonia found over ocean areas is generally believed to be of continental origin *(113)*.

As one of the few alkaline atmospheric trace gases, both anthropogenic and natural ammonia emissions react with acidic pollutants, such as sulfur and nitrogen oxides, to form particulate ammonium salts in air *(114a)*. Studies of the chemical composition of fine (<2.5-μm diameter) and coarse (>2.5-μm and <15-μm) particulate matter in the Detroit, Michigan, metropolitan area showed that the fine fraction accounted for more than 60% of the total mass below 15 μm diameter. Fifty percent of the fine fraction was ammonium sulfate. Nitrate comprised less than 1% of the fine fraction and only 5% of the coarse fraction *(115)*. However, ammonium nitrate has significant volatility in air and has been demonstrated to evaporate from filters, leaving some question as to the actual amounts present in air *(116)*.

D. The Gaseous Halogens

Halogenic substances comprise a large fraction of those atmospheric pollutants having known hazardous effects as well as potential for long-term harmful impacts. They include the elemental and compound forms of fluorine, chlorine, bromine, and iodine. They cause plant, animal, and material damage; bone deformation; kidney and liver damage; and mutagenicity and pose a threat to oceanic life and stratospheric ozone. The gaseous compounds include hydrofluoric and hydrochloric acids, chlorofluorohydrocarbons (freons), fumigants, cleaner solvent, and pesticide vapors (Tables VII and X).

A detailed study of global concentrations and reactions of halogens in the atmosphere has been reported by Cicerone *(117)*. Table XV indicates the average concentrations of gaseous inorganic and organic

Table XV Summary of Available Data on Halogen-containing Gases in the Troposphere (*117*)

Containing	*Organic gases*	*Inorganic gases*
Fluorine	1 ppb	0.1–0.4 ppb
Chlorine	2.5 ppb	1–2 ppb
Bromine	10–25 ppt	0.01–7 ppt
Iodine	1–5 ppt	0.1–3 ppt

halide compounds in the troposphere. Rancher reported similar concentrations of gaseous compounds over the Atlantic, containing chlorine (0.07–0.7 ppb), bromine (1.5–15 ppt), and iodine (2–12 ppt). Organic fractions predominated, and lifetimes found were chlorine 0.5, bromine 1.3, and iodine 2.0 days *(118)*. Antarctica atmospheric measurements by Rasmussen *et al.* found 177–192 ppt CCl_3F (freon-11), 305–311 ppt CCl_2F_2 (freon-12) and 113–161 ppt CH_3CCl_3 (methyl chloroform) *(119, 119a)*.

1. The Fluorides

Hydrogen fluoride is the most common inorganic fluoride gas found in air. It is emitted principally by the steel, coal-burning, aluminum, and phosphate fertilizer industries. Recent estimates place global industrial emissions at 1.8 million tons fluoride per year. Background concentrations are in the 1- to 2-ppb range. Concentrations near emission sources range up to 50 ppb *(120)*. Smith and Hodge conducted an in-depth study of airborne fluorides and their impact on health and the environment *(121)*. Allen *et al.* also reviewed environmental fluorides, phosphate fertilizer industry emissions, and fluoride control legislation *(120)*.

Atmospheric fluoride concentrations are small when compared to intakes from fluoridated water and toothpaste. However, fluorides settling upon and absorbed by certain plants and grasses pose hazards particularly to grazing animals. In the United States many states have adopted ambient air quality standards to prevent fluorosis to animals. Typical standards expressed as average ppb HF in air are 12-hour, 4.5; 24-hour, 3.5; 7-day, 2.0; and 30-day, 1.0. Typical forage standards range from 40 to 80 ppm F, dry weight, depending upon time of the year and length of growing season *(120, 121)*. New source performance standards (NSPS) for the aluminum and phosphate fertilizer industries have been promulgated by the USEPA (*77*, Part 60, Subparts S–X). Many countries of the world have adopted similar ambient air and emission standards (see Vol. VIII, Chapter 5) *(121)*.

2. Chlorine, Hydrogen Chloride, and Chlorinated Hydrocarbons

Chlorine gas (Cl_2) is widely used in manufacturing and water and sewage treatment and in the home. It is irritating and highly reactive in air. Because of its reactivity and low concentrations in relation to other inorganic chlorides, it is not easily measured in ambient air, often being confused with hydrogen chloride. A few studies have indicated tropospheric concentrations in the parts per billion (*117, 122, 123;* see also Vol. I, Chapter 2). Stratospheric concentrations are studied in detail to understand upper atmosphere chemical and photochemical reactions better *(117)*. Hydrogen chloride (HCl), a more common anthropogenic inorganic gaseous chloride pollutant, contributes to acidic rain but is usually considered a minor contributor compared to sulfur and nitrogen oxide emissions. Table VIII shows estimated HCl and Cl_2 pollution emissions to be 4 Tg/year and natural emissions 100 to 200 Tg/year.

Chlorinated hydrocarbons constitute a very broad range of organic compounds, which have been used extensively for more than half a century. They are easily manufactured, inexpensive, stable, excellent solvents and cleaning agents and can be polymerized to make inexpensive plastics *(124)*. Many herbicides, pesticides, and fungicides are chlorinated hydrocarbons. However, it soon became evident that occupational exposures were causing health problems and that pesticides, such as dichlorodiphenyltrichloromethane (DDT), were persistent and not biodegradable, causing potentially serious long-term environmental problems *(125, 126)*. Vinyl chloride has been listed by the USEPA as a hazardous air pollutant *(77)*. Approximately half of the USEPA study list of potential hazardous substances (Table VII), as well as a sizable fraction of the federal extended list of hazardous substances *(79)*, are chlorinated hydrocarbons. The New Jersey project on airborne toxic elements and volatile organic substances found measurable amounts of 14 chlorinated hydrocarbons in the air of three New Jersey cities. Concentrations ranged from detection limits to 0.5 ppb [Table XVI *(126)*]. The USEPA has published "Health Assessment Documents" on carbon tetrachloride, methyl chloroform,

Table XVI Geometric Means of Volatile Organic Compound (VOC) Concentrations at Three Urban Sites in New Jersey, July 6 to August 16, 1981[a] [Reprinted from Harkov *et al.* (126).]

Compound	*Newark*	*Elizabeth*	*Camden*
Vinyl chloride	0	0	0
Vinylidene chloride	0.38	0.35	0.36
Methylene chloride	0.35	0.23	0.72
Chloroform	0.06	0.10	0.04
Ethylene dichloride	0	0	0
Benzene	1.03	1.05	1.11
Carbon tetrachloride	0.01	0.01	0.01
Trichloroethylene	0.50	0.27	0.21
Dioxane	0.01	0.02	0.01
1,1,2-Trichloroethane	0.01	0.01	0.01
Toluene	4.65	2.89	1.82
Ethylene dibromide	0	0	0
Tetrachloroethylene	0.45	0.31	0.24
Chlorobenzene	0.11	0.08	0.07
Ethylbenzene	0.33	0.26	0.17
m,p-Xylene	0.99	0.75	0.49
Styrene	0.13	0.11	0.07
o-Xylene	0.26	0.22	0.15
1,1,2,2-Tetrachloroethane	0.01	0	0
o-Chlorotoluene	0.02	0.02	0.01
p-Chlorotoluene	0.21	0.25	0.22
p-Dichlorobenzene	0.05	0.07	0.04
o-Dichlorobenzene	0.03	0.02	0.01
Nitrobenzene	0.07	0.10	0.07

[a] Concentrations in ppb; undetected quantities set to 0.0025 ppb, which is one-half the detection limit; quantitation limits are set at three times signal to noise ratios or 0.05 ppb for aromatic VOC and 0.10 ppb for chlorinated VOC.

dichloromethane (methylene chloride), perchloroethylene, and trichloroethylene *(127)*, as well as an in-depth review of the fates and impacts of major forest use pesticides *(128)*.

Chlorofluorohydrocarbons are inert and are generally considered to be nontoxic. They make excellent refrigerants and aerosol propellants. Their inertness gives them long residence times in the atmosphere, enabling them to diffuse to the stratosphere, where they interfere with stratospheric photochemical systems and endanger the protective ozone layer (see Section I, J, this chapter).

E. Ozone and Oxidants

Ozone is used as the indicator pollutant for photochemical oxidation products first recognized in what was commonly called "Los Angeles smog." It is formed through a complicated series of photochemical reactions involving sunlight, nitrogen oxides, and volatile organic compounds largely associated with vehicular emissions (see Chapter 2). Besides ozone, the reaction products include a broad range of oxidized and peroxidized organic compounds including peroxyacyl nitrates (PANs) and condensed organic polymers. The total mixture, or "smog," causes eye irritation, lachrymation, respiratory difficulties, and crop damage.

The ambient air ozone concentrations in the United States dropped approximately 14% as indicated by the drop in the annual averages of the second highest maximum daily 1-hour concentrations for 209 nationwide monitoring stations (Fig. 17). All the indicated averages were above the National Ambient Air Quality Standard (NAAQS) of 0.12 ppm. Only 19 of 80 cities having standard metropolitan statistical area populations above 500,000 people had annual averages below the NAAQS *(51)*.

The number of days per station in which the NAAQS was exceeded in the third quarter of the year averaged 6.9 in 1975 and 4.0 in 1981, a 42% drop (Fig. 18). Part of the drop in the number of exceedances can be attributed to automobile catalytic exhaust converters, which re-

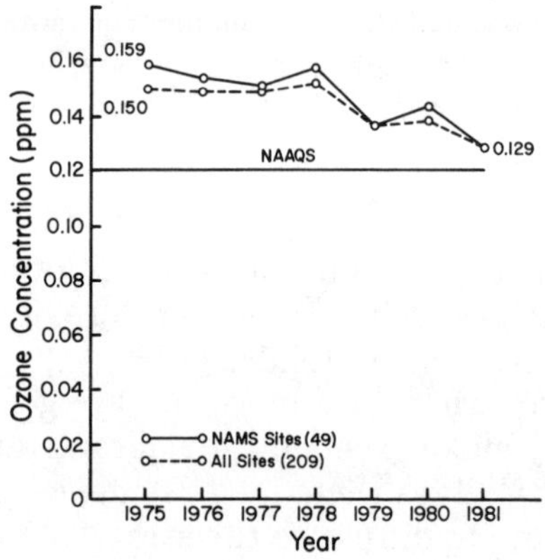

Figure 17. National trend in the composite average of the second-highest daily maximum 1-hour ozone concentration at both National Air Monitoring Systems (NAMS) and all sites, 1975–1981 *(51)*.

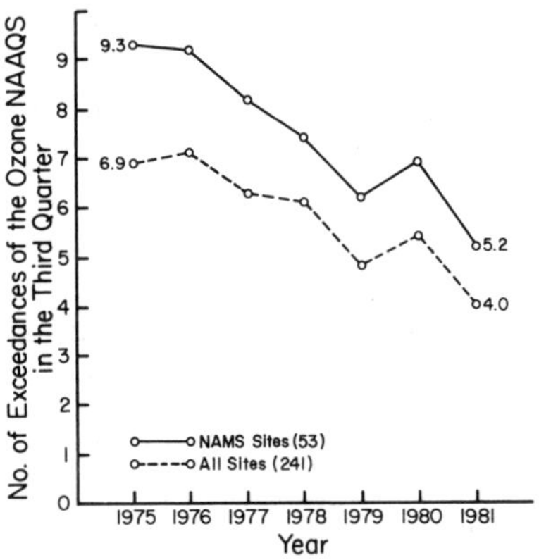

Figure 18. National trend in the composite average of the estimated number of daily exceedances of the ozone National Ambient Air Quality Standard (NAAQS) in the third quarter (July–September) at both National Air Monitoring System (NAMS) and all sites, 1975–1981 *(51)*.

duce hydrocarbon emissions. However, part of the drop was also due to a change in the analytical method used to measure the ozone, and a part was due to the drop in average motor fuel consumption *(51)*.

III. The Primary Air Pollutants—Nonviable Particles

Because of its effect on visibility and its association with soiling, corrosion, and respiratory effects, suspended particulate matter, or smoke, has received the earliest and most universal concern of air pollution clean-up programs. Figure 16 shows the location and Table XIII gives the concentrations of total suspended particulate matter (TSP) for selected cities and countries of the world as measured by the World Health Organization–United Nations Environmental Program global monitoring (GEMS) project *(104–104c)*.

Figure 12 shows the estimated emissions for TSP in the United States for the period 1940–1980 and estimated projections to the year 2000 *(32, 51, 98–100)*. As shown, the emissions of TSP decreased nearly 75% between 1940 and 1980. Projected emissions to the year 2000 predict a slight increase over the 1980 emissions, principally due to increased population effects rather than an easement of control programs.

Figure 15 shows the annual geometric mean values for 24-hour ambient air concentrations of total suspended particulate matter (TSP). The data show a strong 50% decline from 1940 to 1974, followed by a general leveling off for the 1974 – 1981 period. In all probability, the less easily identified, technologically more difficult and possibly more costly to control particulate matter sources are persisting to give relatively high urban area background averages. Such sources include fugitive emissions from general commercial, residential, and vehicular transportation; noninventoried small point and area sources; and resuspension of settled dusts and secondary aerosols formed by thermochemical and photochemical reactions in air.

A. Particle Size Distribution

The size distribution of particulate matter in air determines its potential for adverse effects on health and visibility. The aerodynamic size of the particle dictates whether it falls rapidly to the ground or remains suspended to enter the human respiratory system through the nose or mouth. Again depending upon size and chemical nature, the particle may be filtered out in the upper nasal system, settle in the tracheobronchial system, penetrate into the alveoli, or remain suspended to be expired (Fig. 2, Chapter 7, Vol. II).

Atmospheric particle size distributions show multimodal patterns due in a large part to the mechanisms of their generation. The modes are usually grouped into two broad classifications for air pollution purposes: fine and coarse *(129)*. Fine particles are considered to be those smaller in size than a modal minimum of 2 to 3 μm in diameter (Fig. 15, Chapter 3, Vol. I). The fine (respirable) particles can penetrate into the pulmonary region of the respiratory system. Fine particles are formed by condensation processes from the vapor or gaseous state and range from 0.005 to 0.05 μm in diameter. The condensation process is followed by an agglomeration or "accumulation" process increasing the upper particle size range from 0.05 to about 2.0 μm *(129a)*. Coarse particles are principally formed by mechanical, grinding, or other dispersive forces. Their sizes range from approximately 1 to more than 100 μm in diameter, depending upon the dispersive forces and atmospheric conditions. In practice, fine and coarse fractions are considered to be those collected by the fine and coarse stages of a dichotomous sampler. The fine stage has an upper cutoff point of about 2.5 μm; the coarse (inhalable) stage has an upper cutoff point of either 10 or 15 μm, depending upon design. Present tendencies are to standardize on the 10-μm cutoff point. Figure 19 *(130)* shows the inhalable particle monitoring sites in the United States, and Table

Figure 19. Inhalable particulate network sites established as of 1980 *(130)*.

Table XVII Mean Chemical Composition and Atmospheric Concentrations of Suspended Particulate Matter Sampled by the United States Environmental Protection Agency's Inhalable Particle and National Air Surveillance Networks—μg/m³ and Percentage of Total Mass Sampled *(2)*

Type of sample:	*Urban*						*Rural*	
Number of samples:	*745*				*225*[a]		*133*[b]	
Particle size:	*Coarse, 15–2.5 μm*		*Fine, less than 2.5 μm*		*All less than ~50 μm*			
Particle	*Mean value of 745 values*	*%*	*Mean value of 745 values*	*%*	*Mean value of 225 values*	*%*	*Mean value of 133 values*	*%*
All (total mass)	21.655	100.00	22.680	100.00	74.990	100.00	36.504[a]	100.00
Aluminum	1.797	8.30	0.353	1.56	—	—	—	—
Antimony	0.051	0.24	0.050	0.22	—	—	—	—
Arsenic	0.003	0.01	0.004	0.02	0.005[c]	0.01	0.003[d]	0.01
Barium	0.060	0.28	0.060	0.26	0.273	0.36	0.281	0.77
Beryllium[e]	—	—	—	—	(0.095)	—	(0.084)	—
Bromine	0.019	0.09	0.077	0.34	—	—	—	—
Cadmium	0.006	0.03	0.007	0.03	0.002	0.01	0.001	0.01
Calcium	1.503	6.94	0.340	1.50	—	—	—	—
Chlorine	0.440	2.03	0.155	0.68	—	—	—	—
Chromium	0.008	0.04	0.006	0.03	0.013[c]	0.02	0.015[d]	0.05
Cobalt	—	—	—	—	0.001[c]	0.01	0.001[d]	0.01
Copper	0.019	0.09	0.026	0.12	0.143	0.19	0.136	0.37
Iron	0.743	3.43	0.205	0.90	0.923	1.23	0.254	0.70
Lead	0.083	0.38	0.314	1.38	0.353	0.47	0.066	0.18
Manganese	0.021	0.10	0.013	0.06	0.031	0.04	0.008	0.02
Mercury	0.003	0.01	0.003	0.01	—	—	—	—
Molybdenum	—	—	—	—	0.002	0.01	0.001	0.01
Nickel	0.004	0.02	0.007	0.03	0.007	0.01	0.002	0.01
Phosphorus	0.056	0.26	0.021	0.09	—	—	—	—
Potassium	0.222	1.03	0.156	0.69	—	—	—	—
Selenium	0.001	0.01	0.002	0.01	—	—	—	—
Silicon	2.561	11.83	0.360	1.59	—	—	—	—
Strontium	0.246	0.21	0.051	0.22	—	—	—	—
Sulfur	0.339	1.56	2.056	9.07	—	—	—	—
Tin	0.006	0.03	0.006	0.03	—	—	—	—
Titanium	0.042	0.19	0.015	0.07	—	—	—	—
Vanadium	0.008	0.04	0.010	0.04	0.015	0.02	0.004	0.01
Zinc	0.038	0.18	0.067	0.30	0.147	0.20	0.114	0.31
Nitrate	0.699	3.23	1.071	4.72	4.647	6.20	2.341	6.41
Sulfate	0.706	3.26	5.30	23.37	10.811	14.42	8.675	23.77
Sum of percentages	—	43.82[f]	—	47.34[f]	—	23.20	—	32.64

[a] Except for arsenic, chromium, and cobalt, where the number of samples was 1245.
[b] Except for arsenic, chromium, and cobalt, where the number of samples was 30.
[c] Except for arsenic, chromium, and cobalt, where the mean total mass was 76.647 μg/m³.
[d] Except for arsenic, chromium, and cobalt, where the mean total mass was 30.367 μg/m³.
[e] Nanograms/m³.
[f] Sulfur is counted twice, as sulfur and as sulfate. Some of this sulfur exists as sulfides, sulfites, and forms other than sulfate.

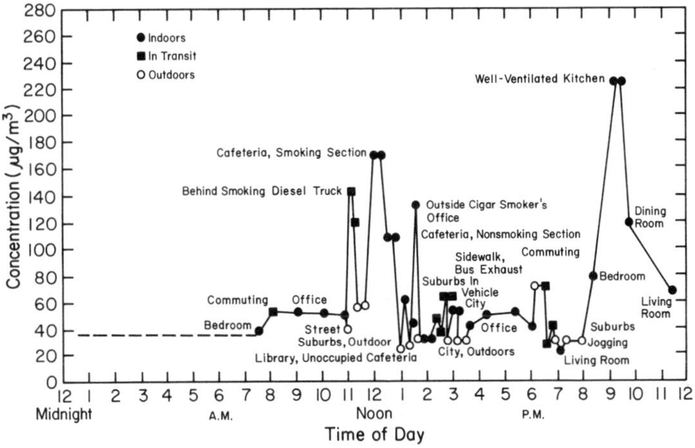

Figure 20. Personal exposure to respirable particles *(130, 131)*.

XVII gives detailed composition data for coarse and fine suspended particulate matter in urban and rural areas in the United States *(97)*. Figure 20 illustrates a representative respirable particle exposure for a person during a typical day *(130, 131)*.

B. Particle Composition

Table XVII shows the major chemical species associated with TSP and its coarse and fine particle fractions. The fine particles in general are secondary-type particles formed by condensation processes. Figure 21 *(130, 131a)* compares the relative composition of fine particulate matter to the total annual stratified arithmetic average of particulate matter in Portland, Oregon. Table XVIII gives the elemental enrichment ratios for fine urban and suburban particles in relation to TSP composition in the Cleveland, Ohio, area *(130)*. A mean 2.8-fold urban enrichment of heavy metal composition in fine particles over suburban areas is shown. Major inorganic ion composition in atmospheric aerosols *(132)* and motor vehicle emissions have been studied *(133)*. Increasing use of incineration to reduce municipal wastes *(52)* or to use refuse as a fuel supplement *(134)* raises questions with respect to the composition of the incinerator emissions and their long-term effects upon the environment.

Particulate ammonium concentrations in the United States have ranged from 0 to 15 $\mu g/m^3$ in urban areas and to 1.2 $\mu g/m^3$ in nonurban areas *(135)*. As much as 33 $\mu g/m^3$ was reported for severe pollution

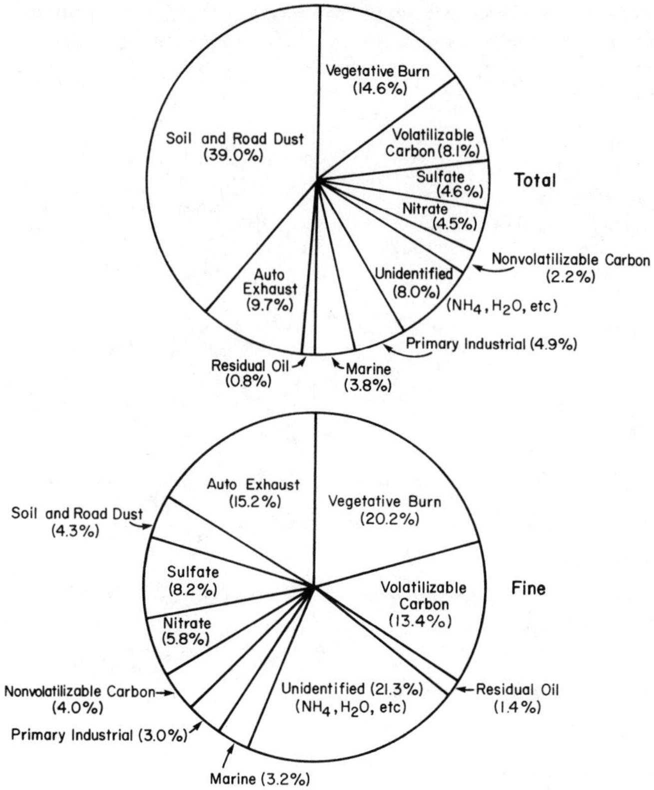

Figure 21. Aerosol sources in downtown Portland, Oregon, annual stratified arithmetic average. Volatilizable and nonvolatilizable carbon approximately correspond to organic and elemental carbon *(130)*.

episodes *(136)*. Vertical profiles of ammonia and ammonium particulate matter follow somewhat similar patterns above 1000 m altitude *(114)*. Worldwide ammonium sulfate atmospheric pollution has been hypothesized to be a major contributor to increasing atmospheric reflectivity *(113–115, 135–137)*.

Organic fraction composition of particulate matter varies considerably, depending upon the types of sources and the atmospheric mixing conditions of an air basin. Lows of less than 1% organic matter in particulate matter for remote areas to as much 45% for the Los Angeles area have been measured. New York City averages 10% to 15% benzene extractable matter *(130)*. The organic fractions contain a large number of compounds, some of which are considered hazardous (see Section I, K, this chapter).

Table XVIII Ratios of Urban (U) to Suburban (S) Concentrations in Air, Cleveland, Ohio, Area *(130)*

Enriched in cities		*Ratio similar to TSP*[a]		*Enriched in suburban particles*	
Element	*U/S*	*Element*	*U/S*	*Element*	*U/S*
Antimony	6.9	Mercury	3.0	Silicon, tin	1.8
Chloride	6.5	Iron	2.8	Copper, vanadium	1.8
Beryllium	6.1	Cadmium	2.5	Aluminum	1.7
Chromium	5.6	Sodium	2.4	Zinc	1.6
Cobalt	3.4	Magnesium	2.4	Arsenic	1.4
Bismuth	3.3	Manganese	2.2	Selenium	1.3
		Calcium	2.0	Bromine	1.2

[a] Mean TSP ratio = 2.8.

IV. The Primary Air Pollutants—Viable Particles

Chapter 4, Volume I, by Jacobson and Morris, defines, illustrates, and discusses the types of viable particulate matter to be found in the atmosphere. Concise descriptions are given for categories within the three broad classes of airborne pollens, microorganisms, and insects. Such substances and organisms have extensive impacts upon the health and well-being of the human population. Increasingly sophisticated research activities in the areas of discovery, identification, evaluation, and prevention are being conducted and reported in the medical and scientific journals. Pedgley describes the meteorology of windborne pests and diseases *(138)*, Hayes describes pesticides studied in man *(139)*, and Couch describes the human viral diseases that are transmitted by air [Table XIX *(140, 140a)*]. The *International Archives of Allergy and Applied Immunology* publishes numerous articles on airborne diseases and allergens *(141)*. Roth has edited a handy, easily understood guide for physicians, the general public, and world travelers on allergenic substances in the air. Listed by

Table XIX Human Viral Diseases Transmitted by the Airborne Route [Reprinted from Couch, *(140)*.**]**

Smallpox	Influenza
Chickenpox	Adenovirus 4 and 7
Measles	Coxsackie A21
Rubella	Lymphocytic choriomeningitis

Table XX Epidemiologic Features of Nine Epidemics of Legionnaires' Disease Associated with Indoor Transmission. [Reprinted from Imperato *(144)*.]

Location	*Date*	*Number of cases*	*Case fatality rate (%)*	*Source*
St. Elizabeth's Hospital	July–August 1965	81	17	Probably soil
Health Department, Pontiac, Mich.	July–August 1968	144	0	Air conditioning
Benidorm, Spain	July 1973	164	18	Not known
Odd Fellows Convention Philadelphia	September 1974	11	10	Possibly air conditioning
Legionnaires' Convention, Philadelphia	July–August 1976	182	17	Possibly air conditioning
Indiana Memorial Union Bloomington, Ind.	February–May 1978	39	10	Possibly air conditioning Possibly soil
Baptist Memorial Hospital, Memphis	August 1978	15	0	Air conditioning
Shopping Center Vasteras, Sweden,	August–September 1979	67	1.5	Unknown
Lido del Savio Hotel, Como, Italy	July–October 1980	21	4.7	Possibly water supply

sections of the United States and countries of the world are the common names of trees, grasses, weeds, molds, and fungi that emit allergy-inducing substances *(142)*.

Increasing recognition that indoor air is a major source for airborne allergies and diseases is reflected in books and sections of books, articles, and symposia devoted to the health aspects of indoor pollution (see Chapters 4 and 7). Solomon and Burge recently reviewed the literature on the role of allergens and pathogens in indoor air *(143)*. Legionellosis, transmitted in a large part by poorly maintained air conditioning systems, has been the subject of a number of studies [Table XX *(144)*]. Airborne indoor viruses *(140)* and allergenic agents *(145)* have been described.

V. The Primary Air Pollutants — Radioactive Gases and Particles

Atmospheric ionizing radiation comes from many sources both natural and artificial (anthropogenic). Principal natural sources include cosmic rays and the radionuclides they produce: radon, thoron, and their respective decay products. Although radon and thoron are produced by trace radioactive minerals in the earth's crust, soil dust contributes a very small amount to the total of natural radioactivity. Anthropogenic sources include effluents from the military and nuclear industries; nuclear wastes; mining; university, medical, and industrial research; and x-rays from applied medical and dental use and industrial and research projects *(57–61, 146–148)*. Figure 22 shows the relative per capita doses received from natural and artificial sources *(59)*.

Atmospheric nuclear weapons testing was discontinued by the United Kingdom in 1958 and the United States and USSR in 1962. However, France continued atmospheric tests through 1974; China has continued atmospheric testing beyond 1978. Table XXI summarizes the number of atmospheric nuclear tests, their fission, and total yields up to 1980 *(57)*.

The generation of nuclear electrical power creates many fission- and neutron-activated radioactive products. Radionuclides released to the atmosphere include the noble gases krypton and xenon; ^{14}C, ^{16}N, ^{35}S, and ^{41}Ar activation gases; tritium; iodine; and particulate matter. Up to 43 radionuclides have been identified in the particulate matter produced. The 1982 report by the United Nations Scientific Committee on the Effect of Atomic Radiation gives detailed inventories of nuclear power plant effluents throughout the world *(57)*.

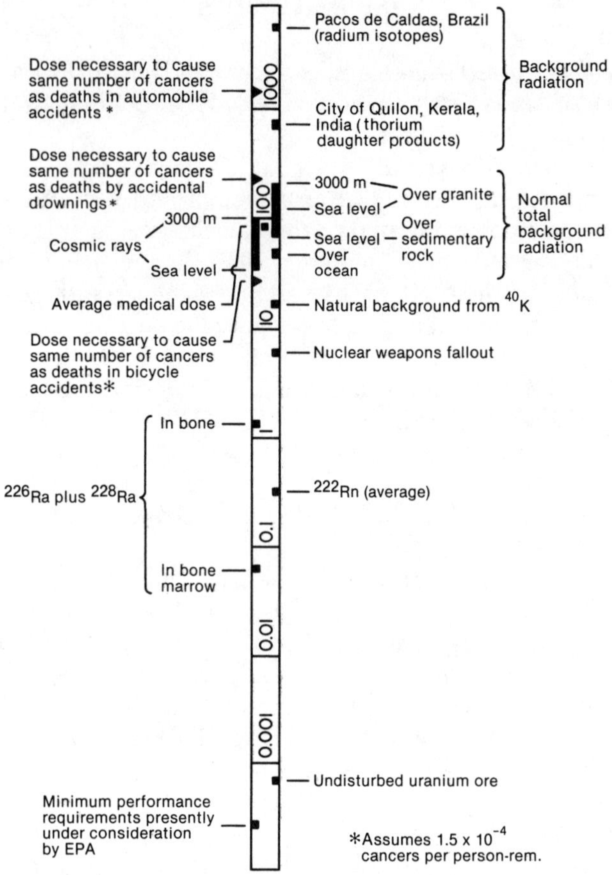

Figure 22. Doses from various radiation sources (mrem per year) (Reprinted with permission from K. R. Hings, G. R. Heath, D. R. Anderson and C. D. Hollister, *Environmental Science and Technology* **16,** 28A [1982]. Copyright 1982 American Chemical Society).

Table XXI Estimated Yields of Atmospheric Nuclear Tests *(57)*

Year	*Country*	*Number of tests*	*Estimated yield (Mt)*	
			Fission	*Total*
1945–1962	U.S.	193	72.1	138.6
1949–1962	USSR	142	110.9	357.5
1952–1953	U.K.	21	10.6	16.7
1960–1974	France	45	10.9	11.9
1964–1980	China	22	12.7	20.7
Total		423	217.2	545.4

Table XXII Annual Per Capita Doses from the Continued Generation of Nuclear Electric Power to the Year 2500 *(57)*

	Year			
	1980	*2000*	*2100*	*2500*
Annual projected nuclear generation [GW(e) a]	80	1000	10000	10000
Annual collective effective dose equivalent (man Sv)	500	10000	200000	250000
World population	4×10^9	10^{10}	10^{10}	10^{10}
Annual per caput effective dose equivalent (μSv)	0.1	1	20	25
Percentage of average exposure to natural sources of radiation (%)	0.005	0.05	1	1

Table XXII from this report projects annual per capita doses from nuclear electric-power-generating plants. The projected annual nuclear generation of 1000 gigawatts of electricity [GW(e)] for the year 2000 from the 80 GW(e) in 1980 seems optimistic in the light of present-day precautionary concerns.

REFERENCES

1. "Acid Rain—Research Summary," EPA-600/8-79-028. U.S. Environmental Protection agency, Cincinnati, Ohio, 1979.
2. A. C. Stern, R. W. Boubel, D. B. Turner, and D. L. Fox, "Fundamentals of Air Pollution," 2nd ed. Academic Press, New York, 1984.
3. E. B. Cowling, *Environ. Sci. Technol.* **16,** 110A (1982).
4. "Acid Rain," EPA-600/9-79-036. U.S. Environmental Protection Agency, Washington, D.C., 1980

4a. R. W. Talbot and A. W. Elzerman, *Environ. Sci. Technol.* **19,** 552 (1985).

4b. R. Lorenz and C. E. Murphey, Jr., *Atmos. Environ.* **19,** 797 (1985).

5. A. E. S. Green and W. H. Smith, eds., "Acid Deposition: Causes and Effects." Government Institutes, Inc., Rockville, Maryland, 1983.
6. A. W. Katzenstein, "An Updated Perspective on Acid Rain." Edison Electric Institute, Washington, D.C., 1981.
7. F. A. Record, D. V. Bubenick, and R. J. Kindya, "Acid Rain Information Book." Noyes Data Corp., Park Ridge, New Jersey, 1982.
8. "The Integrated Lake-Watershed Acidification Study: Proceedings of the ILWAS Annual Review Conference," EPRI EA-2827. Electric Power Research Institute, Palo Alto, California, 1983.
9. "Resource Book." Alliance for Balanced Environmental Studies, Washington, D.C., 1983.
10. "Acid Precipitation Digest," Vol. 1, Nos. 6–11. Center for Environmental Information, Rochester, New York, 1983.

10a. S. B. McLaughlin, *J. Air Pollut. Control Assoc.* **35,** 512 (1985).

10b. G. S. Hart, *J. Air Pollut. Control Assoc.* **35,** 197 (1985).
10c. T. D. Crocker and J. L. Regens, *Environ. Sci. Technol.* **19,** 112 (1985).
10d. R. D. Doctor, C. D. Livengood, J. L. Anderson, D. B. Garvey, and P. S. Farber, *J. Air Pollut. Control Assoc.* **35,** 331 (1985).
11. N. R. Glass, D. E. Arnold, J. N. Galloway, G. R. Hendrey, J. J. Lee, W. W. McFee, S. A. Norton, C. F. Powers, and D. L. Rambo, *Environ. Sci. Technol.* **16,** 162A (1982).
12. "Ecological Effects of Acid Rain," Technical Session Papers, 76th Annual Meeting, *J. Air Pollut. Control Assoc.* **33,** 1031 (1983).
13. G. H. Tomlinson, II, *Environ. Sci. Technol.* **17,** 246A 1983).
14. J. G. Calvert and W. R. Stockwell, *Environ. Sci. Technol.* **17,** 428A (1983).
15. S. Landsberger, R. E. Jervis, G. Kajrys, S. Monaro, and R. Lecomte, *Environ. Sci. Technol.* **17,** 542 (1983).
16. K. W. Ragland and K. E. Wilkening, *Atmos. Environ.* **17,** 935 (1983).
17. D. Golomb, *Atmos. Environ.* **17,** 1387 (1983).
18. A. Eliassen and J. Saltbones, *Atmos. Environ.* **17,** 1457 (1983).
19. J. Harte, *Atmos. Environ.* **17,** 403 (1983).
20. G. M. Hidy, *J. Air Pollut. Control Assoc.* **34,** 518 (1984).
21. M. Havas, T. C. Hutchinson, and G. E. Likens, *Environ. Sci. Technol.* **18,** 176 (1984).
22. D. V. Bubenick, ed., "Acid Rain Information Book," 2nd ed. Noyes Data Corp., Park Ridge, New Jersey, 1984.
23. R. Sequeira, *J. Air Pollut. Control Assoc.* **32,** 241 (1982).
24. J. Wisniewski and E. L. Keitz, *Water, Air, Soil Pollut.* **19,** 327 (1983).
25. Associated Press, Washington, D.C., Sept. 14, 1983.
26. M. L. Newhouse, *in* "Environment and Man" (J. Lenihan and W. W. Fletcher, eds.), Vol. 6, p. 137. Academic Press, New York, 1977.
27. P. Sebastien, M. A. Billon, G. Dufour, A. Gaudichet, G. Bonnaud, and J. Bignon, *Ann. N.Y. Acad. Sci.* **330,** 401 (1979).
28. P. E. Enterline, *J. Air Pollut. Control Assoc.* **33,** 319 (1983).
29. Code of Federal Regulations (CFR), Title 40, Part 61.22. U.S. Environmental Protection Agency, Washington, D.C., 1982.
29a. D. A. Chadwick, R. M. Buchan, and H. J. Beaulieu, *Environ. Res.* **36,** 1 (1985).
29b. B. Singh and J. P. Thouez, *Environ. Res.* **36,** 144 (1985).
30. Council on Environmental Quality, "Global Energy Futures and the Carbon Dioxide Problem." U.S. Superintendent of Documents, Washington, D.C., 1981.
31. B. Hileman, *Environ. Sci. Technol.* **16,** 90A (1982).
32. G. O. Barney, Study Director, "The Global 2000 Report to the President of the United States," Vol. 1. Pergamon, Oxford, 1980.
33. S. W. Matthews, *Nat. Geogr.* **150,** 576 (1976).
34. D. A. Gillette, *Atmos. Environ.* **16,** 2537 (1982).
35. S. B. Idso, *J. Air Pollut. Control Assoc.* **34,** 553 (1984).
36. J. E. Yocum, *J. Air Pollut. Control Assoc.* **32,** 500, 920 (1982).
37. "Indoor Air Pollution: A Serious Health Hazard." National Indoor Environmental Institute, Plymouth Meeting, Pennsylvania, 1983.
38. J. D. Spengler and K. Sexton, *Science* **221,** 9 (1983).
39. J. D. Spengler, C. P. Duffy, R. Letz, T. W. Tibbitts, and B. G. Ferris, Jr., *Environ. Sci. Technol.* **17,** 164 (1983).
40. J. D. Spengler, B. G. Ferris, Jr., D. W. Dockery, and F. E. Speizer, *Environ. Sci. Technol.* **13,** 1276 (1979).

41. B. Meyer, "Indoor Air Quality." Addison-Wesley, Reading, Massachusetts., 1983.
42. B. P. Leaderer, W. S. Cain, R. Isseroff, and L. G. Berglund, *Atmos. Environ.* **18,** 99 (1984).
42a. W. W. Nazaroff, H. Feustal, A. V. Nero, K. L. Revzan, D. T. Grimsrud, M. A. Essling, and R. E. Toohey, *Atmos. Environ.* **19,** 31 (1985).
42b. J. N. Pitts, Jr., T. J. Wallington, H. B. Biermann and A. M. Winer, *Atmos. Environ.* **19,** 763 (1985).
43. G. W. Traynor, J. R. Allen, M. G. Apte, J. R. Girman, and C. D. Hollowell, *Environ. Sci. Technol.* **17,** 369 (1983).
44. B. Hileman, *Environ. Sci. Technol.* **16,** 543A (1982).
45. "Standards for Natural and Mechanical Ventilation;" No. 62-1981. American Society of Heating, Refrigeration and Air Conditioning Engineers, Atlanta, Georgia, 1981.
46. R. C. Bruno, *J. Air Pollut. Control Assoc.* **33,** 105 (1983).
46a. K. Sexton and J. J. Wesolowski, *Environ. Sci. Technol.* **19,** 305 (1985).
46b. I. Turiel, "Indoor Air Quality and Human Health." Stanford Univ. Press, Stanford, California, 1985.
46c. V. J. Konopinski, *Am. Ind. Hyg. Assoc. J.* **46,** 65 (1985).
47. T. F. Gesell, *Health Phys., Spec. Issue* **45**(2), 269 (1983).
48. A. V. Nero, *Health Phys., Spec. Issue* **45**(2), 303 (1983).
49. C. E. Roessler, G. S. Roessler, and W. E. Bolch, *Health Phys., Spec. Issue* **45**(2), 389 (1983).
50. Code of Federal Regulations (CFR), Title 40, Part 50. U.S. Environmental Protection Agency, Washington, D.C., 1982.
51. "National Air Quality and Emission Trends Report, 1981," EPA-450/4-83-011. U.S. Environmental Protection Agency, Research Triangle Park, North Carolina, 1983.
51a. B. E. Saltzman, J. Cholak, L. J. Schafer, D. W. Yeager, B. G. Meiner, and J. Svetlik, *Environ. Sci. Technol.* **19,** 112 (1985).
52. R. L. Bennett and K. T. Knapp, *Environ. Sci. Technol.* **16,** 831 (1982).
53. "Lead," Executive Summary. World Health Organ., Geneva, Switzerland, 1983.
54. A. C. Chamberlain, *Atmos. Environ.* **17,** 677 (1983).
55. I. B. Millar and P. Å. Cooney, *Atmos. Environ.* **16,** 615 (1982).
56. C. Boutron, *Atmos Environ.* **16,** 2451 (1982).
57. "Ionizing Radiation: Sources and Biological Effects," U.N. Scientific Committee on Effects of Atomic Radiation, Report to the General Assembly, United Nations, New York, 1982.
58. "Off-Site Contamination and Radiation Exposures from the Three Mile Island Accident," A review for Atomic Industrial Forum, Washington, D.C., 1979.
59. K. R. Hings, G. R. Heath, D. R. Anderson, and C. D. Hollister, *Environ. Sci. Technol.* **16,** 28A (1982).
60. B. Hileman, *Environ. Sci. Technol.* **16,** 442A (1982).
61. "Environmental Outlook 1980," EPA-600/8-80-003. U.S. Environmental Protection Agency, Washington, D.C., 1980.
62. B. D. Goldstein, *J. Air Pollut. Control Assoc.* **33,** 454, 827 (1983).
63. S. Miller, *Environ. Sci. Toxicol.* **17,** 199A (1983)
64. "Risk Assessment in the Federal Government: Managing the Process." National Academy of Sciences, National Research Council, Washington, D.C., 1983.
65. M. C. Weinstein, *Science* **221,** 17(1983)
66. "Congress and the Clean Air Act: The Critical Issues," Report of the Twelfth Air

Pollution Control Association Government Affairs Seminar, March 22–23, 1984, *J. Air Pollut. Control Assoc.* **34,** 619 (1984).
66a. E. L. Anderson and Carcinogen Assessment Group U.S. EPA, *Risk Anal.* **3**(4), 277 (1983).
67. R. W. Kates, "Risk Assessment of Environmental Hazards." Wiley, New York, 1978.
68. R. C. Schwing and W. A. Albers, eds., "Societal Risk Assessment." Plenum, New York, 1980.
68a. P. F. Ricci and L. S. Molton, *Environ. Sci. Technol.* **19,** 473 (1985).
68b. J. R. Cretney, H. K. Lee, G. J. Wright, W. H. Swallow, and M. C. Taylor, *Environ. Sci. Technol.* **19,** 397 (1985).
69. J. London, *in* "Air Quality Meterology and Atmospheric Ozone" (A. L. Morris and R. C. Barras, eds.) p. 339. Am. Soc. Test. Mater., Philadelphia, Pennsylvania, 1978.
70. H. A. Panofsky, *Pure Appl. Geophys.* **118,** 636 (1980).
71. M. J. Molina and F. S. Rowland, *Nature (London)* **249,** 810 (1974).
72. J. Josephson, *Environ. Sci. Technol.* **16,** 328 (1982).
73. B. W. Boville, *in* "The Ozone Layer" (A. K. Biswas, ed.), p. 1. Pergamon, Oxford, 1979.
74. National Aeronautics and Space Administration, Department of Transportation, "Upper Atmospheric Programs Bulletin," No. 83-2. Federal Aviation Administration, Washington, D.C., 1983.
75. "Causes and Effects of Changes in Stratospheric Ozone: Update 1983," No. ISBN 0-309-03443-4. National Academy Press, Washington, D.C., 1984.
75a. Y. S. Chung and T. Dann, *Atmos. Environ.* **19,** 151 (1985).
76. K. A. Brice, R. G. Derwent, A. E. J. Eggleton, and S. A. Penkett, *Atmos. Environ.* **16,** 2543 (1982).
77. Code of Federal Regulations (CFR), Title 40, Part 61. U.S. Environmental Protection Agency, Washington, D.C., 1982.
78. "Delays in EPA's Regulation of Hazardous Air Pollutants," Report by the Comptroller General, GAO/RCED-83-199. U.S. General Accounting Office, Gaithersburg, Maryland, 1983.
79. "Proposed Rules," *Fed. Regist.* **48**(102), Part 302, Table 302.4, p. 23570 (1983).
80. B. D. Goldstein, *J. Air Pollut. Control Assoc.* **33,** 454, 827 (1983).
81. Joint Work Shop Proceedings, WHO International Institute for Applied Systems Analysis, *Atmos. Environ.* **18,** 487–652 (1984).
82. K. Kawamura and I. R. Kaplan, *Environ. Sci. Technol.* **17,** 497 (1983).
83. H. B. Singh, L. J. Salas, and R. E. Stiles, *Environ. Sci. Technol.* **16,** 872 (1982).
84. P. F. Pankow, L. M. Isabelle, and W. E. Asher, *Environ. Sci. Technol.* **18,** 310 (1984).
85. V. A. Isidorov, I. G. Zenkevich, and B. V. Ioffe, *Atmos. Environ.* **17,** 1347 (1983).
86. D. Grosjean, *Environ. Sci. Technol.* **16,** 254 (1982).
87. E. D. Pellizzari, *Environ. Sci. Technol.* **16,** 781 (1982).
88. R. Harkov, A. Greenberg, F. Darack, J. M. Daisey, and P. J. Lioy, *Environ. Sci. Technol.* **18,** 287 (1984).
89. A. Yasuhara, M. Morita, and K. Fuwa, *Environ. Sci. Technol.* **16,** 805 (1982).
90. J. Josephson, *Environ. Sci. Technol.* **17,** 124A (1983).
91. M. Moller, I. Alfhelm, S. Larssen, and A. Mikalsen, *Environ. Sci. Technol.* **16,** 221 (1982).
92. W. R. Pierson, R. A. Gorse, Jr., A. C. Szkarlat, W. W. Brachaczek, S. M. Japar, F. S. C. Lee, R. B. Zmeldinger, and L. D. Claxton, *Environ. Sci. Technol.* **17,** 31 (1983).

92a. T. Alsberg, U. Stenberg, R. Westerholm, M. Strande *et al.*, *Environ. Sci. Technol.* **19,** 43 (1985).
93. D. G. Bottrell and R. F. Smith, *Environ. Sci. Technol.* **16,** 282A (1982).
94. E. W. Day, Jr., D. G. Saunders, J. W. Mosler, E. M. McKinney, F. L. Powers, R. D. Griggs, and R. Frank, *Environ. Sci. Technol.* **16,** 131 (1982).
95. A. Cavallaro, G. Tebaldi, and R. Gualdi, *Atmos. Environ.* **16,** 731 (1982).
95a. A. A. Boraiko and F. Ward, *Nat. Geogr.* **167**(3), 318 (1985).
96. B. N. Ames, *Science* **224,** 1256 (1983); **204,** 587 (1979).
97. "Inhalable Particle and National Air Surveillance Networks Data." U.S. Environmental Protection Agency, Research Triangle Park, North Carolina, 1980.
97a. M. A. K. Khalil and R. A. Rasmussen, *Atmos. Environ.* **19,** 397 (1985).
98. "The National Air Monitoring Program: Air Quality and Emission Trends," Annu. Rep., Vol. EPA-450/1-73-001a. U.S. Environmental Protection Agency, Research Triangle Park, North Carolina, 1973.
99. "Monitoring and Air Quality Trends Report, 1974," EPA-450/1-76-001. U.S. Environmental Protection Agency, Research Triangle Park, North Carolina, 1976.
100. "National Air Pollutant Emission Estimates, 1970–1981," EPA-450/4-82-012. U.S. Environmental Protection Agency, Research Triangle Park, North Carolina, 1982.
101. P. F. Nelson and S. M. Quigley, *Environ. Sci. Technol.* **16,** 650 (1982).
102. D. Grosjean and K. Fung, *J. Air Pollut. Control Assoc.* **34,** 537 (1984).
103. W. E. Pereira, C. E. Rostad, H. E. Taylor, and J. M. Kein, *Environ. Sci. Technol.* **16,** 387 (1982).
104. "Air Quality in Selected Urban Areas: 1977–1978," Publ. No. 57. World Health Organ., Geneva, Switzerland, 1980.
104a. B. G. Bennett, J. C. Kretzschmar, G. G. Akland, and H. W. deKoning, *Environ. Sci. Technol.* **19,** 298 (1985).
104b. P. V. Hobbs and J. C. Yates, *Atmos. Environ.* **19,** 163 (1985).
104c. K. McGuffie, J. G. Cogley, and A. Henderson-Sellers, *Atmos. Environ.* **19,** 707 (1985).
105. National Research Council, "Hydrogen Sulfide," Subcommittee on Hydrogen Sulfide, Committee on Medical and Biologic Effects of Environmental Pollutants. University Park Press, Baltimore, Maryland, 1979.
106. J. O. Nriagu, ed., "Sulfur in the Environment, Part I. The Atmospheric Cycle." Wiley, New York, 1978.
107. V. P. Aneja, A. P. Aneja, and D. F. Adams, *J. Air Pollut. Control Assoc.* **32,** 803 (1982).
108. V. P. Aneja, J. H. Overton, and A. P. Aneja, *J. Air Pollut. Control Assoc.* **31,** 256 (1981).
109. E. O. Box, *J. Air Pollut. Control Assoc.* **32,** 954 (1982).
110. D. Moller, *Atmos Environ.* **18,** 19, 29 (1984).
111. M. W. Ivanov and J. R. Freney, eds., "Global Biogeochemical Sulfur Cycle." Wiley, Somerset, New Jersey, 1983.
112. S. H. Cadle, R. J. Countess, and N. A. Kelley, *Atmos. Environ.* **16,** 2501 (1982).
113. National Research Council, "Ammonia," Subcommittee on Ammonia, Committee on Medical and Biologic Effects of Environmental Pollutants. University Park Press, Baltimore, Maryland, 1979.
114. J. L. Gras, *Atmos. Environ.* **17,** 815 (1983).
114a. D. Moller and H. Schreferdecker, *Atmos. Environ.* **19,** 695 (1985).
115. G. T. Wolff, M. A. Ferman, N. A. Kelly, D. P. Stroup, and M. S. Ruthkosky, *J. Air Pollut. Control Assoc.* **32,** 1216 (1982).

116. K. Yoshizumi and T. Okita, *J. Air Pollut. Control Assoc.* **33,** 224 (1983).
117. R. J. Cicerone, *Rev. Geophys. Space Phys.* **19,** 129 (1981).
118. J. Rancher, *J. Rech. Oceanogr.* **4**(3), 3 (1979); *Chem. Abstr.* **93,** 76509p.
119. R. A. Rasmussen, M. A. K. Khalil, and R. W. Dalluge, *Science* **211,** 285 (1981).
119a. S. Tassios and D. R. Packham, *J. Air Pollut. Control Assoc.* **35,** 41 (1985).
120. E. R. Allen, P. Urone, M. Hodges, and D. Gancarz, "Atmospheric Fluorides: A Review," Report to Florida Environmental Regulation Department, Gainesville, 1984.
121. F. A. Smith and H. C. Hodge, *CRC Crit. Rev. Environ. Control* **8,** 293 (1979).
122. W. W. Berg and J. W. Winchester, *JGR, J. Geophys. Res.* **82,** 5945 (1977).
123. K. Rahn, R. D. Borys, and R. A. Duce, *Science* **192,** 549 (1976).
124. "Chloroform, Carbon Tetrachloride and Other Halomethanes." National Research Council, Washington, D.C., 1982.
125. R. Harkov, *Sci. Total Environ.* **26,** 67 (1982).
126. R. Harkov, B. Kebbekus, J. W. Bozzelli, and P. J. Lioy, *J. Air Pollut. Control Assoc.* **32,** 1177 (1983).
127. "Health Assessment Documents: Carbon Tetrachloride, 1,1,1-Trichloroethane, Dichloromethane, Tetrachloroethylene, Trichloroethylene," EPA-600/8-82-001–006. U.S. Environmental Protection Agency, Research Triangle Park, North Carolina, 1982.
128. "Environmental Fates and Impacts of Major Forest Use Pesticides," EPA-68-01-3174. U.S. Environmental Protection Agency, Research Triangle Park, North Carolina, 1981.
129. D. A. Lundgren and J. J. Paulus, *J. Air Pollut. Control Assoc.* **24,** 1227 (1975).
129a. R. Toossi and T. Novakov, *Atmos. Environ.* **19,** 127 (1985).
130. "Air Quality Criteria for Particulate Matter and Sulfur Oxides," Vol. II, EPA 600/8-82-029b. U.S. Environmental Protection Agency, Research Triangle Park, North Carolina, 1982.
131. E. P. Perera and A. K. Ahmed, "Respirable Particles." Ballinger Press, Cambridge, Massachusetts, 1979.
131a. S. G. Tuncel, I. Olmez, J. R. Parrington, G. E. Gordon, and R. K. Stevens, *Environ. Sci. Technol.* **19,** 529 (1985).
132. R. M. Harrison and C. A. Pio, *Environ. Sci. Technol.* **17,** 169 (1983).
133. J. M. Ondov, W. H. Zoller, and G. E. Gordon, *Environ. Sci. Technol.* **16,** 318 (1982).
134. D. R. Taylor, M. A. Tompkins, S. E. Kirton, T. Mauney, D. F. S. Natusch, and P. K. Hopke, *Environ. Sci. Technol.* **16,** 148 (1982).
135. APTD-0978. U.S. Environmental Protection Agency, Research Triangle Park, North Carolina, 1972.
136. M. Demuynek, K. A. Rahn, M. Janssens, and R. Dams, *Atmos. Environ.* **10,** 21 (1976).
137. C. Brosset, K. Andreasson, and M. Ferm, *Atmos Environ.* **9,** 631 (1975).
138. D. Pedgley, "Windborne Pests and Diseases: Meterology of Airborne Organisms." Halsted Press, New York, 1982.
139. W. J. Hayes, "Pesticides Studied in Man." Williams & Wilkins, Baltimore, Maryland, 1982.
140. R. B. Couch, *Bull N.Y. Acad. Med.* [2] **57** (10), 907 (1981).
140a. J. Siak, T. L. Chan, T. L. Gibson, and G. T. Wolf, *Atmos. Environ.* **19,** 369 (1985).
141. "International Archives of Allergy and Applied Immunology." Karger, New York.

142. A. Roth, ed., "Allergy in the World." Univ. Press of Hawaii, Honolulu, 1978.
143. W. R. Solomon and H. A. Burge, *in* "Indoor Air Quality" (P. J. Walsh, C. S. Dudney, and E. D. Copenhaver, eds.), Chapter 10. CRC Press, Boca Raton, Florida, 1984.
144. P. J. Imperato, *Bull. N.Y. Acad. Med.* [2] **57** (10), 922 (1981).
145. C. E. Reed, *Bull. N.Y. Acad. Med.* [2] **57** (10), 897 1981).
146. S. Glasstone and W. H. Jordan, "Nuclear Power and its Environmental Effects." American Nuclear Society, LaGrange Park, Illinois, 1980.
147. "Effects on Populations of Exposures to Low Levels of Ionizing Radiation: 1980." National Academy Press, Washington, D.C., 1980.
148. A. Brodsky, "Handbook of Radiation Measurement and Protection." CRC Press, Boca Raton, Florida, 1982.

Lists of books relevant to this chapter published since the lists in Chapter 8, Volume II, 3d edition (1977) can be found in Chapter 1, Table IX, Volume VII.

2

The Transformation of Pollutants

Donald L. Fox

School of Public Health
Department of Environmental Sciences and Engineering
University of North Carolina at Chapel Hill
Chapel Hill, North Carolina

I. Global Sources

The atmosphere is composed principally of nitrogen (~79%), oxygen (~20%), and water vapor (1–4%). All other gases are considered as trace gases. The sources of these gases are based at the earth's surface or in the atmosphere as a transformation product of a surface released gas. By dividing trace gas sources into anthropogenic and natural categories, we can better understand the relative importance of man's activities on a meso- or global scale. Table I contains a summary of sources of trace gases in the atmosphere compiled by Cox and Derwent *(1)*. As can be seen, very few of them are solely of anthropogenic or natural origin. No significant man-made sources of terpenes have been identified, and no significant sources of halogenated compounds such as carbon tetrachloride (CCl_4) or of sulfur hexafluoride (SF_6) have been identified in nature. The impact of upwind ozone on the urban photochemical oxidant problem is an important contemporary problem.

A. Anthropogenic Sources

Anthropogenic, or man-made, sources are dominated by our use of energy for home heating, hot water, and industrial processes such as converting raw materials into consumer products. The major man-made source categories are fossil fuel combustion for power generation and transportation, chemical processes, petroleum recovery, refining, and distribution.

B. Natural Sources

Natural sources may be subdivided into releases from physical or biological processes. Two of the physical processes entail venting of gases from the earth's crust in the form of volcanic eruptions and volatile hydrocarbon seepage. Biological processes include fermentation, denitrification, and terpene release from vegetation. Society can alter natural emission patterns by converting tropical rain forests to agricultural land. Estimates on a global scale compare anthropogenic and natural sources of various trace gases (Chapter 1).

II. Chemical Transformations

A. Trace Gas Cycle

The trace gas cycle involves release to the atmosphere, possible transformation, and eventual removal from the atmosphere. Atmospheric transformations involve physicochemical processes. For example, the oxidation of hydrogen sulfide (H_2S) to SO_2 is a chemical process. The subsequent homogeneous oxidation of gaseous SO_2 to H_2SO_4 is a chemical process followed by a physical transformation from the gaseous H_2SO_4 to sulfuric acid in the aerosol or liquid form.

Depending on its residence time and chemical reactivity, a given trace gas released to the atmosphere may react completely or only partially before removal by the earth's surface. Some gases, such as ozone, have no significant surface-based sources. Ozone is formed from atmospheric chemical processes in two regions of the atmosphere, the stratosphere and the troposphere. Stratospheric ozone formation is discussed in Chapter 1; ozone formation in the troposphere is discussed in this chapter.

B. Classification of Trace Gases

A very large number of individual compounds have been identified in ambient air samples *(2)*. These compounds comprise five groups: carbon-containing, nitrogen-containing, sulfur-containing, halogen-containing, and other. Obviously, these classifications are not mutually exclusive. Table I shows inorganic CO and CO_2 in the carbon-containing compounds along with organic compounds.

C. Chemical Oxidation Processes

Chemical processes in the atmosphere result in oxidation. Volatile hydrocarbons are sequentially oxidized to form carbonyls and, if the residence time is sufficient, to form CO and CO_2. Very little transformation of CO_2 occurs in the atmosphere. Some CO_2 may form carbonic acid in rain water before being removed, but a major atmospheric chemical transformation does not occur. Removal of CO_2 occurs at the earth's surface by plant photosynthesis and absorption by bodies of water. Some volatile organic compounds are oxidized to form oxygenated products with lower vapor pressures. If the vapor pressure of the oxidized product is sufficiently low, the product will condense and form an aerosol or particulate matter in the atmosphere.

Nitrogen-containing species generally enter the atmosphere as NH_3, N_2O, NO, and NO_2. N_2O reacts very little in the troposphere. Ammonia can be oxidized but generally neutralizes the acidic compounds HNO_3 and H_2SO_4. NO and NO_2 do react in the atmosphere: NO is oxidized to NO_2, and NO_2 is oxidized to nitric acid or reacts with organic radicals to form peroxyacyl nitrate (PAN) types of compounds.

Table I Natural and Man-Made Sources of Trace Gases in the Troposphere (1)

Compound	*Natural sources*	*Man-made sources*
Carbon-containing trace gases		
Carbon monoxide (CO)	Oxidation of natural methane;[a] oxidation of natural C_5, C_{10} hydrocarbons; oceans;[d] forest fires[c]	Oxidation of mamade hydrocarbons; incomplete combustion of wood, oil, gas, and coal, in particular motor vehicles;[d] industrial processes:[d] blast furnaces
Carbon dioxide (CO_2)	Oxidation of natural CO; destruction of forests;[e] respiration by plants[f]	Combustion of oil, gas, coal, and wood;[g] limestone burning[g]
Methane (CH_4)	Enteric fermentation in wild animals;[h] emissions from swamps, bogs, *etc.*,[i] natural wet land areas;[j] oceans[h]	Enteric fermentation in domesticated ruminants;[h] emissions from paddies;[i] natural gas leakage;[h] sewerage gas;[h] colliery gas;[h] combustion sources[h]
Light paraffins, C_2-C_6	Aerobic biological source[d]	Natural gas leakage; motor vehicle evaporative emissions; refinery emissions
Olefins, C_2-C_6		Motor vehicle exhaust;[k] diesel engine exhaust[k]
Aromatic hydrocarbons		Motor vehicle exhaust;[k] evaporative emissions:[k] paints, petrol, solvents
Hemiterpenes, C_5H_8	Trees, broadleaves, and conifers; plants[b]	
Terpenes, $C_{10}H_{16}$	Trees, broadleaves, and conifers; plants[b]	
Diterpenes, $C_{20}H_{32}$	Trees, broadleaves, and conifers; plants[b]	
Nitrogen-containing trace gases		
Nitric oxide (NO)[d]	Forest fires; anaerobic processes in soil; electric storms	Combustion of oil, gas, and coal
Nitrogen dioxide (NO_2)[d]	Forest fires; electric storms	Combustion of oil, gas, and coal; atmospheric transformation of NO
Nitrous oxide (N_2O)	Emissions from denitrifying bacteria in soil;[l] oceans[l]	Combustion of oil and coal[m]
Peroxyacetyl nitrate (PAN)	Degradation of isoprene[n]	Degradation of hydrocarbons[o]
Ammonia (NH_3)[d]	Aerobic biological source in soil Breakdown of amino acids in organic waste material	Coal and fuel oil combustion; waste treatment

Compound	*Natural sources*	*Man-made sources*
Sulfur-containing trace gases		
Sulfur dioxide (SO_2)	Oxidation of H_2S;[p] volcanic activity[p]	Combustion of oil and coal;[q] roasting sulfide ores
Hydrogen sulfide (H_2S)	Anaerobic fermentation;[p] volcanoes and fumaroles[p]	Oil refining; animal manure;[r] Kraft paper mills;[r] rayon production;[r] coke oven gas[r]
Carbon disulfide (CS_2)	Anaerobic fermentation[s]	Viscose rayon plants;[t] brick making;[t] fish meal processing[t]
Carbonyl sulfide (OCS)	Oxidation of CS_2;[u] slash and burn agriculture;[c] volcanoes and fumaroles	Oxidation of CS_2;[u] brick making;[t] effluent from Kraft mills;[v] blast furnace gas;[v] coke oven gas;[v] shale and natural gas[v]
Sulfur trioxide (SO_3)		Combustion of S-fuel
Methyl mercaptan (CH_3SH)[s,t]	Anaerobic biological sources	Animal rendering; animal manure; pulp and paper mills; brick manufacture; oil refining
Dimethyl sulfide (CH_3SCH_3)[s,t]	Aerobic biological source	Animal rendering; animal manure; pulp and paper mills
Dimethyl disulfide (CH_3SSCH_3)[s,t]		Animal rendering; fishmeal processing
Other organic sulfur compounds:[s,t] C_2-C_4 mercaptans, dialkyl disulfides, dimethyl trisulfide, alkyl thiophenes, benzothiophenes	Anaerobic biological sources	Animal rendering; fishmeal processing; brick making
Chlorine-containing trace gases[w]		
Hydrogen chloride (HCl)	Volcanoes and fumaroles; degradation of CH_3Cl	Coal combustion; degradation of chlorocarbons
Methyl chloride (CH_3Cl)	Slow combustion of organic matter; marine environment: algae	PVC and tobacco combustion
Methylene dichloride (CH_2Cl_2)		Solvent
Chloroform ($CHCl_3$)		Pharmaceuticals; solvent; combustion of petrol; bleaching of wood pulp; degradation of C_2HCl_3
Carbon tetrachloride (CCl_4)		Solvent; fire extinguishers; degradation of C_2Cl_4

Table I *(Continued)*

Compound	*Natural sources*	*Man-made sources*
Methyl chloroform (CH_3CCl_3)		Solvent; degreasing agent
Trichloroethylene (C_2HCl_3)		Solvent; dry cleaning agent; degreasing agent
Tetrachloroethylene (C_2Cl_4)		Solvent; dry cleaning agent; degreasing agent
Other chlorofluorocarbons: CCl_3F, CCl_2F_2, $C_2Cl_3F_3$, $C_2Cl_2F_4$, C_2ClF_5		Aerosol propellants; refrigerants; foam blowing agents; solvents
Other minor trace gases		
Hydrogen[x]	Oceans; soils; oxidation of methane, isoprene and terpenes *via* formaldehyde	Motor vehicle exhaust; oxidation of methane *via* formaldehyde
HF	Volcanoes and fumaroles	
Ozone	Stratosphere: natural $NO—NO_2$ conversion	Man-made $NO—NO_2$ conversion
H_2O	Evaporation from oceans	Insignificant
SF_6		Electrical insulator
CF_4[y]		Aluminum industry
CH_3Br[z]	Aerobic biological source	Fumigation of soil and grain
CH_3I[z]	Aerobic biological source	Insignificant

[a] W. Seiler, *Tellus* **26,** 116 (1974).
[b] R. A. Rasmussen and F. W. Went, *Proc. Natl. Acad. Sci. U.S.A.* **53,** 215 (1965).
[c] P. J. Crutzen, L. E. Heidt, J. P. Krasnec, W. H. Pollock, and W. Seiler, *Nature (London)* **282,** 253 (1979).
[d] E. Robinson and R. C. Robbins, SRI Proj. PR 6755. Stanford Res. Inst., Stanford, California, 1968.
[e] B. Bolin, *Science* **196,** 613 (1977).
[f] B. Bolin, E. T. Degens, S. Kempe, and P. Ketner, "The Global Carbon Cycle," *SCOPE [Rep.]* **13,** Wiley, New York, 1979.
[g] C. D. Keeling, *Tellus* **25,** 174 (1973).
[h] D. H. Ehhalt, *Tellus* **26,** 58 (1974).
[i] T. Koyams, "Recent Researches in the Field of Hydrosphere, Atmosphere, and Nuclear Geochemistry," p. 143. Maruzen, Tokyo, 1964.
[j] A. Baker-Blocker, T. M. Donahue, and K. H. Mancy, *Tellus* **29,** 245 (1977).
[k] "Vapour Phase Organic Pollutants." Natl. Acad. Sci., Washington, D.C., 1976.
[l] J. Hahn and C. E. Junge, *Z. Naturforsch., B: Anorg. Chem., Org. Chem.* **329,** 190 (1977).
[m] R. F. Weiss and H. Craig, *Geophys. Res. Lett.* **3,** 751 (1976).
[n] R. A. Cox, R. G. Derwent, and M. R. Williams, *Environ. Sci. Technol.* **14,** 57 (1980).
[o] P. A. Leighton, "Photochemistry of Air Pollution." Academic Press, New York, 1961.
[p] L. Granat, H. Rodhe, and R. O. Hallberg, *SCOPE [Rep.]* 7 (1976); *Ecol. Bull.* **22,** 89 (1976).
[q] J. P. Friend, "Chemistry of the Lower Atmosphere," p. 177. Plenum, New York, 1973.
[r] A. C. Stern, "Air Pollution," 3rd ed. Vol. 1. Academic Press, New York, 1976.
[s] J. E. Lovelock, R. J. Maggs, and R. A. Rasmussen, *Nature (London)* **237,** 452 (1972); J. E. Lovelock, *ibid.*
[t] "Report of Working Party on Suppression of Odours from Offensive and Selected Other Trade." Department of Environment, H.M. Stationery Office, London, 1975.

Sulfur-containing compounds enter the atmosphere as SO_2, H_2S, and organosulfides such as dimethyl sulfide. Most of these compounds can undergo photochemical oxidation to form H_2SO_4. SO_2 and H_2S may also be removed by direct contact with the earth's surface.

D. Free Radicals and Other Oxidizers

Carbon-, nitrogen-, and sulfur-containing compounds are oxidized in atmospheric chemical transformations. The species providing all of this oxidizing power are free radicals along with H_2O_2 and O_3. Free radicals are neutral fragments of stable molecules, e.g., hydroxyl radical, OH; hydroperoxy radical, HO_2; atomic oxygen, O^1D; and higher homologs, RO, and RO_2, where R is a hydrocarbon group. Free radicals participate in chain initiation, branching, propagation, and termination reactions in the atmosphere. The $OH-HO_2$ chain is particularly effective in the oxidation of hydrocarbons and nitric oxide. The following set of reactions shows this process.

$$OH + RH \rightarrow R + H_2O \tag{1}$$

$$R + O_2 \rightarrow RO_2 \tag{2}$$

$$RO_2 + NO \rightarrow NO_2 + RO \tag{3}$$

$$RO + O_2 \rightarrow R'CHO + HO_2 \tag{4}$$

$$HO_2 + NO \rightarrow NO_2 + OH \tag{5}$$

This process starts with an OH radical, after one pass through the cycle, two molecules of NO have been oxidized to NO_2, the hydrocarbon is an aldehyde with one less carbon, and the OH radical is available to start the cycle again. Note the pattern: RO (OH) reacts with hydrocarbons to form RO_2 (HO_2), then RO_2 (HO_2) reacts with NO to reform the RO (OH) radical.

---

[u] M. J. Kurylo, *Chem. Phys. Lett.* **58,** 238 (1978).
[v] P. L. Hanst, L. L. Spiller, D. M. Watts, J. W. Spence, and M. F. Miller, *J. Air Pollut. Control Assoc.* **25,** 1220 (1975).
[w] "Fluorocarbons and the Environment," IMOS Rep. U.S. Govt. Printing Office, Washington, D.C., 1975; Pollut. Pap. No. 15. Department of Environment, H.M. Stationery Office, London, 1979; "Chloroform, Carbon Tetrachloride, and other Halomethanes: An Environmental Assessment." Natl. Acad. Sci., Washington, D.C. 1978.
[x] U. Schmidt, *Tellus* **26,** 78 (1974).
[y] R. J. Cicerone, *Science* **206,** 59 (1979).
[z] J. E. Lovelock, *Nature (London)* **256,** 193 (1975); **252,** 292 (1974); H. B. Singh, L. J. Salas, H. Shigeishi, and E. Scribner, *Science* **203,** 899 (1979).

Free radicals can react with hydrocarbons, aldehydes, NO, NO_2, H_2S, SO_2, other free radicals, and many other atmospheric trace gases. Ozone is a good oxidizer for alkenes, and nitric oxide in the gas phase and ozone oxidizes SO_2 in aqueous solutions if the pH is not too low. H_2O_2 is also a good oxidizer for SO_2 in aqueous solutions.

III. Urban Chemical Systems

The cycle of trace gases in the atmosphere may be viewed in an urban setting as, first, release from ground-based sources; second, dispersion into the adjacent layer of the atmosphere; third, physicochemical transformation; fourth, comcomitant removal processes. The urban atmosphere as a chemical reactor is extremely complex in detail but is understood in many general ways. It contains a mixture of carbon, nitrogen, sulfur, and halogen-containing trace gases which undergo transformation. NO is oxidized to NO_2, HNO_3, and organic nitrates and nitrites; hydrocarbons are oxidized to form aldehydes, acids, and alcohols; and SO_2 is oxidized to form sulfates. The availability of suspended particles or water droplets influences the physical gas-to-liquid or solid phase transformations. Adsorption and absorption enhance the conversion of gaseous species to liquid or solid forms.

A. Gaseous Photochemical Smog

Considerable progress has been made in identifying and measuring the chemical reactions in the oxidation of specific hydrocarbons. The interaction of free radicals, hydrocarbons, and NO_x is illustrated in two cases that follow.

1. Propylene Case

Propylene and *trans*-2-butene have been used by researchers for the last 20 years to develop an understanding of the photochemical smog process. Some of the products formed from the atmospheric oxidation of propylene in the presence of oxides of nitrogen and sunlight are formaldehyde, acetaldehyde, CO, formic acid, peroxyacyl nitrate, NO_2, HNO_3, and O_3. These overall transformations occur via a complex set of chemical reactions involving (1) OH and O_3 with propylene, (2) photolysis of NO_2 and reactions of free radicals with NO. These reactions are often classified as organic and nitrogen oxide groups. Consider a series of transformation cycles — one the hydrocarbon oxidation cycle (Eqs. [1] through [5]) and the other the following inorganic nitrogen oxide cycle.

a. Inorganic Nitrogen Oxide Cycle. The inorganic nitrogen cycle establishes the relationship among NO, NO_2, and O_3 and also provides the major pathway for transforming HO_2 radicals to OH radicals. The primary relationship of NO, NO_2, and O_3 is illustrated by

$$NO_2 + h\nu \rightarrow NO + O \qquad (6)$$

where $h\nu$ are photons of light

$$O + O_2 + M \rightarrow O_3 + M \qquad (7)$$

where M is a third body

$$NO + O_3 \rightarrow NO_2 + O_2 \qquad (8)$$

The reaction for converting HO_2 to OH is

$$HO_2 + NO \rightarrow NO_2 + OH \qquad (9)$$

These reactions interact with the hydrocarbon oxidation cycle via the reaction of OH and O_3 with hydrocarbons with the resultant formation of HO_2 radicals and oxygenated hydrocarbons. The HO_2 radicals formed then interact with the inorganic cycle via equation (9) (Fig. 1).

The reaction pathways for propylene with O_3 and OH are shown in

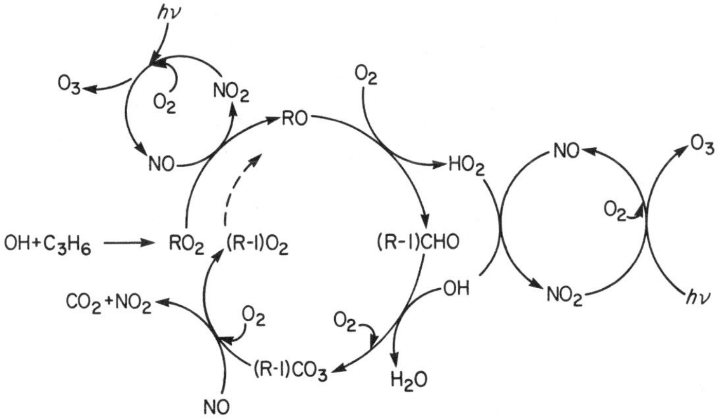

Figure 1. The cyclic interaction of free radicals, hydrocarbons, NO, NO_2, and ultraviolet radiation in photochemical smog. In this example hydroxyl radical reacts with propylene at the left side of the diagram forming RO_2. This cycle interacts with NO and molecular O_2. The inorganic NO_x–O_3 cycle is shown on the right side of the diagram with photolysis of NO_2 eventually forming O_3.

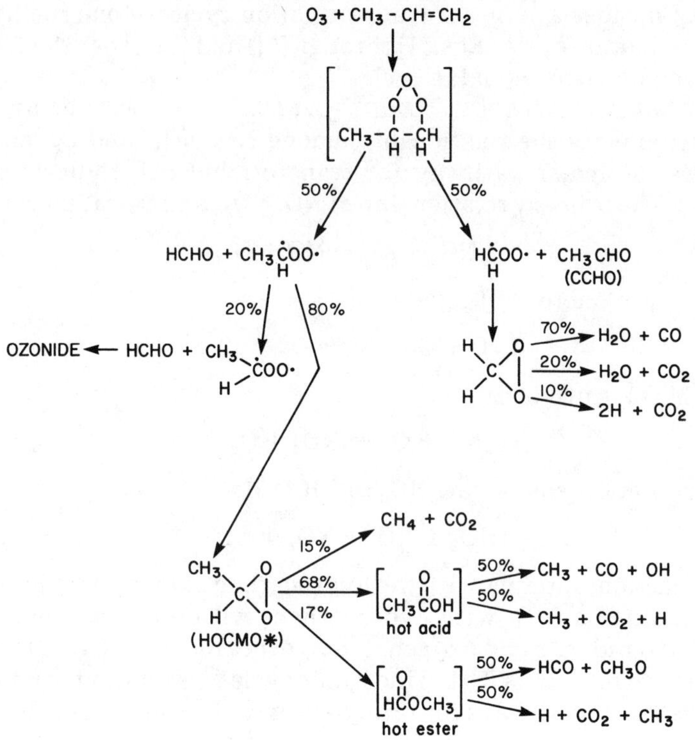

Figure 2. Ozone attack on propylene and subsequent degradation products. When the initial molozonide decomposes, stable aldehydes and other free radicals are formed. Some of these free radicals undergo further decomposition and interact with the NO_x-O_3 cycle.

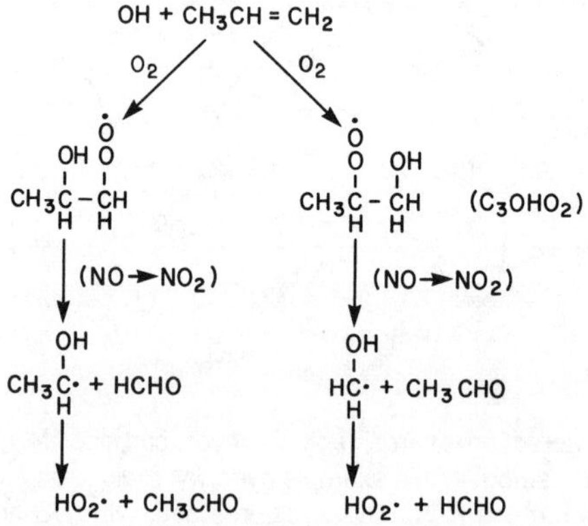

Figure 3. Hydroxyl radical attack on propylene and subsequent degradation products. The OH radical adds to either side of the double bond to form an RO_2 which can oxidize NO to NO_2. Further radical decomposition forms stable aldehydes and hydroperoxyl radicals.

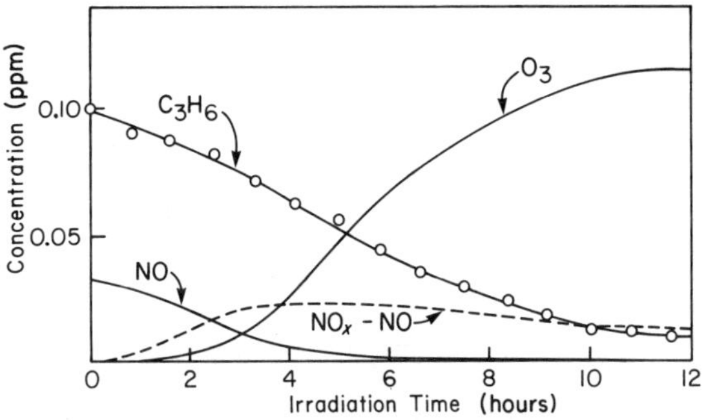

Figure 4. Propylene and NO oxidation under artificial illumination. NO is oxidized to NO_2 and other oxides of nitrogen. Ozone concentrations build up after the ratio of NO_2 to NO increases.

Figures 2 and 3 *(3)*. HO_2 is formed from H because of the very fast reaction of H with molecular O_2 [$H^{\cdot}$ $(+O_2) \rightarrow HO_2{}^{\cdot}$]. The $HO_2{}^{\cdot}$ may participate in the $NO \rightarrow NO_2$ oxidation process, resulting in the accumulation of NO_2 and O_3. Concentration–time profiles for a smog chamber simulation of a propylene–NO_x system with artificial illumination are shown in Figure 4 *(4)*. The patterns are first propylene and NO oxidation, followed by NO_2, PAN, O_3, and aldehyde formation.

2. Toluene Case

A more complex set of reactions has been proposed for the atmospheric reactions of aromatic compounds. Toluene is used to illustrate the types of chemical reactions aromatic compounds undergo in the atmosphere. Killus and Whitten *(5)* have outlined three major differences in the oxidation of toluene from alkenes and alkanes: (1) a high photolysis rate for toluene oxidation products, suggesting the formation of methyl glyoxal; (2) a low rate of peroxyl radical production, as evidenced by the rate of NO to NO_2 conversion in test systems; and (3) a large sink for NO_x at the onset of ozone formation. They have proposed an oxidation mechanism that includes the formation and subsequent reactions of methyl glyoxal and cresol. The hydroxyl radical (OH) is the principal oxidizing agent for aromatic compounds. The OH radical can react with aromatics by hydrogen abstraction from the side chain or by addition to the ring.

a. OH Reaction with Toluene by Hydrogen Abstraction. For toluene, hydrogen abstraction in the presence of oxygen [Equation (10)]:

$$C_6H_5CH_3 + OH \xrightarrow{O_2} \underset{(RO_2{}^{\cdot})}{C_6H_5CH_2O_2} + H_2O \qquad (10)$$

forms the aromatic peroxyl radical, which may oxidize NO to NO_2 [Equation (11)]:

$$C_6H_5CH_2O_2\cdot + NO \rightarrow NO_2 + C_6H_5CH_2O\cdot \tag{11}$$

The benzoyl radical then reacts with oxygen to form HO_2 and benzaldehyde [Equation (12)]:

$$C_6H_5CH_2O\cdot \xrightarrow{O_2} HO_2\cdot + C_6H_5CHO \tag{12}$$

Killus and Whitten (5) also suggest the formation of benzylnitrate [Equation (13)]:

$$C_6H_5CH_2O\cdot + NO_2 \rightarrow C_6H_5CH_2ONO_2 \tag{13}$$

b. Hydroxyl Radical Addition to Toluene. Hydroxyl addition is the more important pathway for OH reaction with toluene because the ratio of hydrogen abstraction to addition is 0.16 at 298°K *(6)*. The OH radical adds to the ring primarily at the ortho position to form an adduct (I) *(7)* [Equation (14)]:

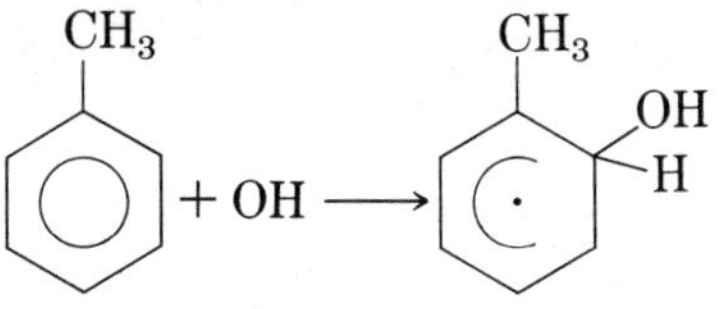

(14)

(I)

This adduct can react by following several pathways: (1) with O_2 to form cresol or the peroxyl radical (II) [Equation (15)]:

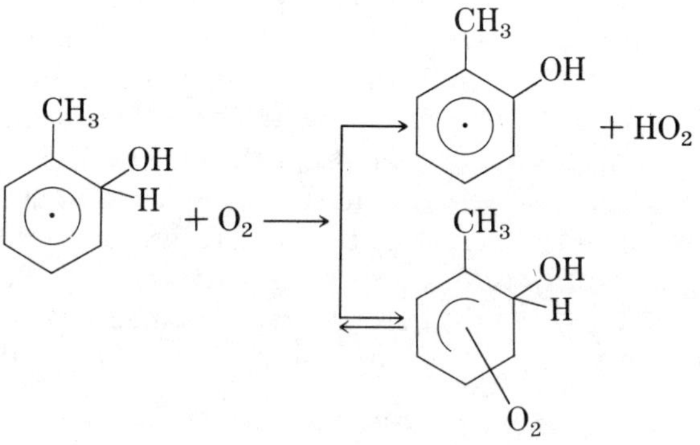

(15)

(II)

or (2) with NO_2 to form nitrotoluene [Equation (16)]:

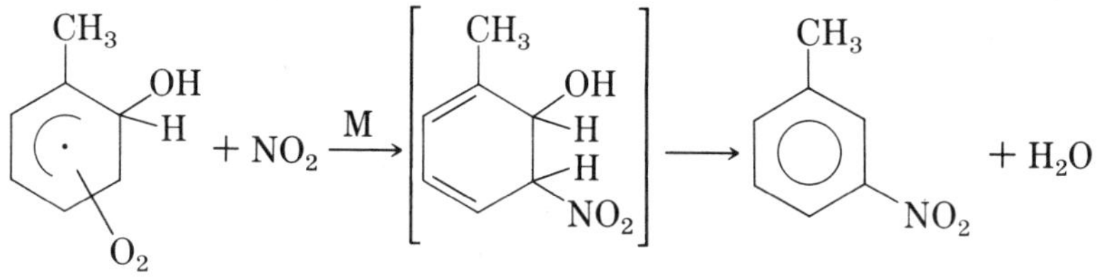

(16)

The peroxyl radical (II) can undergo O_2 addition at the 1, 3, or 5 position of the ring. Atkinson and co-workers *(7)* have proposed reaction steps for the 1 and 3 positions. They omitted the 5 position on the basis of unfavorable thermochemical conditions. The following discussion is based on the O_2 addition at the 3 position (II′) and is intended to illustrate the way the aromatic ring undergoes scission to form methyl glyoxal and *cis*-2-butenedial. The OH-aromatic-O_2 adduct (II′) undergoes rearrangement to form an —O—O— bridge across the aromatic ring [Equation (17)]:

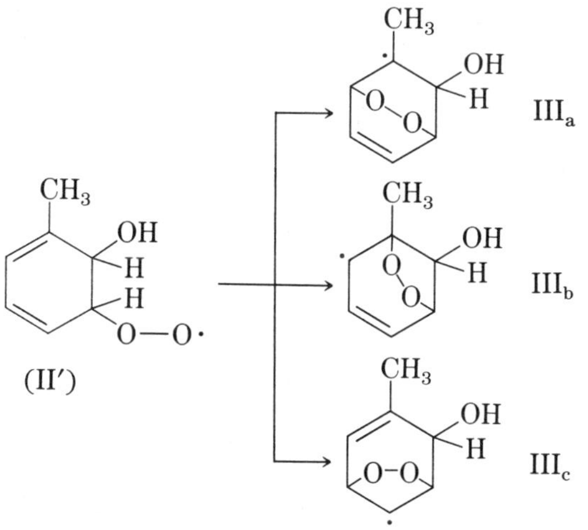

(17)

Although Atkinson and co-workers *(7)* argue that the radicals (IIIa) and (IIIb) are both possible, only (IIIa) is selected for the following discussion. After O_2 addition, this radical can react with NO to form an alkoxy radical (V) or yield a stable nitrate [Equation (18)]:

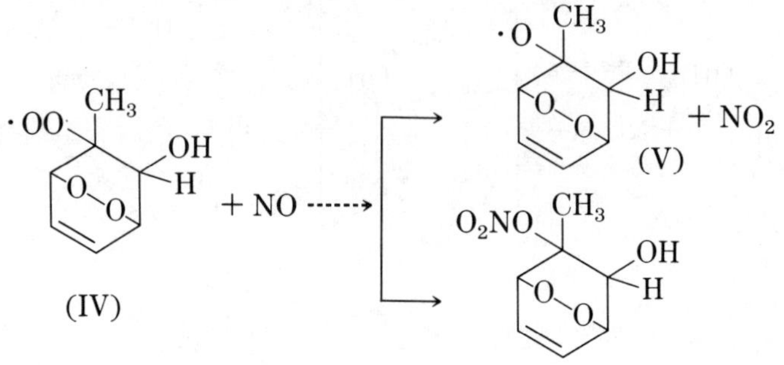

(18)

The bicyclic peroxy alkoxy radicals (V) fragment to form α-dicarbonyl and conjugate γ-dicarbonyl products [Equation (19)]:

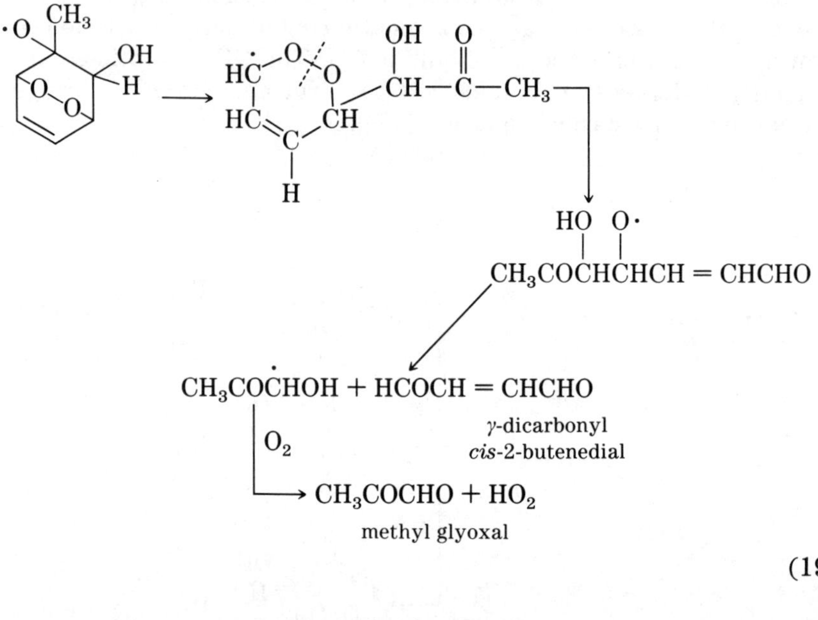

(19)

At this point, the reaction pathways have illustrated the formation of benzaldehyde, *o*-cresol, methyl glyoxal, and γ-dicarbonyl. These constituents can undergo further oxidation in a photochemical system. Benzaldehyde can undergo hydrogen abstraction with OH to form the benzoyl radical. Subsequent addition of O_2 and NO_2 leads to the formation of peroxybenzoyl nitrate (PBzN).

o-Cresol reacts with OH principally by OH addition. The H-abstraction pathway accounts for <10% of the total OH radical reaction and leads to the formation of hydrooxynitrotoluene isomers [Equation (20)]:

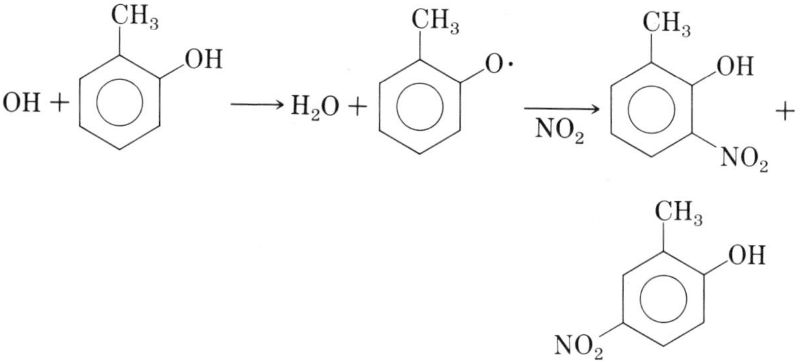

(20)

The addition pathway results in adduct (VI) [Equation (21)]:

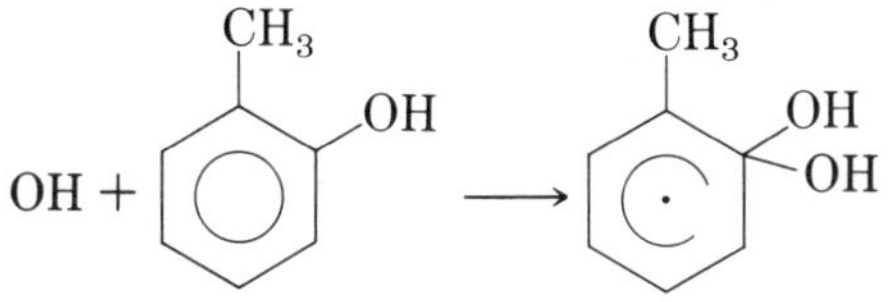

(21)

(VI)

A minor reaction of (VI) leads to the preceding hydrooxynitrotoluene isomers [Equation (20)]. The major reaction of (VI) is with O_2 to form (VII) [Equation (22)]:

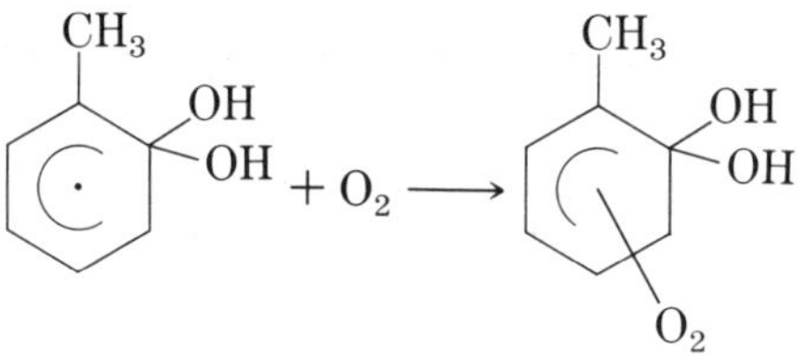

(22)

(VII)

The OH-*o*-cresol-O_2 adduct can react in analogous manner to (II) [Equations (15) and (17)] with the formation of ring fragmentation products.

Methyl glyoxal undergoes photodecomposition, producing HCO and CH_3CO radicals with subsequent formation of HCHO, PAN, and

CH_3ONO_2. Methyl glyoxal may also react with OH forming CO and PAN.

The conjugated γ-dicarbonyls may react with OH to form α-carbonyl peroxy radicals [Equation (23)]

$$\mathrm{HCOCH{=}CHCHO + OH \rightarrow HCOCH{=}CHCO \rightarrow HCOCH{=}CHCO_3\cdot} \qquad (23)$$

which can react with NO or NO_2 to form PAN analogs.

Perry *et al.* *(6)* showed that model calculations based on the types of pathways illustrated were consistent with experimental results on irradiated toluene-benzaldehyde-*o*-cresol-NO_x-air mixtures. With the experimental evidence then available, they were not able to choose among optional sets of equally reasonable reaction sequences.

B. Organic Aerosol Formation

In addition to the role of hydrocarbons in the gas phase photochemical formation of NO_2 and O_3, a subset of the general class of hydrocarbons is transformed into atmospheric particulate matter.

A simple model for the formation and growth of an aerosol at ambient conditions would involve the formation of a gas phase compound (product) by the appropriate chemical oxidation reactions in the gas phase. This product must have a sufficiently low vapor pressure that the gas phase concentration of the oxidized product exceeds its saturation vapor pressure. When this condition occurs, both nucleation and condensation may relieve supersaturation by transferring mass to the condensed phase. The extent of either process depends upon the amount of existing aerosol at a particular time.

The clearest demonstration of this process of homogeneous gas-to-particle conversion is sulfuric acid aerosol formation. Sulfuric acid (H_2SO_4) has an extremely low saturation vapor pressure. Oxidation of relatively small amounts of sulfur dioxide (SO_2) can result in a gas phase concentration of H_2SO_4 exceeding the equilibrium vapor pressure in the ambient atmosphere, with the subsequent formation of sulfuric acid aerosol. In contrast, nitric acid (HNO_3) has a much higher saturation vapor pressure, and, therefore, the gas phase concentration of HNO_3 is not high enough to permit nucleation of nitric acid aerosol in typical atmospheric systems.

In the case of organic aerosol formation, the same product vapor pressure considerations apply. For a particular organic compound, if the gas phase concentration exceeds the saturation vapor pressure gas-to-particle conversion occurs. The organic system is complicated by the large number of organic compounds available in ambient atmospheres. Potential organic aerosol precursor compounds may be identified by physical and chemical criteria. The first of these considerations is the carbon chain length or molecular weight. Oxygenated products with carbon chains $\leq C_4$ have sufficiently high vapor pressures that gas phase concentrations do not result in supersaturated conditions, and, therefore, condensation is not likely at ambient concentrations. Second, the more extensively oxidized compounds have lower vapor pressures for a given carbon chain length. Therefore, difunctional compounds would condense more easily than monofunctional compounds. Third, although the concentration of oxidized species in the atmosphere depends on the properties of the aerosol precursor, the photochemical reactivity of the precursor organically influences the extent of oxidation. This means that alkanes have a lesser role in organic aerosol formation than alkenes because the former are much less photochemically reactive. By these criteria, alkene compounds represent the more chemically reactive class than alkanes and aromatics. A terminally bonded dialkene or a cyclic alkene results in a more extensively oxidized product than a monoalkene. Because of lower vapor pressure of the oxidized product, compounds of C_5 or greater are most likely to condense *(8)*.

For hydrocarbons to act as aerosol precursors they must undergo chemical oxidation. The reactions of the hydrocarbon with the oxidizing species O, O_3, and free radicals (e.g., OH) can yield stable products. Some of the stable products are oxygenated hydrocarbons with low saturation vapor pressures which, when exceeded, result in gas-to-particle conversion by nucleation and condensation. Analysis of ambient organic aerosol samples has indicated compounds containing one or more functional groups such as —COOH, —CHO, —NO_3, or —OH with empirical formulas X—$(CH_2)_m$—Y, where X and Y are functional groups.

C. Application of Chemical Mechanisms to Air Pollution Problems

The use of chemical mechanisms in air pollution problems ranges from the explicit detail of experimental laboratory investigation to the identification of possible oxidation products in the atmosphere. The activities most associated with chemical mechanisms are the function of hydrocarbons and oxides of nitrogen as precursors to atmospheric ozone formation and the role of atmospheric processes in the oxidation and removal of SO_2 in the atmosphere.

The role of hydrocarbons and NO_x in the formation of ozone in the atmosphere is well understood in terms of the formation of free radicals as hydrocarbons are oxidized; participation of free radicals in the NO to NO_2 conversion process; and the buildup of ozone. When this understanding is translated into application, difficulties arise. Researchers have developed chemical mechanisms which accurately calculate the time–concentration profiles for individual and simple mixtures of hydrocarbons studied in laboratory settings *(9)*. The number of mechanistic steps may range from 10 for ethylene to 200 or more for propylene. The potentially large number of chemical reactions required to represent each hydrocarbon in the atmosphere explicitly require methods to simplify the chemical mechanisms. These simplified mechanisms may be included in transport and reactive models used in practical situations.

The vast mixture of hydrocarbon compounds present in urban atmospheres includes alkanes, alkenes, alkynes, aromatics, and oxygenates such as aldehydes or alcohols. Two approaches to simplifying mechanisms are being pursued. The first, apportioning the hydrocarbon mixture by the types of carbon bonding in the organic molecules, is called the *carbon bond approach (10)*. The second is apportioning the hydrocarbon mixture into compounds of similar chemical reactivity or types of reactions, e.g., classifying all aldehydes in one class *(11–13)*.

1. The Carbon Bond Approach

a. Classification Scheme. The carbon bond approach treats carbon atoms with similar bonding in a similar manner regardless of the molecules in which they occur. The simplest form of this approach designates four types of carbon atoms:

C—C:	(PAR)	Includes alkanes and single carbon bonds in alkenes, etc.
Fast C=C:	(OLE)	Primarily alkenes
Slow C=C:	(ARO)	Aromatic ring with 3 double bonds (Ethylene is also in this class)
Carbonyl HC=O:	(CAR)	

With this type of classification, propene, $CH_3—CH=CH_2$, contains 1 PAR and 1 OLE, and toluene contains 3 ARO and 1 PAR. This approach allows mechanisms with many fewer steps than explicit mechanisms. The carbon bond approach avoids the problem of treating the stepwise oxidation of the carbon chain, which requires selection of an average molecular weight of the reacting hydrocarbons. The selection of rate constants for averaging is also much less subject to error. Table II contains a 32-step mechanism based on the carbon bond approach

Table II A Hydrocarbon–NO_x Mechanism Based on the Carbon Bond Approach *(10)*

Number	*Reaction*	*Rate constant*[a]	*Number*	*Reaction*	*Rate constant*
1	$NO_2 + h\nu \rightarrow NO + O$	K_1^b	18	$OLE + O \xrightarrow{2O_2} HC(O)O_2\cdot + CH_3O_2\cdot$	5.3×10^3
2	$O + O_2(+M) \rightarrow O_3(+M)$	2.08×10^{-5}	19[c]	$OLE + O_3 \xrightarrow{O_2} \gamma(HC(O)O_2\cdot + HCHO + OH\cdot)$	1.5×10^{-2}
3	$O_3 + NO \rightarrow NO_2 + O_2$	2.52×10	20	$PAR + OH\cdot \xrightarrow{O_2} CH_3O_2\cdot + H_2O$	1.3×10^3
4	$O + NO_2 \rightarrow NO + O_2$	1.34×10^4	21	$PAR + O \xrightarrow{O_2} CH_3O_2\cdot + OH\cdot$	2.0×10
5	$O_3 + NO_2 \rightarrow NO_3 + O_2$	5×10^{-2}	22	$CAR + OH\cdot \xrightarrow{O_2} HC(O)O_2\cdot + H_2O$	1.0×10^4
6	$NO_3 + NO \rightarrow NO_2 + NO_2$	1.3×10^4 $(2.5 \times 10^4)^d$	23[c]	$CAR + h\nu \xrightarrow{2O_2} \alpha HC(O)O_2\cdot + \alpha HO_2\cdot + (1-\alpha)CO$	$6.0 \times 10^{-3}k_1$
7	$NO_3 + NO_2 + H_2O \rightarrow 2HNO_3$	2.0×10^{-3}	24	$ARO + OH\cdot \xrightarrow{O_2} CAR + CH_3O_2\cdot$	8.0×10^3
8	$HO_2\cdot + NO_2 \rightarrow HNO_2$	2.0×10	25	$ARO + O \xrightarrow{2O_2} HC(O)O_2\cdot + CH_3O_2\cdot$	3.7×10
9	$NO_2 + OH\cdot \rightarrow HNO_3$	9.0×10^3 (1.4×10^4)	26	$ARO + O_3 \xrightarrow{O_2} HC(O)O_2\cdot + CAR + OH\cdot$	2.0×10^{-3}
10	$HNO_2 + h\nu \rightarrow NO + OH\cdot$	$1.9 \times 10^{-1}k_1$	27	$ARO + NO_3 \rightarrow$ products(aerosols)	5.0×10 (1.0×10^2)
11	$NO + OH\cdot \rightarrow HNO_2$	9.0×10^3 (1.4×10^4)	28	$CH_3O_2\cdot + NO \rightarrow NO_2 + CAR + HO_2\cdot$	2.0×10^3 (1.2×10^4)
12	$CO + OH\cdot \xrightarrow{O_2} CO_2 + HO_2\cdot$	2.06×10^2 (4.5×10^2)	29	$HC(O)O_2\cdot + NO \rightarrow NO_2 + CO_2 + HO_2\cdot$	2.0×10^3 (3.8×10^3)
13	$HO_2\cdot + NO \rightarrow OH\cdot + NO_2$	2.0×10^3 (1.2×10^4)	30	$HC(O)O_2\cdot + NO_2 \rightarrow PAN$	1.5×10^2 (6.0×10^2)
14	$HO_2\cdot + HO_2\cdot \rightarrow H_2O_2 + O_2$	4.0×10^3 (1.5×10^4)	31	$CH_3O_2\cdot + HO_2\cdot \rightarrow CH_3OOH + O_2$	4.0×10^3
15	$PAN \rightarrow HC(O)O_2\cdot + NO_2$	2.0×10^{-2}	32	$HC(O)O_2\cdot + HO_2\cdot \rightarrow HC(O)OOH + O_2$	4.0×10^3
16	$H_2O_2 + h\nu \rightarrow OH\cdot + OH\cdot$	$7.0 \times 10^{-4}k_1$			
17	$OLE + OH\cdot \xrightarrow{O_2} CAR + CH_3O_2\cdot$	3.8×10^4			

[a] In units of ppm^{-1} min^{-1} except for photolysis reactions and Reaction 15 (units per minute) and Reaction 7 (units of ppm^{-2} min^{-1}). These rate constants are tentative; they will be changed as dictated by future experimental studies.

[b] The rate constant for NO_2 photolysis, K_1, depends on the light intensity.

[c] $\alpha = 0.5$; $\gamma = 0.67$.

[d] Rate constants in parentheses represent an update to reflect Hampson, R. F., Garvin, D. Natl. Bur. Stand. *(US) Spec. Publ.* **4978,** No. 513.

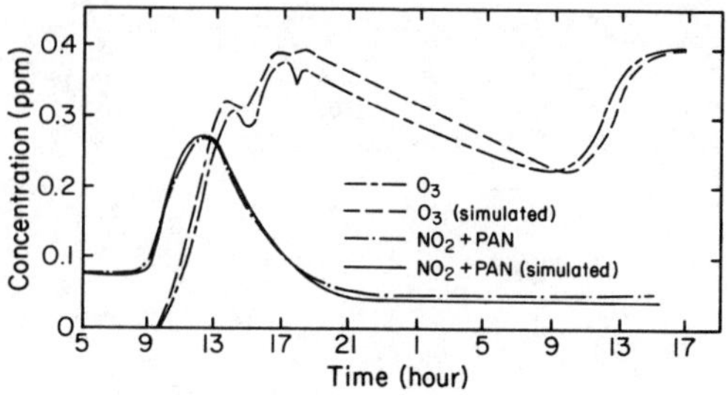

Figure 5. Comparison of experimental results and carbon bond simulation modeling for high hydrocarbon precursor – NO_x mixture in the University of North Carolina outdoor smog chamber *(10)*.

(10). The first 14 steps describe the inorganic chemistry. OLE reactions are in 17 – 19; PAR in 20, 21; CAR in 17, 22 – 24, 26, 28; and ARO in 24 – 27. Other reactions involve the conversion of peroxyalkyl radicals to alkoxy radicals and radical termination steps.

b. Application of the Carbon Bond Approach. The carbon bond mechanism was used to model experimental smog chamber data from two university laboratories and ambient data from Denver, Colorado, and Los Angeles, California. Figures 5 and 6 are model simulations compared to a 36-hour outdoor smog chamber experiment at the University of North Carolina in Chapel Hill North Carolina (UNC) *(10)*. The only modification to the mechanism in Table II was to treat internal and terminal alkenes separately and to change the $N_2O_5 + H_2O$ reaction rate to be consistent with prior data from the UNC chamber. The carbon bond mechanism made predictions in substantial agreement with the measured experimental values; it was also used as part of an urban airshed model which accounted for the interaction of

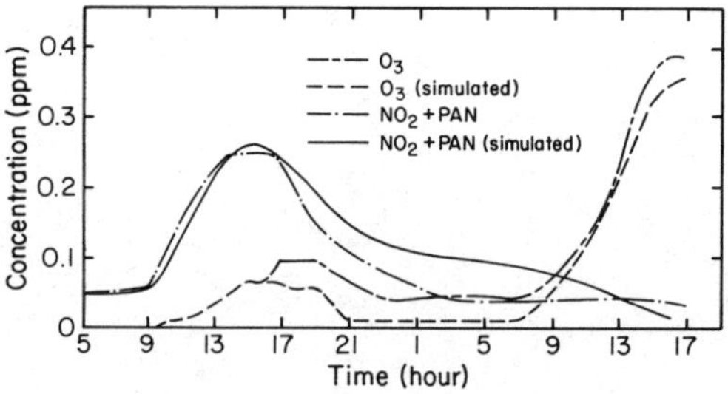

Figure 6. Comparison of experimental results and carbon bond simulation modeling for low hydrocarbon precursor – NO_x mixture in the University of North Carolina outdoor smog chamber *(10)*.

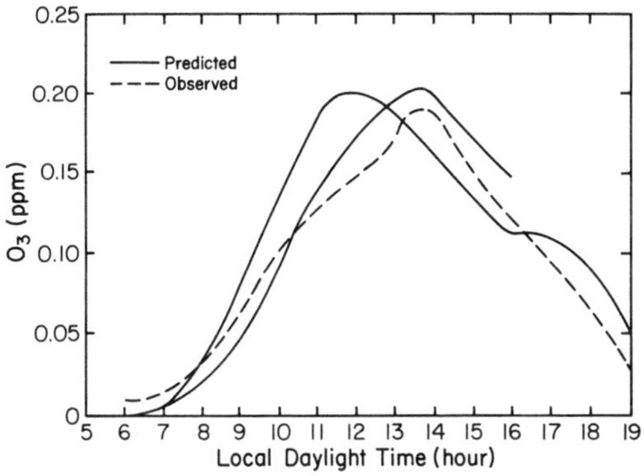

Figure 7. Average ozone concentrations in the Los Angeles basin predicted by the carbon bond mechanism (for two sets of wind fields and initial pollutant concentrations) and measured on June 26, 1974 *(10)*.

emissions, meteorology, and chemistry. Figure 7 contains the results of this carbon bond mechanism applied to the Los Angeles, California, basin *(10)*. Two sets of wind fields and initial pollutant concentrations illustrate the sensitivity to meteorology and ambient measurements. The results reported here from the open literature are for the second formulation of the approach, Carbon Bond II. This modeling approach and mechanism has undergone an additional revision, and at the time of writing, the Carbon Bond III mechanism has been adopted by the United States Environmental Protection Agency for inclusion in air quality modeling activities for compliance with air quality standards *(14)*. Almost all published models and chemical mechanisms are being constantly "improved" by incorporating new information about the physical and chemical phenomena upon which they are based. This applies to the meteorological as well as the chemical components.

2. Lumping by Compound Type

a. CLASSIFICATION SCHEMES. Atkinson and co-workers *(13, 15)* have developed a lumped mechanism based on grouping hydrocarbons into the following general classes: alkanes, 1-alkenes and internal alkenes, aromatics, and aldehydes. For the alkane class, methane and ethane are neglected, propane is considered separately, and $\geq C_4$ alkanes are lumped into a single class; for the alkene class, ethene is considered separately, propene represents the 1-alkenes, and *trans*-2-butene represents the internal alkenes; for the aromatic class, ben-

Table III A Hydrocarbon–NO_x Mechanism Based on Lumping by Hydrocarbon Type (13)

Reaction	*Rate constant (ppm min units)*
Inorganics	
1. $NO_2 + h\nu \xrightarrow{O_2} NO + O_3$	k_1 variable
2. $NO + O_3 \rightarrow NO_2 + O_2$	$k_2 = 1.0 \times 10^6 T^{-1} e^{-1450/T}$
3. $O_3 + h\nu \rightarrow 2OH$	$k_3 = \beta_1 k_1 \times 7.5 \times 10^{-6}[H_2O]$
4. $OH + NO \xrightarrow{M} HONO$	$k_4 = 8.7 \times 10^8 T^{-2}$
5. $OH + NO_2 \xrightarrow{M} HNO_3$	$k_5 = 1.5 \times 10^9 T^{-2}$
6. $HONO + h\nu \rightarrow OH + NO$	$k_6 = \beta_2 k_1$
7. $HO_2 + NO \rightarrow OH + NO_2$	$k_7 = 3.7 \times 10^6 T^{-1}$
8. $HO_2 + NO_2 \xrightarrow{M} HO_2NO_2$	$k_8 = 1.5 \times 10^8 T^{-2}$
9. $HO_2NO_2 \xrightarrow{M} HO_2 + NO_2$	$k_9 = 7.8 \times 10^{15} e^{-10420/T}$
10. $HO_2 + HO_2 \rightarrow H_2O_2 + O_2$	$k_{10} = 3.4 \times 10^4 T^{-1} e^{1100/T} + 5.8 \times 10^{-5} T^{-2} e^{5800/T}[H_2O]$
11. $H_2O_2 + h\nu \rightarrow 2OH$	$k_{11} = \beta_3 k_1$
12. $OH + CO \xrightarrow{O_2} HO_2$	$k_{12} = 1.3 \times 10^5 T^{-1}$
13. $NO_2 + O_3 \rightarrow NO_3$	$k_{13} = 5.3 \times 10^4 T^{-1} e^{-2450/T}$
14. $NO + NO_3 \rightarrow 2NO_2$	$k_{14} = 8.4 \times 10^6 T^{-1}$
15. $NO_2 + NO_3 \xrightarrow{M} N_2O_5$	$k_{15} = 3.1 \times 10^7 T^{-1} e^{-1100/T}$
16. $N_2O_5 \xrightarrow{M} NO_2 + NO_3$	$k_{16} = 3.5 \times 10^{18} e^{-12280/T}$
17. $N_2O_5 + H_2O \rightarrow 2HNO_3$	$k_{17} \sim 1.33 \times 10^{-3} T^{-1}$
18. $NO_3 + h\nu \rightarrow 0.3NO + 0.7NO_2 + 0.7O_3$	$k_{18} = \beta_4 k_1$
19. $OH + O_3 \rightarrow HO_2$	$k_{19} = 7.0 \times 10^5 T^{-1} e^{-940/T}$
20. $HO_2 + O_3 \rightarrow OH$	$k_{20} = 4.8 \times 10^3 T^{-1} e^{-580/T}$
Formaldehyde	
21. $HCHO + h\nu \xrightarrow{O_2} HO_2 + HO_2 + CO$	$k_{21} = \beta_5 k_1$
22. $HCHO + h\nu \rightarrow CO + H_2$	$k_{22} = \beta_6 k_1$
23. $OH + HCHO \xrightarrow{O_2} HO_2 + CO$	$k_{23} = 4.4 \times 10^6 T^{-1}$

Acetaldehyde		
24. $CH_3CHO + h\nu \xrightarrow{O_2} CH_3O_2 + HO_2 + CO$	$k_{24} = \beta_7 k_1$	
25. $OH + CH_3CHO \xrightarrow{O_2} CH_3CO_3$	$k_{25} = 3.0 \times 10^6 T^{-1} e^{250/T}$	
26. $CH_3CO_3 + NO_2 \rightarrow PAN$	$k_{26} = 2.1 \times 10^6 T^{-1}$	
27. $PAN \rightarrow CH_3CO_3 + NO_2$	$k_{27} = 1.2 \times 10^{18} e^{-13543/T}$	
28. $CH_3CO_3 + NO \xrightarrow{O_2} NO_2 + CH_3O_2$	$k_{28} = 3.1 \times 10^6 T^{-1}$	
29. $CH_3O_2 + NO \rightarrow HCHO + HO_2 + NO_2$	$k_{29} = 3.1 \times 10^6 T^{-1}$	
Propane		
30. $OH + PROPANE \rightarrow PO_2$	$k_{30} = 6.6 \times 10^6 T^{-1} e^{-680/T}$	
31. $PO_2 + NO \rightarrow HO_2 + NO_2 + CH_3COCH_3$	$k_{31} = 3.1 \times 10^6 T^{-1}$	
Alkanes		
32. $OH + Alkane \rightarrow AO_2$	$k_{32} = 8.0 \times 10^6 T^{-1} e^{-560/T}$	for explicit *n*-butane
	$k_{32} = 6.6 \times 10^6 T^{-1} e^{-400/T}$	for lumped $\geq C_4$ alkane
33. $AO_2 + NO \rightarrow 1.3NO_2 - 0.4NO + 0.9HO_2 + 0.6CH_3CHO + 0.1RCHO + 0.5MEK$	Explicit *n*-butane	
$\rightarrow 1.7NO_2 - 0.8NO + 0.9HO_2 + 0.15\ HCHO + 0.3\ CH_3CHO + 0.1\ RCHO + 0.3\ CH_3COCH_3 + 0.45MEK$	Lumped $\geq C_4$ alkane mechanism	
	$k_{33} = 3.1 \times 10^6 T^{-1}$	for both systems
Higher aldehydes		
34. $OH + RCHO \xrightarrow{O_2} RCO_3$	$k_{34} = 9.2 \times 10^6 T^{-1}$	
35. $RCO_3 + NO_2 \rightarrow PPN$	$k_{35} = 2.1 \times 10^6 T^{-1}$	
36. $PPN \rightarrow RCO_3 + NO_2$	$k_{36} = 1.2 \times 10^{18} e^{-13543/T}$	
37. $RCO_3 + NO \rightarrow C_2H_5O_2 + NO_2$	$k_{37} = 3.1 \times 10^6 T^{-1}$	
38. $C_2H_5O_2 + NO \xrightarrow{O_2} CH_3CHO + HO_2 + NO_2$	$k_{38} = 3.1 \times 10^6 T^{-1}$	
39. $RCHO + h\nu \xrightarrow{O_2} C_2H_5O_2 + CO + HO_2$	$k_{39} = \beta_8 k_1$	

Table III *(Continued)*

Reaction	Rate constant (ppm min units)
Ketones	
40. $OH + MEK \xrightarrow{O_2} XO_2$	$k_{40} = 4.4 \times 10^6 T^{-1} e^{-330/T}$
41. $XO_2 + NO \rightarrow NO_2 + CH_3CHO + CH_3CO_3$	$k_{41} = 3.1 \times 10^6 T^{-1}$
42. $MEK + h\nu \xrightarrow{O_2} CH_3CO_3 + C_2H_5O_2$	$k_{42} = \beta_9 k_1$
43. $CH_3COCH_3 + h\nu \xrightarrow{O_2} CH_3CO_3 + CH_3O_2$	$k_{43} = \beta_{10} k_1$
Alkenes	
44. $OH + Ethene \xrightarrow{O_2} 2HCHO + NO_2 - NO + HO_2$	$k_{44} = 9.7 \times 10^5 T^{-1} e^{380/T}$
45. $OH + Propene \xrightarrow{O_2} HCHO + CH_3CHO + HO_2 + NO_2 - NO$	$k_{45} = 1.8 \times 10^6 T^{-1} e^{540/T}$
46. $OH + Butene \xrightarrow{O_2} 1.8\ CH_3CHO + 0.9NO_2 + 0.9HO_2 - NO$	$k_{46} = 5.0 \times 10^6 T^{-1} e^{540/T}$
47. $O_3 + Ethene \rightarrow HCHO + 0.4\dot{C}H_2\dot{O}_2 + 0.4CO + 0.12HO_2$	$k_{47} = 4.2 \times 10^3 T^{-1} e^{-2560/T}$
48. $O_3 + Propene \rightarrow 0.5HCHO + 0.5CH_3CHO + 0.2\dot{C}H_2\dot{O}_2 + 0.2CH_3\dot{C}HO\dot{O} + 0.3CO + 0.2HO_2 + 0.1OH + 0.2CH_3O_2$	$k_{48} = 3.1 \times 10^3 T^{-1} e^{-1900/T}$
49. $O_3 + Butene \rightarrow CH_3CHO + 0.4CH_3\dot{C}HO\dot{O} + 0.3HO_2 + 0.2OH + 0.45CH_3O_2 + 0.2CO$	$k_{49} = 3.3 \times 10^3 T^{-1} e^{-1050/T}$
50. $\dot{C}H_2\dot{O}_2 + NO \rightarrow HCHO + NO_2$	$k_{50} = 3.1 \times 10^6 T^{-1}$
51. $\dot{C}H_2\dot{O}_2 + NO_2 \rightarrow HCHO + NO_3$	$k_{51} = 3.1 \times 10^5 T^{-1}$
52. $\dot{C}H_2\dot{O}_2 + SO_2 \rightarrow HCHO + SO_4^{2-}$	$k_{52} = 3.0 \times 10^4 T^{-1}$
53. $\dot{C}H_2\dot{O}_2 + H_2O \rightarrow Product$	$k_{53} = 1.5T^{-1}$
54. $CH_3\dot{C}HO\dot{O} + NO \rightarrow CH_3CHO + NO_2$	$k_{54} = 3.1 \times 10^6 T^{-1}$
55. $CH_3\dot{C}HO\dot{O} + NO_2 \rightarrow CH_3CHO + NO_3$	$k_{55} = 3.1 \times 10^5 T^{-1}$
56. $CH_3\dot{C}HO\dot{O} + SO_2 \rightarrow CH_3CHO + SO_4^{2-}$	$k_{56} = 3.0 \times 10^4 T^{-1}$
57. $CH_3\dot{C}HO\dot{O} + H_2O \rightarrow Product$	$k_{57} = 1.5T^{-1}$
Aromatics	
58. $OH + Benzene \rightarrow 0.25Cresol + 0.25HO_2 + 0.75ADD$	$k_{58} = 5.3 \times 10^5 T^{-1}$

59. $OH + Toluene \rightarrow 0.15ARO_2 + 0.20Cresol + 0.20HO_2 + 0.65ADD$	$k_{59} = 2.7 \times 10^6 T^{-1}$
60. $OH + Xylene \rightarrow 0.25Cresol + 0.25HO_2 + 0.75ADD$	$k_{60} = 7.9 \times 10^6 T^{-1}$ for lumped xylene
	$= 1.05 \times 10^7 T^{-1}$ for explicit *m*-xylene
61. $ADD + NO \rightarrow 0.75NO_2 + 0.75HO_2 + 0.75DIAL + \alpha_1(CHO)_2 + \alpha_2CH_3COCHO$	$k_{61} = 3.1 \times 10^6 T^{-1}$
62. $OH + DIAL \rightarrow El$	$k_{62} = 1.3 \times 10^7 T^{-1}$
63. $El + NO_2 \rightarrow ElNO_2$	$k_{63} = 2.1 \times 10^6 T^{-1}$
64. $ElNO_2 \rightarrow El + NO_2$	$k_{64} = 1.2 \times 10^{18} e^{-13543/T}$
65. $El + NO \rightarrow 3NO_2 - 2NO + \alpha_3HO_2 + \alpha_3(CHO)_2 + \alpha_4CH_3CO_3 + \alpha_4CH_3COCHO + \alpha_3CO$	$k_{65} = 3.1 \times 10^6 T^{-1}$
66. $OH + (CHO)_2 \xrightarrow{O_2} HO_2 + CO$	$k_{66} = 8.8 \times 10^6 T^{-1}$
67. $(CHO)_2 + h\nu \rightarrow HCHO + CO$	$k_{67} = \beta_{11}k_1$
68. $OH + CH_3COCHO \xrightarrow{O_2} CH_3CO_3 + CO$	$k_{68} = 6.6 \times 10^6 T^{-1}$
69. $CH_3COCHO + h\nu \xrightarrow{O_2} CH_3CO_3 + HO_2 + CO$	$k_{69} = \beta_{12}k_1$
70. $OH + Cresol \rightarrow ADD2$	$k_{70} = 1.9 \times 10^7 T^{-1}$
71. $ADD2 + NO \rightarrow 0.75NO_2 + 0.75HO_2 + 0.75DIAL$	$k_{71} = 3.1 \times 10^6 T^{-1}$
72. $NO_3 + Cresol \rightarrow HNO_3 + Phenoxy$	$k_{72} = 6.6 \times 10^6 T^{-1}$
73. $Phenoxy + NO_2 \rightarrow Products$ (*o*-, *p*-nitrophenols)	$k_{73} = 6.6 \times 10^6 T^{-1}$
74. $ARO_2 + NO \rightarrow 0.75NO_2 + 0.75HO_2 + 0.75\ ARCHO$	$k_{74} = 3.1 \times 10^6 T^{-1}$
75. $ARCHO + h\nu \rightarrow Products$	$k_{75} = \beta_{13}k_1$
76. $OH + ARCHO \xrightarrow{O_2} ARCO_3$	$k_{76} = 5.7 \times 10^6 T^{-1}$
77. $ARCO_3 + NO_2 \rightarrow PBZN$	$k_{77} = 2.1 \times 10^6 T^{-1}$
78. $PBZN \rightarrow ARCO_3 + NO_2$	$k_{78} = 1 \times 10^{17} e^{-13025/T}$
79. $ARCO_3 + NO \rightarrow PhO_2 + NO_2$	$k_{79} = 3.1 \times 10^6 T^{-1}$
80. $PhO_2 + NO \rightarrow Phenoxy + NO_2$	$k_{80} = 3.1 \times 10^6 T^{-1}$

[a] Reactions and rate constants used in the basic standard model. Rate constants, etc., from Atkinson *et al.* *(7)* unless otherwise noted.

zene is considered separately, toluene represents the monoalkylbenzenes; and for the aldehyde class, acetaldehyde is considered separately, RCHO represents the higher aldehydes, acetone is considered separately, and higher ketones are represented by methyl ethyl ketone. The basic reaction list has 81 chemical reactions, including 20 inorganic reactions (Table III).

McRae and co-workers *(11, 12)* have developed a lumped chemical mechanism for incorporation in a complex mathematical model for urban air pollution. This model includes source terms, transport processes, and removal terms in addition to a lumped chemical kinetics mechanism based on the four hydrocarbon classes: alkanes, alkenes, aromatics, and oxygenated compounds. They outline their method for selection of rate constants for various lumped reactions, for example,

$$\mathrm{ALK} + \mathrm{OH} \rightarrow \mathrm{RO_2} \qquad (\text{Rate Constant} = k) \tag{24}$$

In this reaction step, ALK is a lumped parameter representing several alkane compounds such as ethane, butane, or isopentane. Each of these individual compounds has a specific rate constant with hydroxyl radicals. The lumped rate constant is composition-dependent, so modelers devise methods to calculate lumped rate constants, sometimes based on comparison with smog chamber experiments.

IV. Regional Chemical Systems

The fate of man-made emissions to the troposphere is threefold: (1) deposition at the earth's surface, (2) accumulation in the troposphere, and (3) transport to the stratosphere. Examples of the latter two pathways are the accumulation of carbon dioxide in the troposphere and the movement of halocarbons to the stratosphere.

Deposition at the earth's surface provides the most direct interaction with man, ranging from health effects in our urban areas to ecological effects in nonurban areas. The pathway of SO_2 from source to ultimate return to the earth's surface *(16)* (Fig. 8) includes (I) transport and mixing; (II) chemical reactions in the homogeneous gas phase (dry reaction); (III) dry deposition; (IV) attachment; (V) chemical reaction in the homogeneous aqueous phase (wet reaction); and (VI) wet deposition. The residence time of a particular pollutant in the atmosphere depends on the meteorology, concentration, and ambient conditions in a given region.

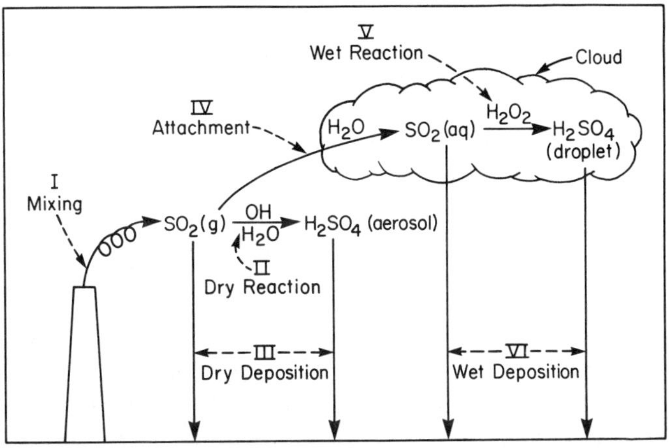

Figure 8. Sulfur dioxide transport, transformation, and deposition processes. Initially SO_2 is mixed into the atmosphere (I). Gaseous SO_2 may undergo oxidation in the gas phase (II) with subsequent formation of H_2SO_4 aerosol. Both gaseous SO_2 and sulfuric acid aerosol may be deposited at the earth's surface (III). Gaseous SO_2 may become dissolved in a water droplet (IV). The dissolved SO_2 can be oxidized in solution (V) to form sulfuric acid aerosol droplets. The sulfuric acid aerosol and the sulfuric acid droplet may be removed to the earth's surface by wet deposition (VI).

A. Acidic Deposition

The problem of greatest interest on the regional scale is acidic deposition. The first major recognition of this problem was in Europe in the 1960s *(17)*. Since then acidic deposition has been recognized in eastern North America. The precursor–product analogy of the hydrocarbon-NO_x-ozone problem is applicable to acidic deposition. The precursors are the gaseous oxides of nitrogen NO and NO_2 and of sulfur, SO_2. The products are nitric acid, HNO_3, and sulfuric acid, H_2SO_4. The gases are emitted by anthropogenic sources, undergo transport and chemical transformation in the atmosphere, and are deposited at the earth's surface as sulfates and nitrates.

1. Transport and Mixing

The first step in the atmospheric cycle of these acidic precursors is transport and mixing in the atmosphere. This process is greatly dependent upon the prevailing meteorological conditions following release to the atmosphere and is discussed in Chapter 3. In some instances, particularly with short stacks, removal at the earth's surface occurs a short time after release and within close proximity of the source. With such short residence times, little chemical transformation occurs in the atmosphere. However, with tall stacks, pollutants are injected into the upper atmosphere, where conditions are more suitable for efficient mixing, and longer residence times permit chemical transformations to occur.

2. Chemical Reactions

a. Homogeneous Gaseous Phase Oxidation. The second step in the cycle are chemical reactions in the homogeneous gas phase (dry reactions). This process is the oxidation in the gas phase of NO and NO_2 to HNO_3 and of SO_2 to H_2SO_4. Several chemical species are capable of oxidizing SO_2 in the laboratory, but the principal gas phase oxidizing agent in daylight conditions is the OH radical, the same agent that was shown to have an important role in urban photochemical smog formation. The source of OH for oxidizing SO_2 depends on the other constituents in the air mass. In urban plumes the OH radicals evolve from the mechanisms described for photochemical smog, i.e., OH reacting with hydrocarbons to form HO_2, then HO_2 reacting with NO to form NO_2 and OH. The OH radicals in this cycle can react with the SO_2 present, forming H_2SO_4. In rural areas where the hydrocarbon ratios are much lower, the source of OH can shift to the photolysis of O_3 where $O(^1D)$ is formed [Equation (25)]:

$$O_3 + h\nu(<305\text{ nm}) \rightarrow O(^1D) + O_2(^1\Delta_g) \qquad (25)$$

$O(^1D)$, an energetic form of atomic oxygen, reacts with water molecules to form two OH radicals [Equation (26)]:

$$O(^1D) + H_2O \rightarrow 2OH \qquad (26)$$

Therefore, there are sources of OH radicals for both urban and rural air masses containing SO_2.

The specific mechanism for the reaction of OH with SO_2 has been subject of much speculation *(18)*. The formation of H_2SO_4 has not been in question, but the specific steps following the initial formation of the $HOSO_2$ adduct have been unknown: Intermediates such as $HOSO_2O_2$, $HOSO_2O$, and $HOSO_2ONO_2$, have been postulated [Equation (27)]:

$$OH + SO_2 \rightarrow HOSO_2 \qquad (27)$$

Because of the lack of information, chemical mechanisms in acidic deposition models have used the following chain terminating step (19):

$$OH + SO_2 \rightarrow H_2SO_4 \qquad (28)$$

The results from this chain terminating assumption support the nonlinear role of SO_2 emission reductions on acidic deposition because OH is removed from the reaction sequence and not regenerated as is the OH + RH sequence described earlier.

i. *IMPORTANCE OF NONLINEAR VERSUS LINEAR MODELS.* Acidic deposition is clearly a function of the amounts of SO_2 emitted into the atmosphere from all sources. The important control strategy question is, Will a fixed percentage reduction in SO_2 emissions cause a proportional reduction in acidic deposition? If there is a linear relationship between the SO_2 mass emitted by sources and the amount of SO_2 converted to sulfate by atmospheric chemical processes, a 50% reduction of SO_2 emissions would result in a 50% reduction in acidic deposition. A nonlinear relationship would result if the amount of atmospheric SO_2 is not the limiting factor in the formation of sulfate. This case could suggest that a 50% reduction in emissions might result in a 10% reduction in acidic deposition and a further 50% emission reduction, another 10% reduction in acidic deposition. An understanding of the relationship between SO_2 and acidic deposition is extremely important in obtaining the desired results from a particular control strategy.

ii. *NONLINEAR MODELS.* In the Rodhe *et al.* model *(19)* of the impact of SO_2 emission reduction, the decrease in acidic deposition is not proportional to SO_2 emission reduction. If the oxidizing species, in this case OH, is the limiting reactant, the formation of H_2SO_4 in Equation (28) is not directly proportional to $[SO_2]$.

iii. *LINEAR MODELS.* Stockwell and Calvert *(20)* suggest that the steps following the OH reaction with SO_2 are not chain terminating. On the basis of their interpretation of SO_2 oxidation in systems containing SO_2, CO_2, CO, HONO, HNO_3, NO, NO_2, H_2O, and O_2, they suggest the following reaction steps:

$$OH + SO_2 \rightarrow HOSO_2 \qquad (27)$$

$$HOSO_2 + O_2 \rightarrow HO_2 + SO_3 \qquad (29)$$

$$SO_3 + H_2O \rightarrow H_2SO_4 \qquad (30)$$

This mechanism is chain-carrying with respect to $OH—HO_2$. This very important modification weakens the nonlinearity argument about the impact of SO_2 emissions reductions on acid deposition.

The OH radical is the principal oxidizing agent in the homogeneous oxidation of NO_2 to nitric acid,

$$OH + NO_2\ (+M) \rightarrow HNO_3\ (+M) \tag{31}$$

Although this reaction is chain-terminating for OH, it represents <10% of the OH—HO_2—radical termination in urban areas and much less in rural areas *(16)*.

b. Homogeneous Liquid Phase Oxidation. A second homogeneous oxidation pathway for formation of HNO_3, potentially important at night, is

$$N_2O_5 + H_2O \rightarrow 2HNO_3 \tag{32}$$

The aqueous phase oxidation of SO_2 (step IV on Figure 8) is preceded by the absorption of SO_2 by the aqueous phase droplet (step III on Figure 8). When SO_2 dissolves in water, a $SO_2 \cdot H_2O$ hydrate is formed,

$$SO_2\ (g) + H_2O\ (aq) \rightarrow SO_2 \cdot H_2O\ (aq) \tag{33}$$

In addition, the following ions are formed: HSO_3^- (bisulfite), SO_3^{2-} (sulfite), and H^+ (hydrogen ion).

$$H_2SO_3 \rightleftharpoons HSO_3^- + H^+ \tag{34}$$

$$HSO_3^- \rightleftharpoons SO_3^{2-} + H^+ \tag{35}$$

The solubility of $SO_2(g)$ in water decreases as the droplet becomes more acidic (pH decreases). This SO_2 solubility behavior affects the role of various proposed oxidizing agents.

c. Liquid Phase Oxidizing Species. Hydrogen peroxide, ozone, manganese, iron, carbon, NO_2, and HNO_2 have been found to oxidize dissolved SO_2 under different conditions. Martin's *(21)* summary of the kinetic rate data for these oxidizing species is shown in Figure 9. Theoretical SO_2 oxidation rates are presented as a function of pH for typical atmospheric concentrations of the various species.

Only H_2O_2 can act as an oxidizing species over a broad pH range. The oxidizing power of all the other species decreases dramatically with lower pH values. Certain circumstances in which these other species would participate in the aqueous phase oxidation of SO_2 could exist in the atmosphere, but because the pH of cloud water is usually less than 5, H_2O_2 is considered the most important aqueous phase oxidizer of SO_2. The NAS document *(16)* could not find any strong evidence to suggest the homogeneous aqueous phase oxidation of NO_2 to form

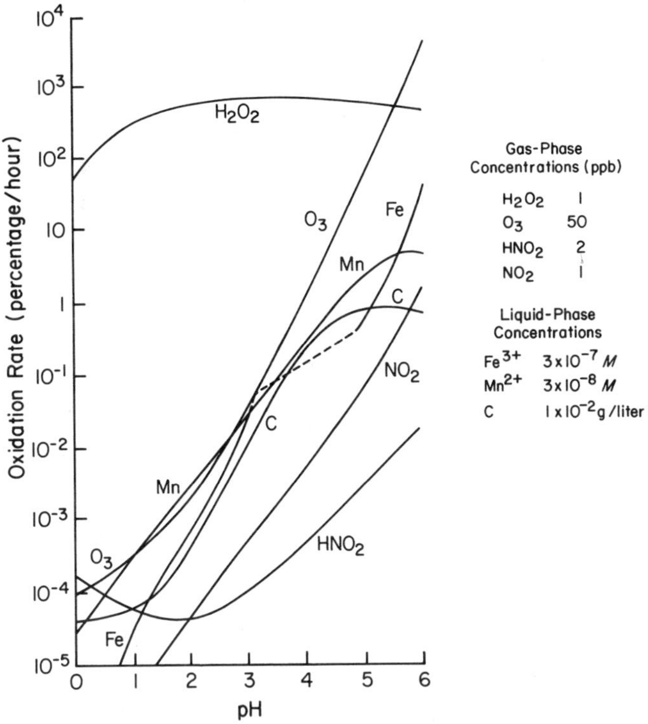

Figure 9. Predicted rates of aqueous phase oxidation of SO_2 by various oxidizing species, assuming 5 ppb of SO_2, 1 ml/m^3 of water in air for conditions shown *(21)*.

HNO_3. The most likely HNO_3 formation pathway results from the oxidation of NO_2 to NO_3 and subsequent formation of N_2O_5, then the dissolution of N_2O_5 in water to form nitric acid.

3. Complex Phase Interaction

The interaction of dry deposition, attachment, and wet deposition steps with the homogeneous aqueous chemical reaction step are shown in the block diagram in Figure 10. The removal process involves several reversible steps. Mixtures of water droplets and gaseous pollutants may form and evaporate (transfer between Box 1 and Box 2). Gaseous pollutants may be absorbed and desorbed from water droplets (transfer between Box 2 and Box 3), chemical reactions may occur in water droplets (transfer between Box 3 and Box 4), and water droplets at Box 3 and 4 of Figure 10 may evaporate, releasing gas-

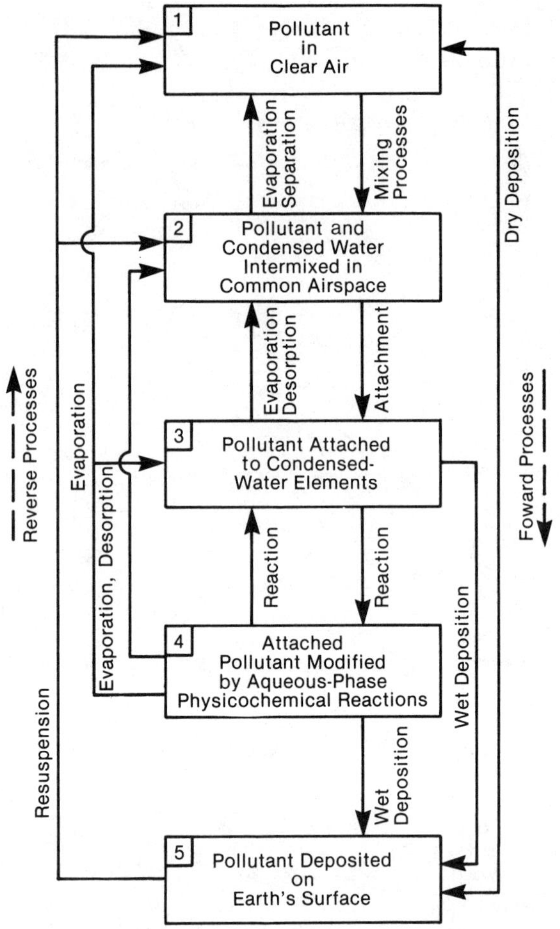

Figure 10. Phase interaction diagram for pollutant scavenging processes. Initially, the pollutant may be in the gas phase (Box 1). The presence of water vapor in the atmosphere provides for the intermixing of gaseous pollutants with aqueous droplets in the same space, i.e., in a cloud. (Box 2). The pollutant gases can become attached to the water droplets, i.e., be absorbed (Box 3), and undergo chemical reaction (Box 4). The gaseous and aqueous pollutants return to the earth's surface (Box 5). Some of these processes have reversible pathways, and others are unidirectional *(16)*.

eous pollutants or submicron aerosols. Wet and dry deposition to the earth's surface represents the removal steps from the atmosphere (transfers from Box 1 to Box 5, Box 3 to Box 5, and Box 4 to Box 5). In the precipitation process, the in-cloud interaction may go through the absorbing–desorbing cycle several times, but chemical reactions alter their equilibrium. For example, a given SO_2 molecule may be absorbed and desorbed easily, depending on the vapor pressure of SO_2 in the surrounding air, but once it is oxidized to sulfate, its desorption as SO_2 cannot occur. More information about the in-cloud processes will assist predicting removal rates and product distributions.

B. Acid Fog

Acidic deposition has been discussed in the context of two categories, wet and dry deposition. The wet deposition component is commonly called *acid rain*. Hoffmann and co-workers *(22, 23)* have taken a new look at an old air pollution problem: polluted fogs. They have collected fog water along the California coast, and their analysis of many samples reveals low pH values. Considerable controversy has evolved over the importance of these findings with respect to health impact *(24–26)*. The issue is whether acid fogs are accompanied by other harmful pollutants such as O_3 or particulate matter. Clearly, the "killer" fogs of London had high levels of other pollutants which contributed to their adverse effects. Results are not conclusive about the frequency of acid fogs containing other pollutants in the 1980s. Several researchers do not believe there is sufficient mass of acid for this type of fog to pose a threat to human health in the western United States *(27, 28)*.

REFERENCES

1. R. A. Cox and R. G. Derwent, *in* "Gas Kinetics and Energy Transfer" (P. G. Ashmore and R. J. Donovan, eds.), pp. 189–234. Burlington House, London, 1981.
2. T. E. Graedel, "Chemical Compounds in the Atmosphere." Academic Press, New York, 1978.
3. A. C. Stern, R. W. Boubel, D. B. Turner, and D. L. Fox, "Fundamentals of Air Pollution," 2nd ed. Academic Press, New York, 1984.
4. H. Akimoto, F. Sakamaki, M. Hoshimo, G. Inoue, and M. Oduda, *Environ. Sci. Technol.* **13,** 53–58 (1979).
5. J. P. Killus and G. Z. Whitten, *Atmos. Environ.* **16,** 1973–1999 (1982).
6. R. A. Perry, R. Atkinson, and J. N. Pitts, Jr., *J. Phys. Chem.* **81,** 296–304 (1977).
7. R. A. Atkinson, W. P. L. Carter, K. R. Darnall, A. M. Winer, and J. N. Pitts, Jr., *Int. J. Chem. Kinet.* **12,** 779–836 (1980).
8. National Academy of Sciences, "Ozone and Other Photochemical Oxidants." National Academy Press, Washington, D.C., 1977.
9. K. L. Demerjian, J. A. Kerr, and J. G. Calvert, *Adv. Environ. Sci. Technol.* **4,** 1–262 (1974).
10. G. E. Whitten, H. Hogo, and J. P. Killus, *Environ. Sci. Technol.* **14,** 690–700 (1980).

11. G. J. McRae, W. R. Goodin, and J. H. Seinfeld, *Atmos. Environ.* **16,** 679–696 (1982).
12. G. J. McRae and J. H. Seinfeld, *Atmos. Environ.* **17,** 501–522 (1983).
13. R. A. Atkinson, A. C. Lloyd, and L. Winges, *Atmos. Environ.* **16,** 1341–1355 (1982).
14. U.S. Environmental Protection Agency, "Guideline for Using the Carbon Bond Mechanism in City Specific EKMA," EPA-450/4-84-005. Office of Air Quality Planning and Standards, USEPA, Research Triangle Park, North Carolina, 1984.
15. R. Atkinson and A. C. Lloyd, *J. Phys. Chem. Ref. Data* **13,** 315 (1984).
16. National Academy of Sciences, "Acid Deposition: Atmospheric Processes in Eastern North America." National Academy Press, Washington, D.C., 1983.
17. C. Brossett, *Ambio* **2,** 2–8 (1973).
18. R. B. Husar, J. P. Lodge, Jr., and D. J. Moore, eds., "Sulfur in the Atmosphere," Special Issue of *Atmos. Environ.,* Vol. 12. Pergamon, Oxford, 1978.
19. H. Rodhe, P. Crutzen, and A. Vanderpol, *Tellus* **33,** 132–141 (1981).
20. W. R. Stockwell and J. G. Calvert, *Atmos. Environ.* **17,** 2231–2236 (1983).
21. L. R. Martin, *in* "Acid Precipitation: SO_2, NO, and NO_2 Oxidation Mechanisms: Atmospheric Considerations" (J. G. Calvert, ed.). Ann Arbor Press, Ann Arbor, Michigan, 1983.
22. J. W. Munger, D. J. Jacob, J. M. Waldman, and M. R. Hoffmann, *JGR, J. Geophys. Res.* **88,** 5109 (1983).
23. B. Hileman, *Environ. Sci. Technol.* **17,** 117A–120A (1983).
24. J. M. Waldman, *J. Air Pollut. Control Assoc.* **34,** 13 (1984).
25. W. J. Deal, *J. Air Pollut. Control Assoc.* **34,** 94 (1984).
26. M. R. Hoffmann, *Environ. Sci. Technol.* **18,** 61 (1984).
27. W. J. Deal, *J. Air Pollut. Control Assoc.* **33,** 691 (1983).
28. W. B. Innes, *Environ. Sci. Technol.* **18,** 61 (1984).

Lists of books relevant to this chapter published since the lists in Chapter 8, Volume II, 3d edition (1977) can be found in Chapter 1, Table IX, Volume VII.

3

The Transport of Pollutants

D. Bruce Turner

Meteorology and Assessment Division
United States Environmental Protection Agency
Research Triangle Park, North Carolina
(on assignment from the National
Oceanic and Atmospheric Administration)

I. The Meteorological Setting

A. Parameterization of the Boundary Layer

Since most of the air pollutants and their subsequent transport and dispersion occur in the planetary boundary layer (PBL), it is important to understand the structure of wind, temperature, and turbulence through the PBL.

1. The Convective Boundary Layer

Wyngaard *(1)* describes the convective boundary layer (CBL) as consisting of three layers: the surface layer, the mixed layer, and the interfacial layer. Typical profiles of potential temperature, wind speed, and direction are shown in Figure 1 *(2)*. Caughey *(3)* shows a fourth layer, the free convection layer, between the surface layer and

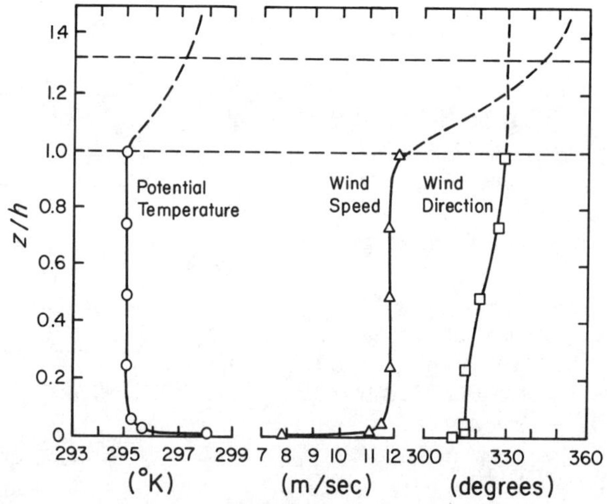

Figure 1. Profiles of wind speed, wind direction, and potential temperature during convective conditions. The near-adiabatic lapse rate and the negligible mean wind shear in the mixed layer are typical for observation periods in this experiment. z = height above ground, h = mixing height. [Reproduced with permission from Kaimal *et al.* *(2)*.]

the mixed layer. The orthogonal components of the wind are u, v, and w with u along the mean horizontal wind direction.

In the surface layer wind shear dominates and Monin–Obukhov similarity applies. The controlling parameters are z, height above ground; τ_0, surface stress $= \rho u_*^2$ where ρ is density and u_* is friction velocity; Q_0, surface heat flux; and g/θ, gravity over potential temperature. Holtslag and Van Ulden *(4)* give a method for determination of the surface heat flux. The scaling velocity of the surface layer is u_*, which can be found from $u = (u_*/k)\ln(z/z_0)$, where u is wind speed, k is Von Karman's constant (about 0.4), and z_0 is roughness length. This is the logarithmic wind profile. The potential temperature at the height z_0 is θ_0. Where both the wind speed and temperature profiles are available, they can be used to estimate u_* *(5)*. The scaling temperature for the surface layer is θ_*, and $w\theta_0 = -u_*\theta_*$. Dimensionless groups are functions of z/L, where L is the Obukhov length, $L = -\theta u_*^3/(gk\overline{w\theta_0})$. The surface layer is less than $|L|$. Irwin and Binkowski *(6)* give a method of iteratively determining L from near surface measurements.

In the free convection layer, τ_0 is no longer important with height. The controlling parameters are z, Q_0, and g/θ. The scaling velocity is u_f, where $u_f = [(Q_0/\rho c_p)z(g/\theta)]^{1/3}$, where c_p is specific heat at constant pressure. The scaling temperature is $T_f = (Q_0/\rho c_p)/u_f$. Dimensionless groups formed with u_f and T_f should be constants for the very unstable conditions where $-z/L > 1$. The upper limit for free convection scaling is about $0.1h$, where h is the boundary layer height. Binkowski *(7)* gives a time-dependent numerical model for h, as well as for the transport flow averaged through h. The convection boundary layer over land can be expected to give $h/|L|$ of greater than 50 to 100.

For the mixed layer the structure of turbulence is insensitive to both z and τ_0. The length scale is h. The scaling velocity is $w_* = [h(g/\theta)\overline{w\theta_0}]^{1/3}$, and the scaling temperature is $T_* = \overline{w\theta_0}/w_*$ *(8)*. A way of estimating T_* from the wind and temperature profiles is given by Nieuwstadt *(5)*. In the mixed layer, dimensionless groups formed with w_* and T_* are functions of z/h. An example of mixed layer scaling is given in Figure 2 from Nieuwstadt *(9)*, where the dimensionless lateral dispersion is shown as a function of $w_*x/\overline{U}h$. Experimental data from Prairie Grass *(10, 11)* along with numerical modeling results of Lamb *(12)* and an empirical function of Deardorff and Willis *(13)* are shown. Ching *et al.* *(14)* discuss the usefulness of three recent field programs for examination of mixed layer scaling.

The entrainment interfacial layer is between about $0.8h$ and $1.2h$. Here turbulence structure may be expected to be dominated by en-

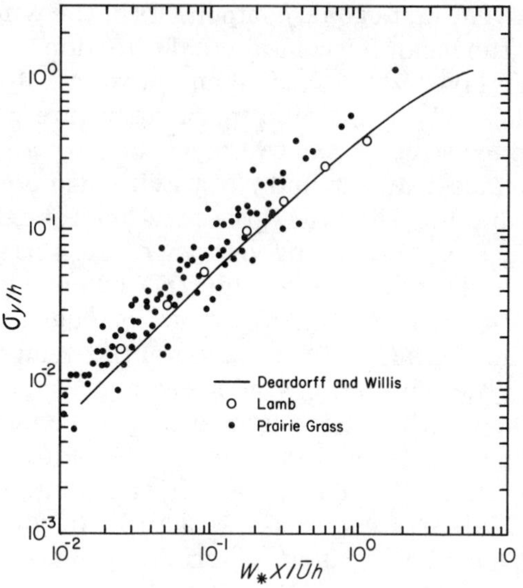

Figure 2. The dimensionless lateral dispersion coefficient σ_y/h as a function of the mixed-layer scaling variable ($w_* x/\bar{U}h$); w_* = convective velocity scale, x = distance downwind, $\bar{U}$ = mean wind speed, h = mixing height, σ_y = horizontal plume standard deviation. [Reproduced with permission from Nieuwstadt *(9)*.]

trainment and the characteristics of the capping inversion and the stable layer above. In the interfacial layer, mixed layer scaling is not expected to apply. Deardorff *(15)* presents a general-structure entrainment model which performs better than earlier models. Scaling expressions for the structure parameters of this layer have been developed by Wyngaard and LeMone *(16)*.

2. The Stable Boundary Layer

Unlike the CBL, the stable boundary layer (SBL) cannot be conveniently separated into various layers. Turbulence in the SBL is produced by mean wind shear and is destroyed both by viscous dissipation and by the negative buoyancy flux created by the positive vertical gradient of potential temperature. Typical profiles of potential temperature, wind speed, and direction are shown in Figure 3 *(17)*. The flux Richardson number, $R_f = [(g/T)\overline{\theta w}]/[\overline{uw}\ (\delta u/\delta z)]$, gives the ratio of the buoyant loss and the shear production. Although the temperature inversion is strongest at the surface, the effects on

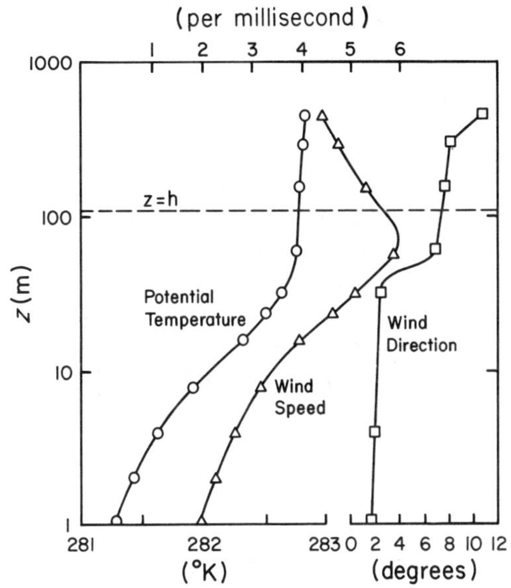

Figure 3. Typical mean profiles of wind speed, wind direction, and potential temperature during stable conditions showing their relationship to boundary-layer depth *h*; *z* = height above ground. [Reproduced with permission from Caughey *et al.* *(17)*.]

the turbulence are weakest there because of the low wind speeds. The stability as indicated by R_f increases farther from the surface. At the SBL top the stability becomes so strong that it completely suppresses the turbulence. The height where this occurs is dependent upon the wind profile, the temperature profile, and the surface energy budget, all of which are time-dependent. Nieuwstadt and Tennekes *(18)* have developed a rate equation for the height of the SBL and, using similarity, have determined the structure of the SBL. The resulting predicted structure agrees well with tower measurements made at Cabauw in The Netherlands *(19)*.

Although the height of the SBL is typically only about 0.1 that of the CBL, the levels of turbulence are so small that study of the SBL by instrumental measurements is extremely difficult in spite of its nearness to the surface. Study of the SBL is also hampered by its unsteady nature, characterized by intermittent periods of turbulence. Also, slight topographic irregularities tend to set up drainage currents which frequently overshadow other factors causing the wind profile.

II. Transport and Dispersion of Stack Effluents

A. Dispersion as Functions of Fluctuations

Although the advice of experts for many years has been to estimate dispersion directly from measurements of atmospheric turbulence or wind fluctuation *(20–22)*, more indirect techniques using standard hourly meteorological observations have frequently been routinely applied. Perhaps this is because such observations represent readily available continuous data sets.

Procedures to use wind fluctuation statistics directly for dispersion estimates have been suggested by Cramer *(23)* and Draxler *(24)*. These can be expressed as

$$\sigma_y = x\sigma_a f_y \tag{1}$$

$$\sigma_z = x\sigma_e f_z \tag{2}$$

where σ_y and σ_z are, respectively, the standard deviations of the horizontal and vertical distributions of concentrations of a continuously emitted plume at the downwind distance x in meters. σ_a is the standard deviation of horizontal wind direction fluctuations in radians, and σ_e is the standard deviation of the fluctuations of the elevation angle of the wind. f_y and f_z are functions which have the value 1 near the source and decrease with distance from the source.

In analyzing available tracer data distributions as functions of wind fluctuations at the point of release, Irwin *(25)* provided simplified expressions with both f_y and f_z having the form

$$f = 1/(1 + 0.9\sqrt{T/T_0}) \tag{3}$$

where T is travel time (distance divided by wind speed) and T_0 has the following values:

$T_0 = 1000$ for f_y

$T_0 = 500$ for f_z for unstable and daytime neutral conditions

$T_0 = 50$ for f_z for stable and nighttime neutral conditions

The dispersion modeling community *(26, 27)* recommend use of fluctuation statistics for dispersion estimates, and models for routine use are becoming available.

B. Buoyancy-Induced Dispersion

Buoyant plumes have a contribution to both vertical and horizontal spread in addition to that caused by ambient turbulence levels. This contribution is due to the entrainment of surrounding air into the expanding plume induced by the vertical motion of the plume in relation to its surroundings. This expansion is related to the amount of rise of the plume *(28)*.

The effective vertical dispersion, σ_{ze}, that due to both buoyancy-induced turbulence and atmospheric turbulence, is found from

$$\sigma_{ze}^2 = (\Delta H/3.5)^2 + \sigma_z^2 \tag{4}$$

where ΔH, the plume rise, and σ_z are evaluated at the distance x from the source. Beyond the distance to final rise, ΔH is constant. At shorter distances it is evaluated for the gradually rising plume.

Since in the initial growth phases of a buoyant plume, the plume is nearly symmetrical about its centerline, the buoyancy-induced dispersion in the crosswind (horizontal) direction is assumed equal to that in the vertical. Thus the effective horizontal dispersion σ_{ye} is found from

$$\sigma_{ye}^2 = (\Delta H/3.5)^2 + \sigma_y^2 \tag{5}$$

The Gaussian plume equations are then used by substituting the value of σ_{ye} for σ_y, and that of σ_{ze} for σ_z.

C. Physical Modeling

Physical modeling and, in particular, wind tunnel studies have been used for many years to help answer questions about building downwash and terrain influences for specific sites. More recently wind tunnels and towing tanks have been used in systemic studies to answer more general questions. The amplification factor is a convenient way of expressing results from physical modeling, whether terrain or building influences are being studied. The amplification factor for a particular position in relation to the source is the ratio of the concentration with the building (or terrain feature) in place to the concentration with no building (or flat terrain) *(29)*.

A number of two- and three-dimensional shapes simulating terrain features have been studied in connection with stack releases. Neutral conditions were simulated in a wind tunnel, and a range of stack heights to height of obstruction and positioning of the stack in relation to the obstruction (upwind, on top of, downwind, etc.) were systematically studied *(30)*. Amplification factors were greatest for two-dimensional (ridgelike) obstructions, as much as 10 for the steepest slope: three slopes of 26°, 16°, and 10° were used. The stack was positioned downwind of the obstruction. For the steepest slope a cavity of reverse flow formed behind the obstruction.

Amplification factors for three-dimensional obstructions are generally less than for two-dimensional obstructions.

Rectangular shapes simulating buildings were studied for neutral conditions *(31)*. Releases from an effective height 2.5 times the building height result in maximum ground-level concentrations 1.2 times as high as those occurring with no building. Releases from an effective height 1.5 times the building height result in maximum concentrations 2.5 times those occurring with no building.

By layering saltwater solutions of varying density and towing a model upside down in the towing tank, one can simulate stable conditions. A critical height directly proportional to wind speed divides streamlines into two classes: those above, which flow over the obstruction, and those below, which flow around the obstruction (Section II,D). Near the critical height the plume impinges on the obstruction surface before flowing around it. Concentrations on the surface can be equal to those at the plume centerline with the obstruction removed *(32)*. Both the direction and speed of simulated flow in the towing tank are steady. In the atmosphere, variation in wind speed can cause the critical height to be above the plume height at one time, and below it somewhat later, so that the pollutant flow is around the obstruction at first and then over the obstruction. Also variations in wind direction will direct the plume centerline to various locations with time so that concentrations averaged over half-hour to hour periods are reduced from that indicated by a steady-state simulation.

D. Complex Terrain Studies

A significant problem in dispersion meteorology is that of estimating surface pollutant concentrations from sources located in complex terrain. A working definition of *complex terrain* is a site where nearby hill or mountain tops are higher than the top of the stack. Because the flow field is altered by the terrain features, the transport and location of the plume centerline are difficult to estimate. Also, the dispersion may be enhanced by flow over increased roughness features or may be decreased as a result of confinement by physical barriers. Estimates of pollutant concentration must be made by making simplifying assumptions about the flow and dispersion, or by first undertaking field investigations on-site (which may include tracer studies) to define further the flow and dispersion regimes under different situations.

In order to bring some new knowledge to making concentration estimates in complex terrain, the United States Environmental Protection Agency began a complex terrain model development program in 1979. A workshop advised that stable plume impaction on terrain should be the principal focus and that field studies in concert with physical modeling studies (tow tank) should proceed from simple cases to more complex topographic cases. Checking the applicability of physical modeling in simple situations would give more confidence in using physical modeling techniques in more complex situations. The first field studies were a three-dimensional conical-shaped obstruction on a flat plane, Cinder Cone Butte, near Boise, Idaho, in 1980; and a two-dimensional ridge, Hogback Ridge, near Farmington, New Mexico, in 1982. A critical dividing streamline height, useful to divide the flow into two regions, was verified in both the Cinder Cone Butte field work and the towing tank simulations. Below the critical height the flow is deformed horizontally and flows around the obstacle; above the critical height the flow goes over the obstruction (Figure 4). The critical height can be determined from the wind profile and the potential temperature structure *(33)*. In addition to estimating the nature of the flow, the resulting concentrations are dependent upon the dispersion. As over flat terrain, measurement of turbulence near the height of release is found to provide the best estimate for horizontal and vertical plume spreading.

Using the preceding facts about two distinct flow regimes, model

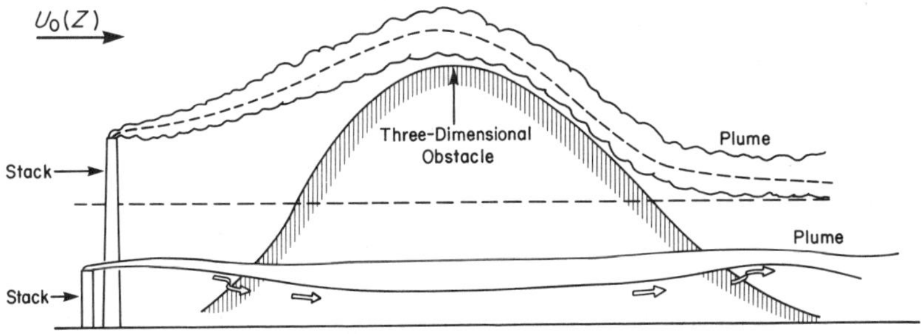

Figure 4. Schematic diagram of plume behavior in stable flow over and around a three-dimensional obstacle. Note different plume behavior above and below the height of the dividing streamline. U = wind speed in the approach flow. [Adapted from Strimaitis *et al.* *(33)*.]

development proceeded in the direction of treating first Neutral and Impingement, then Lift and Wrap, and then the combination of the latter two models. To estimate the critical height, it is necessary to have representative vertical profiles of wind and temperature. To estimate the flow paths and the dispersion, it is necessary to have detailed wind velocity and turbulence data representative of plume heights *(34)*.

Shoreline environments may also be considered as complex terrain. A site-specific study *(35)* in a deep fjord in Norway not only involved a land–water interface but also complex topography, since the mountains surrounding the fjord rise to 1000 m above the water surface. The smelter emissions studied are released near the surface. Concentration estimates for 1-hour periods close to the source (which includes the area of maximum impact) are adequately estimated by using a modified Gaussian model. This was verified through sulfur hexafluoride tracer experiments. However, use of simple Gaussian-type models underestimated 24-hour concentrations on the average by 40%. During the troublesome time of the year studied, summer, pollutants emitted at night were transported by the combination land breeze–down-valley flow out over the fjord with little vertical dispersion, only to return near their point of origin with the daytime on-shore up-valley flow. It was also noted that two wind stations were inadequate to determine the mean flow.

There is a need for additional knowledge of methods to estimate dispersion in complex terrain.

E. Evaluation of Point Source Models

The Electric Power Research Institute *(36)* undertook a project to provide data bases and analyses for rigorous operational validation and diagnostic validation of plume models. The first study was conducted on a simple plain at the Kincaid power plant in Illinois (central United States). The field measurements of the buoyant plume at Kincaid included operation of 30 air quality samplers measuring sulfur dioxide. Data for about 1600 hours from these samplers were compared with model estimates. Sulfur hexafluoride, SF_6, tracer experiments were conducted for about 300 hours using 200 ground-based samplers and two aircraft, one equipped with a particle lidar, the other with continuous analyzers to measure SO_2 and SF_6, and an integrating nephelometer. Indirect measuring equipment included two ground-based lidars, one a particle lidar, the other a differential ab-

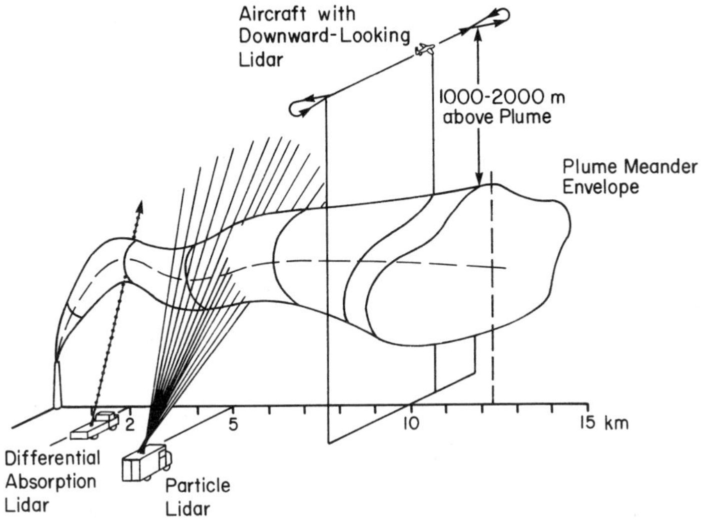

Figure 5. Remote sensing of a plume with lidars. [Reproduced with permission from Electric Power Research Institute *(36)*.]

sorption lidar (DIAL) making remote SO_2 measurements (Fig. 5; see Section V,B).

Objectives of the study were to (1) establish accuracy of the models, (2) assess their performance over a range of conditions, (3) create and archive data bases, and (4) develop and validate improved plume models.

Two of the models chosen for evaluation were CEMQ and TEM. The CEMQ is a CRSTER *(37)* equivalent model with additional features such as the ability to calculate at specific receptor coordinates. CRSTER is a Gaussian plume model widely used in making source impact assessments to meet regulatory requirements. The Texas episodic model (TEM) *(38)* is also a Gaussian model, in which horizontal dispersion is adjusted to the simulated averaging time of the calculations.

The models made estimates for averaging periods from 1 to 24 hours. The highest, 5th highest, and 20th highest concentrations from the entire data set were selected for both models and for the measured concentrations (Figure 6). Although most statistics show underestimation by the models, overestimation is shown by CEMQ for highest and fifth highest 1 hour and highest 3 hour for the tracer data and highest 1 hour for SO_2. Overestimation by TEM occurs only for the highest 3-hour tracer concentration.

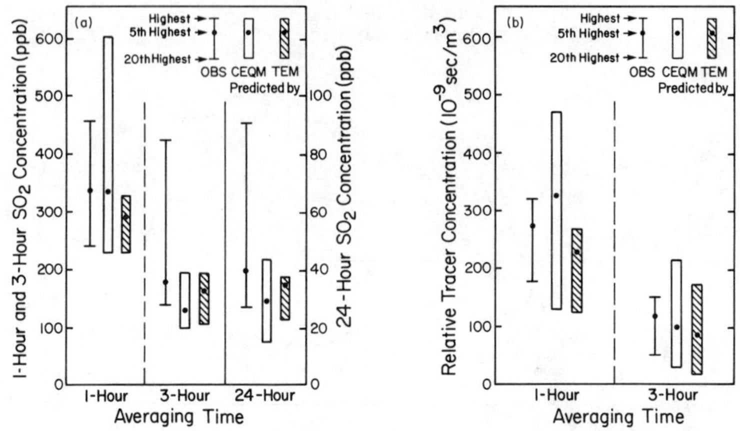

Figure 6. Twenty highest observed and predicted SO_2 values (unpaired) for 1-, 3-, and 24-hour averaging times (left) and 20 highest observed and predicted tracer values (unpaired) for 1- and 3-hour averging times (right). CEQM is a model equivalent to CRSTER (United States Environmental Protection Agency: User's Manual for Single Source [CRSTER] Model. EPA-450/2-77-013. Research Triangle Park, N.C. [NTIS Accession Number PB-271 360] 1977). TEM is the Texas Episodic Model (Texas Air Control Board: User's Guide to the Texas Episodic Model. Austin, Tex. [NTIS Accession Number PB80-227 572] 1980). [Adapted with permission from Electric Power Research Institute *(36).*]

Other results show high model estimates (as shown by the 95th percentile level) to peak 1–3 km from the source and to fall off by a factor of 2 to 4 at a distance of 20 km. However, the measured concentrations show nearly the same high concentrations from 1 to 20 km, i.e., much less change with distance from the source.

Turner and Irwin *(39)*, using data from air monitoring networks at seven power plant sites, made comparisons of model estimates from CRSTER with measurements for second-highest once-a-year 3-hour and 24-hour periods. They noted a dominance of overestimates by the model for receptors that are on elevated terrain with respect to the stack base. For second-highest once-a-year 24-hour concentrations at receptors at elevations higher than 0.5 of the physical stack height, there are 10 overestimates and 3 underestimates; for receptors at elevations below half the stack height there are 9 overestimates and 21 underestimates. For second-highest once-a-year 3-hour concentrations at receptors at elevations higher than 0.5 of the physical stack height, there are 13 overestimates and 2 underestimates; for the lower receptors there are 10 overestimates and 18 underestimates. On the basis of these results, the model tends to underestimate the second-highest once-a-year concentrations for these two averaging periods for receptors representing flat or gently rolling terrain and to overestimate for the elevated receptors.

As a result of a cooperative agreement between the United States Environmental Protection Agency and the American Meteorological Society *(27, 40)*, 10 point-source models were examined by a select group of reviewers and statistics suggested by Fox *(41)* calculated for six monitoring sites near a single power plant.

The reviewers noted that the model features were not as different as expected. Most models replicated features found in models used to address regulatory requirements. Correlations between estimates and measurements for the same times and locations were very low. The sensitivity of concentration to the plume transport direction (as determined by the wind direction input to the model) suggests that wind direction input is not indicative of true plume transport direction as a major factor in the poor correlations.

On the basis of the furnished statistics, the important conclusions of this study summarized by Irwin and Smith *(42)* were "(1) all of the 10 models were poor predictors of observed hourly concentrations when predictions and observations were matched in time and space; and (2) there was no basis for choosing any one model over the others as more suitable for regulatory applications." Using the data generated from this study *(40)*, Irwin and Smith *(42)* examined the variation of maximum concentrations with distance from the source for each Pasquill stability class and noted that by looking at the data in this way, rather than using ensemble statistics, one could detect differences between models.

III. Long-Distance Transport

A. Mesoscale Modeling

The mesoscale includes a distance scale to 500 km and time scales of several days. Modeling on this scale is more difficult than on either small scale (to tens of kilometers) or the global scale, because considerable simplification can be made on the small scale (since wind fields can usually be considered uniform), whereas, on the global scale, vertical exchange processes are of greater concern than transport processes. On the mesoscale, variations in the wind field, both horizontally and vertically, and diurnal variations in stability and removal processes are of importance. Some topographically simple mesoscale areas may be easier to model than those topographically more complex.

1. Simple Topography

Considerable mesoscale modeling activity has centered on The Netherlands and bordering countries, an area of little topographic variation. Although photochemical processes may be of ultimate concern, consideration must be given to transport processes and removal by dry and wet deposition.

For this area Van Dop and de Haan *(43)* have formulated a mesoscale model incorporating a number of features. In order to approximate diurnal behavior, the treatment of vertical diffusion is as follows: At night vertical diffusion is considered within the stable boundary layer. Above this, vertical diffusion is neglected. During daytime, vertical diffusion is considered anytime the unstable boundary layer is less than 600 m. When this layer exceeds 600 m vigorous vertical mixing resulting in uniform concentrations through the boundary layer is assumed. Horizontal advection or transport is considered regardless of condition or pollutant height. Horizontal diffusion is neglected.

Advection equations are solved by using a pseudospectral method. The resulting first-order ordinary differential equations are solved by a fourth-order Runge–Kutta scheme. Solution of the diffusion is by a Crank–Nicolson scheme. Horizontal separation of grid points is 20 km; vertical resolution is 50 m.

Dry deposition uses deposition velocities dependent upon both the pollutant and the vegetative ground cover. Wet deposition is dependent upon precipitation rate.

The wind fields are determined by using 1- to 3-hour-interval 10-m wind and pressure data and 6- to 12-hour-interval radiosonde data. The horizontal wind fields are made nondivergent.

Published model results consist primarily of sensitivity studies. For a period in May 1978, calculations made with no dry deposition increased near-ground concentrations at night about threefold over the calculations with dry deposition; therefore, the influence of dry deposition is considerable. Since there is 50-m resolution in the vertical, the model can well simulate emissions by placing them into the correct vertical layer. The calculated diurnal patterns indicate that fumigation is well simulated. The computer time required is moderate.

Another mesoscale modeling effort *(44)* applies to the same area, The Netherlands and surroundings. The modeling is somewhat simplified so that a minicomputer can be used. Data are hourly averages of wind speed, wind direction, and standard deviation of wind direction at a height of 10 m above ground at 40 stations; hourly solar radiation at 3 stations; and the mixing height as measured by acoustic sounder and occasionally by thermosonde ascents. The principal aims are to represent the spatial variability with a resolution of 20 km (3 km in the urban areas) and the temporal variability of meteorological influences including fumigation.

Emissions are considered by squares 15 km on a side. Individual point sources are treated separately. Emissions are considered in two layers below and above 150 m. Those below are given effective heights of 50 m; those above are given effective heights of 375 m. No temporal variation or variation dependence on temperature of the emissions is considered.

Three modeling layers are used: a 50-m surface layer within which an analytical concentration profile is assumed; a reservoir layer at the top with variable top of from 300 to 700 m, and a mixing layer in the middle whose top can extend through the reservoir layer (Figure 7). Uniform vertical mixing is assumed within both the mixing and reservoir layers. Wind fields are determined separately for these two layers, and each is made nondivergent. Advection is calculated numerically by using a pseudospectral scheme. It has high accuracy and little pseudodiffusion and requires only modest computation time. Six-minute time steps are used. The initial mixing height, the rising speed of the mixing height, the start and end of fumigation, and the hour of afternoon stabilization are given as input (see Chapter 9, Section III,A).

Surface layer characteristics to determine vertical diffusion, dry

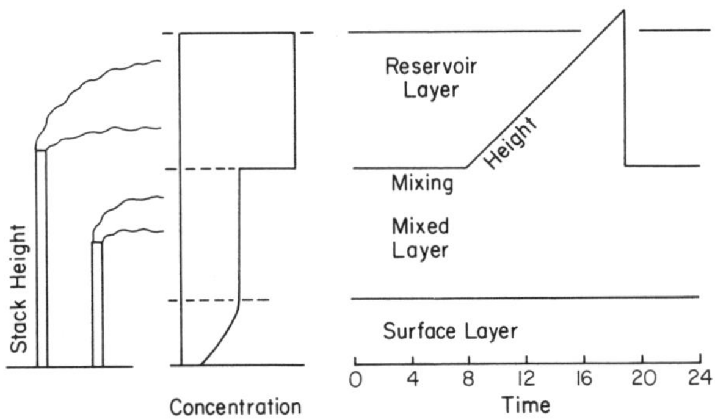

Figure 7. The three layers of the van Egmond and Kesseboom model. [Adapted with permission from van Egmond and Kesseboom *(44)*.]

deposition, and influence of surface cover for the concentration profile are calculated by using friction velocity u_* and Obukhov length L. These latter two are determined hourly from spatially averaged 10-m winds and spatially averaged radiation data. Horizontal diffusion is ignored. Conversion of sulfur dioxide to sulfate is considered as a first-order reaction dependent upon area-averaged solar radiation.

Results from a test case for February 1980 show that diurnal profiles of spatial means, including the fumigation period, appear reasonable. Maximum errors were about 40–45% too low around 1800 hours. An expression of proper spatial patterns is the diurnal variation of the spatial correlation between model and measurements for 100 monitoring stations. This was best at 0300 hours and worst during the fumigation period, 0800–1100 hours. These spatial correlations seem to be greatly affected by errors near the borders of the simulation area, probably due to emissions' taking place beyond the borders and not included in the modeling. Also, modeled concentrations were reduced too much by initial mixing in the squares 15 km on a side. Large elevated sources should be modeled separately. A Lagrangian puff model to do this has been devised by the same authors *(45)*.

2. Difficult Topography

Mesoscale modeling becomes more complicated with difficult topography, such as coastal zones and mountainous terrain. Pielke *et al.* *(46)* suggest techniques to obtain physically consistent flow fields for horizontal areas with scales to several hundred kilometers using 1- to 10-km grid intervals. The resulting flow fields give greater spatial and temporal resolution than simple interpolation between observations. The flow fields can then be coupled with various types of dispersion models, depending on the source type, to yield the field of concentrations *(47)*.

Concentration data collected over a 2-year period at 13 receptor sites at distances from 30 to 140 km from two release sites were used to test performance of 17 models *(48)*. The models included six statistical models using sector-averaged meteorological data, seven Gaussian trajectory models using two-dimensional transport winds, and four three-dimensional models. Statistical measures followed suggestions of an American Meteorological Society workshop *(41)*. No appropriate rules are formulated for handling outliers. An appropriate procedure for outliers must be determined for each model evaluation. It was not surprising that performance varied with averaging time (with annual concentrations estimated well) but twice-daily concentrations were estimated poorly by all statistical measures.

Using a subset of the preceding data base, consisting of 10 data periods, Pepper and Cooper *(49)* compared four numerical algorithms for multiple advection–diffusion on the mesoscale. These algorithms were chapeau function recursion relation (finite element interpolation); method of moments; particle-in-cell; and pseudospectral. For modeling single source releases the particle-in-cell and method of moments show superior accuracy. For distributed sources the pseudospectral and finite element interpolation gave comparable accuracy. Overall the method of moments is felt to be best, although all four models are relatively close in accuracy.

B. Regional Modeling

Much of the interest in long-range (beyond 500 km) modeling has centered on transport of sulfur dioxide and oxides of nitrogen and their transformation products, sulfates and nitrates. Interest includes spatial variation of both air concentrations and surface accumulations due to both dry and wet deposition. Although shorter-period estimates are of interest, longer-period, such as annual, values are useful.

Johnson *(50)* in reviewing regional scale modeling indicates the usefulness of models that can determine the contributions to each of the receptors of individual source areas. Although the transport can be reasonably well handled in a statistical way, deposition cannot.

Most attempts to model on this scale have not used a statistical approach for transport but have used trajectory approaches, although Eliassen *(51)* noted that the long-term deposition patterns from models using statistical transport are similar to those from models with 6-hour time resolution.

Two similar models, EURMAP and ENAMAP *(50, 52)*, use a Lagrangian puff trajectory approach for transport, with a new puff released every 6 or 12 hours from each grid cell. Transformation rates and dry deposition can be reasonably parameterized. However, because of the temporal and spatial variability of precipitation, wet deposition is not so easily handled. Although precipitation data on an hourly basis would be useful for wet deposition estimates, synoptic observations every 3 or 6 hours frequently constitute the only temporal breakdown available.

Comparisons between model estimates and measurements are somewhat more difficult since the model estimates represent an area average and the measurements are for specific points. Averaging several measurements within an area makes them more comparable. Evaluation of both EURMAP and ENAMAP for several monthly periods indicated that estimates gave patterns similar to those derived from the measurements, both for sulfur dioxide concentrations and for sulfate deposition. Eliassen and Saltbones *(53)*, using a simple receptor-oriented trajectory model (96-hour trajectories arrive at each receptor every 4 hours, with mixing height held constant), found reasonable agreement of annual concentrations. However on comparison of performance for two 1-month periods, February and June 1979, the model did not well simulate the decrease in sulfur dioxide from February to June.

Fisher *(54)* indicated that although major transport and removal mechanisms are understood well enough to yield annual estimates within a factor of 2 of measurements, large errors occur for short-period (daily) estimates. He indicated that until the behavior of ammonia and nitrogen compounds is included with that of sulfur dioxide, it will not be possible to predict the effect of emission changes upon acidic deposition with confidence.

There are other long-range problems of interest in addition to those of sulfur and nitrogen compounds. In order for states and provinces to formulate strategies to achieve oxidant air quality standards, it is necessary to assess both the transport of oxidants from outside local areas and the estimated influx of precursors that will create additional oxidant. Lamb and Novak *(55)* give the principal features of the four-layer regional oxidant model (ROM) (Figure 8) designed to simulate photochemical processes over time scales of several days and space scales of 1000 km. Temporal resolution yields hourly concentrations using time steps of 30 minutes and spatial resolution of about 18 km. The processes included are (1) horizontal transport; (2) photochemistry using 35 reactions of 23 species; (3) nighttime chemistry; (4) nighttime wind shear, thermal stratification, and turbulence episodes associated with the nocturnal jet; (5) cumulus cloud effects, including venting from the mixed layer and affecting photochemical reactions by their shadows; (6) mesoscale vertical motion induced by terrain and horizontal divergence; (7) mesoscale eddy effects on trajectories and growth rates of urban plumes; (8) terrain effects on flow and diffusion; (9) subgrid scale chemical processes due to subgrid scale emissions; (10) natural sources of hydrocarbons and oxides of nitrogen; and (11) wet and dry removal processes.

The model was initially applied to the northeastern quarter of the

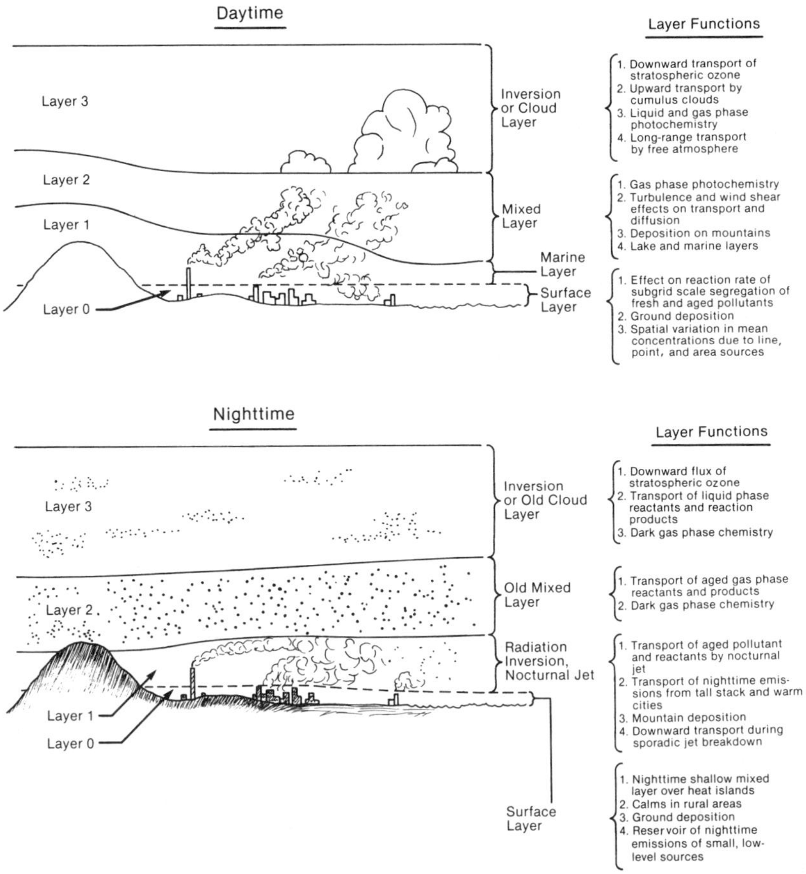

Figure 8. Schematic depiction of the "dynamic" layer structure of the regional model and the daytime and nighttime phenomena each layer is designed to treat. [From Lamb and Novak *(55).*]

United States. A 1980 emissions inventory gathered data on nitrogen oxides, volatile organic compounds, carbon monoxide, sulfur oxides, and total suspended particulate matter. In applying the model the volatile organics are considered as four reactive classes: olefins, paraffins, aldehydes, and aromatics. Application of the model requires acquisition and preparation of emission and meteorological information for the area and a 3- to 4-month commitment by a person having intimate knowledge of the model.

C. Tracers

Dispersion studies such as Prairie Grass *(10, 11)* and Green Glow *(56)*, conducted in the 1950s and 1960s to study the dispersion on the scale of hundreds of meters to several kilometers used sulfur dioxide or fluorescent particles as the tracer. Since then the gas sulfur hexafluoride, SF_6, has frequently been used for studies ranging to tens of kilometers. Draxler *(57)* has provided a summary of experiments from the late 1960s and early to mid-1970s. He divides the studies into seven different categories and gives the principal results from each category.

Johnson *(58)* discusses the various tracers that have been and are being used. Tracers employed in the past such as fluorescent particles and uranine dye particles have the disadvantage of having some deposition as well as requiring long times to analyze the filter samples. More recently used tracers for short to medium ranges such as sulfur hexafluoride and the freons can be collected over convenient sampling periods by gradually filling plastic bags. These are then taken to the laboratory and samples from the bags analyzed for the tracer by gas chromatography with electron capture detection. Particular advantages of bag sampling and gas chromatography are that study design incorporates use of multiple tracers, each analyzed from the sample in a single bag. Since meteorological conditions are difficult to replicate, using multiple tracers allows more than one feature to be studied at one time, e.g., releases made near the ground can be compared with releases made farther aloft.

The most useful tracers for long distances (greater than 100 km) are the perfluorocarbons. There are at least three: perfluoromonomethylcyclohexane (PMCH), perfluorodimethylcyclohexane (PDCH), and perfluorodimethylcyclobutane (PDCB) *(58)*. Their detection sensitivities are four orders of magnitude lower than that of sulfur hexafluoride, and their background concentrations are considerably lower.

Because of difficulties directly sampling in the vertical, e.g., by bag sample collection, the ability to measure remotely is especially attractive (see Section V,B). Lidar can be used to detect oil fog or other smoke plumes. Particulate matter concentrations at a large number of locations above ground can be measured by varying the scanning position of the lidar. Time average concentrations can be approximated by repeating the scanning sequence a number of times over a given time period (such as an hour).

Ten tracer experiments of 1-hour duration using sulfur hexafluoride were carried out by Gryning and Lyck *(59)* in Copenhagen under neutral and unstable conditions. Releases were from a tower at a height of 115 m; ground-level sampling was conducted from 2 to 6 km from the source. Three-dimensional wind velocity fluctuations were measured at the height of release. Both the horizontal tracer variance and the centerline ground-level concentration are better estimated by wind-variance-based methods than by stability-based methods.

IV. Urban Air Quality Simulation Modeling

A. Particulate Matter Modeling

The techniques used to model particulate settling and removal include (1) source depletion, (2) surface depletion, (3) partial reflection, and (4) gradient transfer.

1. Source Depletion

Modification of the emission rate Q has been used to account for dry deposition *(60)*. Frequently the "tilted plume" model has been associated with the source depletion technique. This lowers the plume centerline so that the center of mass of the pollutant plume approaches the ground with transport distance *(61)*. The source depletion model usually is not physically realistic since depletion, as modeled, is through the entire vertical depth of the plume rather than from its lowest part, where removal actually occurs. This model is usable for unstable conditions, where vertical overturning quickly redistributes material, thus coming close to the assumption of depletion throughout the plume, and least applicable under stable conditions when the real atmosphere behaves entirely differently from the model, with the removal occurring near the surface, not throughout the plume.

2. Surface Depletion

Depletion at the surface can be simulated as a negative area source *(62)* extending from the source to the distance of the receptor. Although this concept is physically appealing, it is difficult to apply because of computational complexities.

3. Partial Reflection

Partial reflection models use an "image source" for eddy reflection from the earth's surface. Decreasing the emission rate from the image source results in a partial reflection from the surface. Horst *(63)* indicates that such a source depletion model is easy to use and gives a result close to his surface depletion model at distances near the source, but departs considerably at greater distances.

4. Gradient Transfer Model

Rao *(64)* has provided equations to incorporate the gravitational settling velocity W and deposition velocity V_d for nonreacting gases or particulate matter in operational models; for the simple case of ground-level concentration beneath the centerline of a plume at effective height H the equations take the form

$$C(\hat{x},0,0) = (Q/\pi\sigma_y\sigma_z U)\exp(2\hat{W}\hat{H}\hat{x} - \hat{W}^2\hat{x}^2 - \hat{H}^2) \times (1 - 2\sqrt{\pi}\,\hat{V}_1\,\hat{x}e^{\xi^2}\,\mathrm{erfc}\,\xi) \tag{6}$$

where Q is emission rate, U is wind speed,

$$\hat{x} = x/(\sqrt{2}\,\sigma_z) \tag{7}$$

$$\hat{W} = W/U \tag{8}$$

$$\hat{H} = H/(\sqrt{2}\,\sigma_z) \tag{9}$$

$$\hat{V}_1 = \hat{V}_d - \hat{W}/2 = V_1/U \tag{10}$$

$$\hat{V}_d = V_d/U \tag{11}$$

and

$$\xi = \hat{H} + 2\hat{V}_1\hat{x} \tag{12}$$

The complementary error function is erfc:

$$\mathrm{erfc}(z) = \frac{2}{\sqrt{\pi}}\int_z^\infty e^{-t^2}\,dt \tag{13}$$

Only the vertical dispersion is altered by the deposition. The concentration depends upon the ratio of settling velocity to wind speed and of deposition velocity to wind speed, rather than upon settling velocity and deposition velocity directly. Other equations for off-axis positions and for receptors elevated above the ground are also given.

5. Operational Models

Two models, MPTER *(65)* and PAL *(66)*, applicable only to gases and particles remaining suspended, have been modified to be applicable to particles that settle and removal mechanisms describable by a deposition velocity. The resulting models are MPTER-DS *(67)* and PAL-DS *(68)*.

B. Source Apportionment (Receptor) Modeling

In order to meet air quality standards for particulate matter, it is necessary to determine the origin of the pollutants and the contribution from each source, so that appropriate control strategies can be formulated and implemented. Considerable information about the sources that contribute to particulate matter at a receptor can be learned through chemical analysis of the particulate matter. Specific elements predominate in specific types of sources, such as lead and bromine in auto exhaust and sulfur, nickel, vanadium, and iron in combustion of residual oil *(69–71)*.

The resultant modeling is called *receptor modeling* since the emphasis is centered on information gained from the samples at each receptor. Watson *(72)* gave an overview of receptor model principles. Figure 9 compares the features of receptor models with those of dispersion models. The value of receptor modeling can be greatly en-

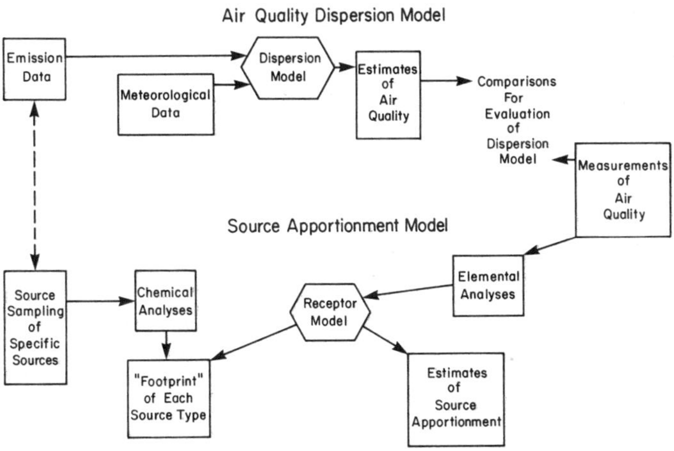

Figure 9. Comparison of features of dispersion and receptor models.

hanced by the collection and analysis of local source samples. This allows more specific "fingerprinting" of various sources by pinpointing the ratios of the principal elements.

The basis for a quantitative source impact determination is chemical mass balance. Assume that b is the total mass received at a receptor from a given source over a given time interval and that a/b is the proportion of this mass due to a particular element. Chemical mass balance assumes that a/b is the same proportion as the mass release of this element to the total mass release at this source. This hypothesis, like atmospheric dispersion modeling, assumes no significant transformation. If there are more elements than source types, one can write simultaneous equations. By using least-squares methods to approximate solutions to these equations, the source-type contributions can be calculated.

Multivariate methods such as factor analysis and pattern recognition can be used in studying variability as measured from a large number of filters. The concentrations of two elements from the same steady-state source increase and decrease together, and their variability with time may be similar. A correlation matrix can show association but does not provide causal relationships.

For small data sets qualitative indications of source contributions can be obtained by using the enrichment factor method. An element characteristic of the source in question is chosen along with another element. Comparisons are made between two samples, one expected to have a contribution by the source, the other representative of background. The enrichment factor E_f is

$$E_f = (C_1/C_2)_{\text{ambient}}/(C_1/C_2)_{\text{background}} \tag{14}$$

The ratio will exceed 1 as the proportion of element 1 is high in the sample.

Time series analyses are useful in evaluating extensive data sets since elements that come from the same source or source type should be correlated. Hourly observations can make use of diurnal patterns. Samples that must be of longer duration may employ seasonal variations. Evaluation of time series may produce strong inferences but cannot quantitate specific source types. This is especially useful in evaluating control measures and in trying to find additional sources.

Examining similarities and differences in samples collected at the same time, but at different locations may yield additional information on sources. Measurement of wind direction may allow the construction of pollution wind roses, if the length of the sampling can be limited to 1 or 2 hours. Longer samples encounter wind from a range of directions and are more difficult to interpret.

Hybrid models incorporating the features of both receptor-oriented models and air quality dispersion models are especially useful in determining effects of particular sources.

C. Photochemical Modeling

Considerable progress in evaluating urban photochemical models has been achieved. Schere and Shreffler *(73)* have compared model estimates from three different models with air quality measurements from 20 high-oxidant days selected from data collected in 1975 and 1976 as part of the Regional Air Pollution Study (RAPS) conducted in St. Louis, Missouri. The three models are briefly described by Shreffler *et al. (74)*.

The *photochemical box model* (PBM) is a single-cell Eulerian model. As applied, the box was 20 km on a side with height dependent upon the mixing height. The photochemistry is described by the 36 reactions for 27 reactive species *(75)* given in Table I. Reactive hydrocar-

Table I Generalized Chemical Kinetic Mechanism in Photochemical Box Model[a] from Demerjian and Schere *(75)*.

1.	$NO_2 \xrightarrow{h\nu} NO + O$	19.	$OLEF + O \longrightarrow RO_2 + ALD + HO_2$
2.	$O + O_2 + M \longrightarrow O_3 + M$	20.	$OLEF + O_3 \longrightarrow RO_2 + ALD + HO_2$
3.	$O_3 + NO \longrightarrow NO_2 + O_2$	21.	$OLEF + HO \longrightarrow RO_2 + ALD$
4.	$O_3 + NO_2 \longrightarrow NO_3 + O_2$	22.	$PARAF + HO \longrightarrow RO_2$
5.	$NO_3 + NO \longrightarrow 2NO_2$	23.	$ALD \xrightarrow{h\nu} 0.5RO_2 + 1.5HO_2 + 1.0CO$
6.	$NO_3 + NO_2 + H_2O \longrightarrow 2HONO_2$	24.	$ALD + HO \longrightarrow 0.5R1O_2 + 0.5HO_2 + HO_2$
7.	$HONO \xrightarrow{h\nu} HO + NO$	25.	$RO_2 + NO \longrightarrow RO + NO_2$
8.	$HO + NO \xrightarrow{(O_2)} HO_2 + CO_2$	26.	$RO + O_2 \longrightarrow ALD + HO_2$
9.	$HO_2 + NO_2 \longrightarrow HONO + O_2$	27.	$R1O_2 + NO_2 \longrightarrow PAN$
10.	$HO_2 + NO \longrightarrow HO + NO_2$	28.	$RO + NO_2 \longrightarrow RONO_2$
11.	$HO_2 + NO_2 + M \longrightarrow HOONO_2 + M$	29.	$RO_2 + O_3 \longrightarrow RO + 2O_2$
12.	$HOONO_2 \longrightarrow HO_2 + NO_2$	30.	$R1O_2 + NO \xrightarrow{(O_2)} RO_2 + NO_2$
13.	$HO + HONO \longrightarrow NO_2 + H_2O$	31.	$PAN \longrightarrow R1O_2 + NO_2$
14.	$HO + NO_2 + M \longrightarrow HONO_2 + M$	32.	$AROM + HO \xrightarrow{(O_2)} R_2O_2 + 2ALD + CO$
15.	$HO + NO + M \longrightarrow HONO + M$	33.	$R_2O_2 + NO \longrightarrow R_2O + NO_2$
16.	$HO_2 + O_3 \longrightarrow HO + 2O_2$	34.	$R_2O + O_2 \longrightarrow ALD + HO_2 + 2CO$
17.	$HO + O_3 \longrightarrow HO_2 + O_2$	35.	$R_2O_2 + O_3 \longrightarrow R_2O + 2O_2$
18.	$HO_2 + HO_2 \longrightarrow H_2O_2 + O_2$	36.	$R1O_2 + O_3 \longrightarrow RO_2 + 2O_2$

[a] Reprinted with permission.

bons are considered by four reactivity groups: olefins, paraffins, aldehydes, and aromatics. Hourly pollutant estimates from the model were compared with the average concentration within the box obtained by averaging concentrations from the 12 measurement stations inside the box. Statistics for the differences between the observed hourly maximum concentration of O_3 for each of the 20 days and the PBM-estimated concentration for the same time are given in Table II. As noted in the statistics the model has a bias toward overestimating concentrations: 17 of the 20 estimated maxima are overestimates. The PBM is useful for assessing urban photochemically reactive pollutants, especially under stagnation conditions.

The second model, the *Lagrangian photochemical model* (LPM), considers an air parcel moving with the air motion which receives emissions into the lowest layer and then mixes vertically and disperses laterally. Initial locations were chosen by using backward trajectories so that parcels would pass the measurement station having the highest ozone concentration at the time of the maximum. The same statistics for the LPM as for the PBM are given in Table II. The LPM is easy to use to explore O_3 production in urban areas.

The third model, the *urban airshed model* (UAM), is a three-dimensional grid-type photochemical model. In applying the model to St. Louis, Missouri, an 80-km by 60-km region was considered, using cells 4 km on a side. Vertically, four layers were used with the bottom two layers simulating the mixed layer. The modeling domain included 21 measurement stations. The other four measurement stations were used for boundary conditions. Statistics similar to those for the other two models are given in Table II. All 20 estimates were slightly lower than the measured concentrations. Intercomparison between models is difficult since the PBM estimated large volume-averaged concentrations, whereas the other two models estimated smaller parcel or

Table II Statistics for Measured and Estimated Ozone Concentrations for the Time of the Measured Maximum Derived from Application of Three Photochemical Models for 20 Selected Days in St. Louis, Missouri[a]

(All statistics in ppm)	*Photochemical box model (PBM)*	*Lagrangian photochemical model (LPM)*	*Urban airshed model (UAM)*
Average difference	−0.012	0.023	−0.062
Standard deviation of differences	0.039	0.058	0.035
Average absolute difference	0.029	0.052	0.062

[a] From Schere and Shreffler *(73)*.

Table III Ranges of Daily Ozone Maxima over 20 Selected Days in St. Louis, Missouri[a]

	Photochemical box model (PBM)	*Lagrangian photochemical model (LPM)*	*Urban airshed model (UAM)*
Largest measured maximum (ppm)	0.183	0.26	0.244
Smallest measured maximum (ppm)	0.072	0.16	0.142
Largest estimated maximum (ppm)	0.223	0.30	0.237
Smallest estimated maximum (ppm)	0.08	0.07	0.059
Twenty-day mean of daily measured maxima (ppm)	0.123	0.199	0.191
Standard deviation of differences divided by the mean measurement	0.317	0.291	0.183

[a] From Schere and Shreffler *(73)*.

cell averages. The ranges of measured and estimated daily ozone maxima, as well as the standard deviation of differences divided by the mean measured concentration, are given in Table III. These vary from around 20% for the UAM to around 30% for the other two models.

The time required to set up a run for each model to perform a 1-day simulation and the central processor unit (CPU) time required for the 1-day run are given in Table IV.

McRae and Seinfeld *(76)* reported on the evaluation of a second-generation mathematical model for urban air pollution *(77)* in simulating a 2-day smog episode (June 1974) in the South Coast Air Basin in Southern California. The model uses 5-km by 5-km grid cells and five cells in the vertical. The photochemistry is accomplished by using 37 reactions for the 15 species. The model accurately reproduces the important features of the ozone and nitrogen dioxide distributions and time variations over the South Coast Air Basin.

Table IV Setup and Execution Times for Photochemical Models[a]

	Setup time for a 1-day simulation (person-months)	*Execution time for a 1-day simulation (CPU minutes)[b]*
Photochemical box model (PBM)	0.15	1.
Lagrangian photochemical model (LPM)	0.20	10.
Urban airshed model (UAM)	0.50	110.

[a] From Schere and Shreffler *(73)*.
[b] Central processor unit.

Table V Statistics for Measured and Estimated 1-hour Ozone Concentrations at 16 Measurement Stations for the Application of the Second-Generation Mathematical Model to the South Coast Basin: 26–27, June 1974[a]

	Value
Average difference (ppm)	0.044
Standard deviation of differences (ppm)	0.035
Average absolute difference (ppm)	0.044
Largest measured maximum (ppm)	0.51
Smallest measured maximum (ppm)	0.23
Largest estimated maximum (ppm)	0.41
Smallest estimated maximum (ppm)	0.21
Mean measured concentration (ppm)	0.331
Standard deviation of differences divided by the mean	0.106

[a] Data used with permission from Table 9 of McRae and Seinfeld *(76)*.

A summary of statistics on the differences between the observed and predicted maximum 1-hour ozone concentrations at the 16 measurement stations having maximum concentrations greater than 0.20 ppm is given in Table V. Except for the two locations having the same measured and estimated concentrations, all estimates are less than measurements. Since these comparisons apply to a number of locations on a single day, these results cannot be compared with those for the maxima for multiple days in St. Louis.

D. Model Evaluation

The usefulness of models is dependent upon their accuracy. Accuracy is evaluated by determining values for measures of performance using suitable data bases. The calculated values for the measures of performance can then be compared with standards of performance. At this writing there are no agreed-upon standards of model performance. Various measures of performance were discussed in a workshop on dispersion model performance held in 1980 *(41)*. Measures of bias, variance, and gross variability were enumerated. The appropriate measure of performance must be consistent with the particular use of the model. For example, statistics designed to compare the top 10% of the estimated distribution of 3-hour concentrations with the top 10% of the measured distribution of 3-hour concentrations are not especially useful if the purpose of the model is to estimate the annual concentration reliably.

One of the major difficulties in evaluating the performance of operational models for short averaging times is that comparison of rare events yields few values for statistical use. For example, in evaluating model performance for the second-highest 24-hour concentration once a year, only one pair of values is available from a year's measured and estimated concentrations for each measurement location. This pair is not very useful in generating meaningful statistics.

1. RAM Model Evaluation

Other than evaluation of photochemical models discussed in the previous section, evaluation of urban models has centered primarily on RAM *(78)*. Kummler *et al. (79)* compared RAM and a modified SAI model, a numerical model, for SO_2 concentrations from a 14-station measurement network in Detroit, Michigan. They concluded "that the numerical model provides a superior predictive tool to evaluate cause and effect relations, but that the RAM model, at far lower cost, predicts the correct magnitude of the worst events."

Hodanbosi and Peters *(80)*, using data for the year 1977, evaluated the RAM in Cleveland, Ohio. SO_2 measurements were available at five continuous monitors and 28 sites using 24-hour bubblers operating from 17 to 94 days of the year. A tabulation (Table VI) of the number of measurement stations having ratios of estimates to measurements in various ranges includes annual, second-high 24-hour, and highest 24-hour values. Since the actual maximum operating conditions were used for the simulation, it is not surprising that the bias is on the side of overestimation, but less for the highest and second-highest 24-hour than for the averages. For examining conformance with regulations, bias in this direction is preferred rather than toward underestimation. No more than four stations had estimates inaccurate by more than a factor of 2.

In another comparison of RAM with observations at four sites in Michigan City, Indiana, and one site in Indianapolis, Indiana *(81)*, maximum 24-hour concentrations were overestimated at all five sites. The ratios of maximum 24-hour estimates to maximum 24-hour measurements at the five sites were 3.36, 2.08, 1.45, 1.30, and 4.20. In Michigan City actual hourly emissions were used for the principal point source, but maximum (design capacity) emissions were used for all other point sources; in Indianapolis maximum emission rates were used for all point sources. Therefore some degree of overestimation

Table VI Numbers of Measurement Stations in Different Ranges of the Ratio of Model Estimates to Measurements (24-hour) in Cleveland, Ohio, 1977[a]

Estimates/measurements	*Average*	*Second-high*	*Highest*
2.8–2.9		1	
2.7–2.8		1	
2.6–2.7			1
1.5–2.6			
2.4–2.5			1
2.3–2.4			
2.2–2.3	1	2	
2.1–2.2	2		
2.0–2.1	1		1
1.9–2.0	2		1
1.8–1.9	1		
1.7–1.8	2	2	1
1.6–1.7	2	1	1
1.5–1.6	4	2	2
1.4–1.5	1	2	1
1.3–1.4	4	2	4
1.2–1.3	2	4	2
1.1–1.2	5	4	7
1.0–1.1	6	4	1
0.9–1.0		5	4
0.8–0.9		2	2
0.7–0.8		1	3
0.6–0.7			1

[a] Data used with permission from Hodanbosi and Peters *(80)*.

would be expected. Although 0.015 ppm was added to all model estimates before comparing with measurements, cumulative frequency distributions of measured concentrations, displayed for three sites, show that 60% to 70% of the concentrations are below this value, implying that this background estimate is too high.

A comparison of RAM, the urban model, with RAMR, the same model with rural dispersion parameters (Pasquill–Gifford values), using 24-hour SO_2 measurements in the vicinity of two Ohio power plants (four sites at Eastlake and three sites at Avon Lake) indicated consistent overestimates by the urban model, but both over- and underestimates for the rural model. The authors *(82)* speculated that in their study, as well as the Michigan City study (discussed previously), the area may not have been sufficiently urbanized to apply the urban model.

Turner and Irwin *(83)*, using the comprehensive data from the Regional Air Pollution Study (RAPS) in St. Louis, Missouri, examined the performance of RAM for SO_2. Table VII shows the ratios of second-high estimates to second-high measurements for SO_2 for four averaging times, and the ratios of annual estimates to measurements for the 13 measurement stations. Although there is considerable station-to-station scatter in the results, there appear to be no overall bias and no particular trend in performance with averaging time.

In a subsequent study, Turner and Irwin *(84)* used the data from the prior study to explore the effect of wind speed and stability on the performance of RAM. They found a significant stability bias. Under stable conditions the model estimated high values (as one might ex-

Table VII Ranked Ratios of Second-highest Estimates to Second-highest Measurements for Four Averaging Times and Ranked Ratios of Annual Estimates to Annual Measurements for 1976 for the 13 Measurement Stations in St. Louis, Missouri[a]

	Averaging time				
Rank	*1-Hour*	*3-Hour*	*8-Hour*	*24-Hour*	*Annual*
1	2.76	2.07	2.23	2.31	2.32
2	1.78	1.60	2.06	1.93	1.45
3	1.63	1.47	1.35	1.21	1.27
4	1.34	1.11	1.27	1.21	1.27
5	1.34	0.98	1.20	0.80	1.08
6	1.03	0.86	1.02	0.77	1.04
7	0.98	0.74	0.90	0.70	1.01
8	0.94	0.74	0.65	0.67	0.95
9	0.93	0.74	0.64	0.67	0.86
10	0.74	0.61	0.60	0.66	0.73
11	0.58	0.55	0.55	0.49	0.70
12	0.50	0.50	0.48	0.41	0.70
13	0.47	0.28	0.21	0.18	0.67
Number of overestimates	6	4	6	4	7
Number of underestimates	7	9	7	9	6

[a]Data used with permission from Turner and Irwin *(83)*.

pect from area sources), whereas the measurement maxima occurred under neutral conditions at three sites, under stable conditions at six sites, and under unstable conditions at four sites. Tables VIII and IX were constructed by ranking the stability classes according to the magnitude of the highest 1-hour concentration in each class for each of the 13 measurement stations. The model estimates display a very regular pattern with stable conditions having highest maxima and unstable conditions having the lowest maxima. The measurements display considerable irregularity except that the extremes of stability, *A* and *G*, have the lowest maxima. Turner and Irwin were unable to prove, but speculated, that the major cause for the behavior noted was the inclusion of many small to moderate point sources as part of the area source inventory.

As these studies show, there is considerable room for improvement in modeling even the less reactive pollutants in urban areas. Collecting sufficiently detailed emission inventories in any urban area is a challenge.

Table VIII Stability Class Ranking According to Maximum 1-hour RAM Estimates Using 1976 Regional Air Pollution Study Data in St. Louis, Missouri[a]

	Stability class rank[b]							
Station	*1 Highest maximum*	*2*	*3*	*4*	*5*	*6*	*7*	*8 Lowest maximum*
101	G	E	F	DN	DD	C	B	A
103	G	E	F	DN	C	DD	B	A
104	F	E	G	DN	DD	C	A	B
105	F	G	E	C	A	DD	B	DN
106	F	E	G	DD	DN	C	B	A
108	G	E	F	DN	DD	C	B	A
113	G	E	F	DN	C	DD	B	A
114	G	E	F	DN	DD	C	B	A
115	F	E	G	DN	DD	C	B	A
116	F	G	E	DN	C	DD	B	A
120	E	F	DD	G	DN	C	B	A
121	G	F	DD	DN	C	E	B	A
122	F	E	C	G	DN	B	DD	A

[a] Data used with permission from Turner and Irwin *(84)*.
[b] Stability classes are unstable, A through C; neutral, DD and DN; stable, E through G.

Table IX Stability Class Ranking According to Maximum 1-hour Measurements Using 1976 Regional Air Pollution Study Data in St. Louis, Missouri[a]

	Stability class rank[b]							
Station	*1 Highest maximum*	*2*	*3*	*4*	*5*	*6*	*7*	*8 Lowest maximum*
101	F	E	B	C	DN	DD	G	A
103	C	B	E	F	DD	DN	A	G
104	E	DN	DD	F	C	G	B	A
105	C	F	DN	DD	E	G	B	A
106	DN	DD	E	F	C	B	G	A
108	E	DN	C	DD	B	F	G	A
113	B	F	E	DN	C	DD	G	A
114	C	DD	E	DN	F	B	G	A
115	E	DN	B	C	A	DD	F	G
116	F	DD	DN	C	B	E	G	A
120	DD	B	E	F	C	DN	G	A
121	E	DN	F	DD	B	C	G	A
122	DD	E	C	DN	B	G	F	A

[a] Data used with permission from Turner and Irwin *(84)*.
[b] Stability classes are unstable, A through C; neutral, DD and DN; stable, E through G.

E. Regulatory Use of Models

Before passage of the 1977 amendments to the United States Clean Air Act *(85)*, air quality simulation modeling was applied to assess the air quality impact of sources, but such assessment was not mandatory. Section 165 of the 1977 amendments on prevention of significant deterioration provides for issuance of permits prior to construction of major emitting facilities. It states that "the required analysis has been conducted in accordance with regulations" and that these regulations "shall specify with reasonable particularity each air quality model or models to be used under specified sets of conditions." The required regulations *(86)*, promulgated on June 19, 1978, specify that "estimates of ambient concentration required shall be based on the applicable air quality models given in the Guidelines on Air Quality Models."

Section 310 of the 1977 amendments called for the "conduct of a conference on air quality modeling" within 6 months of the passage of the act and "at least every 3 years thereafter." The first conference on air quality modeling required by statute was held in Washington,

D.C., in December 1977 and reviewed modeling guidelines *(87)* published in 1978.

The models given in the guidelines are Gaussian-based models and depend heavily on available airport data for meteorological input. Revisions were proposed in August 1980 for a new edition of the guidelines, and public comments were received at three public meetings held in October 1980. However, a new edition was not issued. Instead, revisions were incorporated during 1983 and 1984 with the expectation that public comments would be sought early in 1985.

For the most part, impact assessments submitted to the United States Environmental Protection Agency (USEPA) for permits or other regulatory action are reviewed by personnel in one of the ten regional USEPA offices. In order to obtain most consistency in the application of air quality dispersion models from region to region, especially where the guidelines are not specific, USEPA has done two things: (1) It has held annual workshops on modeling applications attended by regional people, along with representatives from its Office of Air Quality Planning and Standards (OAQPS) and its Office of Research and Development; and (2) it has established a model clearinghouse within OAQPS to assist the regions on modeling application issues for which no earlier precedent had been set. Both of these have succeeded in establishing consistency in model application from region to region.

1. Impact Assessment

Most of the difficulty in applying models for impact assessment has been in estimating values to compare with the short-term standards. At the time of this writing both SO_2 and particulate matter 24-hour standards were in terms of a value not to be exceeded more than once per year, and a secondary standard for SO_2 on the same basis existed for 3-hour concentrations. Therefore, it became necessary for the models to make reasonable estimates for these extreme events. Because routine meteorological data can be obtained on an hourly basis, most of the modeling assumes steady conditions for 1-hour periods. Unfortunately the available meteorological measurements, especially relating to horizontal dispersion, are not good discriminators of winds with little directional variation as opposed to those with considerable directional variation over an hour. Although the maximum concentrations over the hour resulting from the considerably varying wind are much lower (because of the enhanced horizontal dispersion) than those for an hour with fairly steady winds, the routinely available meteorological measurements consider the 2 hours to be meteorologically identical.

Also the resolution of routine airport wind direction measurements to only 10° has led to the placement of each hourly wind direction within a 10° sector according to a randomly generated digit. The hope here is that over the course of a year's data the various stability conditions may have a variety of hour-to-hour wind direction variations similar to the distribution encountered over the year. However, the extremes of these conditions cannot be expected to occur in the directions from the source or at the times which real extremes occur.

One might expect second-highest concentrations to have considerable variation from year to year. Therefore, in regulatory applications, the recommendation has been to use 5 years' data where possible.

It has been considered desirable to give some indications of accuracy along with model estimates given to those responsible for making regulatory decisions. Beyond the typical "within a factor of 2" estimate for accuracy, the lack of extensive field measurements has hampered building a history of accuracies from model evaluations.

A workshop on "uncertainty in air quality modeling" *(88)* was held in 1982 to consider the ways to express uncertainty, to determine it in model evaluations, and to convey it properly to decision makers.

The direction being pursued to overcome many of the modeling difficulties stated is to make measurements more directly related to dispersion and to provide the modeling framework to use these measurements. The measurement of the standard deviations of azimuth angle and elevation angle of the wind at plume centerline height is expected to relate most closely to the horizontal and vertical plume dispersion. Since the plumes from larger sources may be at heights in the vicinity of 600 m above ground or higher, direct measurements of wind fluctuation are difficult, and it is necessary to make measurements by using remote sensing or to devise techniques to make reasonable estimates of wind fluctation at plume level from measurements made at other heights above ground.

V. Meteorological Measurements

A. Microcircuitry and Storage of Information

A common past practice for obtaining meteorological information has been to use a data logger to interrogate each sensor and store each piece of data received on a convenient storage medium, such as magnetic tape, which could periodically be changed and taken to a computer installation to calculate the required statistics for each appropriate time interval. This procedure can result in considerable accumulation of data prior to analysis and can create a data storage problem.

With the current ease of performing numerous calculations with a properly designed silicon chip, calculation of appropriate statistics as the data are received is not only possible but very attractive from the standpoint of data storage, as only the statistics, not the raw data, need to be stored. It becomes a simple matter to cumulate sums of instantaneously sampled variables such as wind speed or temperature and to calculate means over periods of interest, such as hourly periods.

Because wind direction is a circular function with a crossover between 360° and 0°, difficulties are encountered in calculating means and standard deviations. If all values are retained, for example, 3600 instantaneous wind direction values taken each second over an hour, it is a simple matter to calculate the mean and standard deviation, eliminating crossover problems. However, it is highly desirable to have a methodology so that calculations can be made without storage of a large number of values (3600 in the preceding example). Two techniques, those of Verrall and Williams *(89)* and of Ackermann *(90)*, provide methods to avoid crossover and make accurate calculations for means and standard deviations by cumulating only six sums (five for the Verrall–Williams technique) as the wind direction values are read. The processor must be able to calculate the sine and cosine of the wind direction azimuth a and to sum the following values: $\sin a$, $\cos a$, $\sin^2 a$, $\cos^2 a$, N (the number of values over the time period used for the sum, 3600 in the preceding example); additionally, the Ackermann technique calculates $\sin a \cos a$. Details on calculating the statistics from the sums are given in Table X and in the two papers. Note that in the application of the Verrall–Williams technique for the standard deviation, the absolute values of the means are used in obtaining the two vectors.

The wind direction mean and the standard deviation can then be easily calculated from the five or six sums along with other statistics such as mean wind speed and temperature for each time period of interest, such as an hour; they are stored on cassette tape or other storage medium or are transmitted to a central data logger by either dedicated telephone line or radio.

ble X Data Collected and Calculations Made to Obtain Estimates of Mean Wind Direction, $\bar{a}$, and andard Deviation, σ_a

mmations of instaneous values accumulated over the selected time period (for example, 1 hour):

$\Sigma \sin a$, $\Sigma \cos a$, $\Sigma \sin^2 a$, $\Sigma \cos^2 a$, N, $\Sigma \sin a \cos a$ (for Ackermann)

lculations made for both Ackerman *(90)* and Verrall–Williams *(89)* techniques:

$\overline{\sin a} = \Sigma \sin a/N$, $\overline{\cos a} = \Sigma \cos a/N$, $\bar{a} = \arctan(\overline{\sin a}, \overline{\cos a})$

$$\sigma_s^2 = \frac{\Sigma \sin^2 a}{N} - (\overline{\sin a})^2 \qquad \sigma_c^2 = \frac{\Sigma \cos^2 a}{N} - (\overline{\cos a})^2$$

Ackermann

$$\sigma_{sc} = \frac{\Sigma \sin a \cos a}{N} - \overline{\sin a}\,\overline{\cos a}$$

$$\sigma_a = \frac{[(\overline{\cos a})^2\sigma_s^2 + (\overline{\sin a})^2\sigma_c^2 - 2\,\overline{\sin a}\,\overline{\cos a}\,\sigma_{sc}]^{1/2}}{(\overline{\sin a})^2 + (\overline{\cos a})^2}$$

Verrall–Williams

e east and north components of vector 1:

$E_1 = |\overline{\sin a}| - \sigma_s$

$N_1 = |\overline{\cos}| + \sigma_c$

The east and north components of vector 2:

$E_2 = |\overline{\sin a}| + \sigma_s$

$N_2 = |\overline{\cos a}| - \sigma_c$

The magnitude of vectors 1 and 2:

$M_1 = [E_1^2 + N_1^2]^{1/2}$

$M_2 = [E_2^2 + N_2^2]^{1/2}$

The dot product of vectors 1 and 2:

$D = E_1E_2 + N_1N_2$

$\sigma_a = 0.5 \arccos[D/(M_1M_2)]$

B. Remote Sensing

1. Acoustic Remote Sensing

Sound direction and ranging, sodar, is an acoustic sensing technique useful for obtaining information about the boundary layer. Sound waves interact with the thermal and turbulent microstructure of the atmosphere, and reflected sound is detected by a sensitive receiver *(91)*.

A co-located transmitter and receiver is called a *monostatic system*. This system is useful in determining the vertical extent of inversions and mixing depths. Displays can be on facsimile. However, dot matrix printers and color graphics terminals can provide easier data interpretation. Monostatic systems are used to quantify a temperature structure parameter. Errors are caused by excess attenuation of the beam by turbulence, and by beam bending due to horizontal winds. These effects were studied by Haugen and Kaimal *(92)* by comparing estimates of the temperature structure parameter from an acoustic sounder with estimates from measured temperatures on a tower. The resulting errors were found to be functions of both height and stability. To decrease the effect of these errors the sodar can be operated in the bistatic mode (separating the transmitter and receiver). In this mode scattering takes place from small-scale velocity fluctuations as well as from temperature inhomogeneities.

A Doppler system with the capability of analyzing the return signal for its frequency allows determination of the component of wind velocity parallel to the pathway of the sound. Although there are a number of methods of analyzing for the frequency shift, the use of a fast Fourier transform (FFT), which can be accomplished by a chip processor, is common.

In order to measure the three-dimensional wind, it is necessary to use three transmitters or receivers. Two different combinations are (1) trimonostatic, with three monostatic systems pointing in different azimuth and elevation angles, and (2) bistatic, with two separated transmitters and one vertically pointed receiver which also is used as a third transmitter.

An advantage of the trimonostatic system is that the three units can be located near one another, resulting in a relative compact and more movable system. The geometric configuration used by one such system is depicted in Figure 10 *(93)*. Also, because of the coincidence of the transmit and receive beams, greater heights can be reached. However, since the monostatic units depend only on the temperature structure, the return signals may be weaker. Its disadvantages are that the three differently positioned beams are observing different air volumes (an arrangement which causes problems, if the atmosphere is not homogeneous, e.g., near a terrain obstacle or near a

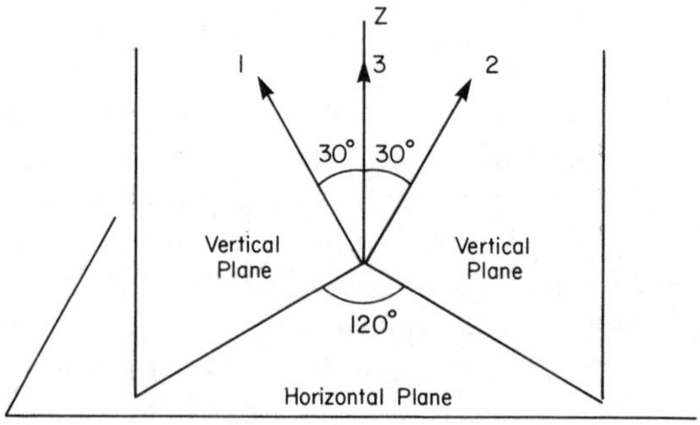

Figure 10. Geometric configuration of a trimonostatic Doppler sodar system. [Reproduced with permission from Weill *et al.* *(93)*.]

coastline), two of the beams are not vertical and receive a lot of environmental noise, and the antennae must have heavy shielding.

Advantages of the bistatic system are that only one air volume is sampled and that the beam is reflected by both temperature discontinuities and air turbulence. Its principal disadvantge is the requirement for two 100- to 200-m baselines resulting in considerable length of cable to link the components. The geometric configuration of a bistatic system is depicted in Figure 11.

Convective activity is easily recognized by acoustic systems. The Doppler systems easily measure horizontal wind profiles to heights around 500 m *(94)*. The standard deviation of errors in wind speed at a height of 100 m was found by Weill *et al. (93)* to be 0.6 m $second^{-1}$. Errors in wind measurement by Doppler sodar are affected by duration of pulse, duration of the receiving gate, and beam width *(95)*. Adjustment of these can decrease the error in wind measurement at the expense of the error in the height measurement and vice versa. Spizzichino *(95)* also pointed out that refraction effects on the acoustic beam are negligible for measurement of phenomena of a few seconds' duration but are important for averaged quantities. Measurement of vertical velocities and of turbulence statistics is more difficult. Weill *et al. (93)* found the standard deviation of vertical velocity errors to be 0.3 m $second^{-1}$.

Weill *et al. (96)* demonstrated that by measuring the vertical veloc-

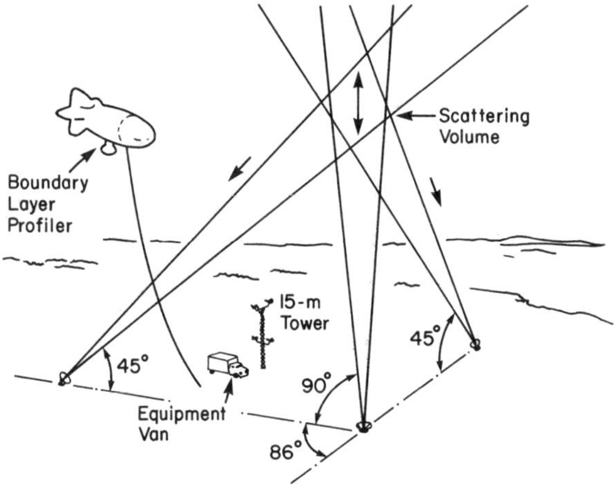

Figure 11. Geometric configuration of a bistatic Doppler sodar system. [Reproduced with permission from Hall *et al. (94)*.]

ity profile with a vertical Doppler sounder, the heat flux profile in the well mixed layer can be determined, also permitting estimation of the surface heat flux.

Using Doppler sodar, Taconet and Weill *(97)* studied the vertical profile of the vertical velocity third-order statistic w'^3, under convective conditions, finding a maximum at 0.3 z_i, where z_i is the convective boundary layer height. This study also showed the vertical velocity distribution to have a bimodel distribution, with the principal peak at small negative velocities (around −0.1 m second^{-1}) and a smaller peak near larger positive velocities (about 0.7 m second^{-1}), confirming expectations.

2. Lidar

Light detecting and ranging, lidar, is an optical remote sensing technique. The light source is usually a laser. Backscatter from aerosols in the atmosphere is the principal form in which the outgoing beam is returned to the receiver. Boundary layer development such as the formation and growth of convective cells can be followed by the movement of the atmospheric aerosols *(98)*, and the depth of the convective boundary layer can be determined. Lidars can also distinquish between ice and water in clouds. Using specialized lidars having two wavelengths, it is possible to obtain information on aerosol particle size.

Wind can be determined by lidars by several methods. By tracking inhomogeneities in the backscatter, horizontal winds can be estimated. This is a second-order method since the winds are measured indirectly.

The radial component of the wind can be determined by detection of the Doppler shift in the return signal. The horizontal components can be determined by scanning with the lidar in a circle (a path like an inverted cone). The resulting *velocity azimuth display* (VAD) can be interpreted to give both speed and wind direction. Vertical winds can be determined from a vertically pointed Doppler lidar.

Lidars using two different wavelengths (with one wavelength absorbed by a particulr pollutant) can determine integrated pollutant concentrations along the lidar path. This is the differential absorption lidar (DIAL).

Lidars are single-ended remote sensing devices determining range-resolved data along the line of sight. The spatial averages seen by the lidar can be varied by proper design. Because the region sampled can be varied by pointing the lidar beam in different directions, a single system can sample many vectors in three dimensions, thus replacing many fixed samplers. The lidar beam does not disturb the air that is being measured, an especially important feature for vertical velocities. Lidars are especially suitable for intensive research campaigns.

3. Comparison of Monostatic Sodar and Lidar

Russell *et al.* *(99)* made comparisons between measurements of atmospheric structure with a monostatic sodar and a lidar using a ruby laser. In the early forenoon as the depth of the mixed layer rose, the two instruments performed equally well. During the strong convection in the afternoon the sodar detected only the lower part of the convective cells, whereas the lidar detected the full extent of the convection. Later as the surface inversion formed it was detected immediately by the sodar but only later by the lidar when a discontinuity in aerosol content in the two layers became obvious.

4. Pulsed-Doppler Radar

Remote sensing by radio detecting and ranging (RADAR) is an additional method of sensing phenomena in the boundary layer. A number of wavelengths can be used. The short wave lengths of 3 cm and 8.6 mm are less affected by ground clutter. Winds can be estimated by using insects as tracers of air motion or by following released chaff. Dual Doppler techniques can obtain the two-dimensional wind velocity and can compute the third component by integration of the equation of mass continuity *(100)*. Wind field accuracies are on the order of 0.2 to 0.4 m $second^{-1}$ for horizontal winds and about 0.5 m $second^{-1}$ for vertical winds. In exploring profiles throughout the boundary layer, the following can be measured: mean wind components, Reynolds stresses, eddy dissipation rate, turbulent kinetic energy, and divergence. Radars cannot measure heat flux or pressure velocity.

C. Quality Assurance

Quality assurance (QA) of meteorological measurements includes both activities to provide quality data (quality control) and activities to provide assurance that the quality control system is performing adequately (quality assurance) *(101)*.

Quality control requirements should be determined by the underlying purpose for the data collection. The necessary program can then be designed toward achieving that data quality. The formal document for quality assurance, the QA plan, lists all necessary procedures and their required frequency. Specific areas are

1. Project personnel responsibilities
2. Data reporting procedures
3. Data validation procedures
4. Audit procedures
5. Calibration procedures
6. Preventive maintenance schedules
7. Quality reports

The quality assurance program implements all items in the plan in such a way that problems and potential problems are detected quickly and corrected before a significant loss of data occurs. The value of lost data determines the level of effort that is cost-effective to limit the loss of data.

Audits should be quantitative and unbiased. Systems audits review the adequacy of the instruments for the desired measurements and the adequacy of the quality control procedures, including logs and control charts. Performance audits provide a physical verification of the instrument calibration. Having audits performed by persons outside the organization collecting the data may be more impartial and is encouraged. Audits are checks and should not be used to determine calibration coefficients or to correct or adjust data.

Data validation by inspection of collected data and screening of values for reasonableness is important for both ensuring data quality and detecting problems that have gone undetected. Screening procedures include checking for proper range, checking against a known statistical distribution, comparing values adjacent in time, and making comparisons based on known physical relationships.

The expense of quality assurance should always be anticipated in planning meteorological data collection. Proper planning and execution should make the quality assurance cost-effective, since the result is a high percentage of data capture and valid data.

VI. Air Pollution Climatology

Because numerous air quality dispersion models use Pasquill *(22)* stability class as input and because the United States National Climatic Center (at Asheville, North Carolina) receives many requests to process hourly data to produce stability wind roses (joint frequencies of stability class, wind direction, and wind speed), Doty *et al.* *(102)*

Table XI Mean (across stations) Frequencies of Occurrence of Seven Stability Classes at All Weather Stations in the United States, Percentage[a]

	Stability class						
	A	*B*	*C*	*D*	*E*	*F*	*G*
Annual	1.0	6.6	11.2	47.7	11.8	13.4	8.7
Winter	0.0	2.6	7.1	57.5	12.1	13.6	8.7
Spring	1.0	6.2	11.2	52.2	11.4	12.0	7.3
Summer	2.6	11.3	16.4	36.1	11.5	13.4	8.6
Fall	0.5	6.2	10.2	44.5	12.5	15.6	11.4

[a] Data from Doty *et al.* *(102)*.

have prepared a climatological analysis of Pasquill stability categories.

Averaged over all stations the percentage frequencies of seven stability classes annually and seasonally are given in Table XI. As one might infer from other climatological knowledge, increased cloudiness and wind speeds in winter increase the frequency of neutral conditions. Conversely, decreased cloud cover and generally lighter winds in summer decrease the neutral frequencies and increase the unstable cases.

Geographical variation of the frequencies of unstable, neutral, and stable conditions in the United States is given by Doty *et al.* *(102)* in maps for these three categories for each season as well as for the year. From the annual maps it is noted that for neutral conditions the highest frequencies, i.e., 40–60%, extend all the way across from the Pacific northwest to New England as well as southward from the Great Plains to Texas. The rest of the country has lower neutral frequencies, around 30% in the southwest and about 40% in the

Table XII Annual Average Effective Stack Height[a]

Physical stack height in meters	*Effective stack height, in meters*	
	At 1115 GMT	*At 2315 GMT*
50	100–200	200–700
400	650–750	700–1100

[a] Data from Holzworth *(103)*.

southeast. Unstable frequencies are highest in the southwest, 20–30%, about 20% in the Pacific northwest and mountain states; they exhibit a north–south gradient in the eastern two-thirds of the country, with about 20% typical of the south and 10% typical of the north. Stable conditions are typically near 40% over a great extent of the country. Exceptions are 20–30% over the Great Lakes to New England and around 30% over the Great Plains and extending to Texas.

Holzworth *(103)* examined calculated plume rise for two source sizes using 5 years' radiosonde data for the 48 contiguous United States. The two source sizes, 50-m and 400-m physical stack heights, were selected to include nearly every possible source size. The annual effective heights for the times of the two radiosonde releases — 1115 GMT (0315 PST, 0615 EST) and 2315 GMT (1515 PST, 1815 EST)— are given in Table XII. Lowest effective heights are through the central plains and highest effective heights over the western mountains.

Of interest as it relates to the phenomena of fumigation is the annual percentage of days that effective heights of the two stacks at 1115 GMT were in or below an inversion and subsequently were

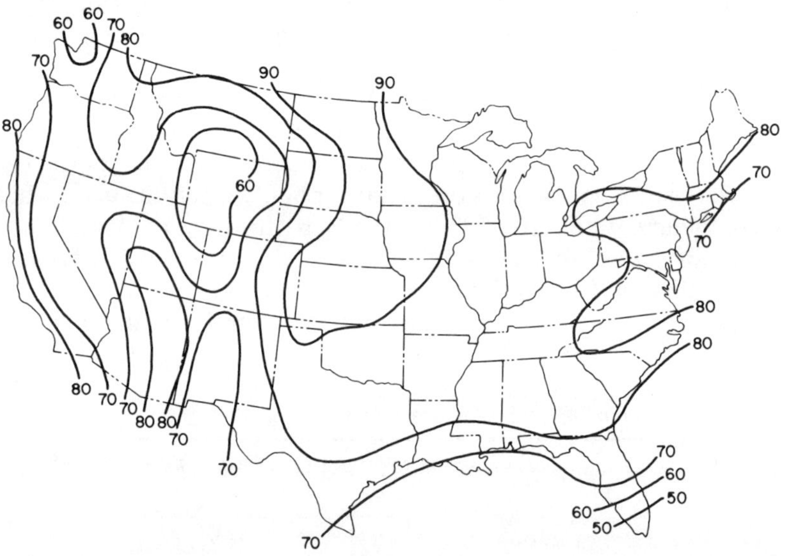

Figure 12. Isopleths of annual percentage that effective heights of 50-m stacks at 1115 GMT were in or below an inversion and subsequently were reached by the afternoon mixing height. [From Holzworth *(103)*.]

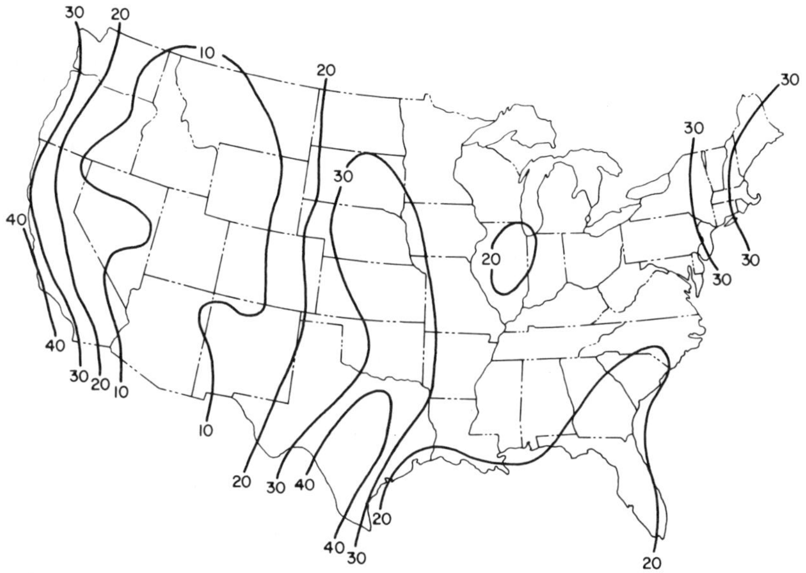

Figure 13. Isopleths of annual percentage that effective heights of 400-m stacks at 1115 GMT were in or below an inversion and subsequently were reached by the afternoon mixing height. From Holzworth *(103)*.]

reached by the afternoon mixing height. For the 50-m stack these percentages ranged from 50% to 90%. For the 400-m stack the percentages ranged from less than 10% to more than 40%. The isopleths are given for the 50-m stack in Figure 12 and for the 400-m stack in Figure 13.

Holzworth and Fisher *(104)*, using 5 years of radiosonde observations, primarily from 1960 through 1964, provide isopleth analyses over the contiguous United States of frequencies of surface-based and elevated inversions at the time of the sounding releases.

Representativeness affects not only air pollution climatology but the collection and use of all meteorological data. Its definition was proposed at a workshop on representativeness *(105)*. "Representativeness is the extent to which a set of measurements taken in a space–time domain reflects the actual conditions in the same or different space–time domain taken on a scale appropriate for a specific application." It was made clear that "the same set of measurements may be deemed representative for some purposes but not for others." It was felt that representativeness will remain a value judgment.

REFERENCES

1. J. C. Wyngaard, *in* "Mesoscale Meteorology—Theories, Observations, and Models" (T. Gal-Chen and D. K. Lilly, eds.), pp. 603–650. Reidel Publ., Dordrecht, The Netherlands, 1983.
2. J. C. Kaimal, J. C. Wyngaard, D. A. Haugen, O. R. Coté, Y. Izumi, S. J. Caughey, and C. J. Readings, *J. Atmos. Sci.* **33,** 2152–2169 (1976).
3. S. J. Caughey, *in* "Atmospheric Turbulence and Air Pollution Modeling" (F. T. M. Nieuwstadt and H. van Dop, eds.), Chapter 4, pp. 107–158. Reidel Publ., Dordrecht, The Netherlands, 1982.
4. A. A. M. Holtslag and A. P. van Ulden, *J. Clim. Appl. Meteorol.* **22,** 517–529 (1983).
5. F. T. M. Nieuwstadt, *Bound.-Layer Meteorol.* **14,** 235–246 (1978).
6. J. S. Irwin and F. S. Binkowski, *Atmos. Environ.* **15,** 1091–1094 (1981).
7. F. S. Binkowski, *Bound.-Layer Meteorol.* **27,** 217–236 (1983).
8. J. W. Deardorff, *J. Atmos. Sci.* **27,** 1211–1213 (1970).
9. F. T. M. Nieuwstadt, *J. Appl. Meteorol.* **19,** 157–162 (1980).
10. M. L. Barad, ed., "Project Prairie Grass, a Field Program in Diffusion," Geophys. Res. Pap. No. 59, Vols. I and II, Rep. AFCRC-TR-58-235. Air Force Cambridge Research Center, 1958 (NTIS, Accession Nos. PB-151 425 and PB-151 424).
11. D. A. Haugen, ed., "Project Prairie Grass, a Field Program in Diffusion." Geophys. Res. Pap. No. 59, Vol. III, Rep. AFCRC-TR-58-235. Air Force Cambridge Research Center, 1959 (NTIS, Accession No. PB-161 101).
12. R. G. Lamb, *Prepr.—Symp. Turbul. Diffus. Air Pollut., 4th, 1979,* pp. 27–33 Boston, Massachusetts (1979).
13. J. W. Deardorff and G. E. Willis, *J. Appl. Meteorol.* **14,** 1451–1458 (1975).
14. J. K. S. Ching, J. F. Clarke, J. S. Irwin, and J. M. Godowitch, *Atmos. Environ.* **17,** 859–871 (1983).
15. J. W. Deardorff, *J. Atmos. Sci.* **36,** 424–436 (1979).
16. J. C. Wyngaard and M. A. LeMone, *J. Atmos. Sci.* **37,** 1573–1585 (1980).
17. S. J. Caughey, J. C. Wyngaard, and J. C. Kaimal, *J. Atmos. Sci.* **36,** 1041–1052 (1979).
18. F. T. M. Nieuwstadt and H. Tennekes, *J. Atmos. Sci.* **38,** 1418–1428 (1981).
19. F. T. M. Nieuwstadt, *Bound.-Layer Meteorol.* **20,** 3–17 (1981).
20. J. S. Hay and F. Pasquill, *in* "Advances in Geophysics, Vol. 6, Atmospheric Diffusion and Air Pollution" (F. N. Frenkiel and P. A. Sheppard, eds.), pp. 345–365. Academic Press, New York, 1959.
21. H. E. Cramer, *Am. Ind. Hyg. Assoc. J.* **20,** (3), 183–189 (1959).
22. F. Pasquill, *Meteorol. Mag.* **90,** 33–49 (1961).
23. H. E. Cramer, *in* "Proceed. 7th Int. Tech. Meet. on Air Pollut. Modeling and Its Applic.," pp. 731–780. NATO/CCMS Pub. No. 51, Brussels, Belgium, 1976 (NTIS, Accession No. PB-270 799).
24. R. R. Draxler, *Atmos. Environ.* **10,** 99–105 (1976).
25. J. S. Irwin, *J. Clim. Appl. Meteorol.* **22,** 92–114 (1983).
26. S. R. Hanna, G. A. Briggs, J. Deardorff, B. A. Egan, F. A. Gifford, and F. Pasquill, *Bull. Am. Meteorol. Soc.* **58,** 1305–1309 (1977).
27. D. G. Fox, D. Randerson, M. E. Smith, F. D. White, and J. C. Wyngaard, "Synthesis of the Rural Model Reviews," EPA-600/3-83-108. U.S. Environmental Protection Agency, Research Triangle Park, North Carolina, 1983 (available only from NTIS, Accession No. PB84-121 037).
28. F. Pasquill, "Atmospheric Dispersion Parameters in Gaussian Plume Modeling.

Part II. Possible Requirements for Change in the Turner Workbook Values," EPA-600/4-76-030a. U.S. Environmental Protection Agency, Research Triangle Park, North Carolina, 1976.
29. W. H. Snyder, "Fluid Modeling of Terrain Aerodynamics and Plume Dispersion—A Perspective View," Ext. Abstr. 6th Symp. Turbul. Diffus., pp. 317–320. Am. Meteorol. Soc., Boston, Massachusetts, 1983.
30. I. P. Castro and W. H. Snyder, *Atmos. Environ.* **16,** 1869–1887 (1982).
31. A. H. Huber and W. H. Snyder, *Atmos. Environ.* **16,** 2837–2848 (1982).
32. W. H. Snyder, R. E. Britter, and J. C. R. Hunt, "Wind Engineering" (J. E. Cermak, ed.), Vol 1, pp. 319–329. Pergamon, Oxford, 1979.
33. D. G. Strimaitis, A. Venkatram, B. R. Greene, S. Hanna, S. Heisler, T. F. Lavery, A. Bass, and B. A. Egan, "EPA Complex Terrain Model Development, Second Milestone Report—1982," EPA-600/3-83-015. U.S. Environmental Protection Agency, Research Triangle Park, 1982.
34. F. A. Schiermeier, T. F. Lavery, D. G. Strimaitis, A. Venkatram, B. R. Greene, and B. A. Egan, EPA Model Development for Stable Plume Impingement on Elevated Terrain Obstacles, *in* "Air Pollution Modeling and Its Application," Vol. IV (C. DeWispelaere, ed.), pp. 637–647. Plenum, New York, 1985.
35. B. Sivertsen, B. Lamb, and K. E. Grønskei, *Atmos. Environ.* **17,** 1915–1922 (1983).
36. Electric Power Research Institute, "Overview, Results, and Conclusions for the EPRI Plume Model Validation and Development Project: Plains Site," EPRI EA-3074. Electric Power Research Institute, Palo Alto, California, 1983.
37. U.S. Environmental Protection Agency, "User's Manual for Single Source (CRSTER) Model," EPA-450/2-77-013. USEPA, Research Triangle Park, North Carolina, 1977 (NTIS, Accession No. PB-271 360).
38. Texas Air Control Board, "User's Guide to the Texas Episodic Model." TACB, Austin, 1980 (NTIS, Accession No. PB80-227 572).
39. D. B. Turner and J. S. Irwin, *Atmos. Environ.* **16,** 1907–1914 (1982).
40. M. E. Smith, *Bull. Am. Meteorol. Soc.* **65,** 554–558 (1984).
41. D. G. Fox, *Bull. Am. Meteorol. Soc.* **62,** 599–609 (1981).
42. J. S. Irwin and M. E. Smith, *Bull. Am. Meteorol. Soc.* **65,** 559–568 (1984).
43. H. van Dop and B. J. de Haan, *Atmos. Environ.* **17,** 1449–1456 (1983).
44. N. D. van Egmond and H. Kesseboom, *Atmos. Environ.* **17,** 257–265 (1983).
45. N. D. van Egmond and H. Kesseboom, *Atmos. Environ.* **17,** 267–274 (1983).
46. R. A. Pielke, R. T. McNider, M. Segal, and Y. Mahrer, *Bull. Am. Meteorol. Soc.* **64,** 243–249 (1983).
47. M. Segal, R. T. McNider, R. A. Pielke, and D. S. McDougal, *Atmos. Environ.* **16,** 1381–1397 (1982).
48. A. H. Weber, M. R. Buckner, and J. H. Weber, *J. Appl. Meteorol.* **21,** 1633–1644 (1982).
49. D. W. Pepper and R. E. Cooper, *Atmos. Environ.* **17,** 1881–1895 (1983).
50. W. B. Johnson, *J. Air Pollut. Control Assoc.* **33,** 563–574 (1983).
51. A. Eliassen, *J. Appl. Meteorol.* **19,** 231–240 (1980).
52. R. M. Endlich, K. C. Nitz, R. Brodzinzky, and C. M. Bhumralkar, "The ENAMAP-2 Air Pollution Model for Long-Range Transport of Sulfur and Nitrogen Compounds," EPA-600/7-83-059. U.S. Environmental Protection Agency, Research Triangle Park, North Carolina, 1984 (available only from NTIS, Accession No. PB84-120 930).
53. A. Eliassen and J. Saltbones, *Atmos. Environ.* **17,** 1457–1473 (1983).
54. B. E. A. Fisher, *Atmos. Environ.* **17,** 1865–1880 (1983).
55. R. G. Lamb and J. H. Novak, *in* "Proceedings of the EPA-OECD International

Conference on Long Range Transport Models for Photochemical Oxidants and Their Precursors," EPA-600/9-84-006, pp. 67–86. U.S. Environmental Protection Agency, Research Triangle Park, North Carolina, 1984.
56. J. J. Fuquay, C. L. Simpson, and W. T. Hinds, *J. Appl. Meteorol.* **3,** 761–770 (1964).
57. R. R. Draxler, "A Summary of Recent Atmospheric Diffusion Experiments," Nat. Oceanic Atmos. Adm. Tech. Memo. ERL ARL-78. NOAA, Silver Springs, Maryland, 1979.
58. W. B. Johnson, *J. Clim. Appl. Meteorol.* **22,** 931–946 (1983).
59. S. -E. Gryning and E. Lyck, *J. Clim. Appl. Meteorol.* **23,** 651–660 (1984).
60. I. van der Hoven, *in* "Meteorology and Atomic Energy 1968" (D. H. Slade, ed.), NTIS TID-24190, Sect. 5-3. U.S. Atomic Energy Commission, Oak Ridge, Tennessee, 1968.
61. S. R. Hanna, G. A. Briggs, and R. P. Hosker, Jr.; *in* "Handbook on Atmospheric Diffusion," DOE/TIC-11223, Chapter 10, pp. 67–73 Technical Information Center, U.S. Department of Energy, Oak Ridge, Tennessee, 1982.
62. T. W. Horst, *Atmos. Environ.* **11,** 41–46 (1977).
63. T. W. Horst, *in* "Atmospheric Sulfur Deposition: Environmental Impact and Health Effects" (D. S. Shriner, C. R. Richmond, and S. E. Lindberg, eds.), pp. 275–283. Ann Arbor Sci. Publ., Ann Arbor, Michigan, 1979.
64. K. S. Rao, "Analytical Solutions of a Gradient-Transfer Model for Plume Deposition and Sedimentation," EPA-600/3-82-079. U.S. Environmental Protection Agency, Research Triangle Park, North Carolina, 1982 (available only from NTIS, Accession No. PB82-215 153).
65. T. E. Pierce and D. B. Turner, "User's Guide for MPTER—A Multiple Point Gaussian Dispersion Algorithm with Optional Terrain Adjustment," EPA-600/8-80-016. U.S. Environmental Protection Agency, Research Triangle Park, North Carolina, 1980.
66. W. B. Petersen, "User's Guide for PAL—A Gaussian-Plume Algorithm for Point, Area, and Line Sources." EPA-600/4-78-013. U.S. Environmental Protection Agency, Research Triangle Park, North Carolina, 1978.
67. K. S. Rao and L. Satterfield, "MPTER-DS: The MPTER Model Including Deposition and Sedimentation—User's Guide," EPA-600/8-82-024. U.S. Environmental Protection Agency, Research Triangle Park, North Carolina, 1982 (available only from NTIS, Accession No. PB83-114 207).
68. K. S. Rao and H. F. Snodgrass, "PAL-DS Model: The PAL Model Including Deposition and Sedimentation—User's Guide," EPA-600/8-82-023. U.S. Environmental Protection Agency, Research Triangle Park, North Carolina, 1982 (available only from NTIS, Accession No. PB83-117 739).
69. G. E. Gordon, *Environ. Sci. Technol.* **14,** (7), 792–800 (1980).
70. J. A. Cooper and J. G. Watson, Jr., *J. Air Pollut. Control Assoc.* **30,** (10), 1116–1125 (1980).
71. U.S. Environmental Protection Agency, "Receptor Model Technical Series." Vol. I. Overview of Receptor Model Application to Particulate Source Apportionment," EPA-450/4-81-016a. USEPA, Research Triangle Park, North Carolina, 1981.
72. J. G. Watson, *J. Air Pollut. Control Assoc.* **34,** 619–623 (1984).
73. K. L. Schere and J. H. Shreffler, "Final Evaluation of Urban-Scale Photochemical Air Quality Simulation Models," EPA-600/3-82-094. U.S. Environmental Protection Agency, Research Triangle Park, North Carolina, 1982.
74. J. H. Shreffler, K. L. Schere, and K. L. Demerjian, *in* "Air Pollution Modeling and

Its Application III" (C. De Wispelaere, ed.), pp. 477–504. Plenum, New York, 1984.
75. K. L. Demerjian and K. L. Schere, *in* "Proceedings, Ozone/Oxidants: Interactions with the Total Environment II," pp. 329–352. Air Pollut. Control Assoc., Pittsburgh, Pennsylvania, 1979.
76. G. J. McRae and J. H. Seinfeld, *Atmos. Environ.* **17,** 501–522 (1983).
77. G. J. McRae, W. R. Goodin, and J. H. Seinfeld, *Atmos. Environ.* **16,** 679–696 (1982).
78. J. H. Novak and D. B. Turner, *J. Air Pollut. Control Assoc.* **26,** 570–575 (1976).
79. R. H. Kummler, B. Cho, G. Roginski, R. Sinha, and A. Greenberg, *J. Air Pollut. Control Assoc.* **29,** 720–723 (1979).
80. R. F. Hodanbosi and L. K. Peters, *J. Air Pollut. Control Assoc.* **31,** 253–255 (1981).
81. P. H. Guldberg and C. W. Kern, *J. Air Pollut. Control Assoc.* **28,** 907–910 (1978).
82. H. M. Ellis and P. C. Liu, *J. Air Pollut. Control Assoc.* **31,** 148–152 (1981).
83. D. B. Turner and J. S. Irwin, *in* "Air Pollution Modeling and Its Application II" (C. De Wispelaere, ed.), pp. 695–706. Plenum, New York, 1983.
84. D. B. Turner and J. S. Irwin, The relation of urban model performance to stability, *in* "Air Pollution Modeling and Its Application, IV," (C. De Wispelaere, ed.), pp. 721–732. Plenum, New York, 1985.
85. U.S. Congress, "Clean Air Act Amendments of 1977," Public Law 95-95. U.S. Gov. Printing Office, Washington, D.C., 1977.
86. Federal Register, "1977 Clean Air Act; Prevention of Significant Air Quality Deterioration. State Implementation Plans; Requirements," Part V, 26380-26410. U.S. Govt. Printing Office, Washington, D.C., 1978.
87. U.S. Environmental Protection Agency, "Guideline on Air Quality Models," EPA-450/2-78-027, OAQPS No. 1.2-080. USEPA, Research Triangle Park, North Carolina, 1978.
88. D. G. Fox, *Bull. Am. Meteorol. Soc.* **65,** 27–36 (1984).
89. K. A. Verrall and R. L. Williams, *J. Appl. Meteorol.* **21,** 1922–1925 (1982).
90. G. R. Ackermann, *J. Clim. Appl. Meteorol.* **22,** 959–961 (1983).
91. W. D. Neff and R. L. Coulter, *in* "A M S Short Course on Instruments and Techniques for Probing the Atmospheric Boundary Layer," Boulder, Colorado, 1983, pp. 486–545. Am. Meteorol. Soc., Boston, Massachusetts, 1986, in press.
92. D. A. Haugen and J. C. Kaimal, *J. Appl. Meteorol.* **17,** 895–899 (1978).
93. A. Weill, F. Baudin, J. P. Goutorbe, P. van Grunderbeeck, and P. Leberre, *Bound.-Layer Meteorol.* **15,** 375–390 (1978).
94. F. F. Hall, Jr., J. G. Edinger, and W. D. Neff, *J. Appl. Meteorol.* **14,** 513–523 (1975).
95. A. Spizzichino, *JGR, J. Geophys. Res.* **79,** 5585–5591 (1974).
96. A. Weill, C. Klapisz, B. Strauss, F. Baudin, C. Jaupart, P. van Grunderbeeck, and J. P. Goutorbe, *J. Appl. Meteorol.* **19,** 199–205 (1980).
97. O. Taconet and A. Weill, *Bound.-Layer Meteorol.* **23,** 133–151 (1982).
98. R. Schwiesow, *in* "A M S Short Course on Instruments and Techniques for Probing the Atmospheric Boundary Layer," Boulder, Colorado, 1983, Am. Meteorol. Soc., Boston, Massachusetts, 1986, in press.
99. P. B. Russell, E. E. Uthe, F. L. Ludwig, and N. A. Shaw, *JGR, J. Geophys. Res.* **79,** 5555–5566 (1974).
100. R. A. Kropfli, *in* "A M S Short Course on Instruments and Techniques for Probing the Atmospheric Boundary Layer," Boulder, Colorado, 1983, pp. 437–485. Am. Meteorol. Soc., Boston, Massachusetts, 1986, in press.
101. P. L. Finkelstein, D. A. Mazzarella, T. J. Lockhart, W. J. King, and J. H. White,

"Quality Assurance Handbook for Air Pollution Measurement Systems: Volume IV. Meteorological Measurements," EPA-600/4-82-060. U.S. Environmental Protection Agency, Research Triangle Park, North Carolina, 1983.
102. S. R. Doty, B. L. Wallace, and G. C. Holzworth, "A Climatological Analysis of Pasquill Stability Categories Based on 'STAR' Summaries." National Climatic Center, Environmental Data Service, National Oceanic and Atmospheric Administration, Asheville, North Carolina, 1976.
103. G. C. Holzworth, *in* "Preprints of Joint Conference on Application of Air Pollution Meteorology," pp. 80–87. Am. Meteorol. Soc., Boston, Massachusetts, 1977.
104. G. C. Holzworth and R. W. Fisher, "Climatological Summaries of the Lower Few Kilometers of Rawinsonde Observations," EPA-600/4-79-026. U.S. Environmental Protection Agency, Research Triangle Park, North Carolina, 1979.
105. C. J. Nappo, J. Y. Caneill, R. W. Furman, F. A. Gifford, J. C. Kaimal, M. L. Kramer, T. J. Lockhart, M. M. Pendergast, R. A. Pielke, D. Randerson, J. H. Shreffler, and J. C. Wyngaard, *Bull. Am. Meteorol. Soc.* **63,** (7), 761–764 (1982).

Lists of books relevant to this chapter published since the lists in Chapter 8, Volume II, 3d edition (1977) can be found in Chapter 1, Table IX, Volume VII.

4

Air Pollution Effects on Physical and Economic Systems

John E. Yocom

TRC Environmental Consultants
East Hartford, Connecticut

Norbert S. Baer

Conservation Center
Institute of Fine Arts
New York University
New York, New York

Elmer Robinson

Mauna Loa Observatory
National Oceanic and Atmospheric Administration
Hilo, Hawaii

I. Effects on Visibility

A. Introduction

The visual quality of the atmosphere, consisting of its color and the clearness with which distant objects can be seen, is one of three measures of environment quality that are available to almost every member of the public whether they be trained scientists or not. The other two readily apparent air quality indicators are odors and large dust or soot particles.

In the United States the importance of the visual quality of the atmospheric environment was recognized in the Clean Air Act Amendments of 1977, Public Law 95-95 *(1)*. A brief discussion of this law with regard to air pollutant impacts was given by Lewis *(2)*, and Horn *(3)* has reviewed some of the early responses of the United States Environmental Protection Agency to this legal requirement for scenic vista protection and visibility improvement where man-made impairment of visibility existed.

Most of the efforts directed toward visibility improvement and protection in the United States have been centered in the western and southwestern regions, where spectacular distant vistas, such as the Grand Canyon, are present and where very large visual ranges are a natural part of the environment *(3a)*. These large natural visibilities are the result of a combination of low concentrations of atmospheric particulate material, low ambient relative humidity, and distant mountains and landmarks to serve as visibility targets. This combination of conditions is not commonly found in other parts of the United

States or in many other countries. In the southern and eastern regions of the United States, high humidity typically causes the natural and anthropogenic atmospheric particles to grow and become important factors in the reduction of visibility. In most of the more heavily populated European and Asian areas a combination of marine air masses with moderate or high relative humidity and significant concentration of hygroscopic marine salt aerosol particles usually precludes visibility even approaching that typical for a remote desert region such as is found in the southwestern United States. The addition of industrial pollutants over wide areas of Europe and Asia augments the background aerosol particles and reduces visibility over broad regions.

It is probably difficult to overemphasize the value to mankind of earth's spectacular vistas not just the Grand Canyon of Arizona, but also the Alps of Europe, the fjords of Scandinavia, the Taj Mahal of India, and the majestic Mt. Fuji of Japan. These and countless other visual treasures can be lost through the degradation of the clarity of the atmosphere caused by air pollution impacts.

This discussion amplifies and brings up to date the presentation in Vol. II 3d ed., Chapter 1 with particular emphasis on the problems presented by long-range scenic vistas and other related concerns in the area of air pollution impacts on the atmosphere.

B. Visibility and Optical Properties of the Atmosphere

Visibility is the ability to distinguish and identify a distant object, and the distance beyond which this cannot be done is the measurement of visibility. Details of this measurement have been described by a number of authors, including McCartney *(4)*, Horvath *(5)*, and the basic reference cited by almost all investigators, W. E. K. Middleton's "Vision through the Atmosphere" *(6)*.

First, it should be recognized that an object can be seen when there is a perceptible degree of contrast between the object and its background. Extensive tests of visual acuity carried out during World War II showed that there was a wide variation in the sensitivity contrast threshold of the human eye from less than 1% to greater than 5% *(6)*. In many calculations or models of visibility it is assumed that the limiting contrast is 2%; although this is a midrange value, it is not necessarily an average or typical value for the general population. Middleton *(6)* presents a good discussion of the idea of a limiting contrast.

1. Visibility Terms and Definitions

The increased attention being given to the problem of visibility has enlarged the scope of visibility studies and has expanded the discussions of a variety of visibility-related topics. This in turn increases the need for a set of terms and definitions tailored for the air pollution visibility problem. To begin to meet these needs, Walther *(7)* prepared a preliminary summary of visibility terminology for the Air Pollution Control Association Visibility Committee (TE-5). This preliminary report has been used here as a guide for the definitions in the following list. This is not intended to be an "official" glossary, but it seems to be a necessary step to take at this stage of development of visibility problems. This list is only a partial compilation of the likely relevant terms: Walther *(7)* had 110 items in his initial survey. The following list places emphasis on terms for the interested but nonexpert technical person. Where references to basic technical sources can be given, they are included.

The meaning of *visibility* and similar expressions needs to be emphasized because, as noted (Vol. II 3d ed., Chapter 1), the terms *visibility, visual range*, and more recently *meteorological range* have been applied by some investigators to the same physical feature of the atmosphere. Because there is a "visibility," indicating a distance along a sight path, the following abbreviated definitions of these three basic terms from the "Glossary of Meteorology" *(8,* pp. 367, 390, 613, 614) are suggested for air pollution investigators.

Visibility: The greatest distance in a given direction at which it is possible to see and identify with the unaided eye (1) in the daytime, a prominent dark object against the sky at the horizon and (2) at night, a known, preferably unfocused, moderately intense light source.

There are inherent difficulties with the conventional requirement that visibility markers be both detected and identified. A more rigorously defined concept of visual range avoids reference to identification; thus, if the identification requirement were dropped, visibility could be so defined. Daytime estimates of visibility are subjective evaluations of atmospheric attenuation of contrast, whereas nighttime estimates represent attempts to evaluate something quite different, namely attenuation of flux density. Thus, visibility data must be regarded as falling into two distinct classes, those obtained by day and those by night.

Daytime Visual Range (8): The distance, under daylight conditions, at which the apparent contrast between a specified type of target and its background becomes just equal to the threshold contrast of an observer; to be distinguished from the night visual range. The visual range is a function of the atmospheric extinction coefficient, the albedo and visual angle of the target, and the observer's threshold contrast at the moment of observation. The formula for the daytime visual range V is

$$V = (1/\sigma)\ln(|C|/c) \tag{1}$$

where there is a contrast C between the horizon target and its background viewed through an atmosphere of extinction coefficient σ by an observer whose momentary threshold contrast is c. In practice, an important application of this formula is to the case of a black target (for which $C = -1$); hence

$$V = \frac{1}{\sigma}\ln\frac{1}{c} \tag{2}$$

Night Visual Range (8): The greatest distance at which a point source of light of a given candlepower can be perceived at night by an observer under given atmospheric conditions. If L is the candlepower of the point source, σ the extinction coefficient of the atmosphere along the light path, and I_t the threshold illuminance for the observer's eyes, then the night visual range v satisfies the equation

$$v = \frac{1}{\sigma}\left(\ln\frac{L}{I_t} - 2\ln v\right) \tag{3}$$

Unlike the daytime visual range, which is limited by atmospheric attenuation of contrast, the night visual range is limited by atmospheric (and geometric) attenuation of luminous flux density. It is important to note that there is no unique night visual range under given atmospheric conditions since this range is also dependent on the source intensity and on the characteristics of the observer's eyes.

Meteorological Range (8): An empirically consistent measure of the visual range of a target; a concept developed to eliminate from consideration the threshold contrast and adaptation luminance, both of which vary from observer to observer. The meteorological range is the distance V' in the black target form of the visual range formula, when c, the threshold contrast, is set equal to 0.02:

$$V' = \frac{1}{\sigma}\ln\left(\frac{1}{0.02}\right) \tag{4}$$

Thus, V' is a function only of the extinction coefficient σ of the atmosphere at the time and place in question and may be written as

$$V' = \frac{3.9}{\sigma} \tag{4a}$$

where both V' and σ are in compatible units, e.g., km and km^{-1}.

It should be noted, especially by investigators working with climatological files of visibility observations, that the tabulated data are not for *visibility* as described but for prevailing visibility *(6, 8)*.

Prevailing Visibility: In United States weather observing practice, the greatest horizontal visibility which is equalled or surpassed throughout half of the horizon circle; it need not be a continuous half. In the case of rapidly varying conditions, it is the average of the prevailing visibility while the observation is being taken. This value is entered in a surface weather observation.

For air pollution climatology studies it is important to recognize that the *prevailing visibility* need not be either the highest or lowest visibility observed at the station; neither is it a calculated average in the statistical sense. Prevailing visibility was designed to be a useful reported observation for aviation operations.

The following additional terms and definitions are taken from the list assembled by Walther *(7)*.

Airlight: light scattered into the sight path and toward the observer from outside the sight path

Apparent contrast: contrast at the observer of the target with respect to an element of horizon sky directly above the target

Blue–red radiance ratio (9): a measure of the discoloration of the background sky caused by a plume, representing the ratio of blue intensities (400 nm) to red intensities (700 nm)

Brightness: attribute of visual sensation by which an area appears to emit more or less light *(10)*

Ground light: visible radiation from the sun and sky reflected by surfaces below the horizon plane *(11)*

Inherent contrast:

$$C_0 = (N_t - N_s)/N_s) \tag{4b}$$

where N_t and N_s are the target and sky radiance (inherent) at the position of the target

Sight path visibility: visibility measured over the entire length of a sight path

2. The Visibility Problem

Figure 1, adapted from Middleton *(6)*, illustrates the visibility problem for a uniformly mixed atmosphere. The observer, *O*, views the target, *T*, along the path influenced by the direct sunlight, diffuse skylight, and light from the ground acting on the infinitessimal elements, *dv*, of the sight path. Within *dv* part of direct and diffuse light is scattered toward the observer. This increases the apparent brightness of the dark target. This scattered light is known as *airlight*. When the total amount of airlight integrated along the path *OT* approaches the intensity of the light from the horizon sky, the target is at the limit of visibility, and the distance *OT* is by definition the visibility. By using the idea of a threshold contrast it is possible to be more precise and state that the visibility is the distance at which the ratio of the intensity or brightness of the airlight to the brightness of the horizon equals the threshold contrast of the observer. In the simplest case it is usually assumed that the target is black, as in the visibility definition. Also the radiance of the horizon is usually considered to be constant with regard to distance.

If the simple, regional visibility situation shown in Figure 1 is made more complex by the interjection of an air pollution plume, the schematic result is shown by Figure 2. This figure and the following discussion of plume impacts on the visibility problem mainly follow that

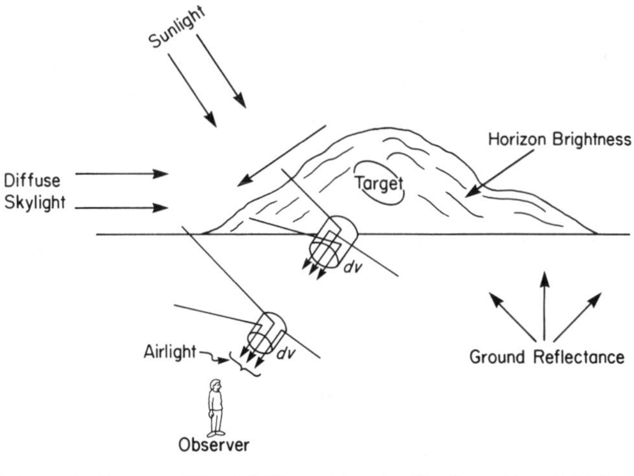

Figure 1. Schematic diagram of the visibility problem and the formation of airlight as a result of sunlight and diffuse skylight scattered into the observers cone of vision [Adapted from Middleton *(6)*].

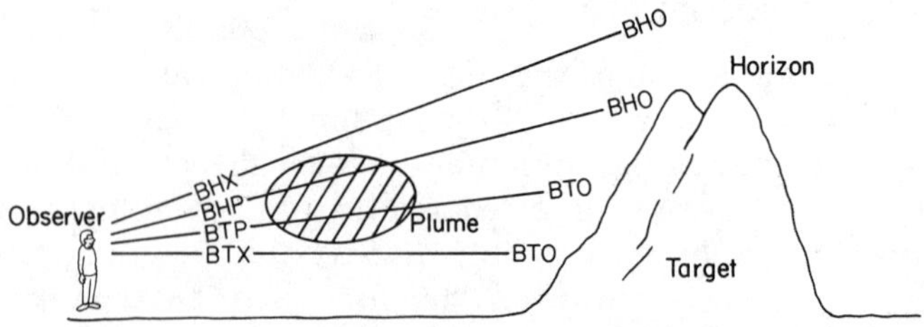

Figure 2. Schematic diagram of the visibility problem with a plume in the observer's cone of vision. Field includes the direct and diffuse illumination as shown in Figure 1, sight path nomenclature: BHO–BHX; brightness of horizon sky at target (O) and observer (X) distance; BTO–BTX, brightness of target at target surface (O) and observer (X) distances; BHO–BHP, brightness of horizon (BHO) at observer viewed through plume (P); BTO–BTP, brightness of target (BTO) at observer viewed through plume (P) [Adapted from Haas and Fabrick *(12)*].

presented by Haas and Fabrick *(12)*. From Figure 2 it is apparent that the change in contrast between the horizon and the target caused by the plume can be expressed as

$$\Delta C = [(B_{\mathrm{Tx}}/B_{\mathrm{Hx}}) - 1] - [(B_{\mathrm{TP}}/B_{\mathrm{HP}}) - 1] \tag{5}$$

In this case the first term on the right-hand side is equivalent to the target contrast equation, and the second term is the effect which is due to the plume. As indicated by Haas and Fabrick *(12)*, C increases with both the fine particle and NO_2 content of the plume, i.e., with the concentration of these plume components that can scatter or absorb visible light.

Another indicator of plume visibility is the contrast of the plume when viewed against the horizon sky *(12)*. In terms of the brightness values illustrated by Figure 2, plume contrast C_p can be expressed as

$$C_{\mathrm{p}} = (B_{\mathrm{Hp}}/B_{\mathrm{Hx}}) - 1 \tag{6}$$

But since the brightness of the horizon is independent of distance

$$B_{\mathrm{Hx}} = B_{\mathrm{HO}} = B_{\mathrm{H}} \tag{7}$$

Equation (6) can be written as

$$C_{\mathrm{p}} = (B_{\mathrm{HP}}/B_{\mathrm{H}}) - 1 \tag{8}$$

In this case, a bright plume may appear brighter than the sky and have a value of C_{p} greater than 1, and a plume with strong absorption characteristics will be darker than the sky and have a contrast value less than 1.

Model calculations of ΔC and C_{p} were made by Haas and Fabrick

(12), who considered the light as white light although it is known that both fine particles and NO_2 can impart color to a plume. The blue–red ratio B/R has been used as a measure of the discoloration of the sky by a plume *(11, 13)*. If the plume contrast Cp is evaluated at two wavelengths, blue ($\lambda = 400$ nm) and red ($\lambda = 700$ nm), this ratio R can be expressed as

$$R = \frac{C_p(400\text{ nm}) + 1}{C_p(700\text{ nm}) + 1} \tag{9}$$

Haas and Fabrick *(12)* point out that a "bright-" or "high blue" appearing plume will have an R value greater than 1, whereas a plume with strong absorption and appearing darker than the sky will look "browner" and have an R value less than 1.

More complex considerations of plume color contrast and changes in color contrast over a distant sky–terrain vista have also been discussed by Haas and Fabrick *(12)*, and the results of plume impacts were evaluated in terms of sulfate aerosol particle content and NO_2 concentration.

3. Visibility in Clean Air Masses

Visibility through the atmosphere is commonly expressed by a form of the Koschmiedier equation

$$V = \frac{1}{\sigma}\ln\left(\frac{1}{c}\right) \tag{10}$$

where σ is the extinction coefficient and c is the contrast threshold. If the contrast threshold is assumed to be 0.02, this expression becomes the meteorological range equation

$$V = \frac{3.9}{\sigma} \tag{11}$$

The units for V and σ must be mutually consistent, e.g., m and m^{-1}. As noted by many authors (see Vol. II, 3d ed., Chapter 1), the extinction coefficient σ is the sum of Rayleigh scattering b_R, particle scattering b_{sp}, particle absorption b_{ap}, and gaseous absorption, b_{ag}. Thus

$$\sigma = b_R + b_{sp} + b_{ap} + b_{ag} \tag{12}$$

If we assume an urban visibility of 20 km, not an especially low value, the value of σ using the meteorological range equation [Equation (11)] is $1.95 \times 10^{-4}\ m^{-1}$. By contrast, the Rayleigh scattering coefficient b_R, which describes the scattering due to atmospheric molecules, has a value of 1.7×10^{-5} m^{-1} for a wavelength of 500 nm at sea level *(4)*. Table I lists the values for the Rayleigh coefficient for several wavelengths in the visible part of the spectrum *(4)*. Since there is at least an order of magnitude difference between the extinction coefficient for a typical urban visibility situation and Rayleigh conditions, it is common practice to neglect Rayleigh scattering in air pollution visibility problems and to concentrate on aerosol particle and NO_2 effects (Vol. II 3d ed., Chapter 1).

However, with the development of a concern for possible visibility problems in areas of very good visibility, Rayleigh scattering may now become an important factor in the visibility problems. For example, in some parts of the western and southwestern United States a visibility of 130 km, or 80 mi, is not unusual. Using the meteorological range expression [Equation (11)], a 130-km visibility corresponds to an extinction coefficient value of 3.0×10^{-5} m^{-1}. Thus b_R with a value of 1.7×10^{-5} m^{-1} accounts for more than half of the total extinction. If the non-Rayleigh part of the extinction, namely about 1.3×10^{-5} m^{-1}, is attributed to the fine particle scattering, b_{sp}, Waggoner *et al.* *(14)* show that this amount of extinction could be accounted for by a fine particle mass concentration of about 4 μgm^{-3}. A reduction in the meteorological range from 130 to 65 km would accompany a change in the extinction coefficient from 3.0×10^{-5} m^{-1} to 6.0×10^{-5} m^{-1}; the Rayleigh component, 1.7×10^{-5} m^{-1}, would account for 28% of the total scattering at this lower meteorological range. The remainder of the scattering coefficient, 4.3×10^{-5} m^{-1}, could be caused by a fine particle concentration of about 14 μgm^{-3}, according to Waggoner *et al.* *(14)*. Rayleigh scattering factors at other wavelengths can be readily calculated from Table I or other more detailed tabulations. Corrections for altitude or ambient temperature can be made easily because

Table I Rayleigh Scattering Coefficient as a Function of Wavelength ($T = 15°C$, $P = 1013$ mb) *(4)*

Wavelength (μm)	b_r *(m^{-1})*
0.45	2.64×10^{-5}
0.50	1.72×10^{-5}
0.55	1.16×10^{-5}
0.60	0.82×10^{-5}
0.65	0.59×10^{-5}

Rayleigh scattering is directly proportional to atmospheric density *(4)*.

These simple arithmetic calculations show clearly that when the ambient visibility is relatively good, Rayleigh scattering accounts for a significant fraction of the total extinction, and large fractional reductions in visibility can result from relatively small changes in total fine particle concentrations. The implications with regard to locating major air pollutant sources in high visibility regions are obvious.

C. Visibility Simulation Models

Considerable effort has gone into the search for one or more models by which downwind visibility patterns could be calculated. However, the complexity of the problem is extreme, as can be seen by noting the components that must be included in a plume visibility model shown schematically by Figure 3. This is a simplified block schematic of the EPA–SAI plume visibility model *(15)*, and it shows that the modeling of visibility requires a combination of at least three separate models each with its own assumptions and possibilities for error. In the EPA–SAI model the plume dispersion model is a Gaussian model and, of course, requires submodels to cover emissions and basic meteorological factors. The plume dispersion model is used to predict the

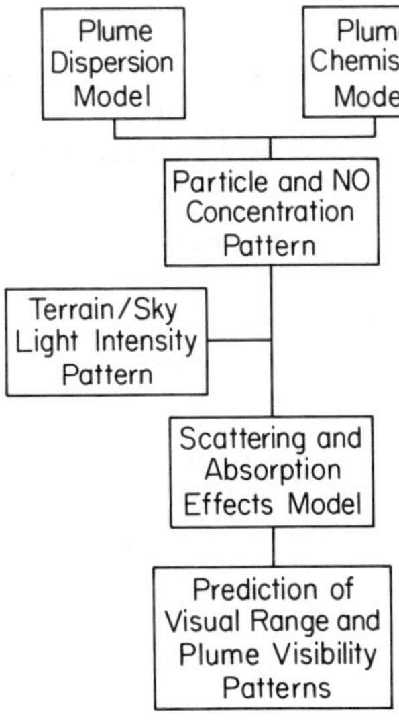

Figure 3. Schematic of the visibility modeling program.

patterns of inert fine particles, e.g., fly ash, and the gaseous plume constituents that are the precursors of NO_2 and secondary aerosol particles, mainly sulfate particles.

The plume chemistry model relates the precursor and ambient atmospheric factors to produce downwind dispersion patterns of NO_2 and secondary fine particles and combines with the primary particle pattern of the Gaussian plume model. The pattern of fine particle and NO_2 concentrations plus ambient light patterns are inputs to the optical effects model which, by a prediction of scattering and absorption effects, leads to a prediction of visual range and plume visibility. The visibility predictions, obviously, are strongly sensitive to the performance of the Gaussian plume model. However, this sensitivity is mitigated to a certain degree because visual qualities are long-path integrated factors rather than point concentration values, and the Gaussian model is probably better able to predict such large area average values.

Bergstrom *et al. (15)* describe a field comparison of the EPA–SAI model using the plume from the Navajo power generating station in Arizona. Their conclusion for this arid, relatively clean background situation was that the overall performance of the model was "good." The greatest uncertainties in the model seemed to be related to the well-known limitations of the Gaussian dispersion model. The plume chemistry model performed "reasonably well" but tended to overestimate the rate of conversion of NO to NO_2 in the plume. For the arid conditions applicable to this Arizona test the formation of a secondary sulfate aerosol was negligible, and thus plume visibility was modeled mainly as an NO_2 effect. Under these specific conditions the optics model performed "reasonably well" but tended to overpredict the visual plume effects to some extent *(15)*.

These results are encouraging in terms of obtaining a useful visibility model in spite of the complexity of the problem, as illustrated by Figure 3. However, the limitations of this desert test situation must be noted, and probably the most important of these was the lack of the necessity for modeling plume SO_2 oxidation to form a secondary sulfate aerosol plume because of the assumed low rate of oxidation in desert regions. Field data describing the particle and chemistry characteristics of a plume relative to its visual impact have been obtained by Hegg *et al. (15a)*.

There have been a number of models developed for visibility simulation in addition to the EPA–SAI model considered previously, and most were described in the visibility meeting proceedings edited by White *(16)*. The Environmental Research and Technology Inc. (ERT) model *(17)* is generally similar to the EPA–SAI model; its output is designed to estimate the regional visibility degradation, i.e., "plume blight," due to dispersion from point sources for use in regulatory programs. Özkaynak *et al. (18)* describe the testing of this model against the analytic theory of atmospheric visibility, i.e., Koschmieder-type relationships, but do not present comparisons with field measurements.

The visibility model developed at the Los Alamos Scientific Laboratory and described by Williams *et al. (19–21)* produces a photographic approximation of the visibility-impaired scene. Basically this modeling procedure is a three-step process. First the pollutant dispersion, chemical reactions, and visibility impacts are calculated in a fashion such as is shown in Figure 3. Then a clean-atmosphere photograph of the impact area is digitized by a microdensitometer, and these point density data are modified according to the results of the visibility model calculations. Finally, the photographic view is reconstructed from the modified densitometer data and used as a visual simulation of the estimated regional plume impact. This technique can provide the estimated numerical visibility relationships, including blue to red ratios and plume–sky brightness values [see Equations (5) and (9)], as in the EPA–SAI model, as well as the more qualitative simulated photographic scene. Williams *et al. (21)* note that when the simulations were tested against actual plume situations the results seemed "reasonable," but that the plume simulations appeared more diffuse and less apparent than the actual scene. Some inadequacies in the dispersion model were indicated *(21)*, as with the EPA–SAI results described by Bergstrom *et al. (15)*.

A regional visibility model has been described by Dabberdt and Eigsti *(22)* and tested against National Weather Service visibility data for the northeastern part of the United States. This visibility model combines a long-range transport model *(23)* which can produce a secondary sulfate aerosol concentration pattern to an optical effects model to estimate a regional visibility pattern. A joint comparison of several visibility models against the same data from an Arizona desert is described by White *et al. (23a)*.

D. Modeling Air Quality Judgments

The visibility models described were directed toward a quantitative analytical goal, namely the prediction of numerical estimates of visibility, although the various investigators were well aware of the fact that the final measure of visibility is a human evaluation. The

Clean Air Act Amendments of 1977 *(1)* clearly state that the protection of scenic vistas is a national environmental goal, and this purpose clearly brings human judgments of air quality to the visibility equation. As a result of this situation, a number of investigators have focused attention on the human evaluation aspects of the overall visibility or "vista preservation" problem *(24–31)*.

Two basic approaches have been used in these several studies. Malm *et al. (28, 29)* used color slides taken of vistas in the southwestern United States shown to panels of individuals, and their responses as to the perceived degree of degradation of the vista were compared to telephotometer data taken at the time of the original photograph. Observers also evaluated the actual Canyonlands National Park vista used in the photographs, and the responses to both color slide and actual vista were compared. Latimer and Daniel *(30)* and Latimer *et al. (26)* used a similar photography and telephotometer experimental design but without an on-site vista evaluation. However, they expanded their vista consideration to include a southwestern United States urban area and eastern United States vistas as well as Grand Canyon, Arizona, views. Slides from this research were evauated by groups with a variety of backgrounds. Another research program, reported by Mumpower *et al. (25)* and Middleton *et al. (24, 24a)*, was designed to determine whether observer evaluations of visual characteristics could be modeled by using air quality estimates and present radiative transfer (i.e., visibility) theory.

The research study by Malm *et al. (29)* showed that there was no statistical difference in the response of observers when perceived visual air quality is evaluated from the actual vista as compared to a color slide of the same view. Observer evaluations were analyzed under a variety of conditions affecting the scene. When the change brought about an apparent increase in vista color, as can occur as a result of changing sun angle, the presence of snow in the distance or foreground, an increase in cloud cover, or a decrease in air pollution, the result was an increase in the perceived visual air quality by the typical observer. Malm *et al. (29)* used their experimental results to develop a model to predict changes in perceived visual air quality (PVAQ) based on changes in the scene. As a result of model calculations, Malm *et al. (29)* concluded that (1) for each isolated scenic element the relationship between PVAQ and apparent target contrast change is linear; (2) when there is more than one scenic component, PVAQ is proportional to the contrast change of each element weighted in proportion to the fractional area subtended by that element; (3) vistas with high inherent scenic beauty are most sensitive to air pollution change; (4) nearby foreground features do not reduce the sensitivity of PVAQ to air pollution impact; and (5) the distance to each of the scenic elements is important in that it establishes which element is most sensitive to air pollution impact.

The results described by Latimer *et al.* *(26)* for their study of vista photographs and observer-perceived air quality summarize the observer evaluations of both air quality and scenic beauty. They determined that for these two elements there is a great deal of agreement between different groups but that ratings differ substantially for different landscapes. For example, in this study *(26)* Grand Canyon vistas were rated much higher than Tucson, Arizona, and eastern United States vistas by the various panels of vista judges used in this research program. As a result, Latimer *et al.* *(26)* concluded that it might be reasonable to set the most stringent visibility goals and priorities in those vista areas with the greatest scenic beauty sensitivity to optical characteristics such as visual range.

The use of color photographic slides as a surrogate for on-site human observers needs to recognize several features of commercial color slide and film as summarized by Henry *(31)*. He notes that the photographic industry designs film color response to produce high customer acceptance of the color reproduction of the perceived scene and not necessarily for an accurate reproduction of the original scene. More specifically, in regard to factors important to evaluations of scenic vistas and pollutant impacts, Henry *(31)* states that (1) the viewer-preferred color of the blue sky in a color reproduction is significantly different from the actual color of the sky, (2) the actual color of the blue sky is even outside the range of preferred sky reproductions, and (3) the color people remember as being the blue sky color is significantly different from either the actual or preferred blue sky color. These conclusions by Henry *(31)* refer to the original studies of Bartleson *(32)* and Bartleson and Bray *(33)*.

The implications of these results for the use of photographs in vista evaluations are obvious and emphasize the need for careful experimental design.

The model results reported by Middleton *et al.* *(24)* for their comparison of on-site observer evaluation of wintertime Denver, Colorado, visual air pollution, i.e., the Denver "Brown Cloud," with the theoretical impacts of pollutant mass on visual parameters showed very poor statistical correlations between observer and theoretical data. This study differs from those described previously in several important ways. First, the area considered was an urban region significantly affected by local pollutant sources and having impaired visibility. Second, the same panel of on-site observers was used to evaluate visual parameters during the month-long field study, and third, the observer data were compared with similar calculated data based on a three-dimensional air quality simulation model. The pollutant input to the model was limited to fine particles (diameter less than 2.5 μm). Contributions of larger particles, secondary particles, and NO_2 to the winter visual air quality in Denver were judged to be of much less importance and were not considered in the simulation. Statistical analysis of observer data and straightforward theoretical simulation model calculations showed that observed target clarity and calculated contrast were significantly correlated; however, the correlation coefficient was low, r equal to 0.365, and accounted for less than 15% of the variability of the data. Other visual parameters included in this study were atmospheric color and the definition of the pollutant cloud border. For these parameters the statistical correlations between observations and the theoretical model were not significantly different from zero. Although several empirical model expressions were tested against the data and better correlations were calculated, the fraction of the variation explained by the various correlations was in all cases less than one-third of the total. The conclusion of Middleton *et al.* *(24)* was that current standard visibility theory cannot be used to predict visual perception of pollutant haze masses. The problems of basing visual air quality management on human judgment have been considered specifically by Middleton *et al.* *(24a)*.

E. Visibility Climatology

Concern about visual air quality has provided the impetus for a relatively large number of studies of visibility climatology. The meeting proceedings for the Plumes and Visibility Conference held at Grand Canyon National Park and edited by White *(16)* contain a dozen or more papers related to plume transport and visibility occurrences. Considerable study has been given, obviously, to the visibility climatology of southwestern United States sites. Space does not permit a summary of visibility climatology, but a few remarks are certainly in order.

The paper by Trijonis *(34)* presents an analysis of 25 years of airport visibility observations at 17 sites over the southwestern United States, including both urban and nonurban locations. Among his conclusions Trijonis stated that visibility in the southwestern United States had deteriorated by 10–30% over the period from the mid-1950s to the early 1970s at almost all of the sites studied. This same southwestern United States region was studied by Macias *et al.* *(35)* using concurrent visibility and wind trajectory data on a case study basis. These authors concluded that periods of reduced visibility at Page, Arizona, and more generally over the southwestern desert areas coincided with transport from the Los Angeles basin of southern California. Aerosol particle chemistry data generally supported this conclusion *(35)*. Turbidity trends at Tucson, Arizona, have been studied for the 27-year period 1956–1983, but no long-term upward or downward trend was identified *(35a)*.

Husar *et al.* *(36)* have used visibility data over periods as long as 80 years to assess changes in visibility and haze over the eastern part of the United States. Among their results, these authors conclude that the strongest increase in haziness occurred in the states adjacent to the Great Smoky Mountains (Tennessee and North Carolina), where the average visibility decreased from 24 to 10 km between 1948 and the 1970s. Sloan *(37–40)* has also analyzed visibility trends in the northeastern region of the United States. Her conclusions *(39, 40)*, based on a more detailed stratification of the visibility record, were somewhat more optimistic than those of Husar *et al.* *(36)* mentioned previously and of Trijonis and Yuan *(41)*. A study of visibility in Illinois over the period 1949–1980 by Vinzani and Lamb *(41a)* shows the complexity of regional visibility by examining records on both an individual station and a regional basis. Visibility may also be related to air mass changes *(41b)*.

Visibility values are a function not only of the gaseous and particulate concentration of the atmosphere but also of the meteorological conditions, as noted by many authors (e.g., Vol. II 3d ed., Chapter 1). It is typical for visibility climatology to be examined without regard to biasing trends in other meteorological parameters such as the relative humidity. Sloan *(39)* in a follow-up of earlier studies *(37, 38)* has examined the climatological trends of both visibility and meteorological factors that can affect the visibility. The region of study was the mideastern United States (Illinois to the East Coast, south of the Great Lakes to Tennessee and North Carolina). A comparison was made between summertime conditions during 1952–1956 and 1971–1975. Summertime visibility in this region was at a 30-year maximum in the mid-1950s and at a relative low in the 1970s, a fact that has often been used to illustrate the importance to regional visibility of increased summertime power production and SO_2 emissions. Sloan's *(39)* analysis of meteorological trends indicated that meteorological conditions adverse to visibility, such as frontal passages, air mass changes and their sources, relative humidity, and temperature, are not randomly distributed throughout the summer months and do not necessarily become randomly distributed as longer time periods are considered. As one example, Figure 4 *(39)* shows the average summertime (June–September) relative humidity values at 1300 EST for three nonurban sites: Lynchburg, Virginia; Williamsport, Pennsylvania; and Roanoke, Virginia. The upward trend from the mid-1950s to the mid-1970s is obvious and statistically significant. A relative humidity of 70% is typically used to indicate the threshold where moisture effects begin to degrade the visibility significantly. In Sloan's *(39)* tabulation of the rural station humidity data, there were 75% more occurences of days with 1300 EST relative humidity over 70% in the period 1971–1975 than there were in 1952–1956.

The extent to which these trends in regional meteorology affected visibility trends was examined in a subsequent paper by Sloan *(40)*. In this analysis she attempted to compensate for the trends in meteorological factors by considering only those visibility observations made under a set of given meteorological conditions considered representative of "typical" conditions for each observational site. After this screening of observational data for 1952–1956 and 1971–1975 the three rural stations showed generally moderate (10–30%) decreases in summertime visibility, but generally large (greater than 30%) improvements in visibility in the other seasons of the year.

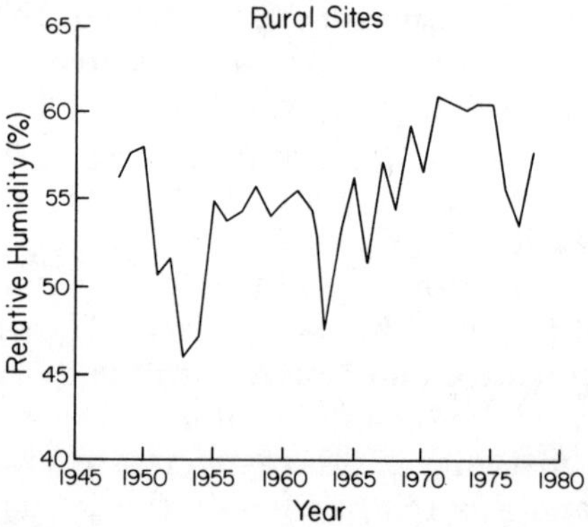

Figure 4. Average summertime relative humidity for three rural sites, Lynchburg, Va., Williamsport, Pa., and Roanoke, Va., based on June–September, 1948–1978, 1300 EST observations [Adapted from Sloan *(39)*].

F. Visibility Measurements

The preceding discussion has emphasized that the basic visibility measurement is a visual observation by a trained observer (see Vol. I 3d ed., Chapter 11 and Vol. II 3d ed., Chapter 1). The Clean Air Act Amendments of 1977 *(1)* increased the need for instrumental means by which visibility could be estimated. The result has been considerable investigation of the nephelometer *(42, 42a)* and telephotometer *(43–45)* as instruments for air pollution visibility assessment. Although transmissometers (Vol. I 3d ed., Chapter 11) can in theory also be used in visibility studies, in practice it has proven very difficult to operate them with suitable accuracy at the ranges of interest in most air pollution studies.

Malm *(27)* points out the way the nephelometer and telephotometer instruments approximate the visibility and the assumptions that are required. The nephelometer measures the integrated scattering coefficient b_{sp} of the atmosphere at a fixed point; the meteorological range can be calculated from

$$V = \frac{3.9}{b_{sp}} \tag{13}$$

as described previously. This approximation of the visibility assumes the following:

1. Absorption coefficient is zero.
2. Extinction coefficient does not vary along the visibility sight path.
3. Visibility target object is black.
4. Sky radiance is the same at the target object as at the observing point.
5. The earth is flat, and the sight path is horizontal.

In addition it must be assumed that the atmospheric aerosol particles do not change as they are drawn through the nephelometer, and, considering the sensitivity of natural and pollutant particulate materials to changes in relative humidity, this may be a very important factor.

As described by Malm *(27)* the basic expression linking the telephotometer to the meteorological range is

$$V = \frac{-3.9x}{\ln C} \tag{14}$$

where x is the distance to the target and C is the apparent contrast of the target as determined by the telephotometer. This expression requires that it be assumed that

1. The atmosphere is homogeneous along the sight path.
2. The sky is cloudless.
3. The earth is flat and the sight path is horizontal.

Telephotometer measurements are frequently approximated by individual photographs, and apparent contrast is determined by densitometer measurements made on the film. This procedure requires that great care be taken in film processing and in the density evaluations. A basic reference for the use of photographic photometry for visibility studies is Steffens *(46)*.

In many cases the use of telephotometer or camera measurements of visibility can be simplified by measuring the contrast of two individual targets at different distances. In this case, as described by Ellestad and Speer *(47)*, the extinction coefficient can be caculated from

$$\sigma = \frac{1}{x_2 - x_1} \ln\left(\frac{C_1}{C_2}\right) \tag{15}$$

where x_1 and x_2 are distances to a near (1) and a far (2) target with photometer-measured contrasts C_1 and C_2. Steffens *(46)* also describes the dual target photographic visibility method.

Nephelometer and telephotometer visibility measurement techniques have been compared in several studies *(45, 47, 48)*. In the data reported by Malm *et al. (45)* taken in Mesa Verde National Park, Colorado, an arid desert area of the southwestern United States, nephelometer-indicated meteorological range had a median value of about 210 km for one test period of a month. Concurrent telephotometer data across a distant vista showed a median value of about 160 km. Malm *et al. (45)* attribute this degree of difference to inhomogeneities along the sight path compared to the single point measurement of the nephelometer.

Tombach and Allard *(48)* describe a 9-month study done at a site in Ohio and one in Pennsylvania in which concurrent visibility estimates were compared using (1) visual observations, (2) telephotometers, (3) photographic photometry, and (4) nephelometers. This study was especially oriented toward the visibility conditions in the eastern part of the United States. Compared to the desert southwestern United States sites, their characteristics include greater aerosol particle concentrations, much higher relative humidity conditions, and different target brightness characteristics.

The initial results of this research *(48)* showed that for "normal" illumination conditions, i.e., excluding cases of cloud shadows and other irregular lighting, agreement between visual observations and both the telephotometer and photographic methods was quite good and without significant bias toward either over- or underestimates by the instrumental methods. The lower scatter in the data comparing prevailing (visual) visibility with the telephotometer than for the camera results was attributed to the accumulation of errors due to film processing and densitometer evaluation *(48)*.

The results of comparing visual prevailing visibility with the nephelometer data showed a great deal of scatter and a strong bias toward indicated visibilities greater than those observed. Figure 5 shows these results; Figure 5(a) presents all of the data, and Figure 5(b) shows only data taken under relatively dry air mass conditions, at a relative humidity of 60% or less. The drier air mass data show less scatter but there is still a marked bias toward values greater than those reported by the visual observations.

The nephelometer comparisons *(48)* in Figure 5 probably represent errors due to at least two of the assumptions that are necessary to relate the nephelometer output to visual range. First, and perhaps

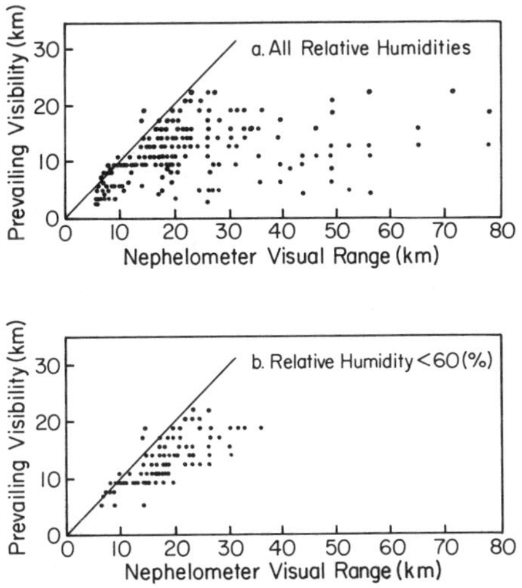

Figure 5. Comparison of observed prevailing visibility and the meteorological range calculated from integrating nephelometer measurements of the scattering coefficient. Combined data taken near Duncan Falls, Ohio, and Scranton, Pa., during June–August 1979 [Adapted from Tombach and Allard *(48)*].

most important, the ambient aerosol is probably heated and dried somewhat as it is drawn into the instrument; this would reduce the size of hygroscopic and deliquescent particles, causing the scattering signal to decrease and the calculated visibility to increase (see Vol. II, 3rd ed., Chapter 1). In addition, the nephelometer results neglect aerosol absorption, and in an area where industrial and urban aerosol particles are common this may be a source of significant error. Waggoner *et al.* *(14)* note that the particle absorption coefficient may be nearly equal to the particle scattering coefficient in urban areas, although absorption is more nearly 10% of the scattering coefficient in arid background areas in the southwestern United States.

G. A Look to the Future

The need for using a quantitative measurement for the visibility and bypassing the human observer will doubtless continue to grow in importance as aesthetic qualities of the environment receive increased attention *(48a)*. However, it seems clear that the development of a practical instrumental system for visibility measurements that equates to the performance of the human observer may be difficult to achieve, since considerable variation is found when similar experiments are made in different locations *(24, 24a)*. One solution to this problem would be to redefine the visibility measurement *for control purposes* on the basis of a specified instrumental measurement, e.g., a telephotometer, as has been done for gaseous and particulate pollutants. There would be obvious assumptions required for such a procedure because of restrictions to daytime data over "perfect" sight paths, and so on. However, visibility described in this manner could be subjected to simulation modeling and changes in visibility dealt with in a quantitative fashion as is done with other atmospheric pollutants.

II. Effects on Economic Materials and Structures

A. Introduction

Damage to materials from air pollution takes a variety of forms, including the corrosion of metals, erosion and discoloration of paints, decay of building stone, and the weakening and fading of textiles (Table II). All of these effects occur to a significant degree as a result of natural environmental conditions, even in unpolluted atmospheres. Moisture, atmospheric oxygen, carbon dioxide, sunlight, temperature fluctuations, and the action of microorganisms all contribute to the deterioration of materials. Quantifying the specific contributions of anthropogenic air pollutants to such damage is a formidable task. Furthermore, distinguishing the relative amount of damage caused by specific pollutant transformation and contact scenarios (for example, acidic deposition) becomes even more elusive.

This section deals with the effects on materials of air pollutants of anthropogenic origin. Although its scope is extremely broad in concept, the literature describing research on any one specific contact and effect scenario may be limited or even nonexistent.

There is a significant body of literature describing the effects of primary air pollutants on materials as determined by both laboratory and field experiments. This literature has been summarized in detail by the United States Environmental Protection Agency (EPA) in its Criteria Documents supporting the establishment of air quality standards for such pollutants as sulfur oxides and particulate matter *(49)*. Other reviews include Yocom and Stankunas *(50)*, Haagenrud *et al.* *(51)*, dealing primarily with effects of sulfur compounds, and Yocom and Baer *(52)*, considering acid deposition effects on materials.

1. Long-Range and Local Effects

Acidic pollutants, whether they are present as primary pollutant gases (e.g., SO_2 and NO_x), as fully oxidized acids, as salts (e.g., sulfates and nitrates), or in the form of acidified precipitation, may have arrived at a material surface from local pollutant sources or may have been transported from distant sources (Table III). The separation of long-range and local characteristics is somewhat artificial since phenomena associated with long-range transport may be generated by local sources under the appropriate conditions. The key point is that the different acidic deposition scenarios are likely to respond differently to alternative control strategies.

2. Complex Mechanisms of Exposure and Deposition

The work that has been done to measure damage to materials from air pollution has not usually considered to any significant degree the specific mechanisms of exposure, deposition, and subsequent damage. Most of the studies which have used laboratory chamber exposure or field exposure in the ambient atmosphere are not able to isolate specific deposition mechanisms from the many interrelated chemical and physical processes involved. The following list presents

Table II Air Pollution Damage to Materials

Materials	*Types of damage*	*Principal damaging pollutants*	*Other environmental factors*	*Methods of measurement*	*Mitigation measures*
Metals	Corrosion, tarnishing	Sulfur oxides and other acid gases	Moisture, air, salt, particulate matter	Weight loss after removal of corrosion products, reduced physical strength, change in surface characteristics	Surface plating or coating, replacement with corrosion-resistant material, removal to controlled environment
Building stone	Surface erosion, soiling, black crust formation	Sulfur oxides and other acid gases	Mechanical erosion, particulate matter, moisture, temperature fluctuations, salt, vibration, CO_2, microorganisms	Weight loss of sample, surface reflectivity, measurement of dimensional changes, chemical analysis	Cleaning, impregnation with resins, removal to controlled environment
Ceramics and glass	Surface erosion, surface crust formation	Acid gases, especially fluoride-containing	Moisture	Loss in surface reflectivity and light transmission, change in thickness, chemical analysis	Protective coatings, replacement with more resistant material, removal to controlled environment
Paints and organic coatings	Surface erosion, discoloration, soiling	Sulfur oxides, hydrogen sulfide	Moisture, sunlight, ozone, particulate matter, mechanical erosion, microorganisms	Weight loss of exposed painted panels, surface reflectivity, thickness loss	Repainting, replacement with a more resistant material
Magnetic storage media	Loss of signal, tape failure	Particles	Moisture, abrasion, wear	Signal quality, physical and chemical analysis	Removal to controlled environment, periodic cleaning and recopying

Paper	Embrittlement discoloration	Sulfur oxides	Moisture, physical wear, acidic materials introduced in manufacture, microoganisms	Decreased folding endurance, pH change, molecular weight measurement, tensile strength	Synthetic coatings, storing in a controlled atmosphere, deacidification, encapsulation, impregnation with organic polymers
Photographic materials	Microblemishes	Sulfur oxides	Particulate matter, moisture	Visual and microscopic examination.	Removal to controlled environment
Textiles	Reduced tensile strength, soiling	Sulfur and nitrogen oxides	Particulate matter, moisture, light, physical wear, washing	Reduced tensile strength, chemical analysis (e.g., molecular weight), surface reflectivity	Replacement, use of substitute materials, impregnation with polymers
Textile dyes	Fading, color change	Nitrogen oxides, ozone	Light, temperature	Reflectance and color value measurements	Replacements, use of substitute materials, removal to controlled environment
Leather	Weakening, powdery surface	Sulfur oxides	Physical wear, residual acids introduced in manufacture	Loss in tensile strength; chemical analysis	Removal to a controlled environment, consolidation with polymers, replacement
Rubber	Cracking	Ozone	Sunlight, physical wear	Loss in elasticity and strength, measurement of crack frequency and depth	Addition of antioxidants to formulation, replacement with more resistant materials

Table III Characteristics of Local Air Pollution and That Transported from Distant Sources

Pollutant or effect	*Transported from distant sources*	*Local*
Pollutant concentration patterns	Low concentrations and uniform distribution	High to moderate concentrations and strong gradients in time and space
Sulfur oxides	Tendency of SO_2 to be oxidized to particulate sulfates	Exists primarily as SO_2; however, under light winds and stable atmospheric conditions conversion to particulate sulfate possible
Nitrogen oxides	Significant conversion to particulate nitrates	Exist primarily as NO and NO_2; under low wind speed, stable conditions, and sunlight, possible conversion to organic or inorganic nitrates in particulate form
Particulate matter (includes aerosols)	Only the smallest primary particle sizes persisting; large component of material converted from gases and vapors to particulate form, such as sulfates	Exists in wide range of sizes which may be bimodal; particles capable of producing surface soiling and participating in the formation of corrosion layers (e.g., black crust on stone)
Ozone and other oxidants	Ozone and other oxidants produced from hydrocarbons and NO_x over moderate to long-range transport in presence of sunlight	Formation of ozone and other oxidants likely only under low winds and sunlight if precursors are present
Dry acidic deposition	Dry deposition of acidic particles (for example, sulfates) possible	Dry deposition of acidic particles possible, especially under stable conditions, often enhanced by moist surfaces
Acidic precipitation	Acidic rain formation may be predominantly through droplet condensation around acidic particles	Acidic rain formation may be predominantly through rain washout of acidic particles and pollutant gases
Acidic fog (includes liquid aerosols)	Acidic fog may be formed by drop condensation around small acidic particles or other acidic condensation nuclei	Same as that transported from distant sources

a series of simplified mechanisms that the authors believe occur in one form or another. These mechanisms are based upon the presence of acidic gases such as SO_2 and NO_2, their transformation products, and moisture in some form.

Dry gas, dry surface: An acid gas is absorbed on a material surface (for example, building stone) and later exposure to moisture as dew or rain forms acids which attack the material.

Dry gas, wet surface: An acid is absorbed in moisture (condensed dew or collected precipitation) already on surfaces and results in acid attack.

Large dry particle, dry surface: Large particles containing acid components fall on the material's surface and lead to damage directly. An example would be acid-containing soot from an oil-fired boiler.

Small particle, dry or wet surface: A small particle containing acidic compounds such as sulfuric or nitric acid salts capable of reacting with moisture to form acids settles on or impacts on a dry or wet surface and subsequently leads to acid attack.

Acidic precipitation: Rain or snow containing acidic components falls on the material surface and leads to damage directly.

The result of each of these scenarios is acidic deposition capable of damaging materials.

3. Laboratory versus Field Studies

The effects of air pollution on materials have been studied under both laboratory and field conditions. In laboratory studies, the conditions of exposure can be controlled, and the specific effects of a single pollutant or environmental parameter can be isolated. However, in order to produce measurable material damage in a reasonable time period, the material is often exposed continuously to severe environmental conditions (e.g., extremely high pollutant concentrations and high humidity), completely unrepresentative of field conditions. Furthermore, the exposure conditions are programmed through predetermined cycles which may only remotely resemble the complex interaction of temperature, humidity, surface wetness, sunlight, pollutant concentration, and other environmental factors occurring in the ambient atmosphere. Because of this, laboratory experiments have thus far been unable to present a true picture of the effects of pollutants under conditions of long-range transport, where such transformation would have ample opportunity to take place.

Field studies normally consist of exposing samples of materials to ambient atmospheres representing various combinations of pollutant concentrations and other environmental factors. By comparing damage level (e.g., loss of surface material) with pollutant concentration and other environmental factors (e.g., humidity, "time of wetness," or pH of rainwater) statistical models may be developed for the damage. The principal difficulties with this approach are the following:

- Materials exposed may not represent materials in actual use.
- In normal use materials are found in combination. Field studies seldom include interactions of other materials in contact with test materials.
- Damage is a complex function of many environmental conditions, and the effect of one condition is difficult to isolate.
- Measured variables may be interrelated (e.g., pH of rain may be dependent upon SO_2 level).

4. Measurement of Materials Damage

Material damage is usually measured by noting quantitative changes in some physical or chemical feature of the material (e.g., weight or thickness of a sample, surface color or reflectivity, chemical analysis, and characterization of corrosion products) (Table II).

B. Damage to Materials

A wide range of sensitive materials can be affected by air pollution. Metals, masonry, paint and other coatings, other materials (paper, photographic materials, textiles, and leather), and cultural property (historically and culturally valuable structures and objects) form classes of materials that are widely used and are economically important.

1. Metals

The atmospheric corrosion of metals is generally an electrochemical process governed by diffusion of moisture, oxygen, and acidic pollutants (e.g., SO_2) to the surface. The United States Environmental Protection Agency (EPA) Draft Criteria Document for Sulfur Oxides and Particulate Matter *(49)* provides a review of the mechanisms governing the corrosion of metals and other materials in the presence of SO_2 and moisture.

The position of metals in the electrochemical series determines their relative reactivity. However, the solubility of the particular metal salt and the stability of the metal oxide coatings that tend to form in the atmosphere determine their ability to corrode as a result of acidic deposition. For example, aluminum is high in the series, but aluminum oxide coatings that form in the atmosphere resist corrosion even in the presence of significant amounts of acidic deposition. However, even aluminum may be pitted in atmospheres containing sea salt or large acidic particles.

Thermodynamic considerations governing electrochemical corrosion are conveniently examined with the help of Pourbaix potential-pH diagrams (Figure 6) *(52a, 52b)*. By plotting electrical potential against solution pH, one can indicate regions of stability for various chemical species. In simplified form, when reactions to form soluble species occur, one has "corrosion." When the free metal is stable, the region is designated "immune" to corrosion. When a chemically stable oxide or salt film forms on the surface, leaving the metal resistant to subsequent attack, the region is "passive." Caution must be exercised in interpreting these diagrams since kinetic factors with nonequilibrium behavior may govern corrosion.

Moisture is always required for metal corrosion, each metal tending to have a critical humidity above which corrosion tends to accelerate.

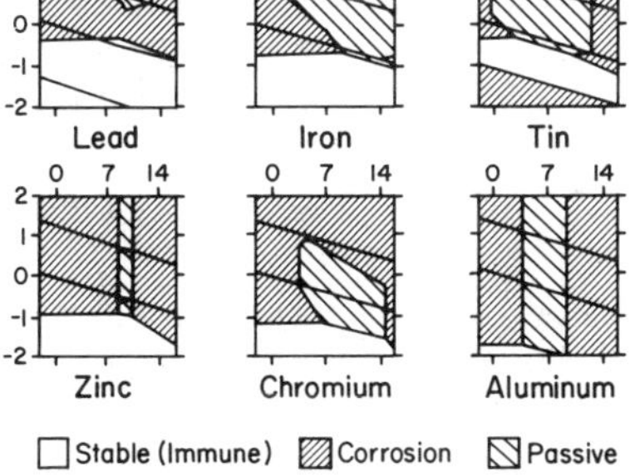

Figure 6. Pourbaix diagrams for various metals. The ordinate is in volts (electron potential versus standard hydrogen electrode), and the abscissa is in units of pH. The upper thin diagonal line is the O_2 evolution line; the lower line is that for H_2 evolution [After Pourbaix *(52a)* and Kruger *(52b)*].

Depending on the specific metal, these critical humidities are in the range of 60–80% relative humidity (RH). The relative length of time a metal surface is wet (time of wetness) is the single most important variable affecting the acceleration of corrosion by acidic deposition. Some workers *(49)* have found that hygroscopic corrosion products (e.g., iron sulfate) cause metal surfaces to remain wet at lower RH than if these products were not present.

a. *FERROUS.* Corrosion of iron and steel in polluted atmospheres has received a great deal of attention over the years. Steel, unless it is an alloy designed for unprotected exposure, is usually coated by painting or plating (e.g., zinc) when used in outdoor exposures. Data on unprotected iron and steel corrosion provide valuable information on the relative importance of acidic deposition components and the mechanisms causing damage. The Pourbaix diagram for the iron system is given in Figure 6.

(1) Deterioration Mechanisms. Many theories on the principal chemical reactions that describe iron and steel corrosion in the presence of SO_2 and moisture have been advanced. A variety of chemical routes are possible, and it is likely that several of them operate in actual atmospheric exposures.

According to Nriagu *(53)*, once corrosion has been initiated, the progress of the reaction is controlled largely by sulfate ions produced from the oxidation of absorbed or adsorbed SO_2. However, the actual mechanism of SO_2 oxidation on the surface is poorly understood. The work of Johnson *et al.* *(54)* appears to show that sulfur and sulfates are only minor constituents of the corrosion products of steel. They exposed mild steel samples in two urban areas near Manchester, England. One area was heavily polluted, and the other was lightly polluted. Scanning electron microscopy, energy dispersive x-ray analysis, and x-ray diffraction analysis of corrosion products showed them to be predominantly γ-$Fe_2O_3 \cdot H_2O$, α-$Fe_2O_3 \cdot H_2O$, and α-FeOOH. Some minor amounts of sulfur were found in a few of the samples. The relative amount of corrosion produced was strongly dependent on whether or not the sample was initially wet at the beginning of the exposure. It is possible that any sulfates formed were soluble and washed away.

A set of laboratory experiments attempting to approximate field conditions was conducted by Haynie *et al.* *(55)*. Various materials were exposed to controlled pollutant concentrations and moisture conditions at levels that encompased those found in ambient urban atmospheres. Sunlight and the formation of dew were also simulated.

Steel corrosion was determined in terms of weight loss of the steel panels by chemically removing the corrosion products, and the results showed that there was a strong, statistically significant relationship between steel corrosion and SO_2 concentration, together with high humidity.

(2) Field Studies and Damage Functions. A problem inherent in field studies is that iron and steel corrosion occurs even in unpolluted atmospheres, and the impact of specific acidic deposition scenarios is difficult to isolate completely. Therefore, the effects of a given pollutants can only be inferred by statistical treatment of the data.

Haynie and Upham *(56, 57)* in a series of experiments exposed steel samples of varying compositions for 1 and 2 years in field exposure studies. Pollutants of interest were SO_2, total suspended particulate matter, and the sulfate and nitrate content of the particulate matter. An empirical function was developed to relate sulfate in particulate matter, and humidity, to corrosion. However, the authors believed that SO_2 rather than sulfate was the causative agent in producing corrosion and have transformed the relationship from a linear regression between sulfate and SO_2 to the following one based on SO_2:

$$\text{cor} = 325\sqrt{t}\ e^{[0.0027 SO_2 - (163.2/RH)]} \tag{16}$$

where cor = depth of corrosion, μm; t = time, years; SO_2 = SO_2 concentration, $\mu g/m^3$; and RH = average relative humidity, %.

The relationship between pseudocorrosion rate (cor/$\sqrt{t}$), relative humidity, and SO_2 concentrations shows that the corrosion rate is much more sensitive to humidity than to SO_2, especially at levels of SO_2 normally experienced in urban areas.

For example, if one were comparing relative corrosion at 55% relative humidity in two areas with average SO_2 levels of 100 and 150 $\mu g/m^3$ (a very significant difference in relative air pollutant levels), the difference in relative corrosion would be approximately 3 pseudocorrosion units. On the other hand, if one were comparing relative corrosion at a constant SO_2 level of 100 $\mu g/m^3$ between two areas with a moderate difference in average relative humidity (55% and 65%), the difference in relative corrosion rate would be approximately 15 pseudocorrosion units.

This damage function shows that the sensitivity of corrosion to humidity is far greater than that for SO_2, especially at SO_2 levels normally experienced in urban areas. A number of damage functions relating steel corrosion to SO_2 and humidity (or time of wetness) developed by other workers have been summarized by EPA *(49)* and Haagenrud *et al. (51)*.

A study of material damage in the St. Louis, Missouri, area in 1974–1975 by Mansfield *(58)* included the use of special atmospheric corrosion monitors which measure the length of time that a corrosion panel is wet enough for electrochemical corrosion to take place (time of wetness). His sample exposure array included weathering steel, galvanized steel, house paint, and Georgia marble. Concentrations of SO_2 measured in this study were an order of magnitude lower than those measured by Upham's earlier study *(56)*. Mansfield was not able to show any significant correlation between corrositivity and pollutant levels.

Some of the experiments of Vernon *(59)* showed that moist air polluted with SO_2 and particles of charcoal produced corrosion much more rapidly than air containing SO_2 and moisture alone. He reasoned that the effect of the particles was primarily physical in that they increased the SO_2 concentration. Sanyal and Singhania *(60)* stated that particulate matter had a "profound" effect on corrosion rates. They believed that the influence of particulate matter on corrosion was related to its electrolytic, hygroscopic, and acidic properties and its ability to absorb corrosive pollutant gases. Although these laboratory studies appear to show a strong influence of particulate matter with corrosion, field studies have not confirmed this effect.

Haynie *(61)* has attempted to address the effects of small particles on materials. Lacking a significant body of experimental data, he has dealt with the question theoretically by utilizing data on deposition velocities. He considered four species of small particles: carbon, sulfuric acid, ammonium sulfates, and ammonium nitrate. Experimental data from one study *(62)* supported the chemical models for damage. Furthermore, in terms of calculated pollutant fluxes, SO_2-induced damage will tend to exceed H_2SO_4 effects in most urban areas.

In summarizing several years' work in Norway, Haagenrud *(63)* states that monthly corrosion rates for carbon steel are strongly influenced by long-range transport of both acidic precipitation and SO_2. However, episodes of precipitation of $<$pH 4 occur so seldom that these episodes do not strongly influence long-term corrosion rates. Similarly, episodes of high SO_2 concentration also affected monthly corrosion rates but had little effect on long-term values because they occurred so seldom.

b. *NONFERROUS METALS.* The corrosion rates of commercially important nonferrous metals in polluted atmospheres are generally less than those for steel but cover a wide range.

(1) Aluminum. Aluminum is quite resistant to SO_2-related acidic

deposition. However, deposited particulate matter may produce a pitted or mottled surface in the presence of SO_2 and moisture. In view of the reductions in emissions of SO_2 and particulate matter, especially larger particles or agglomerates that could act as centers for corrosion initiation, SO_2-related acidic deposition and surface corrosion of aluminum do not appear to be significant problems *(64)*.

(2) Copper. Copper and copper alloys in most atmospheres develop thin, stable surface films, which inhibit further corrosion. Initial atmospheric corrosion is a brown tarnish of predominantly copper oxides and sulfides which can thicken to a black film. After a few years, the familiar green patina forms. Analysis of this surface layer indicates it to be either basic copper sulfate or, in marine atmospheres, basic copper chloride. However, in coastal urban areas, the sulfate may still predominate (e.g., the Statue of Liberty) because of the continuous availability of SO_2 over many years. Nevertheless, both the sulfate- and chloride-based patinas are generally resistant to further attack.

In a series of studies, Graedel *et al. (64a, 64b)* demonstrated that carbonyl sulfide (COS) in wet air corroded copper at a rate similar to that for hydrogen sulfide. The rate was approximately linear with total exposure. Also observed were the rate-enhancing roles of solar radiation and ozone. Graedel concluded that the high rates of corrosion, i.e., copper sulfidation, observed in urban areas were the result of a complex sequence of multicomponent photochemical processes.

(3) Zinc. Zinc is used primarily for galvanizing steel to make it resistant to corrosion in the atmosphere and as an alloying metal with copper to produce brass. Zinc as a coating on steel is anodic with respect to steel such that when zinc and steel are in contact with an electrolyte, the current flow protects the steel from corrosion at the expense of some oxidation of zinc.

Because of its economic importance, the behavior of zinc in the presence of acidic deposition has been studied intensively by a number of workers. Guttman *(65)* carried out long-term measurements of atmospheric corrosion of zinc from which he developed an empirical damage function for zinc corrosion in relation to SO_2 concentrations and time of wetness.

Time of wetness was measured by means of a dew detector. SO_2 was measured by lead peroxide sulfation candles and conductiometric SO_2 measurements. Guttman's empirical damage function is

$$Y = 0.005461(\mathrm{A})^{0.8152} \times (\mathrm{B} + 0.02889) \qquad (17)$$

where Y = corrosion loss, mg, for a 3- by 5-in. panel; A = time of wetness, hours; and B = atmospheric SO_2 content during the periods that the panels were wet, ppm.

Haynie and Upham *(66)* carried out a zinc corrosion study in eight cities, where zinc panels were exposed and where data on SO_2, temperature, and humidity were collected concurrently. They developed the following empirical damage function relating zinc corrosion to SO_2 levels and relative humidity.

$$y = 0.001028\,(\mathrm{RH} - 48.8)\,\mathrm{SO_2} \tag{18}$$

where y = corrosion rate, m/year; RH = average annual relative humidity; and SO_2 = average SO_2 concentration, μg/m^3.

The damage coefficients for these two functions plus functions developed from four other studies were compiled by EPA *(49)*. These coefficients are compared in Table IV.

Benarie *(68a)*, attempting to rationalize the many damage functions proposed for zinc, used the well-known corrosion expression

$$M = at^b \tag{18a}$$

where M is the metal loss during exposure time t. The two constants a and b are assigned physical significance and are derived statistically. If a is given in terms of time-of-wetness (t_w) where

$$t_w = f \cdot t \tag{18b}$$

then

$$a = f^b A \tag{18c}$$

Table IV Experimental Regression Coefficients with Estimated Standard Deviations for Small Zinc and Galvanized Steel Specimens Obtained from Six Exposure Sites

Study	*Time-of-wetness coefficient (μm/year)*	*SO_2 coefficient[a] (μm/year)/(μg/m³)*	*Number of data sets*	*Reference*
Field Studies				
CAMP	1.15 ± 0.60	0.081 ± 0.005	37	*66*
ISP	1.05 ± 0.96	0.073 ± 0.007	173	*67*
Guttman	1.79	0.024	<400	*65*
Guttman and Sereda	2.47 ± 0.86	0.037 ± 0.008	136	*68*
St. Louis	2.36 ± 0.13	0.022 ± 0.004	153	*58*
Chamber Study				
Haynie *et al.*	1.53 ± 0.39	0.018 ± 0.002	96	*55*

[a] 1 ppm SO_2 = 2620 μg/m^3 SO_2.

where A may be approximated as a linear function of the SO_2 and Cl^- deposition. For zinc

$$A \text{ (weight loss per wetness year)} = 4.8 + 0.53[SO_2 + Cl^-] \quad (18d)$$

The exponent b characterizes the diffusivity of the corrosion product layer is given by

$$b = 1.55 - 0.142(\text{average rain pH}) \quad (18e)$$

2. Stone

The term *masonry* is applied to a large number of building and decorative materials exhibiting a broad range of surface activities in response to physical and chemical stresses imposed by the environment. The importance of acidic deposition to this class of materials may be related to the effect produced directly on a single material (e.g., limestone or marble) or to direct or indirect damage to composite masonry systems. An example of direct damage to composite systems is the rusting of steel reinforcing bars in concrete, which expand and crack the concrete. Indirect damage includes damage to brick–mortar systems in which the relatively reactive mortar is damaged directly by acidic materials and rainfall, and the salts released by these reactions then diffuse into the brick, causing stress and subsequent spalling.

Comparative photographs taken over extended periods of time for monuments exposed to polluted atmospheres (Fig. 7) are often said to illustrate the accelerated rate of deterioration occurring in this century. However, caution must be exercised in interpreting such photographs since they do not provide information on the degree to which chemical transformation without loss of material may have occurred (see Sec. II,B,2a). Indeed, for the example cited above, there is clear evidence of black crust formation with attendant spalling, despite the superficially good condition of the figure in 1913.

a. Deterioration Mechanisms. Stones composed almost entirely of calcium carbonate (limestone, marble, travertine, etc.) or those whose cementing material is calcium carbonate are particularly vulnerable to damage from acidic deposition.

The attack of sulfur dioxide on such carbonate stones has been studied for over a century. Yet, no quantitative relationship has been developed between ambient SO_2 levels and resulting materials damage.

The general decay mechanism includes aerodynamic factors con-

Figure 7. Sixteenth Century marble sculpture from the Renaissance facade of Castle Horst, Federal Republic of Germany, photographed in 1913 and 1984 *(68b)*.

trolling delivery of SO_2 to the stone surface, oxidation of SO_2 to sulfate and the subsequent reaction with the carbonate surface, mechanical stress by which reaction products destroy the stone structure, and removal of the stone and its alteration products by rainfall and other weathering phenomena *(69, 70)*.

Although the primary air pollutants causing damage to stone are sulfur compounds, a comprehensive decay mechanism must include the roles of nitrogen compounds, carbon dioxide, and water. For the carbonate stone–sulfur compound system three general modes of attack obtain:

Gaseous SO_2:

$$SO_2 + CaCO_3 \rightarrow CaSO_3 + CO_2 \tag{19}$$

$$CaSO_3 + 1/2O_2 \rightarrow CaSO_4 \tag{20}$$

Wet deposition:

$$H_2SO_4 + CaCO_3 \rightarrow CaSO_4 + H_2O + CO_2 \tag{21}$$

Dry deposition, the reaction between sulfates in particulate matter and calcium carbonate, either in the form of sulfuric acid and as in wet deposition or in the form of ammonium sulfates *(71)*:

$$(NH_4)_2SO_4 + CaCO_3 \rightarrow CaSO_4 + (NH_4)_2CO_3 \tag{22}$$

$$NH_4HSO_4 + CaCO_3 \rightarrow CaSO_4 + NH_4HCO_3 \tag{23}$$

The anhydrous $CaSO_4$ is hydrated to form gypsum, which is highly susceptible to surface erosion.

Humidity plays a key role in all aspects of the interactions of SO_x with carbonaceous stone. In autoradiographic experiments using sulfur-35, Spedding *(72)* showed surface saturation of olitic limestone samples by SO_2 at 81% RH occurring in less than 10 min. However, at the same concentrations, but at 11% RH, only a few distinct sites showed reaction after 20-min exposure, with approximately 25% of the total SO_2 uptake measured for the high humidity case.

b. *Field Studies.* The accelerated decay of stone buildings and monuments in highly industrialized areas has been documented by comparison of current condition with historic photographs and plaster casts. Photographs taken in 1908 and 1969 of a sandstone sculpture carved in 1702 in Westphalia, West Germany, demonstrate a dramatic loss of material in the past 60 years with virtual obliteration of the object *(73)*. Similarly, comparison of a plaster cast made in 1802 with a photograph taken in 1938 demonstrates substantial deterioration of a sculpture from the west frieze of the Parthenon in Athens, Greece *(74)*. A detailed account of the restorations on the Acropolis in Athens, Greece, and measurements of the thickness of gypsum layers formed on its exposed marble surfaces is presented by Skoulikidis *(75)*. The deteriorating condition of the Caryatids of the Erechthion on the Acropolis led to their replacement with replicas and their removal to the controlled environment of the Acropolis Museum *(76)*.

Few quantitative studies of air pollution damage to stone have been reported, although the increased rate of erosion of marble tombstones in the urban environment of Edinburgh, Scotland, was observed as early as 1880 *(77)*. In a study of tombstones in United States National Cemeteries, Baer and Berman *(78)* developed methodology for measuring damage to marble headstones exposed to the environment for 1–100 years. Their data base consists of measurements of some 3900 stones in 21 cemeteries distributed throughout the United States. The factors affecting damage rates include grain size, total precipitation, and local air quality.

In the United States, measured rates of marble deterioration have generally been small, on the order of 2.0 mm per 100 years *(73)*. This is substantially below values reported for stones exposed in urban areas in Europe, although direct comparison is difficult because the stones exposed in Europe are generally more reactive.

Comparing the condition of similar samples of sandstone exposed in different areas of Germany for about 100 years, Luckat associated large differences in observed deterioration with trends in local air quality *(79, 80)*.

A series of measurements made at St. Paul's Cathedral, London, England, on the Portland stone (biosparite limestone) balustrade demonstrate a high rate of weathering *(81)*. Using lead plugs filled in 1718 in openings in the stone as base level references revealed a mean rate of lowering of 0.078 mm $year^{-1}$ for the period 1718–1980. The balustrades represent conditions of exposed rain flow. Similarly, use of a microerosion meter (dial micrometer gauge mounted on reference studs) obtained a current erosion rate of 0.139 mm $year^{-1}$ for six sites on the cathedral. These sites represented drip erosion zones. Though the two sets of data are not strictly comparable, both represent substantially higher rates of loss than observed for marble in the United States. The results of these and other measurements of stone damage have been reviewed and tabulated by Livingston and Baer *(82)*.

3. Concrete

World production of concrete amounts to some 3 billion m^3/year. Many important structures, e.g., bridge decks, highways, military installations, and naval shore structures, suffer from severe durability problems *(83)*.

The highly alkaline nature of cement–concrete leaves such surfaces vulnerable to acidic deposition. Spedding *(72)*, reporting on the contamination–decontamination of laboratory surfaces accidentally exposed to sulfur–sulfur dioxide, observed that good decontamination was obtained by simple water washing. This finding suggests that the reaction products of the deposition of SO_2 on concrete are water-soluble. The high volume of water flow through rain collection and distribution culverts in drinking water systems raises questions about the possible release of toxic materials leached from the concrete matrix.

Similar concerns have been expressed over the erosive effects of acidified streams on concrete bridge piers. The literature reveals only limited research on the effects of acidified water runoff on concrete durability.

Cements used in dams and culverts require a special formulation for sulfate resistance when exposed to concentrations in excess of 200 ppm of sulfate in water *(53)*.

Specialized concretes in which sulfur replaces cement as the binding agent have been developed by the United States Bureau of Mines for resistance to acid and salt attack and damage to freeze–thaw cycling *(84)*.

The role of chloride ion as a major contributor to corrosion of reinforced concrete is well established *(85, 86)*. The alkalis in the hardened cement passivate the reinforcing steel, and penetrating chlorides depassivate the iron. Other factors in corrosion of the steel include the development of electrolytic corrosion cells and the penetration of atmospheric O_2 through the concrete to the steel. The reaction of SO_2 and SO_4^{2-} with cement involves the formation of calcium sulfate and calcium sulfate aluminum hydrate (ettringite).

A survey of the effects of acid deposition on Portland cement concrete (PCC) revealed very little qualitative or quantitative information. However, there appeared to be an increase in the rate of deterioration of reinforced PCC structures in polluted areas *(86a)*.

4. Ceramics and Glass

Although enamels and glasses are quite resistant to chemical attack by air pollutants, damage has been observed. In a 3-year exposure study of porcelain enamels placed in seven United States cities, some change in surface condition of the enamel was observed although the base metal was protected *(87)*.

Fluorides, especially HF, are capable of attack on a wide variety of ceramic materials and glasses. Legislation restricting fluoride emissions has, for the most part, eliminated new fluoride-induced damage.

Glass weathering is the process of removing alkali cations (e.g., Na^+ and K^+) from glass by reaction with water or sulfur dioxide. The reaction with water involves the exchange of sodium ions by hydrogen ions with the rate of reaction limited by the diffusion of sodium ions to the surface. The reaction with sulfur dioxide in the range 20–100°C in gas saturated with SO_2 involves the same process at approximately the same rate as with water alone *(88)*.

Perhaps the most serious glass damage problem is that associated with the decay of medieval stained glass windows. The unique composition of these glasses combined with their open exposure to the atmosphere makes them particularly susceptible to deterioration. This problem is discussed in detail in Section II,C,3.

Properly fired brick is highly resistant to attack by air pollutants, but poorly fired brick is highly susceptible to chemical attack. Acidic solutions accelerate such damage, increasing the rate of reaction 10-fold over that of water alone. Residual sulfates from decay of mortars can combine with other salts to produce failure in brick *(89)*.

5. Organic Materials

Deterioration of organic materials exposed to the atmosphere is determined primarily by the effects of atmospheric oxygen, UV light, and atmospheric oxidants such as ozone. Painted surfaces are the most widely employed organic surfaces exposed to the atmosphere and are susceptible to some degree to acidic deposition. However, these effects are relatively minor when compared with natural environmental factors such as sun and precipitation *(55)*.

Paint damage from acidic deposition is strongly related to the paint formulation. Such factors as the ratio of pigment and extenders to film-forming ingredients determine the hardness, flexibility, and permeability of the surface. It has been shown that the presence of extremely high concentrations of SO_2, a reducing gas, can interfere with the drying process, which is an oxidation–polymerization reaction *(90)*.

The degradation of paper and textiles is dominated by three factors: light, humidity, and acidity. Paper and other cellulosic materials (e.g., cotton, linen, and rayon) are highly susceptible to acid hydrolysis at the glucosidic linkage in the cellulose chain. Among proteinaceous textile materials, silk is most susceptible to damage by light. In bright light, silks may lose 60% of their strength in as little as 8 weeks of exposure *(91)*.

a. PAINT. Paint consists of pigment and vehicle. Pigments, such as titanium dioxide and zinc oxide, provide color, hiding power, and durability. Sometimes fillers such as calcium carbonate and inorganic silicates are also added. The vehicle provides the film-forming properties of the paint and contains resin binders, solvents, and additives. Together, the pigment (along with fillers) and vehicle protect the underlying surface and enhance the appearance of the exposed surface. Air pollution may limit both of these functions by damaging the protective coating, thus exposing the underlying surface to attack or spoiling the appearance of the surface. The most important potential effects of SO_2 on paints are interference with the drying process and acceleration of the normal erosion process.

The primary effect of particulate matter on paint is soiling. Soluble salts such as iron sulfate contained in deposited particles can also produce staining. Chemically active large particles such as acid smut (or soot) from oil-fired boilers, mortar dust near building demolition sites, and iron particles from grinding operations can severely damage automotive paint *(92)*. The effects range from discoloration of the paint film to corrosion of the substrate in the vicinity of individual particles. Large particles becoming embedded in a freshly painted surface can act as wicks to transfer moisture and corrosive pollutants such as SO_2 to the underlying material surface.

Holbrow *(90)* has reported a number of experiments to determine effects of sulfur dioxide on newly applied paints. Drying time for various oil-based paints exposed to extremely high concentrations of SO_2 (1–2 ppm) was increased 50–100%. Thus far no experiments have been carried out on the effect of SO_2 on drying time of water-based latex paints.

(1) Field Studies. Campbell *et al. (93)* carried out an extensive study of paint erosion for a variety of paint types and exposure conditions (including exposure to SO_2 and O_3). Both chamber and field experiments were conducted. The researchers evaluated four important types of paint: acrylic latex and oil-based house paints, urea–alkyd coil coating; nitrocellulose–acrylic automotive refinishing paint, and alkyd industrial maintenance coating.

Generally, exposures to high concentrations (1 ppm of both SO_2 and ozone) produced statistically significant erosion rate increases compared to clean air (zero pollution) conditions. Oil-based house paint experienced the largest erosion rate increases. The greater susceptibility of oil-based house paint to SO_2 was attributed to the use of extenders such as $CaCO_3$ or metal silicates. Latex and coil coatings experienced moderate increases, and the industrial maintenance coating and automotive refinish experienced the smallest increases. In general, exposures to sulfur dioxide produced higher erosion rates than exposures to ozone. Unshaded panels eroded more than shaded panels. Exposures to 0.1 ppm pollutants did not produce significant erosion rate increases over clean air exposures. It should be noted that even these lower concentrations are high when compared with average concentrations found in the ambient air of urban areas.

In the field portion of this same study, painted panels were exposed at four locations with different environments: rural–clean air (Leeds, North Dakota); suburban (Valparaiso, Indiana); urban–sulfur dioxide-dominant (Chicago, Illinois); and urban–oxidant-dominant (Los Angeles, California).

In most cases, southern exposures produced somewhat larger erosion rates, a finding which agreed with the unshaded versus shaded results of the laboratory study. Oil-based house paint again experienced by far the largest erosion rate increases, followed in order by the urea–alkyd coil coating, latex house paint, industrial maintenance paint, and automotive refinish. Generally, the field exposures showed that the relative paint erosion rate was about the same for the sulfur dioxide-dominant as for the oxidant-dominant location, which appeared to contradict the chamber studies somewhat. However, the authors believed that differences in the pollutant mix at the two locations, and especially the presence of nitrogen oxides at the oxidant-dominant site, could have enhanced the erosion rate at this location, bringing it up to the level of damage at the sulfur dioxide-dominated location.

It is noteworthy that the oil-based house paint and urea–alkyd coil coating experienced the largest erosion rate increases in both the field and laboratory sulfur dioxide exposures. These coatings were the only ones that contained a calcium carbonate extender, a substance that is sensitive to attack by acidic materials.

(2) Damage Function. Spence *et al.* *(94)* summarized the results of paint exposure to several gaseous pollutants from the full-scale chamber studies reported by Haynie *et al.* *(55)* and discussed earlier in relation to metal exposures. Four classes of painted surfaces were evaluated: oil-based house paint, vinyl–acrylic latex house paint, vinyl coil coating, and acrylic coil coating. Paint erosion for the oil-based house paint, SO_2, and humidity were highly correlated. The vinyl and acrylic coil coating were unaffected, but blistering was noted on the latex house paint. It was not certain whether the blistering was the result of SO_2 or of moisture.

A multiple regression relationship was developed for the joint influence of SO_2 and relative humidity on the oil-based house paint:

$$E = 14.3 + 0.0151\ \mathrm{SO_2} + 0.388\ \mathrm{RH} \qquad (24)$$

E = erosion rate, μm/year; SO_2 = concentration of SO_2, μg/m^3; RH = mean annual relative humidity, %.

This relationship implies that paint erosion is significantly more sensitive to changes in humidity than to SO_2. However, one must be careful in using models based on accelerated chamber tests for actual exposures since they would predict that in an atmosphere with no SO_2 present, and an average relative humidity of 50%, the paint erosion rate would be about 34 μm/year. Assuming a typical paint thickness of 50 μm, the paint film would be completely eroded within 1.5 years. One may assume that the SO_2 and RH coefficients are correct but that the intercept may be in question.

b. PAPER. In addition to coatings, a wide range of organic materials are found to be susceptible to attack by atmospheric pollutants. These materials, including paper, photographic materials, textiles, and leather, were not considered in the Criteria Assessment Documents but are considered here even though the indoor locations in which they are normally found dictate gaseous transport mechanisms for deposition.

The role of sulfur dioxide in the deterioration of paper has been accepted since the 1930s. Early experiments *(95)* relied on unrealistically high SO_2 concentrations of 5000 ppm interacting with damp paper. Hudson and Milner *(96)* used sulfur-35 as a radioactive tracer to demonstrate that measurable amounts of SO_2 were rapidly deposited into paper. Working with concentrations of 10 ppm, Grant *(97)* showed that SO_2 deposition increased with increasing aluminum sulfate–resin sizing of the paper.

A comparative study of identical copies of twenty-five 17th- and 18th-century books in two English libraries, one in an unpolluted rural atmosphere in Chatsworth, the other in the badly polluted urban atmosphere of Manchester, revealed a significant increase in paper acidity in the Manchester library *(98)*. This acidity was greatest at the page edges and decreased greatly toward the center of the page, which might be considered the initial sheet acidity.

Wallpapers form an important part of the indoor surface area available for SO_2 sorption. Spedding and Rowlands *(99)* measured the sorption characteristics of polyvinyl chloride and conventional wallpapers on exposure to maximum initial SO_2 concentrations of 150 $\mu g/m^3$. Sorption depended largely on surface finish and design pattern with greater sorption by conventional wallpapers. The researchers suggested that SO_2 sorption accelerates the deterioration of wallpaper.

c. Photographic Materials. Under normal conditions of temperature and relative humidity, paper, acetate film, and other photographic materials are oxidized at a very slow rate. One of the most serious factors in the preservation of photographic materials is the presence of large quantities of oxidizing gases: hydrogen sulfide, sulfur dioxide, and to a lesser extent NO_x, peroxides, formaldehyde, and ozone *(100)*.

The effect of these pollutants is usually yellowing and fading of the silver of the image. The paper base may also be degraded and stained. Acidic gases degrade gelatin, paper, and the film base of negatives *(100)*.

Agfa produces a colloidal silver test strip which is 8 to 10 times more sensitive to gaseous pollutants than are ordinary photographic materials. In a survey of major libraries and archives using this technique, many examples of significant air quality problems were observed *(101)*.

d. Textiles and Textile Dyes. Sulfur oxides are capable of causing deterioration to natural and synthetic fibers. Cotton, like paper, a cellulosic fiber, is weakened by sulfur dioxide. Under circumstances in which sulfuric acid comes into contact with a cellulosic surface, the product of the reaction is water-soluble, with little tensile strength *(102)*. In field stress tests in St. Louis, Missouri, cotton duck exposed to varying SO_x levels showed a direct relationship between loss in tensile strength and increasing SO_x concentration *(103)*. Zeronian *(104)* exposed cotton and rayon fabrics under accelerated aging conditions of light and water spray with and without 0.1 ppm SO_2. Loss in strength was 13% in the absence of SO_2 and 22% in the presence of SO_2. In a study of nylon fabrics exposed to 0.2 ppm SO_2 under similar conditions, he found that they lost 40% of their strength under the SO_2-free conditions and 80% of their strength in the presence of SO_2 *(105)*.

The degradation of nylon-66 by exposure to light and air is increased by the addition of 0.2 ppm of SO_2 to the air. Chemical properties and yarn tensile properties both reflect this damage *(106)*. Results demonstrated that the mode of degradation is not changed, although SO_2 accelerates the rate of reaction.

Among proteinaceous textiles, silk is most vulnerable to the effects of light, acidity, and sulfur dioxide, demonstrating much greater loss in strength than wool *(91)*.

Damage to textiles has been attributed to NO_x *(107)*. Such damage has been caused by both loss of fiber strength and fading of textile dyes. Significant reduction in breaking strength and increase in cellulose fluidity were observed for combed cotton yarns exposed in Berkeley, California, to unfiltered air when compared to exposure to carbon-filtered air *(108)*. Both sets of samples were unshaded and exposed at a 45° angle facing south. Though the authors did not isolate the effects of individual pollutants, they implied that compounds associated with photochemical smog, especially NO_x, were the probable cause of increased damage.

In an EPA chamber study of the effects of individual pollutants on 20 dyed fabrics, it was demonstrated that NO_2 at 0.1–1.0 ppm produced appreciable dye fading, and SO_2 at 0.1–1.0 ppm caused visible fading on wool fabrics *(109)*. It was also concluded that higher temperatures and relative humidities increase dye fading and that the rate of fading as a function of exposure time appears to be nonlinear.

e. *Leather.* According to Parker *(110)*, Michael Faraday is attributed with having established in 1843 a link between the rotting of leather armchairs in the London (England) Atheneum Club and sulfur dioxide emitted by its gas illumination. Plenderleith *(111)*, Innes

(112), and Smith *(113)* describe the sequence of chemical deterioration for leather and consider possible mitigative actions.

It has been observed that leather initially free of sulfuric acid accumulates up to 1% acid by weight per year if exposure to an atmosphere containing SO_2. The mechanism is thought to involve the metal ion catalyzed conversion to sulfuric acid of the SO_2 absorbed by the collagen of the leather. Using sulfur-35-labeled SO_2, Spedding *et al.* *(114)* showed that it is sorbed evenly over the leather surface, with the limiting factor in uptake being gas-phase diffusion to the surface.

C. Cultural Property

The United States has over 6000 museums, historical societies, and related institutions; more than 10,000 entries on the National Register of Historic Places; and in excess of 26,000 libraries and archives of substantial size *(115)* and an estimated historic building stock of 200,000 structures *(115a)*. Light, oxidation, fluctuations in humidity, and chemical pollutants threaten this precious cultural heritage.

Damage to cultural property cannot be quantified in simple dose–response terms. Just as an electrical component may require replacement because of corrosion of a fraction of its mass or stress–corrosion fracture may lead to failure of a mechanical system, damage to texture of sculpture or the surface of a fresco exposed to the environment diminishes its aesthetic importance to a greater degree than the amount of material damage. Still more critical is the circumstance that, for most cultural property, replacement is impossible. What is lost is lost.

1. Architectural Monuments

Historic and artistic structures represent the single most visible aspect of our history and culture. For the United States, legislation providing a mandate for preservation began with the Antiquities Act of 1906, followed most recently by the Historic Preservation Act Amendments of 1980. In Canada, the Archaeological Sites Protection Act and the Historic Sites and Monuments Act were adopted in 1953.

Architectural monuments are universally threatened by the effects of pollution and urbanization, as well as by weathering cycles and other natural phenomena *(116)*. Although damage to these monuments is frequently attributed to acidic precipitation, no clear evidence of a cause and effect relationship between acidic precipitation and damage to a specific monument exists. In general, it appears that although acidic deposition can effect significant damage to cultural property, the sources are predominantly of local origin.

In 1984, the Environment Committee of the House of Commons (England) agreed with witnesses who attributed the "destruction of carvings and other important features of priceless buildings" to SO_2 emissions and recommended that "the Government give urgent and immediate consideration to the cost-benefit of preventing the avoidable erosion of both historic and modern buildings *(117)*."

2. Museums, Libraries, and Archives

As discussed previously, the sorption of SO_x and NO_x by organic materials in the indoor environment is well established. In some cases, as in paper and leather embrittlement, dye fading, and "red–ox" blemishes on microfilm, a direct relationship between pollutant sorption and damage has been established. This has led major museums, libraries, and archives to install scrubbers for the removal of acid gases.

Among the systems in use are activated charcoal and Purafil (activated alumina impregnated with $KMnO_4$) dry scrubbers and wet scrubbers. Such systems have been introduced as part of new construction or retrofitted at the National Gallery (London), the Library of Congress (Washington, D.C.), the Newberry Library (Chicago), and the National Gallery (Washington, D.C.). Many other collections of cultural artifacts are preparing for the eventual retrofitting of their air handling systems to use scrubbers for removing air pollutants. In Table V the results of air quality measurements in archives, libraries, and museums are presented *(118–121)*.

The universal nature of concern for the effects of polluted air on cultural property is reflected in a Japanese study of ambient and indoor SO_x and NO_x concentrations for buildings where important screen and panel paintings are housed *(122)*. Six sites in Kyoto were investigated. Average concentrations of SO_x and NO_x were found to be about one-third of those in Tokyo. Seasonal concentrations for SO_x peaked in winter and were highest for a site near a dyeing factory whose liquid wastes emitted SO_2. The NO_2 concentrations were found to be more evenly distributed throughout the city. Tight buildings showed higher NO_x levels indoors than were found for ambient conditions. Although they did not cite specific examples of damage, the authors called for protective measures to prevent air pollution damage to paintings.

Cass and co-workers *(123, 124)*, studying the effects on watercolor

Table V Measured Indoor–Outdoor Pollution Levels for Archives, Libraries, and Museums

			Concentration		
Institution[a]	*Pollutant*	*Dates Measured*	*Exterior*	*Interior*	*Filtration System (Reference)*
National Archives and Records Service[a] (Archives Building)	SO_2	Nov. 1977	32–40 ppb	<3 ppb (10× reduction)	Particulate *(118)*
National Archives and Records Service[a] (Archives Building)	SO_2	Dec. 1982–Jan. 1983	7–34 ppb[b] daily avg.	2–25 ppb	Particulate *(118)*
National Gallery[a] (East Building)	SO_2	Feb. 1983	7–34 ppb[b] daily avg.	≤1 ppb	Water wash *(118)*
Library of Congress[a] (Madison Building)	SO_2	Jan. 1983	7–34 ppb[b] daily avg.	≤0.5 ppb	Purafil *(118)*
Tate Gallery (London, U.K.)	SO_2	1980–1983	12–80 ppb	0–4 ppb	Activated carbon *(119)*
Victoria and Albert (London, U.K.)	SO_2	Jan.–Feb. 1983	22–60 ppb	3–42 ppb	None *(119)*
National Archives and Records Service[a] (Archives Building)	NO_x	Sept. 1977	41 ppb average	20–80 ppb	Particulate *(118)*
National Archives and Records Service[a] (Archives Building)	NO_x	Dec. 1982–Jan. 1983	10–527 ppb[b]	10–252 ppb	Particulate *(118)*
National Gallery[a] (East Building)	NO_x	Feb. 1983	40–92 ppb[b]	7–50 ppb	Water wash *(118)*
Library of Congress[a] (Madison Building)	NO_x	Jan. 1983–Feb. 1983	46–318 ppb[b]	4–154 ppb	Purafil *(118)*
National Archives and Records Service[a] (Archives Building)	O_3	Sept. 1977	97 ppb	0–42 ppb	Particulate *(118)*
National Archives and Records Service[a] (Archives Building)	O_3	Dec. 1982–Jan. 1983	1–21 ppb[b]	≤0	Particulate *(118)*
National Gallery[a] (East Building)	O_3	Feb. 1983	1–21 ppb[b]	≤0	Water wash *(118)*
Library of Congress[a] (Madison Building)	O_3	Jan. 1983	1–21 ppb[b]	≤0	Purafil *(118)*
Los Angeles County Museum (Los Angeles, California)	O_3	July 1982	200 ppb average	<10 ppb	Activated carbon *(120)*
Sainsbury Centre (Norwich, U.K.)	O_3	Sept. 1981	56 ppb maximum	40 ppb maximum	Particulate *(121)*

[a] Washington, D.C.
[b] Measured at 24th and L Streets, N.W., by the District of Columbia.

pigments, concluded that ozone at the concentrations found in photochemical smog can fade or alter the color of pigments, in particular alizarin-based lakes and yellow pigments used in Japanese woodblock prints and in watercolors, and that ozone could pose a threat to the preservation of works of art.

The effects of indoor air pollutants on cultural property are reviewed by Baer and Banks *(124a, 124b)*. Various attempts have been made to develop air quality standards for libraries and archives (Table VI) *(125–129)*. These standards are currently under American National Standards Institute review.

Table VI Air Quality Criteria for Archives, Libraries, and Museums

Authority installation	*Pollutant* SO_x	NO_x	O_3	*Particulate matter*	*Reference*
ASHRE[a]	Cannister-type filters or spray washers of chemical pollutants in outdoor air			85% DSM[i]	*125*
BML[b]	0	0	0	0	*125*
CCI[c]	Should not exceed 10 ppb.			85% ≥1 μm, 50% 0.5–1 μm	*126, 127*
LC[d]	Purafil system in use			85%	*125*
NBS[e]	1 μgm^{-3} (0.4 ppb)	5 μgm^{-3} (2.5 ppb NO_2)	25 μgm^{-3} (13 ppb)	75 μgm^{-3} TSP[k] (Hi Vol)	*125*
N-PNB[f]	*4 ppb*	≤10 μgm^{-3}	≤2 μgm^{-3}	High rating DSM[i]	*128*
ROM-C[g]	Charcoal or equivalent filtration to remove SO_x, NO_x, O_3			99% ≥10 μm 95% ≥1 μm	*125*
T[h]	≤10 μgm^{-3}	≤10 μgm^{-3}	0–2 μgm^{-3}	60–80 MBT[j]	*129*

[a] ASHRE American Society of Heating, Ventilating, and Air Conditioning Engineers.
[b] BML British Museum Libraries, London, U.K.
[c] CCI Canadian Conservation Institute, Ottawa, Canada.
[d] LC Library of Congress (Madison Building), Washington, D.C.
[e] NBS National Bureau of Standards, Washington, D.C.
[f] N-PNB Newberry Library, Chicago–P.N. Banks Planning Study.
[g] ROM-C Royal Ontario Museum Conference, Ontario, Canada.
[h] T G. Thomson *(129)*.
[i] DSM Dust spot method.
[j] MBT Methylene blue test.
[k] TSP Total suspended particulate matter.

3. Medieval Stained Glass

Some evidence exists that medieval stained glass exposed to the atmosphere has deteriorated more rapidly since World War II than in previous centuries. This accelerated deterioration has been attributed to the effects of air pollution *(130–133)* because gypsum and syngenite ($CaCO_4 \cdot K_2SO_4 \cdot H_2O$) are found in the weathering crust. However, such crusts are found even in locations with low SO_2 concentrations, suggesting that background SO_2 levels are sufficient to produce the sulfates observed. An alternative mechanism of decay that has been suggested is that storage of the windows under damp conditions during the war permitted the formulation of a fissured hydrated layer that led to enhanced corrosion after reinstallation of the windows. The sulfates found in the weathering crusts are thought to be by-products of the deterioration process *(134, 134a, 134b)*. In West Germany, a catalog of impacted stained glass windows has been published with detailed photographs of the types of damage observed on individual pieces of glass in the inventory of stained glass windows *(135)*.

A 1985 research report describes accelerated tests on lacquerlike coatings incorporating colorless inorganic fillers (mica or glass flakes) to reduce water or gaseous pollutant diffusion to the surface of the painted glass left *in situ (135a)*.

A broad range of preservation techniques has been employed, including lamination, coating with inorganic and organic materials, and isothermal glazing. In the latter process, the ancient glass is moved forward, to just inside the building, and colorless modern glass takes its place in the window space.

D. Conservation and Mitigation Costs

While noting the difficulties inherent in valuing damage to cultural property, a 1984 Dutch study provided estimates of replacement and prevention costs to various classes of air pollution impacted cultural property (Table VII) *(135b)*.

Further indication of the problem's magnitude is given by cost for mitigative actions taken for cultural property in West Germany (Table VIII). Similar cost estimates exist for national preservation programs in the United Kingdom, Greece, France, Italy, and the United States. For example, the Italian parliament designated an amount equal to $200 million in 1980 for a 5-year program to restore and maintain the ancient monuments in Rome *(136)*, and a British

Table VII Estimated Costs Associated with Air Pollution Damage to Cultural Property in the Netherlands[a]

Type of cultural property	*Basis for cost estimate*	*Period*	*Costs in Guilders (1984)*
Monuments	Restoration	Annual	15,000,000–30,000,000
Monuments	Damage	Total	120,000,000–200,000,000
Stained Glass	Prevention	Total	20,000,000
Carillions	Damage	Annual	250,000
Textiles	Damage	Total	20,000,000
Paper	Prevention	Annual	10,000,000
Paper	Damage	Total	620,000,000

[a] Damage costs from Feenstra *(135b)*. Total damage costs are based on man-year restoration estimates.

parliamentary committee estimates that restoration of the fabric of the Houses of Parliament will cost up to £5 million *(137)*.

1. Economic Implications

The possibility of determining the economic costs of air pollution's damaging effects has long attracted environmental policy makers. If reliable cost estimates could be developed for such effects in relation to the pollutant levels which produced them, it then might be possible to compare the costs for achieving various levels of air quality control through emission control with the cost savings from reduced damage, a significant step toward developing cost–benefit relationships for air pollution control. The many attempts to estimate costs associated with air pollution-induced material damage have been summarized by Yocom and Stankunas *(50)*. Without exception, all of the generalized estimates of material damage costs related to all types of air pollution existing at the time of this review are of questionable value. There are a number of reasons for this imprecision, which include the following:

- The difficulty of isolating the specific portion of damage and therefore the associated costs created by a given air pollution effect (discussed previously)
- Improper assumptions and inaccurate estimates of the quantities of materials in place and exposed to pollutants
- Unrealistic or improper scenarios of use, repair, and replacement of materials susceptible to air pollution damage, together with improper or inaccurate assignment of costs to the scenarios

Table VIII Estimated Costs Associated with Air Pollution Damage to Cultural Property in Federal Republic of Germany[a]

Location	*Objects*	*Measures*	*Period*	*Costs, DM*
Federal Republic of Germany	All municipal bronze monuments and sculptures	Desirable cleaning	Annual	4,000,000
	All metal sculptures in museums and open air	Desirable cleaning, conservation	Annual	1,000,000
	All medieval stained glass	Desirable conservation	10-Year cost estimate	200,000,000–300,000,000
	Artifacts in museums	Air conditioning with air improvement	During construction	15% of construction costs
Münster	Castle facade	Cleaning restoration, conservation	1965–1973	1,000,000
Cologne	Cathedral stained glass windows	Conservation	1978	448,000
Cologne	Cathedral facade	Cleaning, restoration, conservation	Annually 1977–1997	3,000,000–60,000,000 (estimated)
Freiburg	Cathedral stained glass windows	Restoration, conservation	1978	150,000
Ulm	Cathedral stained glass windows	Desirable restoration	Total cost	3,000,000

[a] After Schreiber *(80)*.

• Incomplete knowledge of substitution scenarios in which more expensive material systems may replace more susceptible materials
• Inadequate knowledge of the exposure conditions of susceptible materials, for example, coexistence of pollutants with other environmental effects such as moisture and temperature, and the physical aspects of exposure such as orientation and degree of sheltering

A study of Stankunas *et al.* *(138)* attempted to surmount many of these difficulties. In this study the quantities of potentially susceptible materials were determined within 357 randomly selected 100- by 100-ft square areas covering the Boston metropolitan area. Teams of observers using survey techniques determined the areas of various types of exposed painted surface; bare metal of several types; brick, stone, and concrete; and several other types of surfaces. Of the 357 areas selected, 183 were found to contain man-made structures. The total areas of each material found at the survey sites were extrapolated to the entire Boston metropolitan area. Then, by using air quality records for SO_2 in the Boston area, together with humidity data and published air pollution damage functions for given materials, the areawide damage was calculated. The excess painting costs for the Boston metropolitan area attributable to SO_2 damage for the year 1978 were estimated to be $31.3 million. This is equivalent to a per capita cost between $11 and $12. Costs for damage to zinc-coated materials were two orders of magnitude lower.

Haynie *(139)* estimated costs for damage to zinc-coated transmission towers and to galvanized roofing, siding, and guttering. By using SO_2–moisture damage functions for zinc corrosion and an estimate (based primarily on changes in wind speed with height) of the effect of height above ground on SO_2 deposition velocity, estimates were made of change in zinc thickness with time for the group of transmission towers. Then, by using several scenarios of painting, repair, and replacement, annualized costs, based on local SO_2 and humidity level, were estimated for mitigating the effects of the damage. Since the Tennessee Valley Authority (TVA) owned the towers, such costs could be internalized and were estimated to be 0.0028 mills/Kwh $\pm$ 0.0011 to be added to customers' electric bills. These estimates were based on an SO_2 concentration of 17 $\mu g/m^3$. If SO_2 levels were allowed to reach the ambient air quality standard of 80 $\mu g/m^3$, the annualized extra maintenance cost would rise to an estimated 0.0132 mills/Kwh $\pm$ 0.0052. This is a small fraction of the estimated costs of 5.5 mills/Kwh $\pm$ 2.4 for SO_2 emission control by flue gas desulfurization.

Cost estimates in the United States for nationwide damage to galvanized roofing, siding, and guttering required taking into account the relative quantity of these materials in place. One of the complicating factors in making this determination was the trend in recent years of replacing bare galvanized materials exposed to the outdoor atmosphere with coil coated galvanized steel or bare aluminum. Various models were used to convert data on shipments of the materials in question and anticipated use of alternate materials to a realistic picture of the amount of bare galvanized materials in these categories in 1980. Damage functions for the effects of SO_2 and moisture on zinc, together with estimates for the thickness of zinc coatings and various maintenance scenarios and their costs, were used to estimate per capita costs. These per capita costs were computed to be in the range of \$0.60 to \$1.50 with the best estimate being \$1.05 at an annual average SO_2 concentration of 30 $\mu g/m^3$. At the primary standard of 80 $\mu g/m^3$, the best estimate of per capita costs would be \$1.80.

2. Mitigative Measures

Assuming that air pollution produces significant damage to materials, the primary mitigative measure is to reduce the concentrations of pollutants in the ambient atmosphere. However, as has been implied in the foregoing discussion, the relative amount of damage and associated costs from individual pollutants are not known. Therefore, at present there is no basis for a given degree of control of air pollution and precursors that will eliminate or bring to acceptable levels the potential damage from a given source.

Given that there is some degree of damage to materials from air pollution, there are a wide range of mitigative actions that may be taken in response to damage. Table II lists several of these in relation to various material categories. The particular mitigative measure and likelihood of its implementation depend on many factors, including the following:

- Physical and chemical nature of the material
- Age and state of repair of the materials system
- Availability and cost of substitute materials
- Feasibility of isolating the object or surface of concern from the ambient environment
- Importance of aesthetics in the appearance of the material
- Impact of damage on structural integrity
- Attitudes of those responsible for the objects made of the materials in question regarding the relative importance of the damage

As stated earlier, material damage from air pollution is often indistinguishable from damage caused by the natural environment. However, chemical analysis of corrosion or damage products can often distinguish various damage mechanisms. In general, superimposing air pollution on these natural phenomena tends only to shorten the time before some mitigative measure must be considered. It does not change the mitigative actions themselves. Thus, mitigative measures taken to protect, replace, repair, and maintain materials exposed to the ambient environment generally will not change whether or not acidic deposition has an effect. Only the frequency of implementing these measures will change.

III. Effects on Indoor Air Quality

A. Introduction

The interest in indoor air quality, although slow to develop, has increased rapidly in recent years. This accelerated interest may be measured by the number of major reviews and books *(140–148)* on the subject which have been published just since 1981. The five volumes of proceedings from the Third International Conference on Indoor Air Quality and Climate held in Stockholm, Sweden, in August 1984 is another excellent source of information on indoor air quality *(149)*. A newsletter covering "items of interest to the indoor air quality community" was being published by the Lawrence Berkeley Laboratory of the University of California at the time of this writing *(150)*.

It has been long recognized that indoor exposures to air contaminants are far more important than outdoor exposures in terms of the time spent in each environment. However, except for occupational standards applicable primarily to the industrial work room, the principal environmental standards designed to protect human health from air pollutant exposure in the United States have been developed by the United States Environmental Protection Agency (EPA) for the outdoor ambient air. Paradoxically, the United States Clean Air Act does not preclude consideration of the indoor environment as part of the "ambient air," but regulations developed by the EPA *(151)* define ambient air as "that portion of the atmosphere, external to buildings to which the general public has access" *(146)*. The closest approximation to indoor air quality standards in the United States are the guidelines published by the American Society of Heating, Refrigeration, and Air Conditioning Engineers (ASHRAE) for building ventilation to produce acceptable indoor air quality. Such guidelines are sometimes incorporated in building design specifications *(140)*.

In northern Europe, where the need for energy-efficient housing developed earlier than in the United States, the problem of indoor air quality has been addressed much more squarely with air exchange standards and indoor air quality standards for such contaminants as formaldehyde and radon *(140)*.

Indoor air quality is a complex function of a number of variables, the most important of which are outdoor air quality, air exchange rate, indoor sources, and pollutant decay rates. A typical single compartment conservation of mass equation for a single pollutant shows the interrelation of these factors:

$$\frac{dC_i}{dt} = aC_o - aC_i - DC_i + S \tag{25}$$

where C_i and C_o are indoor and outdoor concentrations of the pollutant, a is the air exchange rate, D is the decay or deposition rate for the pollutant, and S is the source strength for the pollutant per unit volume. More complex models have been developed; for example, Özkaynak *et al.* *(152)* have developed multicompartment models with reactive chemistry. However, Equation (25) is adequate to illustrate the most important variables controlling indoor air quality.

From Equation (25) it is apparent that anything that reduces air exchange a increases indoor concentrations of pollutants generated indoors and reduces indoor concentrations of pollutants generated outdoors. Dietz and Cote *(153)* have developed a unique perfluorocarbon tracer–passive monitor system for measuring air exchange rate that has received wide acceptance among indoor air quality researchers. Air exchange rate is affected by a number of factors, the most important of which are building design and meteorological effects.

1. Building Design

The integrity of the building envelope is the most important variable affecting air exchange. The quality of construction and materials used determine in large measure the relative permeability of a structure. Specific features affecting air exchange rate are types of windows and doors, presence of a fireplace, use of vapor barriers, utility accesses, and types of heating, cooling, and ventilation systems and their modes of operation.

2. Meteorological Effects

The wind field acting upon a building provides the driving force for air infiltration. If openings such as windows and doors are located on the side facing the wind, the positive pressure forces infiltration on that side of the building. Blomsterberg and Harrje *(154)* believe that the locations of building openings can produce a factor of 2 range in infiltration rates.

Temperature differences between indoors and outdoors can provide the motive force for enhanced air exchange based on differences in air density. In the winter, with warm air indoors and cold air outdoors, the indoor air space produces a stack effect with warm air escaping through openings in the upper part of the building and air infiltrating at the bottom. Changes in humidity can cause wood alternately to swell and shrink, thus changing the dimensions of cracks and joints and the potential air exchange rate.

Weatherization techniques that reduce air infiltration, such as use of storm windows, weather stripping, and vapor barriers, affect indoor air quality. However, such retrofits reduce air infiltration on the average only about 30%. Thus, the weatherization of older buildings usually has only a limited effect on indoor air quality. Although indoor air quality data are sparse for pre- and postretrofitting of weatherization measures, such data, except for respirable particulate matter from smoking, tend to show only minor increases in indoor pollutants as a result of weatherization.

B. Categories of Indoor Pollutants

Table IX, based on the United States National Academy of Sciences (NAS) review *(140)*, summarizes categories of indoor air pollutants in terms of their principal source locations. This type of categorization is especially helpful when considering the relative importance of indoor and outdoor sources in relation to relative tightness of buildings as measured by air exchange rate.

As might be expected, all of the pollutants which are the basis of ambient air quality standards in the United States (SO_2, CO, particulate matter, O_3, NO_2, and Pb) have outdoor sources. However, several of them also have important indoor sources: CO and NO_2 from indoor combustion such as occurs in gas stoves, and particulate matter from smoking, cleaning, and other indoor activities. Several authors *(142, 144)* have commented on the illogic of basing ambient air quality standards in the United States primarily on outdoor exposures although people spend most of their time indoors.

Table IX Typical Sources of Some Pollutants Grouped by Origin *(140)*

Pollutants	*Sources*
Group I: Sources predominantly outdoor	
Sulfur oxides (gases, particles)	Fuel combustion, smelters
Ozone	Photochemical reactions
Pollens	Trees, grass, weeds, plants
Lead, manganese	Automobiles
Calcium, chlorine, silicon, cadmium	Suspension of soils or industrial emission
Organic substances	Petrochemical solvents, natural sources, vaporization of unburned fuels
Group II: Sources both indoor and outdoor	
Nitric oxide, nitrogen dioxide	Fuel burning
Carbon monoxide	Fuel burning
Carbon dioxide	Metabolic activity, combustion
Particles	Resuspension, condensation of vapors and combustion products
Water Vapor	Biologic activity, combustion, evaporation
Organic substances	Volatilization, combustion, paint, metabolic action, pesticides, insecticides, fungicides
Spores	Fungi, molds
Group III: Sources predominantly indoor	
Radon	Building construction materials (concrete, stone), water
Formaldehyde	Particleboard, insulation, furnishings, tobacco smoke
Asbestos, mineral, and synthetic fibers	Fire-retardant, acoustic, thermal, or electric insulation
Organic substances	Adhesives, solvents, cooking, cosmetics, solvents
Ammonia	Metabolic activity, cleaning products
Polycyclic hydrocarbons, arsenic, nicotine, acrolein, etc.	Tobacco smoke
Mercury	Fungicides, in paints, spills in dental care facilities or laboratories, thermometer breakage
Aerosols	Consumer products
Viable organisms	Infections
Allergens	House dust, animal dander

Indoor air quality studies in recent years have placed more emphasis on pollutants generated indoors whether or not they also have outdoor sources. For example, many studies *(140, 142, 155–165)* of indoor-generated NO_2 from gas stoves have been carried out and are still in progress. Some *(162–164)* have been part of larger epidemiological studies.

C. *Indoor Sources of Pollutants*

Spengler and Sexton *(144)* have pointed out that a number of pollutants from indoor sources are capable of affecting human health: tobacco smoke, radon and its decay products, asbestos fibers, fiberglass, formaldehyde, combustion by-products (e.g., polycyclic aromatic hydrocarbons, NO_2, CO, HCN, and SO_2), aeropathogens, and allergens capable of producing a wide range of physiological effects.

1. Combustion Sources

Combustion sources such as furnaces and water heaters totally vented to the atmosphere seldom make a direct contribution to indoor air pollution unless the vent system has been improperly installed, leaks in the system have developed, or significant negative pressures occur indoors. Such situations, although occurring infrequently, may cause elevated indoor concentrations of combustion gases (e.g., CO) and create health effects and even deaths. Unfortunately few systematic studies have provided details on indoor air pollution under these circumstances.

Unvented or incompletely vented gas stoves are the most important indoor combustion source since they emit significant quantities of NO, NO_2, CO, and CO_2 and lesser amounts of formaldehyde and particulate matter. Table X presents data from Girman *et al. (160)* on pollutant emissions from gas ranges.

The emission factors for NO, NO_2, and CO are in the same range as those found earlier by Wade *et al. (155)*, who also found small but measurable effects on emissions from different cooking pan materials and determined that typical kitchen exhausts were only about 40% efficient in exhausting combustion products to the outdoors.

Several workers *(165–167)* have measured emissions from unvented kerosene heaters (Table XI). NO_2 emission rates are in about the same range as gas stoves, but because of the extremely efficient combustion in kerosene heaters, CO and formaldehyde emissions are much lower than for gas stoves. Note, however, that the small residual sulfur component in kerosene produces a significant amount of indoor SO_2.

Except in some Third World countries, fireplaces and wood stoves are always vented to the atmosphere and therefore should have only a minimal effect on indoor air quality. However, poor design of fireplace flues, faulty installation of wood-burning stoves, logs rolling out of fireplaces, downdrafts, and indoor negative pressures may

Table X Summary of Pollutant Emission Rates from Gas Ranges *(160)*

	Oven[a]				Top burners[b]			
Pollutant	*Mean ($\mu g/kJ$)*	*Standard deviation ($\mu g/kJ$)*	*Range ($\mu g/kJ$)*	*No. of runs*	*Mean ($\mu g/kJ$)*	*Standard deviation ($\mu g/kJ$)*	*Range ($\mu g/kJ$)*	*No. of runs*
Gases								
CO	226	17	214–238	2	200	34	172–249	5
CO_2	42,700	2,300	38,380–46,360	9	45,320	1,700	43,900–47,600	4
NO_x(as N)	6.17	0.67	4.98–7.03	6	9.0	1.3	6.9–10.5	6
NO[c]	6.61	—	—	—	9.7	—	—	—
NO_2[c]	10.14	—	—	—	14.8	—	—	—
SO_2	0.11	0.02	0.08–0.13	5	0.16	0.02	0.14–0.18	2
HCHO	2.73	0.41	2.36–3.39	5	1.7	1.1	0.9–2.5	2
O_2	—	—	—	—	78,200	7,300	73,000–83,300	2
Particles (2.5 μm)								
Carbon	0.02	—	—	2	0.22	0.01	0.21–0.23	3
Sulfur (as SO_4)	0.001	—	—	2	0.22	0.01	0.001–0.017	4
Mass[d]	0.05	—	—	2	0.41	0.19	0.24–0.62	3
Mass ($<0.5\ \mu m$)[e]	0.015	0.001	0.014–0.015	2	0.50	0.36	0.21–1.03	4

[a] Operated at 180°C (350°F) for 1 hour. Fuel consumption rate is 8400 kJ/hour.
[b] Two burners operated for 16 minutes. Fuel consumption rate is 9200 kJ/hour per burner.
[c] Assuming a volumetric NO_2-to-NO ratio of 1.0. Our measurements varied from 0.4 to 2.0.
[d] Based on gravimetric analysis of filters.
[e] Based on electrical mobility analysis and assuming a particle density of 2.0 g/cm^3.

Table XI Comparison of Pollutant Emissions from Kerosene Heaters

		Emission Rate, μg/kJ					
Researchers	*Type heater*	*CO*	*CO_2*	*NO*	*NO_2*	*HCHO*	*SO_2*
Traynor *et al.* *(165)*	New convective	14.5	70,100	23.7	14.1	0.01	—
	New convective	10.3	72,500	25.2	12.5	0.08	—
	New radiant	60.2	70,300	1.4	5.2	0.63	—
Caceres *et al.* *(166)*	Radiant	102	—	0.71	3.1	0.10	19.4–55.2[a]
	Radiant	51.1	—	1.5	4.7	0.09	—
	Wick	13.8	—	4.4	4.1	0.06	—
Leaderer *(167)*	Radiant (normal setting)[b]	58.2	67,100	0.08	5.0	—	16.1
	Convective (normal setting)[b]	22.3	71,600	11.6	14.5	—	16.1

[a] Results reported as SO_x.
[b] Results converted from milligrams per gram (mg/g) and based on heating value of kerosene of 43.5 kJ/g.

permit combustion products from such devices to enter the indoor environment. Fortunately, wood smoke is highly irritating and usually provides a warning that combustion products are present to those occupying spaces heated by such means.

Duncan *et al.* *(168)* have determined emissions of various pollutants from wood stoves. Results from this work are presented in Table XII, which shows the wide range of organic compounds emitted from such heating systems. Knight *(169)* has shown that some of the more recent developments in wood heaters reduce emissions of CO and organic compounds.

Moschandreas *et al.* *(170)* measured benzo[*a*]pyrene (BaP) both indoors and outdoors in a group of dwellings that contained fireplaces and wood stoves. In one of the houses with a fireplace, indoor concentrations of BaP were always greater than those outdoors. On one occasion when the fireplace was in operation, the indoor concentration reached 12 ng/m^3, although the outdoor concentration was less than 1 ng/m^3. In a house with a wood stove, indoor BaP concentrations averaged 4.7 ng/m^3, nearly five times typical outdoor levels in urban areas.

2. Tobacco Smoking

Over 2000 organic and inorganic compounds have been identified in cigarette smoke, many of which are odorous, irritating, and capable of

Table XII Emission from Residential Wood-fired Stoves[a]

	Emission, lb/cord[b]	
Substance emitted	*Range*	*Average*
Criteria pollutants		
Particles	3–93	30.3
SO_x	0.5–1.5	0.7
NO_x	0.7–2.6	1.6
Hydrocarbons	1–146	41.6
Carbon monoxide	300–1,220	598.3
Noncriteria pollutants		
Polycyclic organic materials	0.6–1.22	0.9
Formaldehyde	0.3–1	0.8
Acetaldehyde	0.1–0.3	0.4
Phenols	0.3–8	3.3
Acetic acid	5–48	21.1
Aluminum	—	1.3
Calcium	—	10.2
Chlorine	—	0.1
Iron	—	0.7
Magnesium	—	2.0
Manganese	—	1.6
Phosphorus	—	1.0
Potassium	—	3.6
Silicon	—	1.6
Sodium	—	0.7
Titanium	—	0.02

[a] Based on Duncan *et al. (168).*
[b] The relationship used to convert from pounds per ton (lb/ton) to pounds per cord (lb/cord) was 1 lb/ton = 1.65 lb/cord.

causing adverse health effects *(140)*. Particulate matter in cigarette combustion products contains nicotine plus a wide range of polycyclic organic materials, including BaP. These materials are part of what is commonly called *tar*. In addition to pollutants in particulate form, emissions from tobacco combustion include gases and vapors containing CO, CO_2, HCHO, acrolein, NO_2, and many other compounds.

Almost all individuals are exposed to cigarette smoke at some point in their lives, whether or not they have ever been smokers. It is important to distinguish between "Mainstream" smoke (that inhaled by the smoker) and "sidestream" smoke (that escaping to the atmosphere from the burning tobacco between puffs). Smokers are exposed to both types of smoke; nonsmokers are exposed to sidestream smoke and exhaled mainstream smoke. Lebowitz *(170a)* has assessed the potential for lung cancer from passive smoking.

The composition and relative quantities of various compounds released by tobacco combustion vary significantly between mainstream and sidestream tobacco smoke largely because of differences in the combustion dynamics. Sidestream cigarette smoke is a much more important source of indoor air contamination than mainstream smoke because of the longer period over which sidestream smoke is emitted and the higher burn temperature. Although the amount of tobacco burned during sidestream smoke production is only about 20% greater than that during mainstream smoke production, the amount of pollutant production for many compounds on a per cigarette basis is much greater for sidestream smoke than for mainstream smoke. For example, the ratio of sidestream to mainstream production for BaP is between 3.9 and 4.5, for CO it is 4.7, for NO_2 it is 3.6, for ammonia it is 46.3, and for numbers of particles it is 3.3 *(140)*.

The effect of smoking on indoor air quality varies widely depending upon size of room, number of cigarettes smoked, and air exchange rate. Spengler *et al.* *(171)* have shown from a study of 80 homes that a "pack-a-day" smoke contributes approximately 20 $\mu g/m^3$ to the 24-hour average indoor concentrations of respirable particulate matter. Repace and Lowrey *(172)* have measured CO and particulate matter in bars, conference rooms, airplanes, offices, and other enclosed areas where smoking occurs. They found concentrations between 2 and 35 ppm for CO and 10 and 1000 $\mu g/m^3$ for particulate matter.

3. Consumer Products, Furnishings, and Construction Materials

Many of the materials used in construction of buildings and the products used by their occupants emit a wide range of contaminants. The contaminant receiving the most attention in recent years is formaldehyde (HCHO), suspected of being a carcinogen *(173)*. One of its most significant sources, urea–formaldehyde (UF) resin, is used in many building products such as plywood, particle board, paneling, and a variety of furnishings (carpets and draperies). It also is a component of urea–formaldehyde foam insulation (UFFI). This type of insulation was popular several years ago, but because of high rates of HCHO released under certain environmental conditions (high ambient temperature and humidity) and improper installation methods, this material is being severely controlled in several countries and outlawed in several of the states in the United States *(140, 142)*. Formaldehyde is also emitted by other indoor sources such as gas cooking (Table X) and tobacco smoking.

Matthews *et al.* *(174)* have measured HCHO emission rates for a

Table XIII Formaldehyde Emission from Selected Products *(175)*

Product	*Range of emission rates, μg/g-day*
Particle board	0.4–8.1
Plywood	0.03–9.2
Paneling	0.84–2.1
Fiberglass insulation	0.3–2.3
Clothing	0.2–4.9
Draperies	ND[a]–3.0
Paper products	0.03–0.36
Carpet	ND[a]–0.06

[a] ND = not detectable.

number of different building material surfaces using both a newly developed surface monitor and an environmental test chamber. Emission rates varied between essentially zero for standard interior and exterior plywood to almost 1 mg/m^2-hr for decorative paneling. Good agreement was found between the two emission measurement methods. Gupta *et al.* *(175)* have summarized formaldehyde emissions taken from a variety of products on the basis of laboratory tests (Table XIII). Meyer *(176)* has developed mathematical models that show the release rate of formaldehyde with time for various types of paneling; from these models one can compute equilibrium indoor concentration for formaldehyde, given the air exchange rate.

Formaldehyde is only one of the organic compounds released by this category of sources. Use of solvents, adhesives, and accelerators in building materials and furnishings causes indoor contamination from many organic materials. Personal care and home maintenance products such as hair spray, cleaning materials, and insecticides add to the list of organic indoor pollutants. Some of these products emit inorganic compounds such as ammonium hydroxide and other caustic materials. Such products are constantly changing and being improved in terms of their primary use, but not necessarily in terms of their general effect on indoor air quality. Information on emissions of organic pollutants from this category is extremely sparse, and since it is a "moving target," specification of sources, characterization of emissions, and determination of their impact on indoor air quality are formidable, continuing tasks.

Cote *et al.* *(177)* conducted a survey of consumer products used in a group of households, identified the active ingredients in each, and estimated emission rates and effects on indoor air quality based on hypothetical use patterns. Moschandreas *et al. (178)* performed episodic releases of spray oven cleaner and antiperspirants in two dwellings and measured their effect on indoor air quality at several indoor locations. Yocom *et al. (179)* have measured emissions from office building materials, furnishings, and office equipment by using direct reading mass spectrometry. This method is sensitive enough to identify and quantify organic species near their odor thresholds. Molhave *(180)* has studied the emissions of organic compounds from 42 commonly used building materials and has evaluated the effect of these emissions on indoor air quality.

4. Radon and Progeny

Radon-222 is the alpha-decay product of radon-226 which occurs naturally in the earth's crust. It is a noble gas with a half-life of 3.8 days. Radon may move from the site of its formation and escape the earth's surface. The rate of radon release varies from place to place on the earth's surface as a function of the composition and geological structure of the earth's crust.

The primary effect of radon on humans is not from the gas itself, since after inhalation it is mostly expelled by the lungs. Rather, it is the alpha-active short half-life decay products, or progeny, principally polonium-218 and polonium-214, which are of concern. These decay products exist as solids that may become attached to respirable particulate matter and ultimately to the lungs. Lung cancer in uranium miners is most likely associated with radon progeny.

Outdoor concentrations of radon and progeny are usually insignificant because of the rapid dispersal and dilution by air movement. Since radium-226 is a component of soil and rock, radon is emitted from such natural materials and is present in spring and well water, natural gas, concrete, brick, and building stone. Therefore, there are many routes by which radon may enter a structure. As with other pollutants generated or trapped indoors, indoor concentrations tend to rise when the structure in which the radon is trapped is sealed from the outdoors as part of energy conservation measures. As one might expect, indoor radon concentration is usually higher in basements and enclosed crawl spaces than in other parts of a house in view of its source in stone, soils, concrete foundations and basement walls, and ground water.

The common units of measure of radon concentrations are nanocuries per cubic meter (nCi/m^3), which is equivalent to picocuries per liter (pCi/liter), or becquerels per cubic meter (Bq/m^3), where 1 nCi = 37 Bq. The unit used for radon progeny concentration, the *working level* (WL), is any combination of short half-life radon progeny in 1 liter of air that results in the ultimate emission of 1.3×10^5 MeV of alpha-particle energy. Although there are no established ambient indoor or outdoor air quality standards for radon or its progeny, the United States Environmental Protection Agency has set 4 nCi/m^3 as the upper limit in houses built on reclaimed phosphate lands in Florida. This radon concentration is approximately equivalent to 0.02 WL for radon progeny. Bruno *(181)* has reported "normal" outdoor radon levels to be about 0.25 nCi/m^3 and outdoor levels over "mineralized" areas to be about 0.75 nCi/m^3.

Sachs *et al. (182)* studied radon levels in a large number of houses in Pennsylvania and found that indoor radon concentrations varied more than two orders of magnitude. The greatest variability in radon source was found to be water, followed, in descending order, by geological substrates and building materials.

Nazaroff and Nero *(183)* reviewed indoor radon data from several countries and pointed out that variability in indoor concentrations from dwelling to dwelling is caused by structural differences. Further, they suggest that in many situations, soil adjacent to the foundation is the chief source of radon and that radon penetrates the structure by pressure-driven flow rather than through molecular diffusion.

D. Monitoring Indoor Air Quality

Monitoring methods designed for ambient air quality and occupational air quality are both potentially applicable to the measurement of indoor air quality. However, some of the special requirements of indoor air quality monitoring impose problems in this technology transfer. These problems are related to such factors as sensitivity, short-term response, multiple point monitoring at several locations in a structure, ability to depict human exposure, noise and other intrusions in occupied dwellings, prevention of contamination or change of the indoor environment by monitoring equipment exhaust or by removing excessive quantities of sample air from the indoor space with external samplers, and finally, as in any monitoring program, cost.

There are a number of good reviews of indoor air quality sampling methods. The books by NAS *(140)*, Sandia National Laboratories *(143)*, and Diamond and Grimsrud *(147)* provide good overviews of general sampling approaches and the situations in which they are most applicable. The book by Meyer *(146)* provides descriptions of a wide range of sampling and analytical methods and includes an extensive bibliography. This chapter will not attempt to review all indoor air quality monitoring methods. The reader is directed to the references cited here and later in this section.

Before embarking on an indoor air quality study, one must clearly define the objective of the study and develop a monitoring protocol that fits the time, budget, and manpower constraints. Typical objectives for indoor air quality monitoring programs might include characterizing indoor–outdoor relationships for a particular type of dwelling, studying the effect of an indoor source on indoor air quality, determining total human exposure to a given pollutant, and providing exposure data for epidemiological studies. A monitoring protocol indicates the types of monitoring equipment and analytical and data analysis support needed. There are five basic monitoring approaches. However, some methods are not available for some pollutants.

Continuous monitors provide real-time monitoring with the capability of measuring short-term fluctuation and calculating long-term averages. Samplers must usually be operated at fixed locations either inside or outside the structure with sampling lines run to the sampling points. Through the use of sampling tubes and manifolds and suitable valving and electronic control, such devices can be operated as multipoint samplers. This technique is especially useful in characterizing indoor–outdoor air quality relationships and indoor air quality patterns.

Dynamic time-integrated samplers collect pollutants on adsorbents, absorbents, filters, or impaction plates over fixed periods representing a fixed volume of air sampled. Such devices, which are usually cheaper than continuous monitors, can be ganged together in a sequential sampling device to collect a series of time-averaged samples over an extended period. Such samples usually require chemical or physical analysis subsequent to sample collection.

Passive time-integrated samplers do not require a pump to draw air through a collection medium. Rather, the sensing of the pollutant is accomplished by the physical characteristics and dimensions of the sampler and the physical properties of the pollutants. Examples of such devices are the diffusion tubes for NO_2 monitoring developed by Palmes *et al. (184)* and the Track-Etch samplers *(185)* used for measuring total alpha activity.

Grab sampling is used for indoor sampling when single, short-term samples can adequately describe the indoor pollutant concentrations. Such an approach would not normally be used when indoor air quality concentrations are varying rapidly unless a large number of grab samples can be taken at regular intervals. Examples of this method include filling evacuated flasks or cylinders or inert plastic bags with indoor air samples for subsequent laboratory analysis, and using a hand pump to draw measured amounts of air over adsorbents in glass tubes.

Personal monitoring devices are used when there is a need to establish total exposure to a pollutant by individuals as they move about both indoors and outdoors. Such devices may be "dynamic," with air drawn by battery-powered pumps through the sampling system, or passive, of the type described previously. An example of the dynamic type is that used by the Harvard Group in their six cities study *(186)*. The Palmes tube *(184)* has also been widely used as a personal sampler for indoor NO_2 studies in relation to gas stove use. Wallace and Ott *(187)* have conducted an in-depth review of personal monitors.

E. Indoor Concentrations of Pollutants

The National Academy of Sciences Report *(140)* summarizes semi-quantitatively the sources, possible indoor concentrations, indoor–outdoor concentration ratios, and typical locations for the concentrations of the most important indoor air pollutants and pollutant categories. This summary is presented as Table XIV. In a critical review of indoor–outdoor air quality relationships, Yocom *(142)* summarized available information as follows:

Sulfur dioxide: Outdoor concentrations overpredict indoor exposure by a factor of about 2.

Total Particulate matter: Indoor exposures bear little relationship to outdoor concentration patterns. Furthermore, the chemical and physical nature of the particulate matter indoors is different from that outdoors.

Carbon monoxide: In the absence of indoor sources, the outdoor concentration pattern over long averaging times is a reasonably good predictor of indoor exposure. Indoor sources such as unvented stoves and heaters may cause indoor CO exposure to exceed outdoor concentration levels by a factor of 2 or more.

Nitrogen dioxide: Without indoor sources, outdoor NO_2 concentration patterns overpredict indoor exposures by a factor of about 2. With indoor sources such as unvented stoves and heaters, outdoor levels underpredict indoor exposures by a factor of 2 or more in portions of the house where such devices are operated.

Table XIV **Sources, Possible Concentrations, and Indoor-to-outdoor Concentration Ratios of Some Indoor Pollutants** *(140)*

Sources of				
Pollutant	*Indoor pollution*	*Possible indoor concentration*[a]	I/O concentration ratio	Location
Carbon monoxide	Combustion equipment, engines, faulty heating system	100 ppm	$\geqslant 1$	Skating rinks, offices, homes, cars, shops
Respirable particles	Stoves, fireplaces, cigarettes, condenstation of volatiles, aerosol sprays, resuspension, cooking	100–500 $\mu g/m^3$	$\geqslant 1$	Homes, offices, cars, public facilities, bars, restaurants
Organic vapors	Combustion, solvents, resin products, pesticides, aerosol sprays	NA	>1	Homes, restaurants, public facilities, offices, hospitals
Nitrogen dioxide	Combustion, gas stoves, water heaters, dryers, cigarettes, engines	200–1000 $\mu g/m^3$	$\geqslant 1$	Homes, skating rinks
Sulfur dioxide	Heating system	20 $\mu g/m^3$	<1	Removal inside
Total suspended particles without smoking	Combustion, resuspension, heating system	100 $\mu g/m^3$	1	Homes, offices, transportation, restaurants
Sulfate	Matches, gas stoves	5 μ1.0 ppm	>1	Homes, offices
Formaldehyde	Insulation, product binders, particleboard	0.05–1.0 ppm	>1	Homes, offices
Radon and progeny	Building materials, ground water, soil	0.1–30 nCi/m^3	$\geqslant 1$	Homes, buildings
Asbestos	Fireproofing	<1 fiber/cm^3	1	Homes, schools, offices
Mineral and synthetic fibers	Products, cloth, rugs, wallboard	NA	—	Homes, schools, offices
Carbon dioxide	Combustion, humans, pets	3000 ppm	$\geqslant 1$	Homes, schools, offices
Viable organisms	Humans, pets, rodents insects, plants, fungi, humidifiers, air conditioners	NA	>1	Homes, hospitals, schools, offices, public facilities
Ozone	Electric arcing, UV light sources	20 ppb 200 ppb	<1 >1	Airplanes, Offices

[a] Concentrations listed are only illustrative of those reported indoors. Both higher and lower concentrations have been measured. No

Ozone: In view of the extreme reactivity of ozone and the lack of indoor sources, outdoor concentration overpredicts indoor exposure by a factor of about 5.

Nonmethane hydrocarbons: Since there are many possible indoor sources of hydrocarbons and other organic gases and vapors, outdoor concentrations underpredict indoor exposures by a factor of about 1.5.

Lead: Since there are no significant indoor sources of airborne lead, outdoor concentrations overpredict indoor exposures by a factor of about 1.5.

Respirable particulate matter (RPM): In the absence of indoor sources, outdoor RPM tends to underpredict indoor exposures. The chemical and physical nature of indoor RPM may be quite different from that outdoors. Indoor sources, principally smoking, may produce indoor–outdoor (I/O) ratios in excess of 5.

Sulfates: Outdoor concentrations tend to overpredict indoor exposures by a factor of about 1.5.

1. Criteria Pollutants

a. SULFUR DIOXIDE. Since SO_2 is generated outdoors and is quite reactive, indoor concentrations are usually much lower than those outdoors when no indoor sources are present. Spedding *(188)* points out the importance of indoor furnishings (especially wool rugs and leather) in absorbing SO_2. There is evidence *(170)* that the indoor–outdoor relationship for SO_2 is inversely proportional to outdoor concentration. The results from the Six Cities Study *(161)* also confirm this observation. In the cities with modest to high outdoor SO_2 levels, I/O ratios are in the range of 0.3 to 0.5, while in cities where SO_2 levels are lowest, I/O ratios are in the range of 0.7 to 0.9. In Kingston, Tennessee, where all the monitored homes were air-conditioned, the I/O ratio was 0.1.

One can deduce, therefore, that during the winter, when outdoor SO_2 levels tend to be highest and the home is tightly closed against the cold, indoor SO_2 concentrations are only a small fraction of those outdoors. Under these conditions, the home provides an effective refuge from exposure to outdoor SO_2 concentrations. The most important conclusion to be drawn, however, is that outdoor measurements for SO_2 are poor indicators of the exposure of humans to SO_2.

b. PARTICULATE MATTER. Very few indoor measurements for particulate matter can be compared with outdoor levels of total suspended particulate matter (TSP) since the standard high-volume sampler is usually an inappropriate method for measuring indoor concentrations because of noise and excessive volume flow rate. A variety of other methods have been used for measuring indoor particulate matter *(142)*.

Alzona *et al.* *(189)* carried out an experimental study in which 10 different indoor spaces (rooms in university buildings and houses, and the interiors of automobiles) were first cleaned of particulate matter, and then indoor airborne particulate matter was collected on filters at various times thereafter. Equilibrium between indoor and outdoor air quality was reached within less than 1 day. The I/O relationships for particulate matter were evaluated in terms of the I/O ratios for various elements in the particulate matter. With all windows and doors closed, I/O ratios for various elements for all the indoor spaces were Ca, 0.10; Fe, 0.24; Zn, 0.41; Pb, 0.42; and Br, 0.36. These results (Table XV) appear to be consistent with the relative particle size of compounds containing these elements and their ease of penetration. In one of the rooms tested, data were collected under widely differing conditions of sealing openings, opening windows, and covering indoor surfaces with plastic. By comparing the results of cases 1, 5, and 6, the authors believed that these results showed that dust resuspension is much less important in controlling I/O ratios than particulate matter penetration through openings.

c. Carbon Monoxide. CO is an unreactive gas and, in the absence of indoor sources, indoor concentrations tend to follow outdoor concentrations, given a sufficiently long averaging time. Most of the recent indoor air quality studies *(140)* have dealt with the effects of combustion sources (including smoking) on indoor air quality. Brune-

Table XV Indoor–Outdoor (I/O) Relationships for Elements in Particulate Matter in Relation to Methods of Sealing a Room to Prevent Dust Reentrainment *(189)*

			I/O				
Case	*Number of runs*	*Conditions*	*Ca*	*Fe*	*Zn*	*Pb*	*Br*
1	3	Normal	0.10	0.17	0.52	0.49	0.33
2	2	Plastic over windows	0.10	0.15	0.71	0.17	0.17
3	1	Window wide open	0.52	0.81	0.93	1.2	1.0
4	2	Window cracked open	0.20	0.16	0.69	0.55	0.53
5	6	All surfaces plastic-covered	0.02	0.12	0.24	0.15	0.20
6	3	All but windows plastic-covered	0.10	0.15	0.58	0.57	0.32

kreef *et al.* *(190)* measured CO concentrations in 254 houses in the Netherlands to determine the effect of unvented gas water heaters (geisers) and cook stoves on indoor air quality. In 17% of the homes with geisers, CO levels greater than 50 ppm were found in the kitchens. Ziskind *et al.* *(191)* used personal CO monitors of several types to measure exposure of nine members of the general public. They found that the United States 8-hour air quality standard for CO was exceeded on 22 person-days. These elevated CO exposures were frequently associated with periods spent in commuting. Ott and Flachsbart *(192)* also studied indoor and outdoor CO concentrations by using personal monitors. Average concentrations measured by personal monitors were usually greater than those measured by nearby outdoor ambient air monitoring stations.

d. *NITROGEN DIOXIDE.* Many studies of indoor NO_2 concentrations have been carried out. Most *(140, 142, 155–161)* have been related to the use of indoor combustion devices, especially gas cooking stoves. At the time of writing this chapter, new projects in the United States were being carried out by the Gas Research Institute, Chicago, Illinois; Southern California Gas Company, Los Angeles, California; and the New York State Energy Research and Development Agency, Albany, New York.

It has been shown conclusively in a number of studies that indoor NO_2 concentrations with unvented or partially vented gas stoves are significantly higher than concentrations outdoors. Table XVI summarizes the results of selected studies. The data also show that in homes with gas stoves maximum NO_2 concentrations are found in the kitchen and that NO_2 concentrations in the other parts of the house are usually higher than outdoor concentrations.

Indoor NO_2 decay rates have been measured by a number of workers by measuring exponential decay in relation to CO concentrations since CO is considered unreactive. Workers in the United States *(140, 155, 178, 193)* found NO_2 decay rates in the range of 0.83 to 1.4 $hour^{-1}$. These same workers found NO decay in the range of -0.004 to 0.082 $hour^{-1}$. Yamanaka *(194)* has compared NO_2 decay rates to CO decay in a typical Japanese livingroom. Decay rates were measured in relation to relative humidity and surface modifications, which included covering all or portions of the room with polyethylene film and covering the floor mats with acrylic fiber rugs. At low humidity ($<$50% RH), the decay rate was 0.99 ± 0.19 $hour^{-1}$ and was independent of surface covering. Above 50% RH, the decay rate increased linearly with RH, but the slope was steeper when the room was not covered with polyethylene than when it was covered.

Table XVI Summary of Selected Indoor and Outdoor Nitrogen Dioxide Concentrations, μg/m³

		Gas stove homes				Electric stove homes			
			Reported NO_2 levels				Reported NO_2 levels		
Reference	*Season or location*	*Number of homes*	*Kitchen*	*Nonkitchen*	*Outdoors*	*Number of homes*	*Kitchen*	*Nonkitchen*	*Outdoors*
155	Spring	1	100	61, 52	44				
	Fall	1	67, 60	55	50				
	Fall	1	110, 67	49	46				
	Spring	1	114	75, 58	51				
	Fall	1	53, 47	28	35				
	Fall	1	180, 140	70	32				
	Winter	1	213, 120	71	39				
158	Spring	10	92	47	—	9	16	13	—
161	Annual means								
	Portage, WI	3	—	20	11	9	—	8	11
	Topeka, KS	4	—	44	21	9	—	20	21
	Kingston, TN	—	—	—	—	10	—	12	19
	Watertown, NY	9	—	53	49	2	—	47	49
	St. Louis, MO	7	—	47	36	3	—	20	36
	Steubenville, OH	6	—	45	41	6	—	31	41
157	June	5	59	37	17	4	16	15	17
159	Summer	9	67 same floor as kitchen; 23 different floor from kitchen	38	5	10	8 same floor as kitchen; 6 different floor from kitchen	6	11
162	Winter	428 (107)[a]	211	57	35	87 (18)[a]	34	26	35

[a] Number of nonkitchen samples in parentheses.

e. OZONE. Ozone (O_3) is an extremely reactive pollutant and one of the most powerful oxidizers in nature. Its source in the indoor environment is outdoor air, where most of it, at least in urban areas, is formed by photochemical reactions. Indoor sources include electrostatic air cleaners, fluorescent lights, and certain types of office copying machines. Operation of electrical motors produces relatively small quantities of ozone.

Sabersky *et al.* *(195)* determined indoor–outdoor relationships for O_3 in two campus laboratory–office buildings and a private residence and then investigated the mechanism of O_3 decay indoors. The pattern of indoor O_3 followed that of outdoor concentrations, but concentrations were lower inside than outside and displaced somewhat later in time.

In a private home, two experiments were carried out to estimate decay rates for O_3. In the first experiment, after the house was filled with outside air, all windows and doors were closed. The ventilation system was turned off, and persons walked through various rooms in "normal" activity patterns. The indoor concentration of O_3 dropped from 0.23 ppm (450 $\mu g/m^3$) to 0.04 ppm (79 $\mu g/m^3$) in 30 minutes. In the second experiment, the house was again filled with outside air and all openings closed. Then air was mechanically circulated through glass–fiber filters by the internal ventilation system. The indoor O_3 levels decayed exponentially from 0.23 ppm (450 $\mu g/m^3$) to 0.03 ppm (59 $\mu g/m^3$) in 20 minutes. In the second phase of this work the first-order heterogeneous decomposition of O_3 was studied in a test chamber in which panels of various materials were exposed. It was shown that the decomposition rate of O_3 is higher for surfaces of rubber, plastics, and fabrics than for surfaces of metals and glass.

Mueller *et al.* *(196)* measured O_3 decay in rooms of several sizes and established that the reaction is clearly first-order and that the rate is directly related to the surface-to-volume ratio of the interior space. These workers determined that humidity increases decay with certain types of surfaces, and in a typical bedroom the half-life of O_3 was only 5.7 minutes. The half-life in an office was found to be 11 minutes.

Moschandreas *et al.* *(170)* measured the following average indoor–outdoor ratios for O_3 for houses and office buildings in the Boston, Massachusetts, metropolitan areas:

Residences with gas stoves	0.19
Electric residences	0.20
Office buildings	0.29

Berk *et al.* *(197)* determined indoor–outdoor relationships for O_3 in an elementary school for different energy-efficient ventilation rates.

It was found that although the concentration of ozone outdoors reached as high as 98 $\mu g/m^3$ (50 ppb), the O_3 levels never exceeded 19 $\mu g/m^3$ in any of the rooms of the school. From a review of the concentration frequency distributions indoors and outdoors, it is evident that I/O ratios of O_3 undoubtedly decrease with decreasing outdoor airflow rates.

In other work, Berk *et al.* *(198)* measured indoor concentrations for O_3 and other pollutants in a weatherization program of houses in the Medford, Oregon, area. Indoor–outdoor ratios for O_3 were in the range of 0.10 to 0.25, and no significant change in the I/O ratio could be seen after installation of weatherization measures.

From the preceding review it is clear that indoor concentrations of O_3 almost invariably are significantly lower than those outdoors and that the indoor environment is an effective refuge from outdoor exposure to O_3.

f. *HYDROCARBONS.* The United States National Ambient Air Quality Standard for hydrocarbons (HC), before it was deleted in 1983, applied to nonmethane hydrocarbons (NMHC) and in reality was not a direct health-related standard in the same mold as the standards for SO_2, CO, and NO_2. The standard for NMHC was 160 $\mu g/m^3$ (as CH_4) averaged over the 3 hours between 6 A.M. and 9 A.M. NMHC is a precursor of photochemical smog. The strategy in setting the standard was that if levels were below this early morning average, the undesirable manifestations of photochemical smog (such as oxidant formation, eye irritation, and visibility reduction) would be reduced later in that day. Thus, the NAAQS for NMHC had very little meaning for the indoor atmosphere, where photolysis of reactive pollutants normally does not take place. Nevertheless, the term *hydrocarbons* as applied to air quality measurements is still widely used and includes not only true hydrocarbons but also encompasses a wide range of organic gases and vapors. These substances are of possible concern in the indoor atmosphere as a result of their toxic, irritating, or odorous characteristics. Although the indoor air quality literature on hydrocarbons *(142)* has been reviewed in the context of its relationship to outdoor air quality standards, discussion of this category will be included in Section 4.

g. *LEAD.* The United States Ambient Air Quality Standard for lead (Pb) is 1.5 $\mu g/m^3$ averaged over a 3-month period. Lead may enter the indoor environment as a component of atmospheric particulate matter or as part of soil tracked into the house and then made airborne by various indoor activities. Lead may also be present in indoor particulate matter in older buildings through peeling and cracking of lead-based paint.

Indoor and outdoor concentrations of Pb were determined by the General Electric Company *(199)* at two high-rise apartment buildings in New York City. One building was built over a highway to utilize its "air rights" while the other was a "street canyon" building (Table XVII). Note that most of these concentration levels (indoor and outdoor) approach or exceed the 3-month average ambient standard for Pb. The results for the air rights building show that the maximum outdoor concentrations were measured on the lower floor of the structure. The indoor–outdoor ratio for Pb at this sampling level was 0.56 during the heating season and 0.80 during the nonheating season. At the upper sample level (32d floor) the indoor–outdoor ratio for Pb was 1.07 during the heating season and 1.09 during the nonheating season. The apparent increase in the indoor–outdoor ratio of Pb with height in the building is similar to that which the authors noted for CO. These pollutants are distributed through the building in such a way that their indoor–outdoor ratios may exceed 1.0 on the upper floors.

Since the indoor and outdoor sampling points were not at the same building elevations at the canyon structure, it was not feasible to evaluate indoor–outdoor Pb relationships at specific building heights. However, a cursory inspection of Table XVII suggests that at upper floors the indoor–outdoor ratio for Pb would significantly exceed 1.0.

The indoor and outdoor particulate Pb levels measured by Moschandreas *et al. (178)* varied over an extremely wide range based on data from 4-day sampling cycles. Unexplained high outdoor and indoor levels of between 4.0 and 12.5 $\mu g/m^3$ at one house accounted for the extremes at the upper end of the range. If data for this one house are excluded, the outdoor range of Pb values is 0.1 to 2.8 $\mu g/m^3$, and the indoor range is 0.1 to 1.1 $\mu g/m^3$. From these ranges, it appears that

Table XVII Indoor–Outdoor Relationships for Lead Concentrations at Two New York City High-Rise Apartment Houses *(199)*

	Mean particulate lead concentration, $\mu g/m^3$			
Building and season	*Outdoors*		*Indoors*	
Air-rights structure	*2d floor*	*Roof*	*2d floor*	*Roof*
Heating season	3.45	1.47	1.42	1.58
Nonheating season	1.89	1.30	1.51	1.42
Street canyon building	*3d floor*	*Roof*	*11th floor*	*18th floor*
Heating season	1.15	1.23	1.45	1.53
Nonheating season	1.27	1.62	1.91	2.19

Pb levels both indoors and outdoors during this study are significantly lower than those found by the General Electric Study *(199)*. Such differences could be the result of innumerable factors such as the removal of Pb in some types of gasoline between 1970 and 1975, sampling site differences, and sampling and analytical methods.

Moschandreas *(178)* also included an elemental analysis of particles from continuous samplers which collected particles on Nuclepore filter material drawn continuously across an orifice connected to a vacuum device. Samples were collected simultaneously at outdoor and several indoor sites and provided a continuous record of particulate elemental analysis. At one site the indoor levels of Pb fluctuated synchronously with outdoor levels. At another site there was evidence of indoor sources of Pb at certain times which corresponded with indoor activities that could reentrain house dust. This result tends to contradict the conclusions of Alzona *et al.* *(189)* that reentrainment of particulate matter is not important in establishing I/O ratios for the elements in particulate matter. At other times the lead–bromine ratio indicated that the outdoor contributions of auto exhaust predominated.

In another study by Moschandreas *et al.* *(170)*, the numerical indoor–outdoor ratios for Pb were estimated as follows:

Residences with gas stoves	0.65
Residences with electric stoves	0.68
Office buildings	0.42

Halpern *(200)* measured indoor and outdoor concentrations of particulate Pb in four building settings in New York City (Table XVIII)

Table XVIII Indoor–Outdoor (I/O) Relationships for Lead in Four New York City Locations *(200)*

	Mean particulate lead concentration, μg/m³		
Location	*Indoor range*	*Outdoor range*	*I/O ratio*
New air-conditioned apartment building	0.12–0.40	0.13–0.50	0.82
Air-conditioned public building	0.15–0.79	0.33–1.18	0.63
Non-air-conditioned storeroom in public building	0.45–0.98	0.38–1.05	0.81
Older non-air-conditioned apartment building	0.14–0.51	0.17–0.64	0.87

and used the results to show that a "homeostatic balance between indoor and outdoor pollutant levels" does not exist, at least in terms of this pollutant. A unique sampler consisting of two Nuclepore filters in series was used for collecting particulate matter. The pore size of the first filter was 8 μm, and the second was 0.8 μm. The material collected on the second filter was defined as "respirable particulate matter" and was the basis for determining indoor–outdoor Pb relationships. Several 2-hour concurrent indoor and outdoor samples were collected sequentially at each sampling site so that a time series of indoor–outdoor relationships could be examined. The time series plots showed that indoor Pb concentrations paralleled those outdoors, but usually at a lower level. However, in some instances, indoor concentrations appeared to lag behind those outdoors.

The air-conditioned public building (Table XVIII) showed the lowest indoor–outdoor ratio for Pb, but the ratios for the other sites (including the air-conditioned apartment building) were all higher, all in the range of 0.81 to 0.87. In light of the differences in indoor–outdoor ratios between buildings, the author stated that "the results of this study demonstrated the importance of considering parameters other than outdoor pollutant levels in determining indoor air quality."

Tosteson *et al.* *(201)* studied the metals content of indoor, outdoor, and personal monitoring samples in Topeka, Kansas (Table XIX). Eighteen nonsmoking individuals took part in the personal monitoring portion of the study. The collected respirable particulate matter was analyzed for several metals. These results show that there are significant indoor sources of respirable particulate matter, some indoor sources of aluminum, and no significant indoor sources of lead. Certainly the 0.74 I/O ratio is in the range of earlier work under

Table XIX Mean, Indoor, Outdoor, and Personal Monitoring Results for Iron, Lead, and Aluminum in Particulate Matter in Topeka, Kansas *(201)*

	Concentrations, μg/m³			
Monitoring location	*Respirable particulate matter*	*Fe*	*Pb*	*Al*
Personal	29.6	0.0499	0.1605	0.0276
Indoor	30.5	0.0328	0.0919	0.0305
Outdoor	12.2	0.0349	0.1242	0.0122
I/O ratios (dimensionless)	2.5	0.94	0.74	1.08

real-life conditions in spite of a major difference in the sampling method and the range of particle size collected. The I/O ratio of 0.49 determined by Alzona *et al. (189)* cannot be compared with these results in view of the unique conditions of their experiments. It was concluded from the work in Topeka that the outdoor level of Pb is not a good predictor of the personal level even though the overall concentrations are about the same.

Berk *et al. (202)* included measurement of indoor and outdoor Pb concentrations in their studies of indoor air quality in relation to conservation measures. Studies of energy conservation retrofits on several houses in the Pacific Northwest of the United States *(198)* found that the pre- and postretrofit I/O ratios for Pb were 0.56 and 0.38, respectively, for one house and 0.87 and 0.37, respectively, for another. This finding indicates that energy conservation measures provide additional isolation from outdoor exposures to Pb.

2. Respirable and Inhalable Particulate Matter

The United States Environmental Protection Agency *(203)* has proposed revision to the National Ambient Air Quality Standards for particulate matter in which health-related standards would be based upon selected sizes of particles rather than the broad spectrum of particle sizes collected by such devices as the High Volume Air Sampler. The proposal is for a standard for inhalable particulate matter (IPM), with a cut size of about 10 μm, and respirable particulate matter (RPM), with a cut size of about 2.5 μm. Few data have been collected for indoor IPM, but there is a significant data base for indoor concentrations of RPM. Several different methods have been used for monitoring RPM in indoor–outdoor air quality studies. Therefore, comparisons of results from various studies should be made with caution.

Moschandreas *et al. (178)* measured RPM with dichotomous samplers. The RPM fraction represented particles essentially 3.5 μm and less in diameter. The data from 14 different residential settings were divided into two distinct groupings: those with indoor–outdoor (I/O) ratios for RPM 2.0 or greater (range 2.0 to 3.5) and those with I/O ratios of 1.0 or less (range 0.4 to 1.0). The six residences with I/O ratios >2 all had children 7 years old or less, and all but one of the homes had smoking occupants. The one home with no smoking but with small children had an I/O ratio of 2.8. The eight homes with I/O ratios <1.0 had no young children and had little or no smoking (<10

cigarettes/day). These results show not only an apparent influence on I/O RPM ratios of smoking, but also indoor activity or other conditions related to small children.

In another study, Moschandreas *et al.* *(170)* again sampled for RPM by using dichotomous samplers. Numerical values for I/O ratios were not reported, but average I/O ratios were estimated by the scaling of histograms for average indoor and outdoor values and were then stratified in terms of smoking and nonsmoking residences and office buildings, as follows:

Residences with smokers	4.4
Residences without smokers	1.4
Office buildings	1.1

Whereas all of the residences with smokers had gas stoves, one home had a gas stove and no smokers and its I/O ratio for RPM was 1.4, which was in the range for the homes with electric stoves and no smoking. On the basis of this one case, gas stoves appear to have little effect on indoor RPM levels. Use of fireplaces in two of the homes had a strong transient impact on indoor RPM. No significant impact on indoor RPM was noted from the use of a basement wood stove in one of the homes.

As part of the Harvard Six Cities Study, Dockery and Spengler *(204)* assessed the importance of smoking, air conditioning, and heating systems on indoor–outdoor relationships for RPM. Although indoor concentrations of RPM are related to outdoor concentrations, indoor sources such as smoking tend to overshadow the influence of the outdoor ambient air quality. In areas of high outdoor pollution levels there was a tendency to close up the houses to isolate occupants from outdoor pollutants. Thus, with no large indoor sources, this condition produces low I/O ratios for RPM. Except for Steubenville, Ohio, which exhibited high outdoor RPM levels, the average RPM values for air-conditioned homes were higher than those outdoors, implying that the greater isolation of the air-conditioned home from the outdoor environment magnifies the impact of indoor sources.

In another section in the Six Cities Study, Spengler *et al.* *(205)* summarized long-term measurements for indoor and outdoor RPM for all six of the cities monitored in this program. There was considerable similarity in outdoor RPM in all the cities except Steubenville, Ohio, with its heavy contribution of industrial sources, but differences in indoor RPM were significant. Smoking was one of the most important contributions to these differences. The following mean values were developed from data for all six cities.

Location	*Mean concentration* ($\mu g/m^3$)	*I/O ratio*
Outdoor	21.1	—
Indoor: no smokers	24.4	1.16
Indoor: 1 smoker	36.5	1.73
Indoor: 2+ smokers	70.4	3.34

Repace and Lowrey *(172)* have studied indoor–outdoor relationships for RPM in relation to smoking, using a sampler based on the piezobalance principle that is reported to collect particles in the range of 0.01 and 3.5 μm. Sampling was conducted in a wide range of indoor settings with and without smoking. In nonsmoking settings (residences, libraries, restaurants, etc.), indoor RPM was in the range of 24 to 57 $\mu g/m^3$. On those occasions when parallel outdoor samples were taken, the I/O ratio for RPM was reasonably close to unity. On the other hand, a wide range of indoor values was obtained in the smoking situations. In one lodge hall, an indoor value for RPM of 697 $\mu g/m^3$, which was equivalent to an I/O ratio of 11.6, was measured.

Berk *et al.* *(197)* measured indoor and outdoor RPM at an elementary school by using a dichotomous sampler (2.5-μm, 50% cut size). Sampling was carried out at damper settings representing normal ventilation conditions and two reduced-ventilation (energy-conserving) modes. The following data were obtained:

Ventilation rate m^3*/hour/occupant*	*Mean RPM concentration,* $\mu g/m^3$		
	Indoors	*Outdoors*	*I/O ratio*
10.8	19.0	26.3	0.73
9.1	11.7	15.3	0.77
2.5	15.5	26.3	0.59

Although these results are based on too few data to produce definitive conclusions about the quantitative relationship between reduced ventilation and I/O ratio for RPM, the authors were assured that indoor RPM concentrations did not increase in this building with decreased ventilation and that such buildings offer additional protection from RPM in a reduced-ventilation mode.

In a study of the effect of energy conservation retrofits on houses in the northwestern United States, Berk *et al.* *(198)* measured indoor and outdoor RPM concentrations at two houses. House 1 had smoking occupants, but data were collected during both smoking and nonsmoking periods. House 2 had no smoking occupants. The retrofits reduced air exchange rates from 20% to 30% with the ventilation fan turned on. Table XX summarizes the RPM data for these houses.

Table XX Summary of Indoor–Outdoor (I/O) Respirable Particulate Matter (RPM) Levels in Two Houses Before and After Energy Conservation Retrofits *(198)*

Monitoring condition	*Preretrofit* RPM concentrations (μg/m³) *Indoor*	*Preretrofit* RPM concentrations (μg/m³) *Outdoor*	*Preretrofit* *I/O ratio*[a]	*Postretrofit* RPM concentrations (μg/m³) *Indoor*	*Postretrofit* RPM concentrations (μg/m³) *Outdoor*	*Postretrofit* *I/O ratio*[a]
House no. 1						
Smoking	31	10	3.44	36	11	5.42
Nonsmoking	9	14	0.64	8	13	0.39
House no. 2						
Nonsmoking occupants	12	9	1.94	10	10	1.33

[a] I/O ratios are averages of daily values which may differ from the I/O ratios of average concentrations.

From this table the striking impact on smoking on RPM levels in house 1 is evident. The energy conservation retrofit increased the I/O ratio for the smoking situation and decreased it for the nonsmoking situation. Although house 2 appears to have indoor sources other than smoking, on the basis of the I/O ratio of 1.94, the energy conservation retrofit caused a decrease in the I/O ratio to 1.33. No explanation was given for this result. These workers concluded that the energy conservations retrofit procedures at these houses produced a minimal effect on indoor air quality.

3. Radon and Progeny

In a study of indoor air quality in energy-efficient houses, Berk *et al. (202)* measured indoor radon levels between 0.5 and 1.0 nCi/m³ in an experimental house at Iowa State University. In an energy-efficient house in Maryland radon levels of 20 to 30 nCi/m³, well above the United States EPA guidelines for Florida phosphate lands, were measured. In another study of the effect of energy conservation retrofits on indoor air quality, Berk *et al. (198)* measured indoor radon levels before and after retrofit in one house to be <1 and 1.2 nCi/m³, respectively. In other house in this same study, indoor radon levels were <1 nCi/m³ both before and after retrofits.

As a result of the high radon levels found in the Maryland energy-efficient house, Moschandreas and Rector *(205)* surveyed 52 houses for indoor radon concentrations in concentric zones around the study house to determine whether these high radon values were specific to the study house or were characteristic of all houses in the general area. Table XXI summarizes the results. Statistical analysis of the data showed that basement radon concentrations in the neighborhood and town zones were from like distributions and were significantly higher than basement concentrations in the rural zone, and that the main floor concentrations were not significantly different across the three zones. Furthermore, 55% of the basements and 30% of all main floors displayed radon concentrations in excess of the 4.0 nCi/m^3 EPA limit applied to houses built on Florida phosphate land. Although the authors did not report it directly, it appears that the high readings noted by Berk *et al.* *(202)* at the energy-efficient house were not anomalies, and that house location is an extremely important factor in determining indoor radon levels.

The proceedings of the Third International Conference on Indoor Air Quality and Climate held in Stockholm, Sweden, in August 1984 *(149)* contain over 20 papers on the subject of indoor radon source concentrations and risk. Those wishing to pursue this subject in greater detail are directed to these volumes. In general the data on indoor radon show great variability from home to home in respect to structural features and from location to location in terms of geographical factors. Table XXII, based on Oswald and Alter *(206)* data collected in several countries by using integrating Track Etch® samplers,

Table XXI Mean Indoor Radon Levels in Concentric Zones around an Energy-Efficient House in Maryland *(205)*

Zone or house	*Radon concentrations (nCi/m³)* Basement	*Radon concentrations (nCi/m³)* Main floor	*Radon progeny (WL)* Basement	*Radon progeny (WL)* Main floor	*Infiltration rate (air changes per hour)*
Neighborhood zone (≤0.5 km from house)	19.2	8.7	0.195	0.038	0.25
Town zone[a] (≤2 km from house)	9.2	2.8	0.047	0.025	0.45
Rural zone[b] (≤25 km from house)	3.4	2.4	0.023	0.019	0.55
Energy-efficient house	12.1	10.8	—	—	0.1 (nominal)

[a] Exclusive of neighborhood.
[b] Exclusive of town and neighborhood.

Table XXII A Summary of Indoor Radon Concentration Using Track Etch[R] Integrating Monitors *(206)*

Location	*Number of measurements*	*Highest reading (PCi/liter)*	*Percentage greater than 4 PCi/liter*
United States	2496	133	12
Canada	546	34	21
Switzerland	634	729	62
Norway	293	288	58
Italy	67	60	28
Sweden	47074	1140	81

illustrate the kind of variability one finds in broad spectrum radon surveys. Many of the measurements were made in "problem" homes and may not be representative of all homes sampled in that country. Nevertheless, the authors believe that the results indicate that indoor radon poses a potentially serious health problem.

Another aspect of indoor radon that may tend to increase the health risk is the joint presence of radon and radon progeny with respirable particles. Radon progeny are known to attach to aerosol particles. Bergman *et al. (207)* describe experiments that seem to indicate that indoor exposure to radon progeny may increase considerably in the presence of cigarette smoke, since newly formed, electrically charged progeny tend to attach to smoke particles rather than to walls, furnishings, etc., and thus be drawn deeply into the lungs.

4. Organic Gases and Vapors

As pointed out in Section C,3, an incredibly wide range of organic gases and vapors is released indoors from building materials, furnishings, consumer products, and human activity. The organic material receiving the most attention is formaldehyde, since it is widely used in resins and adhesives and as a binder for a variety of products.

a. FORMALDEHYDE. There has been extensive work on measuring indoor formaldehyde levels in northern European countries where particle board is used extensively in construction. The study of formaldehyde concentrations indoors in the United States has been aggressively pursued only relatively recently.

Moschandreas *et al. (178)* monitored total aldehydes indoors and outdoors in a number of dwellings. The range of mean values indoors over 14-day sampling periods as 47–470 $\mu g/m^3$ (approximately 0.04–

0.4 ppm). Highest mean values were found in one of the mobile homes. Concurrent outdoor values were in the range of one-fifth to one-thirtieth the indoor values. The maximum 4-hour value of 934 $\mu g/m^3$ (0.76 ppm) was also found in one of the mobile homes. The interior of mobile homes is usually constructed with a significant amount of plywood and laminated paneling.

The potential for high indoor concentrations of formaldehyde in mobile homes has been confirmed by Dally *et al. (208)*, who found maximum 30- to 60-minute average formaldehyde concentrations of 3.68 ppm (4500 $\mu g/m^3$) in a mobile home, an indication that the 8-hour United States occupational standard for formaldehyde might be approached or exceeded in such settings. Elevated indoor formaldehyde concentrations were also found in homes with UF foam insulation, but these concentrations were lower than those found in mobile homes.

In a study of energy-efficient homes, Berk *et al. (202)* monitored both aliphatic aldehydes and formaldehyde. Formaldehyde values were about 60–70% of total aliphatic aldehyde values. Indoor and outdoor formaldehyde concentrations at two of these houses are presented as follows:

	Concentrations, $\mu g/m^3$		
	Indoors		
House	*Range*	*Average*	*Outdoor average*
1	34–75	42	<9
2	54–182	120	<12

Neither house had extremely high indoor formaldehyde levels, but one had average levels at or exceeding typical European standards.

In a study of energy conservation ventilation modes on indoor air quality at a school, Berk *et al. (197)* showed that total aldehyde levels indoors were extremely low, never exceeding 25 $\mu g/m^3$. Berk *et al. (198)*, in a study of the effect of energy-conserving retrofits, showed that formaldehyde levels did not change significantly after the retrofits and that they were below typical European standards of 120 $\mu g/m^3$.

In a study of 40 homes in Tennessee, Hawthorne *et al. (209)* found that 25% of the homes had formaldehyde concentrations exceeding 100 ppb (123 $\mu g/m^3$). In homes that had high concentration (e.g., those with UFFI), formaldehyde concentrations usually increased with temperature. Newer homes had higher indoor formaldehyde concentrations than older homes. Niemelä and Toppila *(210)* found that ventilation rates of at least 0.5 $hour^{-1}$ were needed to keep indoor concentrations of formaldehyde below 150 $\mu g/m^3$, the Finnish indoor standard.

b. Other Organic Compounds. The broad class of other organic compounds has been overshadowed by emphasis on formaldehyde. Now it is generally recognized that among the many organic compounds found in the indoor atmosphere are many with significant potential to create adverse health effects, and at the very least these compounds may produce odors and irritation to the eyes, nose, and throat. Many of these compounds have been implicated in the "sick building syndrome."

Moschandreas *(170)* monitored for benzo[*a*]pyrene (BaP), one of a group of polycyclic organic compounds suspected of being strongly carcinogenic. In a house with a fireplace, indoor BaP concentrations were always above those outdoors; however, the house also had smoking occupants. Nevertheless, on one occasion when the fireplace was in use, the indoor concentration reached a concentration of almost 12 ng/m^3, showing an unmistakable contribution from the fireplace. Outdoor concentrations of BaP at this house were always less than 1 ng/m^3. In a house with a wood stove and no smokers, indoor BaP concentrations averaged 4.7 ng/m^3, nearly five times typical outdoor levels in urban areas.

Miksch *et al. (211)* investigated trace organics in offices, using a variety of analytical techniques. Comparative gas chromatograms for indoor and outdoor samples showed qualitatively, but strikingly, the extreme complexity of the mixture of organic compounds in the indoor office atmosphere as compared with the outdoor samples. A wide range of alkanes (benzene, toluene, and xylene) and chlorinated compounds (e.g., trichloroethylene and tetrachloroethylene) were found. Outdoor concentrations for most of these compounds were negligible. This study also identified a wide range of organic gases and vapors in the head space above a variety of building materials and furnishings.

At the Third International Conference on Indoor Air Quality and Climate held in Stockholm, Sweden, in August 1984 *(149)*, over 20 papers were presented on sources, concentrations, and problems presented by organic gases and vapors in the indoor environment. Clearly this category of pollutant is coming of age and is receiving the international attention that it deserves.

Pellizzari *et al. (212)* have tentatively identified over 359 chemicals in indoor air. Table XXIII shows the classes of compounds and numbers within each class identified in a home for the elderly. Samples were collected on Tenax and were analyzed by GC/MS. In other

Table XXIII Chemicals Identified in Indoor Air of Home for the Elderly *(212)*

Chemical group	*No. identified*	*Examples*
Alkanes/alkenes alkynes	179	Methylcyclobutane *n*-Pentane 4-Nonyne
Alcohols	21	*n*-Butanol 2-Ethyl-4-methyl-1-pentanol
Aldehydes	6	*n*-Nananal 3,3-Dimethylhexanal
Esters	1	Ethyl acetate
Ethers	12	1,4-Dioxane 2-Furancarboxaldehyde
Ketones	22	Acetone 4-Penten-2-one
Nitriles	1	Benzonitrile
Aromatics	69	Benzene Styrene
Halogenated alkanes	9	Methylene chloride Tetrachloroethylene
Halogenated aromatics	4	*o*-Dichlorobenzene
Aromatic aldehydes	2	α-Methylbenzaldehyde
Sulfur compounds	3	6-Methylbenzo[*b*]thiophene
Miscellaneous	30	
Total	359	

work by Pellizarri *et al. (213)* a design of the sampling and quality assurance protocol is presented. Wallace *et al. (214)* used personal monitors and breath analysis to determine exposure to several of the categories of organics identified by Pellizarri *(212)*. On the basis of the results for both of these methods, total (indoor and outdoor) exposure to the organic gases and vapors was invariably greater than that contributed by outdoor concentrations alone. The authors concluded that breath analysis is a useful means of determining previous exposure to such compounds, both indoors and outdoors, and to associated activity patterns.

Yocom *et al. (179)* used a real-time, mobile mass spectrometer system to identify and measure concentrations of a wide range of organic compounds in office buildings where complaints of poor indoor air quality had been received from workers. Table XXIV presents the data obtained in one of the buildings before and after adjustments to the ventilation system were made. Many of the compounds measured were above the odor or irritation threshold before adjustments, but, as can be seen from the results, the concentrations were orders of magnitude below United States occupational standards.

Table XXIV Comparison of Contaminant Concentrations in an Office Building before and after Ventilation adjustments *(179)*

Contaminant	*Before adjustment (ppb)*	*After adjustment (ppb)*	*Lower detection limit (ppb)*	*United States Occupational Standard (ppb)*
Ethylamine	55–400	0.030–0.95	0.00039	100,000
Diethylamine	8.0–25	0.60–0.83	0.0035	5,000
Dimethyl Acetamide	3.0–1,300	<0.0011–9.3	0.0011	10,000
Xylene	<3.0–470	13–35	2.7	100,000
Butyl acrylate	<0.10–9.0	1.3–3.1	0.10	10,000
Acetic acid	44–200	3.9–6.3	0.0092	10,000
Nicotine	<0.050–2.0	<0.0034	0.0034	—
Pyruvic acid	<0.024–1.6	0.57–1.4	0.024	—
Lactic acid	<0.0010–0.50	0.090–0.16	0.0010	—

Lebret *et al. (215)* used charcoal tubes and then high-resolution capillary gas chromatography and flame ionization detection to determine concentrations of 45 volatile hydrocarbon in 134 houses in Ede, Netherlands. Most indoor concentrations were above those outdoors, and homes with smokers showed higher indoor concentrations of these compounds than homes without smokers. The authors believe that modeling indoor concentrations in terms of emission rates from building materials has serious drawbacks because consumer products also emit some of the same materials emitted by building materials.

Vedel and Nielsen *(216)* have measured phthalate esters in Danish homes. These materials are used in a wide variety of synthetic building materials and furnishings (e.g., plasticizers in coatings and floor tiles). Virgin *(217)* points out that phthalates may cause chlorophyll deficiency in house plants.

Van der Kolk *(218)* has measured indoor concentrations of wood preservatives in homes in the Netherlands. Although indoor concentrations of such materials as pentachlorophenol may reach concentrations in the range of 50–200 $\mu g/m^3$ in the period up to 3 months after application, these concentrations tend to decay rapidly thereafter.

F. Control of Indoor Air Quality

Indoor air quality can be maintained at acceptable levels by preventing contaminants from being released into the indoor environment, removing the contaminants that have been released, or diluting them with less contaminated air.

The control of indoor air quality can be looked at within two categories: preventative control, which can be applied prior to or incorporated in the construction or major modification of a building, and after-the-fact control, in which controls are applied to an existing building, after indoor air quality problems have occurred. Control of indoor air quality is particularly critical when attempts to conserve energy are being implemented.

Examples of preventative control include the following:

- Location of building to minimize impact of outdoor sources and radon penetration of basement areas
- Use of low-emission construction materials
- Isolation or venting of indoor sources
- Careful design and control of ventilation system to optimize thermal comfort and indoor air quality

Some of the same options are available for after-the-fact control but are invariably less-cost effective than preventative controls:

- Provide additional ventilation, assuming that outdoor air is of acceptable quality
- Remove or encapsulate high-emission construction materials and furnishings.
- Change building use patterns to minimize time spent in areas of poor indoor air quality
- Pass indoor air through an air cleaning device

Spengler and Sexton *(144)* have summarized the principal concepts for indoor air quality control and typical pollutants to which they apply and have provided several examples (Table XXV). Woods *(219)* proposes a method which can be used to evaluate the performance of a building in terms of indoor air quality control, energy consumption, and cost-effectiveness.

The Natural Academy of Sciences Report *(140)* and the "*Manual on Indoor Air Quality*" by Sandia National Laboratories *(143)* provide detailed lists of approaches and specific systems and equipment for control of indoor air quality. Billings and Vanderslice *(220)* present a detailed discussion of indoor air quality control options along with models for ventilation control.

Janssen *et al.* *(221)* designed an indoor air quality control system based on fresh makeup air controlled by both CO_2 and temperature. This work showed that it was possible to save up to 20% on energy

Table XXV Control Measures for Indoor Air Pollutants *(144)*

Control measure description	*Pollutant*	*Example*
Ventilation: dilution of indoor air with fresh outdoor air or recirculated filtered air, using mechanical or natural methods to promote localized, zonal, or general ventilation	Radon and radon progeny; combustion by-products; tobacco smoke; biological agents (particles)	Local exhaust of gas stove emissions; air-to-air heat exchangers; building ventilation codes
Source removal or substitution: removal of indoor emission sources or substitution of less hazardous materials or products	Organic substances; asbestiform minerals; tobacco smoke	Restrictions on smoking in public places; removal of asbestos
Some modification: reduction of emission rates through changes in design or process; containment of emissions by barriers or sealants	Radon and radon progeny; organic substance; asbestiform minerals; combustion by-products	Plastic barriers to reduce radon levels; containment of asbestos; design of buildings without basements to avoid radon; catalytic oxidation of CO to CO_2 in kerosene burners
Air cleaning: purification of indoor air by gas adsorbers, air filters, and electrostatic precipitators	Particulate matter; combustion by-products; biological agents (particles)	Residential air cleaners to control tobacco smoke or wood smoke; ultraviolet irradiation to decontamirate ventilation air; formaldehyde sorbent filters
Behavioral adjustment: reduction in human exposure through modification of behavior patterns; facilitated by consumer education, product labeling, building design, warning devices, and legal liability	Organic substances; combustion by-products; tobacco smoke	Smoke-free zones; architectural design of interior space; certification of formaldehyde concentrations for home purchases

while maintaining acceptable indoor air quality. Södergren *(222)* also showed that a CO_2-controlled ventilation system is an energy-efficient method of maintaining acceptable indoor air quality.

The proceedings of the Third International Conference on Indoor Air Quality and Climate *(149)* contain a large number of papers on the effectiveness of various control concepts in improving indoor air quality. Those interested in pursuing this area are directed to these volumes and their extensive lists of references. The following paragraphs provide brief summaries of some of the work reported at this conference that illustrate the spectrum of potential control methods.

Sextro *et al.* *(223)* evaluated 11 portable air cleaners for their effectiveness in removal of respirable particulate matter. They also evaluated the effects of particle removal on levels of radon progeny. In general, air cleaners (electrostatic precipitators and high-efficiency filters) were quite effective, whereas panel filters, an ion generator, and mixing fans were relatively ineffective in removing both respirable particulate matter and radon progeny. Nitschke *et al.* *(224)* found that activated charcoal was of little value in controlling indoor radon concentrations. Sealing cracks in basement walls and ventilating the subslab were the most effective control techniques. In very tight houses, whole house ventilation with heat exchange was also shown to be effective.

Lehti *(225)* evaluated a smoke hood to control emissions from a smoking area in a Finnish squash club. The electrostatic precipitator in the hood system removed 96% of the 0.5- to 1-μm smoke particles. Ueno *(226)* compared the effectiveness of a ventilator, an air filter, an electrostatic precipitator, and a humidifier in reducing radon concentrations of tobacco smoke aerosol. It was found that the humidifier was surprisingly effective, and the author recommends changes in humidifier design to improve aerosol collection efficiency.

Levin and Hahn *(227)* have shown that sealing of surfaces of wood treated with pentachlorophenol is an effective method of controlling indoor concentrations of vapors of this material. Pederson and Fisk *(228)* measured formaldehyde removal efficiencies, using air washers and then a refrigeration system downstream to remove excess humidity. Control efficiencies were in the range of 30–63%.

G. *Indoor Air Quality Regulations*

The indoor environment is an extension of the outdoor environment. However, providing a comfortable indoor environment necessitates modifying the temperature and often the humidity of the air significantly during certain seasons of the year. These modifications have a profound effect upon air exchange and indoor concentrations of pollutants. Because of the uniqueness of indoor air quality and the complex interrelationship between the factors that control it, regulations and specific guidelines to achieve acceptable air quality have been slow to develop. This development has been particularly slow in the United States, where air quality criteria and standards for the general public have dealt exclusively with outdoor ambient air, and where regulators and politicians are reluctant to develop standards that may take away independence of action by the home owner in determining the kind of indoor environment he or she wants. These issues are discussed in some detail by Spengler and Sexton *(144)*. The only standards for indoor air quality in the United States are an ozone standard whereby devices used indoors produce ozone and standards for radon in homes built over uranium mine tailings or phosphate mining land. Several countries have developed standards for indoor radon concentrations *(140)*. Some local governments in the United States are beginning to attack the issue of indoor air quality regulations by establishing controls on smoking in public buildings. In Section II, indoor air quality criteria for archives, libraries, and museums are discussed.

The National Academy of Sciences report *(140)* summarizes a number of indoor air quality standards and guidelines as of 1981. Table XXVI is a summary of formaldehyde standards from this report.

One indoor air quality guideline that has been and will continue to be widely used in the United States as an engineering guideline for ventilation systems, indoor air quality, and energy conservation in buildings is that published by the American Society of Heating, Refrigerating and Air Conditioning Engineers (ASHRAE). The first version appeared in 1973 *(229)* and an updated version, which is still under review, in 1981 *(230)*. Janssen *(231)* has prepared a status report on the 1981 version. Table XXVII from Janssen's work compares the criteria for acceptable indoor air quality in both versions of the standard. At this writing (1985) the principal issues being addressed are the amount of ventilation air per occupant required, the question of whether these criteria should consider the impact of indoor air quality on visitors, and the way to deal with smoking. In view of the attention that indoor air quality and energy conservation are receiving, this standard when finally adopted will be widely used by engineers and architects when planning new buildings.

Table XXVI A Summary of Formaldehyde Standards *(140)*

Country	*Concentration, parts per million (ppm)*[a]	*Status*
Indoor air:		
United States	[b]	[b]
Denmark	0.12 ppm maximum	Recommended
Netherlands	0.1 ppm maximum	Recommended by Ministers of Housing and Health
Sweden	0.1 ppm maximum, new buildings 0.4 ppm maximum, old buildings[c] 0.7 ppm maximum, old buildings[c]	Proposed by National Board of Health and Welfare
Federal Republic of Germany	0.1 ppm maximum	Recommended by Ministry of Health
Occupational air:		
United States	3 ppm, 8-hour time-weighted average	Promulgated by the Occupational Safety and Health Administration
	5 ppm, ceiling	Promulgated by the Occupational Safety and Health Administration
	1 ppm, 30-minute maximum	Recommended by the National Institute of Occupational Safety and Health

[a] 0.1 ppm ≃ 120 g/m^3.
[b] Several states have proposed indoor standards in the range of 0.2–0.5 ppm.
[c] 0.4–0.7 ppm is a border range. Concentrations higher than 0.7 ppm do not meet the standard. Those lower than 0.4 ppm do meet the standard. Those within the range do not meet the standard if dwellers complain. In recently built houses, 0.7 ppm should be acceptable during first 6 mo.

Table XXVII Comparison of Criteria for Acceptable Indoor Air Quality in the 1973 and 1981 Versions of ASHRAE Ventilation Standard 62 *(231)*

	Annual average		*Short-term level*			
	1973	*1981*	*1973*		*1981*	
Contaminant	*μg/m³*	*μg/m³*	*μg/m³*	*Averaging time*	*μg/m³*	*Averaging time*
Carbon monoxide	20,000		30,000	8 Hours	40,000	1 Hour
Lead	—	1.5 for 3 months	—		—	
Nitrogen dioxide	200	100	—		—	
Oxidants (ozone)	100		500	1 Hour	235	1 Hour
Particulates	60	75	150	24 Hours	260	24 Hours
Sulfur dioxide	80	80	400	24 Hours	365	24 Hours
Hydrocarbons	1800	—	4,000	3 Hours		

REFERENCES

1. The Clean Air Act as amended August 7, 1977. Public Law 95-95. U. S. Govt. Printing Office, Washington, D.C.
2. W. H. Lewis, Jr., *J. Air Pollut. Control Assoc.* **30,** 118 (1980).
3. J. Horn, *J. Air Pollut. Control Assoc.* **30,** 120 (1980).
3a. R. D. Rowe and L. G. Chestnut, "Managing Air Quality and Scenic Resources at National Parks and Wilderness Areas." Westview Press, Boulder, Colorado, 1983.
4. E. J. McCartney, "Optics of the Atmosphere," Chapter 1. Wiley, New York, 1976.
5. H. Horvath, *Atmos. Environ.* **15,** 1785 (1981).
6. W. E. K. Middleton, "Vision Through the Atmosphere." Univ. of Toronto Press, Ottowa, Canada, 1952.
7. E. G. Walther, "Visibility Terminology," Preliminary report to Visibility Committee (TE-5). Air Pollut. Control Assoc., Pittsburgh, Pennsylvania, 1984.
8. R. E. Huschke, ed., "Glossary of Meteorology." Am. Meteorol. Soc., Boston, Massachusetts, 1959.
9. D. A. Latimer, R. W. Bergstrom, S. R. Hayes, M.-K. Lui, J. H. Seinfeld, G. Z. Whitten, M. A. Wojcik, and M. J. Hillyer, "The Development of Mathematical Models for the Prediction of Anthropogenic Visibility Impairment," Vol. 1, EPA-450/3/78-110a. U. S. Environmental Protection Agency, Research Triangle Park, North Carolina, 1978.
10. International Commission on Illumination, "International Lighting Vocabulary," 3rd ed., CIE No. 17 (E-1.1). U. S. National Committee, National Bureau of Standards, Washington, D.C., 1970.
11. Illuminating Engineering Society of North America, "IES Lighting Handbook Reference Volume." IESNA, New York, 1981.
12. P. J. Haas and A. J. Fabrick, *Atmos. Environ.* **15,** 2171 (1981).
13. D. A. Latimer and G. S. Samuelson, "Visual Impact of Plumes from Power Plants," UCI-ARTR-75-3. University of California, Air Quality Laboratory, Irvine, 1975.
14. A. P. Waggoner, R. E. Weiss, N. C. Ahlquist, D. S. Covert, S. Will, and R. J. Charlson, *Atmos. Environ.* **15,** 1891 (1981).
15. R. W. Bergstrom, C. Seigneur, B. L. Babson, H.-Y. Holman, and M. A. Wojcik, *Atmos. Environ.* **15,** 2135 (1981).
15a. D. A. Hegg, P. V. Hobbs, and J. H. Lyons, *Atmos. Environ.* **19,** 1147 (1985).
16. W. H. White, ed., "Plumes and Visibility: Measurements and Model Components — Symposium Proceedings," *Atmos. Environ.*, Vol. 15, No.'s 10/11 and 12, pp. 1785–2646, Pergamon, Oxford: 1981.
17. P. J. Drivas, A. Bass, and D. W. Heinhold, *Atmos. Environ.* **15,** 2179 (1981).
18. H. Özkaynak, C. Berkley, A. Curreri, and D. Heinhold, "Multiple- and Single-source Visibility Modeling on Regional Scales," Proc.: View on Visibility, APCA Spec. Conf., p. 137. Air Pollut. Control Assoc., Pittsburgh, Pennsylvania, 1979.
19. M. D. Williams, E. Treiman, and M. Wecksung, "Plume Blight Visibility Modeling with a Simulated Photograph Technique, "Proc.: View on Visibility, APCA Spec. Conf., p. 108. Air Pollut. Control Assoc., Pittsburgh, Pennsylvania, 1979.
20. M. D. Williams, E. Treiman, and M. Wecksung, *J. Air Pollut. Control Assoc.* **30,** 131 (1980).
21. M. Williams, L. Y. Chan, and R. Lewis, *Atmos. Environ.* **15,** 2151 (1981).
22. W. F. Dabberdt and S. L. Eigsti, *Atmos. Environ.* **15,** 2055 (1981).

23. W. B. Johnson, D. E. Wolf, and R. 1. Mancuso, *Atmos. Environ.* **12,** 511 (1978).
23a. W. H. White, C. Seigneur, D. W. Heinold, M. W. Eltgroth, L. W. Richards, P. T. Roberts, P. S. Bhardwaja, W. D. Conner, and W. E. Wilson, *Atmos. Environ.* **19,** 515 (1985).
24. P. Middleton, T. R. Stewart, and R. L. Dennis, *Atmos. Environ.* **17,** 1015 (1983).
24a. P. Middleton, T. R. Stewart, and J. Leary, *J. Air Pollut. Control Assoc.* **35,** 11 (1985).
25. J. Mumpower, P. Middleton, and R. L. Dennis, *Atmos. Environ.* **15,** 24433 (1981).
26. D. A. Latimer, H. Hago, and T. C. Daniel, *Atmos. Environ.* **15,** 1865 (1981).
27. W. Malm, *J. Air Pollut. Control Assoc.* **29,** 1042 (1979).
28. W. C. Malm, K. K. Leiker, and J. V. Molenar, "Human Perception of Visual Air Quality," Proc.: View on Visibility, p. 36. Air Pollut. Control Assoc., Pittsburgh, Pennsylvania, 1979.
29. W. Malm, K. Kelley, J. Molenar, and T. Daniel, *Atmos. Environ.* **15,** 1875 (1981).
30. D. A. Latimer and T. C. Daniel, "Preliminary Results of a Study of Human Judgments of Visual Air Quality," Proc.: View on Visibility, p. 11. Air Pollut. Control Assoc., Pittsburgh, Pennsylvania, 1979.
31. R. C. Henry, "The Human Observer and Visibility—Modern Psychophysics Applied to Visibility Degradation," Proc.: View on Visibility, p. 27. Air Pollut. Control Assoc., Pittsburgh, Pennsylvania, 1979.
32. C. J. Bartelson, *Photogr. Sci. Eng.* **5,** 327 (1961).
33. C. J. Bartleson and C. P. Bray, *Photogr. Sci. Eng.* **6,** 19 (1962).
34. J. Trijonis, *Atmos. Environ.* **13,** 833 (1979).
35. E. S. Macias, J. O. Zwicker, and W. H. White, *Atmos. Environ.* **15,** 1987 (1981).
35a. R. J. Szymber and W. D. Sellers, J. Climate Appl. Meteor **24,** 725 (1985).
36. R. B. Husar, J. M. Hollaway, D. E. Patterson, and W. E. Wilson, *Atmos. Environ.* **15,** 1919 (1981).
37. C. S. Sloan, *Atmos. Environ.* **16,** 41 (1982).
38. C. S. Sloan, *Atmos. Environ.* **16,** 2309 (1982).
39. C. S. Sloan, *Atmos. Environ.* **17,** 763 (1983).
40. C. S. Sloan, *Atmos. Environ.* **18,** 1217 (1984).
41. J. Trijonis and K. Yuan, "Visibility in the Northeast," EPA-600/3-78-075. U. S. Environmental Protection Agency, Research Triangle Park, North Carolina, 1978.
41a. P. G. Vinzani and P. J. Lamb, *J. Climate Appl. Meteor* **24,** 435 (1985).
41b. R. Hitzenberger, *Atmos Environ.* **19,** 835 (1985).
42. N. C. Ahlquist and R. J. Charlson, *J. Air Pollut. Control Assoc.* **17,** 457 (1967).
42a. M. G. Ruby, *J. Air Pollute Control Assoc.* **35,** 244 (1985).
43. H. Horvath, *Appl. Opt.* **19,** 2651 (1980).
44. H. Horvath, *Atmos. Environ.* **15,** 2537 (1981).
45. W. Malm, A. Pitchford, R. Tree, E. Walther, M. Pearson, and S. Archer, *Atmos. Environ.* **15,** 2547 (1981).
46. C. Steffens, *Ind. Eng. Chem.* **41,** 2396 (1949).
47. T. G. Ellestad and R. E. Speer, *Atmos. Environ.* **15,** 2443 (1981).
48. I. Tombach and D. Allard, *J. Air Pollut. Control Assoc.* **30,** 134 (1980).
48a. R. D. Rowe and L. G. Chestnut (eds.), "The Value of Visibility: Economic Theory and Applications for Air Pollution Control," Abt Books, Cambridge, Massachusetts, 1982.
49. U. S. Environmental Protection Agency, "Air Quality Criteria for Sulfur Oxides and Particulate Matter (Draft)." USEPA, Research Triangle Park, North Carolina, 1981.
50. J. E. Yocom and A. R. Stankunas, "A Review of Air Pollutant Damage to Mate-

rials," A report to EPA Environmental Criteria and Assessment Office (NATO-CCMS Program). U. S. Environmental Protection Agency, Research Triangle Park, North Carolina, 1980.

51. S. E. Haagenrud, R. W. Lanting, and G. Santomauro, "Draft Report on Effects of Sulphur Compounds on Materials, Including Historical and Cultural Monuments." United Nations Economic and Social Council, Interim Executive Body for Convention on Long-Range Transboundary Air Pollution, Geneva, Switzerland, 1982.
52. J. E. Yocom and N. S. Baer, *in* "The Acidic Deposition Phenomenon and Its Effects: Critical Assessment Review Papers, Chapter E-7, Sect. 7.1, EPA 600/8-83-016BF. U. S. Environmental Protection Agency, Washington, D.C., 1984.
52a. M. Pourbaix, "Atlas of Electrochemical Equilibrium in Aqueous Solutions." Pergamon, Oxford, 1966.
52b. J. Kruger, *in* "Corrosion and Metal Artifacts: A Dialog Between Conservators and Archaeologists and Corrosion Scientists," Spec. Publ. No. 479, pp. 59–65. National Bureau of Standards, Washington, D.C., 1977.
53. J. O. Nriagu, *in* "Sulfur in the Environment. Part II. Ecological Impacts" (J. O. Nriagu, ed.), pp. 1–59, Wiley, New York, 1978.
54. J. B. Johnson, P. Elliot, M. A. Winterbottom, and G. C. Wood, *Corros. Sci.* **17,** 691–700 (1977).
55. F. H. Haynie, J. W. Spence, and J. B. Upham, "Effects of Gaseous Pollutants on Materials—A Chamber Study," EPA-600/3-76-015. U. S. Environmental Protection Agency, Research Triangle Park, North Carolina, 1976.
56. J. B. Upham, *J. Air Pollut. Control Assoc.* **17,** 398–402 (1967).
57. F. H. Haynie and J. B. Upham, *Mater. Prot. Perform.* **10,** 18–21 (1971).
58. F. B. Mansfield, "Regional Air Pollution Study. Effects of Airborne Sulfur Pollutants on Materials." EPA-600/4/80-007. U. S. Environmental Protection Agency, Research Triangle Park, North Carolina, 1980.
59. W. H. J. Vernon, *Trans. Faraday Soc.* **31,** 1668–1700 (1938).
60. B. Sanyal and G. K. Singhania, *J. Sci. Ind. Res., Sect. B* **15,** 448–55 (1956).
61. F. H. Haynie, "Theoretical Model of Soiling of Surfaces by Airborne Particles," *Proc. 2nd U.S.–Dutch International Symposium: Aerosols.* Williamsburg, Virginia, May 23, 1985. U.S. Environmental Protection Agency, Research Triangle Park, North Carolina.
62. A. B. Harker, F. B. Mansfield, D. R. Strauss, and D. D. Landis, "Mechanism of SO_2 and H_2SO_4 Aerosol Zinc Corrosion," EPA-600/3-80-018. U.S. Environmental Protection Agency, Research Triangle Park, North Carolina, 1980.
63. S. E. Haagenrud, "The Effect of Long Range Transport of Air Pollutants on the Atmosphere Corrosion Rate of Carbon Steel and Zinc as Measured by Weight Loss," Proc. 8th Scand. Corros. Conf. 1978.
64. F. W. Fink, F. H. Buttner, and W. K. Boyd, "Technical-Economic Evaluation of Air Pollution Corrosion Costs on Metals in the U.S.," APTD-0654. U. S. Environmental Protection Agency, Research Triangle Park, North Carolina, 1971.
64a. T. E. Graedel, G. W. Kammlott, and J. P. Franey, *Science* **212,** 663–665 (1981).
64b. T. E. Graedel, J. P. Franey, and G. W. Kammlott, *Science* **224,** 599–601, (1984).
65. H. Guttman, "Metal Corrosion in the Atmosphere," ASTM STP 435, pp. 223–239. Am. Soc. Test. Mater., Philadelphia, Pennsylvania, 1968.
66. F. H. Haynie and J. B. Upham, *Mater. Prot. Perform.* **9** (8), 35–40 (1970).
67. J. H. Cavendar, W. M. Cox, M. Georgevich, N. Huey, G. A. Jutze, and C. E. Zimmer, "Interstate Surveillance Project: Measurement of Air Pollution Using Static Monitors," APTD-0666. U. S. Environmental Protection Agency, Research Triangle Park, North Carolina, 1971.

68. H. Guttman and P. J. Sereda, *in* "Metal Corrosion in the Atmosphere," ASTM STP 435, pp. 223–239. Am. Soc. Test. Mater., Philadelphia, Pennsylvania, 1968.
68a. M. Benarie, Metallic Corrosion as Functions of Atmospheric Pollutant Concentrations and Rain pH. BNL 35668, Brookhaven National Laboratory, Upton, New York, 1984.
68b. Steinzerfall, Deutsches Nationalkomitee für Denkmalschutz, Bonn, Federal Republic of Germany, 1985.
69. R. Rossi-Manaresi, ed., "The Conservation of Stone: *Part A*. Deterioration." Cento per la Conservatione della Sculpture All 'Aperto, Bologna, 1981.
70. R. A. Livingston and N. S. Baer, "Mechanisms of Air Pollution-Induced Damage of Stone," Proc. 6th World Congr. Air Qual., pp. 33–40. International Union of Air Pollution Prevention Associations, Paris, 1983.
71. R. K. Stevens, T. G. Dzubay, R. W. Shaw, W. A. McClenny, C. W. Lewis, and W. E. Wilson, *Environ. Sci. Technol.* **14,** 1491–1498 (1980).
72. D. J. Spedding, *Atmos. Environ.* **3,** 341–346 (1969).
73. E. M. Winkler, *in* "Conservation of Historic Stone Buildings and Monuments," pp. 108–119. National Academy Press, Washington, D.C., 1982.
74. H. J. Plenderleith and A. E. A. Werner, "The Conservation of Antiquities and Works of Art," 2nd ed., pp. 316–317. Oxford Univ. Press, London and New York, 1971.
75. T. N. Skoulikidis, *in* "Atmospheric Corrosion" (W. H. Ailor, ed.), pp. 807–825. Wiley (Interscience), New York, 1982.
76. J. E. Yocom, *J. Air Pollut. Control Assoc.* **29,** 333–338 (1979).
77. A. Geike, *Proc. R. Soc. Edinburg,* pp. 518–532 (1880).
78. N. S. Baer and S. M. Berman, *Proc., Annu. Meet.—Air Pollut. Control Assoc.* **76,** Pap. 83-5.7 (1983).
79. S. Luckat, *Staub—Reinhalt. Luft* **41,** 440–442 (1981).
80. H. Schreiber, *in* "Impact of Air Pollutants on Materials. Report of Panel 3 of the NATO-CCMS Pilot Study on Air Pollution Control Strategies and Impact Modeling," pp. 26–27 (Engl. Transl.). NATO, 1982.
81. A. D. Sharp, S. T. Trudgill, R. V. Cooke, C. A. Price, R. W. Crabtree, A. M. Pickler, and D. L. Smith, *Earth Surf. Processes Landforms* **7,** 387–389 (1982).
82. R. A. Livingston and N. S. Baer, *Proc., Annu. Meet.—Air Pollut. Control Assoc.* **77,** Pap. 84-83.2 (1984).
83. National Materials Advisory Board, "The Status of Cement and Concrete R & D," NMAB Rep. No. 361, p. 23. National Academy of Sciences, Washington, D.C., 1980.
84. Sulphur Institute, *Sulphur Res. Dev.* **2,** 2–14 (1979).
85. R. D. Browne, "Corrosion of Steel in Reinforced Concrete in Marine and other Chloride Environments," Prepr., 2nd Int. Conf. Durability of Build. Mater. Components, pp. 210–224. National Bureau of Standards, Washington, D.C., 1981.
86. A. Volkwein and R. Springenschmid, "Corrosion Reinforcement in Concrete Bridges at Different Ages Due to Carbonation and Chloride Penetration," Prepr., 2nd Int. Conf. Durability Build. Mater. Components, pp. 199–209. National Bureau of Standards, Washington, D.C., 1981.
86a. R. P. Webster and L. E. Kuracka, "Effects of Acid Deposition on the Properties of Portland Cement Concrete State-of-Knowledge," Project Summary EPA-600/53-85/011, U.S. Environmental Protection Agency, Research Triangle Park, North Carolina, 1985.
87. D. G. Moore and A. Potter, *NBS Monogr. (U.S.)* **44,** 13 (1962).
88. R. W. Douglas and J. O. Isard, *Soc. Glass Technol.* **33,** 289–335 (1949).

89. G. C. Robinson, *in* "Conservation of Historic Stone Buildings and Monuments," pp. 145–162. National Academy Press, Washington, D.C., 1982.
90. G. L. Holbrow, *J. Oil Colour Chem. Assoc.* **45,** 701–718 (1962).
91. J. E. Leene, L. Demeny, R. H. Elema, A. J. de Graaf, and J. J. Surtel, "Artificial Aging of Yarns in Presence as Well as in Absence of Light and Under Different Atmospheric Conditions: Condensed Final Report," 4th Trienn. Meet., Prepr., International Council of Museums Committee on Conservation (ICOM),75/10/2: 1–11, Paris, 1975.
92. J. E. Yocom and J. B. Upham, *in* "Air Pollution" (A. C. Stern, ed.), 3rd ed., Vol. 2, pp. 65–116. Academic Press, New York, 1977.
93. G. G. Campbell, G. G. Surr, D. E. Slawikowski, and J. W. Spence, *J. Paint Technol.* **46,** 59–71 (1974).
94. J. W. Spence, F. H. Haynie, and J. B. Upham, *J. Paint Technol.* **47,** 57–63 (1975).
95. W. H. Langwell, The permanence of papers. Part II. *Process and Technical Section of the British Paper and Board Makers Assn.* **29,** 21 (1952); **30,** 170 (1953).
96. F. L. Hudson and W. D. Milner, *Pap. Technol.* **2,** 155 (1961).
97. R. L. Grant, PhD. Thesis, Victoria University of Manchester, England, 1963.
98. F. L. Hudson, *Pap. Technol.* **8,** 189–190 (1967).
99. D. J. Spedding and R. P. Rowlands, *J. Appl. Chem.* **20,** 143–146 (1970).
100. Eastman Kodak, "Preservation of Photographs," Kodak Publ. No. F-30. pp. 25–26. Eastman Kodak, Rochester, New York, 1979.
101. C. Weyde, *Phot. Sci. Eng.* **16,** 283–286 (1972).
102. T. C. Petrie, *Smokeless Air* No. 67, p. 62 (1948).
103. R. J. Brysson, B. J. Trask, J. B. Upham, and S. G. Borras, *J. Air Pollut. Control Assoc.* **17,** 294 (1967).
104. S. H. Zeronian, *Text. Res. J.* **40,** 695–698 (1970).
105. S. H. Zeronian, J. W. Alger, and S. T. Omaye, *in* "Clean Air Congress" (H. M. Englund and W. T. Beery, eds.), pp. 468–476. Academic Press, New York, 1972.
106. S. H. Zeronian, K. W. Alger, and S. T. Omaye, *Text. Res. J.* **43,** 228–237 (1973).
107. D. Harrison, "Who Pays for Clean Air: The Cost and Benefit Distribution of Federal Automobile Emission Controls." Ballinger, Cambridge, Massachusetts, 1975.
108. M. A. Morris, M. A. Young, and T. A.-W. Molnig, *Text. Res. J.* **34,** 563–564 (1964).
109. N. J. Beloin, *Text. Chem. Color.* **5,** 128–133 (1973).
110. A. Parker, "The Destructive Effects of Air Pollution on Materials." National Smoke Abatement Society, London, 1955.
111. H. J. Plenderleith, "The Preservation of Leather Bookbindings." British Museum, London, 1946.
112. F. R. Innes, *Smokeless Air* No. 68, p. 23 (1948).
113. R. D. Smith, "The Preservation of Leather Bookbindings from Sulfuric Acid Deterioration," MLA Res. Pap., University of Denver, Denver, Colorado, 1964.
114. D. J. Spedding, R. P. Rowlands, and J. E. Taylor, *J. Appl. Chem. Biotechnol.* **21,** 68–70 (1971).
115. "Conservation of Cultural Property in the United States." pp. 3–4. National Conservation Advisory Council, Washington, D.C., 1976.
115a. "Historic Buildings: A Study on the Magnitude of Architectural Conservation Needs in America," p. 4. National Institute for the Conservation of Cultural Property, Washington, D.C., 1984.
116. Proposal for a CCMS Pilot Study on the Conservation/Restoration of Monuments, *in* "Conservation of Historic Stone Buildings and Monuments," pp. 341–345. National Academy Press, Washington, D.C., 1982.

117. "Fourth Report from the Environment Committee Session 1983–1984 Acid Rain," Vol. 1, Paragraph 37, p. xxii. H. M. Stationery Office, London, 1984.
118. E. E. Hughes and R. Meyers, "Measurement of the Concentration of Sulfur Dioxide, Nitrogen Oxides and Ozone in the National Archives Building," NBSIR 83-2767. National Bureau of Standards, Washington, D.C., 1983.
119. S. Hackney, *Stud. Conserv.* **29,** (3), 105–116 (1984).
120. C. L. Shaver, G. R. Cass, and J. R. Druzik, *Environ. Sci. Technol.* **17,** 748–752 (1983).
121. T. D. Davies, B. Ramer, G. Kaspyzok, and A. C. Delany, *J. Air Pollut. Control Assoc.* **31,** 135–137 (1984).
122. T. Kadokura and Y. Emoto, *Sci. Conserv.* **12,** 33–34 (1974).
123. C. L. Shaver, G. R. Cass, and J. R. Druzik, *Environ. Sci. Technol.* **17,** 748–752 (1983).
124. K. Drisco, G. R. Cass, and J. R. Druzik, Pap. *Proc., Annu. Meet.—Air Pollut. Control Assoc.* **77,** 84-83.6 (1984).
124a. N. S. Baer and P. N. Banks, *Int. J. Museum Manage. and Curator.* **4,** 9–20, 1985.
124b. N. S. Baer and P. N. Banks, Pap. *Proc., Annu. Meet. —Air Pollut. Control Assoc.,* **78,** 85–88 (1985).
125. National Bureau of Standards, "Air Quality Criteria for Storage of Paper-Based Archival Records," NBSIR 83-2795. NBS, Washington, D.C., 1983.
126. R. H. LaFontaine, *Tech. Bull. (CCI)* **3,** 1–22 (1978).
127. R. H. LaFontaine, *Tech. Bull. (CCI)* **5,** 1–4 (1979).
128. P. N. Banks, "Addendum to Planning Report 7: Preliminary Statement on Environmental Standards for Storage of Books and Manuscripts. Newberry Library, Chicago, Illinois, 1980.
129. G. Thomson, "The Museum Environment." Butterworth, London. 1978.
130. G. Frenzel, *Kirche Kunst* **49,** 58–60 (1971); *Scientific American,* May 1985, pp. 126–135 (revised English version).
131. G. Frenzel, *Spektrum Wiss.,* June, pp. 30–43 (1984); *Scientific American,* May 1985, pp. 126–135 (revised English version).
132. E. Froedel-Kraft, *Öesterr. Z. Kunst Denkmalpflege* **25,** 70–73 (1971).
133. U. D. Korn, *Dtsch. Kunst Denkmalpflege* **19,** 58–74 (1971).
134. R. G. Newton, *Art Archaeol. Tech. Abstr.* **10,** 131–178 (1973).
134a. R. G. Newton, The Deterioration and Conservation of Stained Glass: A Critical Bibliography and Three Research Papers. *Corpus Vitrearum Medii Aevi* Great Britain-Occasional Papers 1, published for the British Academy by Oxford Univ. Press, 1977.
134b. R. G. Newton, The Deterioration and Conservation of Painted Glass: A Critical Bibliography *Corpus Vitrearum Medii Aevi,* 1982.
135. S. Fitz, E. Fitz-Ulrich, G. Frenzel, R. Krüger, and H. Kühn, "Die Einwirkung von Luftunreinigungen auf Ausgewahlte Kunstwerke Mittelaltericher Glasmalerei." Bundesminister des Innern, West Germany (ISBN 3-924183-01-5).
135a. H. Patzelt, G. Tünker, and H. Scholze, *Studies of the Protection of Medieval Glass Windows: Further Experiments,* Translation by B. Turner and R. G. Newton of Umweltbundesamt Research Report 106 08 005/01, 1985.
135b. J. F. Feenstra, *Cultural Property and Air Pollution,* Ministry of Housing, Physical Planning and Environment, Netherlands, 1984.
136. P. Hoffmann, "Rome's Monuments Go Under Wraps," *New York Times,* February 15. N.Y. Times, New York, 1981.
137. *International Herald Tribune,* September 2, p. 5 (1980), Paris.
138. A. R. Stankunas, D. F. Unites, and E. F. McCarthy, "Air Pollution Damage to

Manmade Materials: Physical and Economic Estimates," EPRI Proj. No. RP 1004. Final Report Electric Power Research Institute, Palo Alto, California, 1981.
139. F. H. Haynie, *in* "Atmospheric Corrosion" (W. Ailor, ed.), pp. 3–17. Wiley, New York, 1982.
140. "Indoor Pollutants." National Research Council, National Academy Press, Washington, D.C., 1981.
141. *Indoor Air Pollut., Spec. Issue; Environ. Int.* **8** (1–6), 1–531 (1982).
142. J. E. Yocom, *J. Air Pollut. Control Assoc.* **32,** 500 (1982).
143. Sandia National Laboratories, "Indoor Air Quality Handbook," SAND82-17773. Sandia Natl. Lab., Albuquerque, New Mexico, 1982.
144. J. D. Spengler and K. Sexton, *Science* **221** 9 (1983).
145. R. A. Wadden and P. A. Scheff, "Indoor Air Pollution." Wiley, New York, 1983.
146. B. Meyer, "Indoor Air Quality." Addison-Wesley, Reading, Massachusetts, 1983.
147. R. C. Diamond and D. T. Grimsrud, "Manual on indoor Air Quality," EM-3469, Proj. 2034-3. Electric Power Research Institute, Palo Alto, California, 1984.
148. P. J. Walsh, C. S. Dudney, and E. D. Copenhaver, eds., "Indoor Air Quality." CRC Press, Boca Raton, Florida, 1984.
149. "Indoor Air," Proc. 3rd Int. Conf. Indoor Air Qual. Clim. Swedish Council for Building Research, Stockholm, 1984.
150. "Indoor Air Quality Research Notes." Lawrence Berkeley Laboratory, University of California, Berkeley, 1985.
151. United States Code of Federal Regulations, 40 CFR 50.1. U.S. Government Printing Office, Washington, D.C.
152. H. Özkaynak, P. B. Ryan, G. A. Allen, and W. A. Turner, *Environ. Int.* **8,** 461 (1982).
153. R. N. Dietz and E. A. Cote, *Environ. Int.* **8,** 419 (1982).
154. A. K. Blomsterberg and D. T. Harrje, *ASHRAE Trans.* **85,** Part I (1979).
155. W. A. Wade, III, W. A. Cote, and J. E. Yocom, *J. Air Pollut. Control Assoc.* **24,** 933 (1975).
156. J. D. Spengler, C. P. Duffy, R. Letz, T. W. Tibbits, and B. G. Ferris, Jr., *Environ. Sci. Technol.* **17,** 164 (1983).
157. D. W. Dockery, J. D. Spengler, M. P. Reed, and J. Ware, *Environ. Int.* **5,** 101 (1981).
158. E. A. Palmes, C. Tomczyk, and J. DiMattio, *Atmos. Environ.* **11,** 869 (1977).
159. J. J. Quackenboss, M. S. Kanarek, J. D. Spengler, and R. Letz, *Environ. Int.* **8,** 249 (1982).
160. J. R. Girman, M. G. Apte, G. W. Traynor, J. R. Allen, and C. D. Hollowell, *Environ. Int.* **8,** 213 (1982).
161. J. D. Spengler, B. G. Ferris, Jr., D. W. Dockery, and F. E. Speizer, *Environ. Sci. Technol.* **13,** 1276 (1979).
162. B. D. Goldstein, R. J. W. Melia, S. Chinn, C. Du V. Florey, D. Clark, and H. H. John, *Int. J. Epidemiol.* **8,** 339 (1979).
163. R. J. W. Melia, C. Du V. Florey, S. C. Darby, E. D. Palmes, and B. D. Goldstein, *Atmos. Environ.* **12,** 1379 (1978).
164. F. E. Speizer, B. G. Ferris, Jr., Y. M. M. Bishop, and J. D. Spengler, *Am. Rev. Respir. Dis.* **121,** 3 (1980).
165. G. W. Traynor, J. R. Allen, M. G. Apte, J. R. Girman, and C. D. Hollowell, *Environ. Sci. Technol.* **17,** 369 (1983).
166. T. Caceres, H. Soto, E. Lissi, and R. Cisternas, *Atmos. Environ.* **17,** 1009 (1983).
167. B. Leaderer, *Science* **128,** 1113 (1982).
168. J. R. Duncan, K. M. Morkin, and M. P. Schmierbach, *Proc., Annu. Meet.—Air Pollut. Control Assoc.* **73,** Pap. 80-7.2 (1980).

169. C. V. Knight, "New Technology versus Conventional Airtight Wood Heater Emissions," Proc. Spec. Conf. Meas. Monit. Non-Criteria (Toxic) Contaminants Air. Air Pollut. Control Assoc., Pittsburgh, Pennsylvania, 1983.
170. D. J. Moschandreas, J. Zabransky, and D. J. Pelton, "Comparison of Indoor and Outdoor Air Quality," Rep. EA-1733. Electric Power Research Institute, Palo Alto, California, 1981.
170a. M. Lebowitz, "The Potential of Lung Cancer from Passive Smoking," Proc. 3rd Inte. Conf. Indoor Air Qual. Clim., Vol. 1. Swedish Council for Building Research, Stockholm, Sweden, 1984.
171. J. Spengler, D. W. Dockery, W. A. Turner, J. M. Wolfson, and B. G. Ferris, Jr., *Atmos. Environ.* **15,** 23 (1981).
172. J. L. Repace and A. H. Lowrey, *Science* **208,** 464 (1980).
173. Committee on Aldehydes, "Formaldehydes and other Aldehydes." National Research Council. National Academy Press, Washington, D.C., 1981.
174. T. G. Matthews, A. R. Hawthorne, C. R. Daffron, and T. J. Reed, "Surface Emission Monitoring for Formaldehyde Source Strength Analysis," Proc. Spec. Conf. Meas. Monit. Non-Criteria (Toxic) Contaminants Air. Air Pollut. Control Assoc., Pittsburgh, Pennsylvania, 1983.
175. K. C. Gupta, A. G. Ulsamer, and P. W. Preuss, *Environ. Int.* **8,** 349 (1982).
176. B. Meyer, "Formaldehyde Release from Building Products," Proc. 3rd Int. Conf. Indoor Air Qual. Clim., Vol. 3. Swedish Council for Building Research, Stockholm, Sweden, 1984.
177. W. A. Cote, W. A. Wade, III, and J. E. Yocom, "A Study of Indoor Air Quality," EPA 650/4-74-042, U. S. Environmental Protection Agency, Research Triangle Park, North Carolina, 1974.
178. D. J. Moschandreas, J. W. C. Stark, J. E. McFadden, and S. S. Moore, "Indoor Air Pollution in the Residential Environment," EPA 600/7-78/229A. U. S. Environmental Protection Agency, Research Triangle Park, North Carolina, 1978.
179. J. E. Yocom, N. Hijazi and J. J. Zoldak, "Use of Direct Analysis Mass Spectrometry to Solve Indoor Air Quality Problems," Proc. 3rd Int. Conf. Indoor Air Qual. Clim., Vol. 4. Swedish Council for Building Research, Stockholm, Sweden, 1984.
180. L. Molhave, *Environ. Int.* **8,** 117 (1982).
181. R. C. Bruno, "Sources of Indoor Radon in Houses," Proc. Int. Symp. Indoor Air Pollut., Health, Energy Conserv. University of Massachusetts, Amherst, 1981.
182. H. M. Sachs, T. L. Hernandez, and J. W. Ring, *Environ. Int.* **8,** 97 (1982).
183. W. W. Nazaroff and A. V. Nero, "Transport of Radon From Soil into Residences," Proc. 3rd Int. Conf. Indoor Air Qual. Clim., Vol. 2. Swedish Council for Building Research, Stockholm, Sweden, 1984.
184. E. D. Palmes, A. F. Gunnison, J. DiMattio, and E. Tomczyk, *Am. Ind. Hyg. Assoc. J.* **37,** 570 (1976).
185. H. W. Alter and R. L. Fleischer, *Health Phys.* **40,** 693 (1981).
186. D. W. Dockery and J. D. Spengler, *J. Air Pollut. Control Assoc.* **31,** 153 (1981).
187. L. A. Wallace and W. R. Ott, *J. Air Pollut. Control Assoc.* **32,** 333 (1982).
188. D. J. Spedding, *Atmos. Environ.* **7,** 376 (1973).
189. J. Alzona, B. L. Cohen, H. Rudolph, H. N. Jow, and J. O. Frohliger, *Atmos. Environ.* **13,** 55 (1979).
190. B. Brunekreef, H. A. Smit, K. Biersteker, J. S. M. Boleij, and E. Lebret, *Environ. Int.* **8,** 193 (1982).
191. R. A. Ziskind, K. Fite, and D. T. Mage, *Environ. Int.* **8,** 283 (1982).
192. W. Ott and P. Flachsbart, *Environ. Int.* **8,** 295 (1982).

193. G. W. Traynor, M. G. Apte, J. F. Dillworth, C. D. Hollowell, and E. M. Sterling, *Environ. Int.* **8,** 447 (1982).
194. S. Yamanaka, *Environ. Sci. Technol.* **18,** 566 (1984).
195. R. H. Sabersky, D. A. Sinema, and F. H. Shair, *Environ. Sci. Technol.* **7,** 342 (1973).
196. F. X. Mueller, L. Loeb, and W. H. Maper, *Environ. Sci. Technol.* **7,** 342 (1973).
197. J. V. Berk, C. D. Hollowell, R. A. Young, I. Turiel, and J. H. Pepper, "The Effects of Energy-Efficient Ventilation Rates on Indoor Air Quality in an Ohio Elementary School," LBL-10223, UC-95d, EEB-Vent 80-9. Lawrence Berkeley Laboratory, University of California, Berkeley, 1980.
198. J. V. Berk, R. A. Young, S. R. Brown, and C. D. Hollowell, *Proc., Ann. Meet.—Air Pollut. Control Assoc.* Pittsburgh, Pennsylvania, **74,** LBL-12189 (1981).
199. General Electric Company, "Indoor-Outdoor Carbon Monoxide Pollution Study," Final Rep. EPA-RA-73-020. U. S. Environmental Protection Agency, Washington, D.C., 1972.
200. M. Halpern, *J. Air Pollut. Control Assoc.* Pittsburgh, Pennsylvania, **28,** 689 (1978).
201. T. D. Tosteson, J. D. Spengler, and R. A. Weker, *Environ. Int.* **8,** 265 (1982).
202. J. V. Berk, C. D. Hollowell, J. H. Pepper, and R. A. Young, "Indoor Air Quality Measurements in Energy-Efficient Residential Buildings," LBL-8894, UC-95d, EEB-Vent-80-8. Lawrence Berkeley Laboratory, University of California, Berkeley, 1980.
203. U. S. Federal Register, 40 CFR, Parts 50, 53, and 58, March 20, 1984.
204. D. W. Dockery and J. D. Spengler, *Proc., Annu. Meet.—Air Pollut. Control Assoc.* **71,** Pap. 78-60.5 (1978).
205. D. J. Moschandreas and H. E. Rector, *Environ. Int.* **8,** 77 (1982).
206. R. A. Oswald and H. W. Alter, "Localization of Indoor Radon Sources Using Integrating Track Etch® Detector," Proc. 3rd Int. Conf. Indoor Air Qual. Clim., Vol. 2. Swedish Council for Building Research, Stockholm, Sweden, 1984.
207. H. Bergman, E. Edding, and O. Alexson, "Indoor Radon Daughter Concentrations and Passive Smoking," Proc. 3rd Int. Conf. Indoor Air Qual. Clim., Vol. 2, Swedish Council for Building Research, Stockholm, Sweden, 1984.
208. K. A. Dally, M. S. Kanarek, L. P. Hanrahan, and M. A. Woodbury, *Arch. Environ. Health* **36,** 277 (1981).
209. A. R. Hawthorne, R. B. Gammage, C. S. Dudney, T. G. Matthews, and D. J. Erdman, "Formaldehyde Levels in Forty East Tennesse Homes," Proc. 3rd Int. Conf. Indoor Air Qual. Clim., Vol. 3. Swedish Council for Building Research, Stockholm, Sweden, 1984.
210. R. Niemelä and E. Toppila, "Concentrations of Airborne Formaldehyde in Modern Dwellings with Low Rates of Ventilation," Proc. 3rd Int. Conf. Indoor Air Qual. Clim., Vol. 3. Swedish Council for Building Research, Stockholm, Sweden, 1984.
211. R. R. Miksch, C. D. Hollowell, and H. E. Schmidt, *Environ. Int.* **8,** 129 1982.
212. E. D. Pellizzari, L. S. Sheldon, C. M. Sparacino, and J. T. Bursey, "Volatile Organic Levels in Indoor Air," Proc. 3rd Int. Conf. Indoor Air Qual. Clim., Vol. 4. Swedish Council for Building Research, Stockholm, Sweden, 1984.
213. E. D. Pellizzari, H. S. Zelon, J. T. Bursey, C. C. Leininger, T. D. Hartwell, and J. Breen, "Sampling and Analysis Design for Volatile Halocarbons in Indoor and Outdoor Air," Proc. 3rd Int. Conf. Indoor Air Qual. Clim., Vol. 4. Swedish Council for Building Research, Stockholm, Sweden, 1984.
214. L. Wallace, E. Pellizzari, and T. Hartwell "Analyses of Exhaled Breath of 355 Urban Residents for Volatile Organic Compounds," Proc. 3rd Int. Conf. Indoor Air

Qual. Clim., Vol. 4. Swedish Council for Building Research, Stockholm, Sweden, 1984.
215. E. Lebret, H. J. van de Wiel, H. P. Bos, D. Noij, and J. S. M. Boleij, "Volatile Hydrocarbons in Dutch Homes," Proc. 3rd Int. Conf. Indoor Air Qual. Clim., Vol. 4. Swedish Council for Building Research, Stockholm, Sweden, 1984.
216. A. Vedel and P. A. Nielsen, "Phthalate Esters in the Indoor Environment," Proc. 3rd Int. Conf. Indoor Air Qual. Clim., Vol. 3. Swedish Council for Building Research, Stockholm, Sweden, 1984.
217. H. I. Virgin, "Effects of Di-N-Butylphthalate on the Chlorophyll Formation in Green Plants," Proc. 3rd Int. Conf. Indoor Air Qual. Clim., Vol. 3. Swedish Council for Building Research, Stockholm, Sweden, 1984.
218. J. van der Kolk, "Wood Preservatives in the Indoor Air, Experience in the Netherlands," Proc. 3rd Int. Conf. Indoor Air Qual. Clim., Vol. 1. Swedish Council for Building Research, Stockholm, Sweden, 1984.
219. J. E. Woods, "Air Quality Control Strategies for Health, Comfort, and Energy Efficiency," Proc. 3rd Int. Conf. Indoor Air Qual. Clim., Vol. 1. Swedish Council for Building Research, Stockholm, Sweden, 1984.
220. C. E. Billings and S. F. Vanderslice, *Environ. Int.* **8,** 497 (1982).
221. J. E. Janssen, T. J. Hill, J. E. Woods, and E. A. B. Maldonado, *Environ. Int.* **8,** 487 (1982).
222. D. Södergren, *Environ. Int.* **8,** 483 (1982).
223. R. G. Sextro, F. J. Offerman, W. W. Nazaroff, A. V. Nero, K. Revzan, and J. Yater, "Evaluation of Indoor Aerosol Control Devices and Their Effects on Radon Progeny Concentrations," Proc. 3rd Int. Conf. Indoor Air Qual. Clim., Vol. 5. Swedish Council for Building Research, Stockholm, Sweden, 1984.
224. I. A. Nitschke, J. B. Wadach, W. A. Clarke, G. W. Traynor, G. A. Adams, and J. E. Rizzento, "A Detailed Study of Inexpensive Radon Control Techniques in New York State Houses," Proc. 3rd Int. Conf. Indoor Air Qual. Clim., Vol. 5. Swedish Council for Building Research, Stockholm, Sweden, 1984.
225. H. Lehti, "Smoke Hood," Proc. 3rd Int. Conf. Indoor Air Qual. Clim., Vol 5. Swedish Council for Building Research, Stockholm, Sweden, 1984.
226. Y. Ueno, "The Effects of Ventilation, Air Cleaner, and Humidifier on Indoor Air Pollution Caused by Cigarette Smoking in a House," Proc. 3rd Int. Conf. Indoor Air Qual. Clim., Vol. 5. Swedish Council for Building Research, Stockholm, Sweden, 1984.
227. H. Levin and J. Hahn, "Pentachlorophenol in Indoor Air," Proc. 3rd Int. Conf. Indoor Air Qual. Clim., Vol. 5. Swedish Council for Building Research, Stockholm, Sweden, 1984.
228. B. S. Pederson and W. J. Fisk, "Air Washing for Control of Formaldehyde in Indoor Air," Proc. 3rd Int. Conf. Indoor Air Qual. Clim., Vol. 5. Swedish Council for Building Research, Stockholm, Sweden, 1984.
229. "Standards for Natural and Mechanical Ventilation," ASHRAE Stand. 62-73, ANS. B194.1-1977. Am. Soc. Heat., Refrig., Air Cond. Eng., Atlanta, Georgia, 1973.
230. "Ventilation for Acceptable Indoor Air Quality," ASHRAE Stand. 62-1981. Am. Soc. Heat., Refrig., Air Cond. Eng., Atlanta, Georgia, 1981.
231. J. E. Janssen, "The ASHRAE Ventilation Standard 62-1981: A Status Report," Proc. 3rd Int. Con. Indoor Air Qual. Clim., Vol. 5. Swedish Council for Building Research, Stockholm, Sweden, 1984.

Lists of books relevant to this chapter published since the lists in Chapter 8, Volume II, 3d edition (1977) can be found in Chapter 1, Table IX, Volume VII.

5

Effects on Vegetation: Native, Crops, Forests

Walter W. Heck

United States Department of Agriculture/Agricultural Research Service
North Carolina State University
Raleigh, North Carolina

Allen S. Heagle

United States Department of Agriculture/Agricultural Research Service
North Carolina State University
Raleigh, North Carolina

David S. Shriner

Oak Ridge National Laboratory
Oak Ridge, Tennessee

I. Introduction

The intent of this chapter is to present a fairly detailed update of research on the effects of air pollutants on vegetation since the publication of Vol. II, 3d ed., Chapter 12 (1977). We have scanned several thousand documents published since 1974–1975. The 573 selected references provide a representative cross section of research activities during the past 10 years. This introduction (Section I) includes an overview of research results that develops the rationale for the remaining sections of the chapter, gives a brief evaluation of key reviews, and highlights several major field studies. Section II discusses the primary data base for the two pollutants of primary concern to terrestrial ecosystems (ozone and sulfur dioxide). The intent is to lead the reader through the types of reported research and the intensity of effects reported. There is synthesis in several areas, but synthesis is not the primary goal of this section. Section III highlights other research areas, discusses the relative importance of other pollutants, and covers some research areas that are currently receiving little attention. Sections IV and V focus on two areas of critical worldwide interest. Section IV is a synthesis of our understanding of the impact of acidic deposition on terrestrial systems. It is not an exhaustive review of the subject; thus, some excellent research papers are omitted from the discussion. Section V addresses the state of the art in regional and national assessment methodology, using an ongoing program as an example. The section focuses on ozone and crop production but addresses research with sulfur dioxide and research on forest systems. The section is a synthesis rather than a comprehensive coverage of available information; some growth and yield data that could be included are given in earlier sections. Section VI is meant to give direction; it is not a comprehensive discussion of research needs.

A. Overview of Research on Air Pollution Effects on Vegetation

Since Volume II, 3d ed., Chapter 12 (1977) was published, several in-depth reviews have addressed specific aspects of the effects of air pollutants on terrestrial plant ecosystems (crops, forests, native) or have discussed specific pollutants. None has covered all aspects because complete coverage is no longer manageable in a single article.

Little additional information and few new concepts are available for several areas that were well reviewed in 1977: (1) Visible symptomatology was detailed in that review. The classical acute injury symptoms are still found and recognized. However, there is general acceptance that chronic injury, much of it in the form of early leaf senescence that resembles normal leaf aging, is more important in the field than acute injury. In industrialized countries these chronic symptoms are associated with repetitive low-level exposures and can result in losses in crop and forest productivity. (2) There has been excellent research on gaseous pollutant uptake but no new thinking that furthers our understanding of plants as air pollutant sinks. (3) The protection from losses due to air pollution by cultural or management practices has received little attention. (4) Land use is probably not a helpful control tool except in local situations.

1. Research Areas of Less Concern

Although there are many phytotoxic air pollutants, most research has tended to emphasize those that have the greatest national or international importance. Research on the effects of hydrogen fluoride has decreased as a result of improved control technologies, a fairly solid data base, and a perception that the fluoride problem is well understood *(1)*. Little has been done on the effects of such compounds as chlorine, hydrogen sulfide, ammonia, pesticides, heavy metals (as air pollutants), ethylene, suspended and deposited particulate matter, or combustion products from automobiles.

The effects of elevated carbon dioxide (CO_2) on plant growth and productivity are not discussed in this chapter, although considerable information is available *(2, 3)*. Elevated CO_2 levels increase vegetative growth and productivity and may increase plant water use efficiency. However, possible adverse effects of CO_2 on climate (e.g., changing rainfall patterns) could offset these positive effects. Research on hydrochloric acid (HCl) and aluminum oxide (Al_2O_3), associated with space shuttle exhaust, have proved interesting but do not have regional or national importance *(4, 5)*. The identification of aniline as a substance especially toxic to loblolly pine was an interesting piece of deductive research *(6)* but was a local problem and will not be discussed further.

2. Air Pollutants of National and International Concern

In the order of importance the air pollutants of national and international concern are ozone (O_3), sulfur dioxide (SO_2), nitrogen dioxide (NO_2), and the transformed products of SO_2 and NO_2 (SO_4^{2-}, NO_3^-) that are largely responsible for acidic precipitation. The effects of these pollutants, singly or in mixtures, will be discussed.

In the United States, O_3 is generally accepted as the atmospheric contaminant having the greatest effect on agricultural and forestry production because it is found at damaging concentrations in most sections of the country. This perception is gaining credibility among European scientists, few of whom have previously considered it the primary concern. Within the next few years, O_3 will probably be accepted in all industrialized nations as the pollutant of primary concern. In the United States and Canada, SO_2 is associated mainly with scattered point sources, whereas in Europe it is more of a regional multiple-source problem. Research suggests that SO_2 at low concentrations over long periods of weeks or months can injure sensitive vegetation, but that its primary impact may occur when it is associated with O_3 or NO_2. When NO_2 occurs alone in ambient air, it is not considered an important phytotoxicant except near point sources at which its concentration exceeds 1.0 ppm. The importance of NO_2 in terrestrial ecosystems is twofold: (1) as an important ingredient in photochemical reactions (forming O_3), and (2) when present in association with SO_2 or O_3. Ozone and SO_2 have been the two most intensively studied gaseous pollutants during recent years; NO_2 in combination with O_3 or SO_2 has received some attention.

These three pollutants (O_3, SO_2, NO_2) are also important because they are critical components in the formation of acidic precipitation. The concern for possible effects of acidic precipitation (often incorrectly referred to as *acid rain*, but encompassing all forms of precipitation) on crops and forestry was highlighted in the United States in an international symposium *(7)*. Although the term *acidic deposition* is often applied to this phenomenon, acidic deposition includes both wet and dry deposition of acidifying substances (including SO_2 and NO_2). The term *acidic precipitation* will be used throughout our discussion to separate the transformed products (SO_4^{2-} and NO_3^-) from gaseous SO_2 and NO_2; *acid rain* denotes only that specific form of acidic precipitation. The importance of acidic precipitation to terrestrial ecosystems is unknown. It may have some direct effects on terrestrial vegetation and thus on terrestrial ecosystems, but the primary effect may be indirect: increasing plant sensitivity to other air pollutants and environmental stresses. Vegetation growing where acidic fogs or clouds are common may be more directly affected.

3. Research, an International View

Through the mid-1970s, most of the research on air pollution effects on agricultural and forest plant systems was done in the United States, Canada, the United Kingdom, the Netherlands, and Germany. Other industrialized countries have initiated research since 1975. This increased interest has led to international communication, but a lack of coordination in research of concern to the international community still exists. Within this context, there are certain philosophical differences between European and American approaches to air pollution research. (1) European scientists have only recently acknowledged that O_3 may be a major factor in the air pollution problems they are recognizing. (2) European experiments have usually involved 24-hour continuous exposures of plants to SO_2 or NO_2 at constant concentrations because these gases are almost constantly present in the industrialized areas of Europe. European scientists recognize that gas concentrations vary during a 24-hour period and are starting to perform variable pollutant exposures. Americans are trying to characterize fluctuations in ambient SO_2 and NO_2 concentrations but generally are performing experiments with uniform pollutant additions during a set time period (1–12 hours per day). (3) Europeans have generally studied forest systems and biochemical–physiological responses, whereas Americans have emphasized agricultural crops and large field exposure systems in assessing economic losses. (4) Several European countries (i.e., the Netherlands and Germany) have developed major programs in biological monitoring to determine the extent and impacts of pollutants on agricultural and forest systems. Americans have done little with these monitoring systems since 1974. The primary objectives of current research are to assess the impact of air pollutants on terrestrial vegetation and to understand the mechanism(s) of plant responses.

B. Evaluation of Key Reviews

For a historical perspective, a discussion of knowledge available by the mid-1970s, and the major reviews to that time, see Vol. II, 3d ed., Chapter 12 (1977). Reference *8* also gives a good overview of information through 1976 but is not readily available.

Excellent coverage of the effects of O_3, SO_2, and NO_2 on plant systems is found in publications of the United States National Academy of Sciences [O_3 *(9, 10)*, SO_2 *(11)*, and NO_2 *(12)*] and of the United States Environmental Protection Agency [O_3 *(13, 14)*, SO_2 *(15)*, and NO_2 *(16)*]. The primary weakness of references *13* to *16* is a lack of depth in the coverage of physiological and biochemical areas that is needed for an understanding of response mechanisms.

A number of books are compilations of papers presented at symposia or special meetings. The best of these include references *1*, and *17* through *20*. Table I lists references that provide excellent coverage of major subjects *(6–55)*.

C. Evaluation of Major Field Studies

Many field surveys have been performed around point sources and in large urban areas. Surveys involve considerable planning and much observation, but most are unpublished. Their utility is discussed in Vol. II, 3d. ed., Chapter 12 (1977).

Reports of several major field studies, published since 1974, have increased our understanding of the effects of air pollutants on crop and native terrestrial systems (Table II) *(56–70)*. Unlike the field surveys, these studies have considerable depth, are interdisciplinary, have usually examined mechanisms of response, and are published.

II. Update of Research on Ozone and Sulfur Dioxide

The extensive literature available since 1974 attests to the importance attached to O_3 as an air pollutant. In the United States and

Table I Recommended Books and Review Articles

Subject	*References*[a]
General reviews	*8, 21* (1)
Books	*1, 3, 7, 8, 17, 22–28*
Criteria documents	*13–16*
Ozone	*9, 10, 13, 14, 17* (9, 14), *24* (p. 47), *29, 30*
Sulfur dioxide	*11, 15, 17* (4, 11, 12), *26, 31–35*
Nitrogen oxides	*12, 16, 17* (5)
Fluoride	*1, 17* (7, 13), *24* (pp. 17, 31)
Stomatal physiology	*17* (3), *18* (10, 11), *23* (11), *24* (p. 5), *36*
Metabolic function	*17* (8, 9), *18, 24* (p. 105)
Environmental–biological factors	*17* (16, 17), *18* (24), *23* (12, 13), *37–39*
Crops	*9, 13* (10), *14* (7), *17, 21, 25, 34, 40–42*
Forests	*10, 13* (11), *14* (8), *18* (25), *23, 32, 43–45*
Acidic precipitation	*7, 20, 27, 46, 47*
Genetic variability	*17* (18), *24* (p. 135), *41*
Research direction	*17* (19)
Mixtures	*17* (8, 15), *48–50*
Growth and reproduction	*17* (10–14)
Lichens, algae, moss	*29*
Research highlights	*22, 24, 51, 52*
Methodology	*19, 22* (1)
Biological indicators	*17* (2)
Mechanisms	*17* (6), *18* (27)
General physiology	*17* (4, 7), *18, 26, 29, 33, 42, 53–55*
Yield	*34*

[a] See reference citations at the end of this chapter. A number or numbers in parentheses indicate a chapter number or page numbers within the referenced book. Many references also include research by the author(s).

Canada, O_3 has been the most extensively studied pollutant since the last review [Vol. II, 3d ed. Chapter 12 (1977)]. In the rest of the world, O_3 has only recently received significant attention. From the early 1950s to the 1970s very little research on crop and forest species was done on SO_2 effects. This was due both to the generally accepted perceptions that the effects of SO_2 were understood and that the importance of SO_2 on a national and international level was decreasing because of control measures. New interest in SO_2 was stimulated by increased power generation and the subsequent increased SO_2 emissions, and by the results of long-term assessments around point sources. The scientific community began to realize that earlier perceptions of SO_2 thresholds were flawed and would have to be reexamined.

The discussion in this section is organized around four topical

Table II Reports of Major Field Studies of Air Pollution–Ecological Effects Initiated Since 1971

Name	*Purpose, pollutants, ecological system*	*Importance*[a]	*References*
San Bernardino (1971–1977)	Determine effects of oxidants (primarily O_3) on the Ponderosa pine ecosystem of southern California.	High	*56–58*
White Court (1973–1978)	Determine effects of SO_2 from oil shale development on forests in Alberta, Canada.	Medium	*59, 60*
Colstrip (1974–1979)	Determine effects of SO_2 from a coal-fired power plant on grasslands in the northwestern United States.	Medium–low	*61, 62*
Norwegian-SNSF (1972–1980)	Determine the effects of acidic precipitation from long-distance transport on aquatic and terrestrial ecosystems in Norway (SNSF is a Norwegian acronym).	High	*63, 64*
NCLAN (1980–continuing)	The National Crop Loss Assessment Network (NCLAN) is a coordinated program to determine the effects of O_3, SO_2, and NO_2 (primary emphasis on O_3) on crop production in the United States.	High	*65–68*
NAPAP (1982–continuing)	The National Acidic Precipitation Assessment Program (NAPAP) is a coordinated program across many United States governmental agencies concerned with aquatic and terrestrial effects of acidic precipitation.	High	*27, 69, 70*

[a] Importance is a relative value; the design and objectives of all the studies were excellent; the results, in terms of adding information, varied.

areas: (1) physiological and biochemical systems; (2) factors affecting plant response; (3) growth, biomass, and yield; and (4) dose–response functions. Although we have done some synthesis, the primary goal is to address the research accomplished through the use of selected references. Where helpful, we have separated research on crops and forests (or forest species); we have discussed results with O_3 and SO_2 separately in all except the dose–response subsections.

A. Physiological and Biochemical Systems

Two objective indicators for use in quantifying plant response to air pollutants, including O_3 and SO_2, are under continued investigation: changes in chlorophyll content and release of "stress" ethylene. Chlorophyll loss is related to O_3 concentration in plants exposed under controlled conditions. Knudson *et al.* *(71)* have shown an inverse relationship between increasing O_3 concentration and foliar loss of chlorophyll; a strong correlation between loss of chlorophyll and visible injury was also demonstrated *(71, 72)*. Ethylene release from plants under controlled conditions also correlates with O_3 concentration *(73, 74)*. Both chlorophyll reduction and the release of stress ethylene are examples of biochemical responses of many plants to different stresses. Thus, under controlled conditions, correlations between these responses and O_3 concentration is high. However, these responses are also caused by other stresses. Thus, measurement of chlorophyll loss or ethylene production as a specific test for a single pollutant or group of pollutants is not possible under field conditions. Visible injury from acute exposures can be symptomatic of a specific pollutant and thus, for trained observers, is a good field indicator of cause–effect relationships. However, for chronic exposures, the injury may resemble normal senescence of leaves. Both acute and chronic pollutant injury symptoms can be mimicked by other environmental stress factors. From a practical standpoint, however, visible injury is still the only way to identify air pollution injury in the field conclusively. Research has shown relationships between visible injury and growth and yield for many plant species *(9)*. We believe that such relationships will show high correlations if experiments are designed to study them specifically, but they will not necessarily correlate well if the experiments are designed for other purposes.

The following two sections consider the separate effects of O_3 and SO_2 on crops and forests. Where information is available, these effects are discussed in relation to the following: (1) entrance of the gas into leaf tissues, the stomatal effect, and gas exchange characteristics; (2) membranes as the primary site of action; (3) changes in plant chemical constituents resulting from exposures; and (4) speculation on the mechanisms of plant response.

1. Crop Species

a. OZONE. The entrance of O_3 into plant leaves through stomata has been widely studied. Leuning *et al.* *(75)* found that flux of O_3 into leaves was measured more effectively by using leaf resistance techniques than micrometeorological techniques because the latter did not separate flux to soils and other surfaces from flux to the leaves. A number of studies have addressed the relationship between the rate of flux as influenced by the presence of O_3. Butler and Tibbitts *(76)* found that a resistant bean (*Phaseolus vulgaris*, L.) cultivar had fewer stomata than did a sensitive cultivar. During exposure to O_3, stomata on the resistant cultivars exhibited partial closure, whereas those on the sensitive cultivar did not show closure. The effect of O_3 on stomatal closure was more important than stomatal number in determining sensitivity. Stomatal closure was reported by Faensen-Thiebes *(77)* for bean at low O_3 concentrations. Partial stomatal closure was reported for bean exposed to a mixture of O_3 plus SO_2 *(72)*. However, no relationship was found between O_3 sensitivity and the number of stomata or the rate of gas exchange for 10 azalea cultivars *(78)*, 2 sweet corn cultivars *(79)*, 2 soybean cultivars *(80)*, and a tobacco cultivar *(77)*. No single factor appears related to stomatal control of plant response to O_3 *(80–82)*. However, any stress (e.g., soil moisture) that causes stomatal closure will reduce O_2 uptake and thus protect the plant from O_3 injury. Stomatal closure is a resistance mechanism that is often referred to as *avoidance*, as opposed to *tolerance:* the plant's ability to withstand some level of pollutant within the cell system.

The exact fate of O_3 after entry into the substomatal cavity of plant leaves is not known. Taylor *et al.* *(80)* suggest that O_3 gas- and liquid-phase pathways into the leaf are different from those of water vapor and CO_2. Mudd *et al.* *(83)* suggest that some fraction may pass through the cell membrane. However, the O_3 may react directly with protein or lipid membrane components, or free radical products of O_3 activity within the substomatal cavity may react with membrane components *(53, 84)*. Regardless of the exact mechanism of action, ample evidence indicates that O_3 causes membrane effects and that the cell membrane is probably the site of initial O_3 reaction. This evidence includes leakage of potassium and other electrolytes *(85, 86)*, ultrastructural changes *(87)*, and evidence of membrane repair *(88)* and of induction of senescence *(89)* by membrane effects due to O_3 exposure. Rhoads and Brennan *(90)* reported that chloroplast electron transport was inhibited more in the leaves of O_3-sensitive Bel-W_3 tobacco than in the more tolerant Bel-B, but that isolated chloroplasts from the two cultivars reacted similarly, when tested *in vitro;* this finding suggests that resistance is related to the structure or physiology of the cell membrane that affects O_3 movement into the interior of the cell.

Metabolite levels in plants change after O_3 exposure. Some reports show differences between susceptible and resistant cultivars; protection associated with changed metabolism has been suggested. Pell and Pearson *(91)* found lower concentrations of ribulose-1,5-bisphosphate carboxylase in alfalfa cultivars that were related to cultivar sensitivity and to sensitivity of leaves on a given plant. Seasonal exposure of two potato cultivars to O_3 resulted in lower total glycoalkaloids (in milligrams per 100 g fresh weight) in tubers on a fresh-weight basis but not on a dry-weight basis *(92)*. Rubin *et al.* *(93)* found that isoflavonoid production in bean increased after O_3 exposure, increased with increasing plant injury, and varied with cultivar sensitivity. Effects of O_3 include decreased carbohydrate and mineral content of clover *(94)*, decreased nitrate reduction in soybean *(95)*, decreased phospholipid composition of bean *(96)*, decreased photosynthetic pigments and lipids in spinach *(97)*, and increased peroxidase release *(98–100)*; these changes confirm O_3-induced chemical changes within plants. Increased peroxidase activity was not useful in identifying O_3 sensitivity in tobacco selections *(99)* but may prove useful as a marker for O_3 sensitivity in sweet corn *(100)*.

Mechanisms of O_3 action have been hypothesized on the basis of the results of several experiments. Purvis *(95)* found that sucrose additions eliminated O_3 depression of nitrate reduction in soybean, a result that suggests that depression of nitrate reduction was due to a decrease in available photosynthate and not to O_3 inactivation of nitrate reductase or other direct effects on enzyme systems. Athanassious *(101)* related O_3 resistance in young radish leaves to high concentrations of metabolites and a compact cell arrangement but admitted that these factors did not explain resistance in older leaves. Satoh *et al.* *(102)* reported that sensitive snap bean cultivars had lower concentrations of reducing sugars, higher osmotic potential, and higher transpiration rates than resistant cultivars; the level of reducing sugars was suggested as a good indicator of sensitivity. Hofstra *et al.* *(103)* reported a decreased respiration in bean roots before visible O_3 injury appeared in leaf tissue. Guri *(104)* reported lower reduced glutathione in two sensitive bean cultivars after exposure to O_3 than in two resistant cultivars. He suggested that higher glutathione reductase activity was associated with resistance and that the specific activity of glutathione reductase could be directly related to bean sensitivity. Sakaki *et al.* *(97)* reported that O_3-induced loss of superoxide dismutase and L-ascorbate in spinach leaves resulted in higher levels of active oxygen radicles that subsequently led to loss of chlorophyll *a* and carotenoids. Chanway and Runeckles *(105)* found O_3-decreased levels of superoxide dismutase in snap bean which they associated with increased susceptibility of bean pre-exposed to O_3. Lee and Bennett *(106)* found that a sensitive snap bean cultivar was protected from O_3 injury by addition of the chemical EDU (see Section II,B,3,a) that correlated with increases in superoxide dismutase and catalase activities in the leaves; higher levels of superoxide dismutase were also found in younger, more tolerant leaves.

Research generally supports the conclusion that primary plant effects are associated with direct effects of O_3 (or oxygen radicals) on plasma membrane integrity. Secondary effects may be associated with a variety of oxygen radicals that overpower the plant's defense mechanisms. These two basic phenomena can be viewed as separate mechanisms of O_3 response. The membrane response may be responsible for acute injury symptoms, and both responses may be responsible for chronic symptoms that resemble normal leaf senescence. In summary, both stomatal control of gas exchange and plant physiological factors control plant response to O_3.

b. Sulfur Dioxide. Unsworth and Black *(36)* presented an excellent overview of available information on stomatal responses to air pollutants. Strong evidence suggests that low concentrations of SO_2 can injure epidermal and guard cells, leading to increased stomatal conductance and greater entry of SO_2 into the plant *(107, 108)*. Other data suggest that, in some species, stomatal conductance is reduced, and that this reduction might be one mechanism of resistance *(72, 109, 110)*. Stomatal closure may be related to high SO_2 concentrations and opening to low concentrations *(36)*. Rao *et al.* *(111)* found no effect of 0.5-ppm SO_2 on stomatal conductance in pea during a 3-hour exposure but did report a decrease in carboxylase and dehydrogenase activity in epidermal cells; stomatal conductance decreased with continuing exposure.

Different plant species, SO_2 concentrations, and methodologies have been used for estimates of SO_2 fluxes into leaves *(112–117)*. Flux varies with species, conditions of exposure, SO_2 concentration, and possible entry routes (stomatal, cuticular) into the leaf. Wesley and Hicks *(118)* discussed meteorological and plant canopy conditions that affect SO_2 flux (deposition) to plant tissues and suggested the use of an average stomatal resistance *(115)*. Investigators have used $^{35}SO_2$ to determine uptake rates and distribution of the ^{35}S in plant tissues *(119–121)*. The ^{35}S was rapidly translocated through the plant, including the roots, and was found in organic compounds.

Different methodologies have been used to study effects of SO_2 on photosynthesis, transpiration, or respiration in a number of plant species *(122–126)*. Generally, SO_2 reduced net photosynthesis in all plants at all concentrations; dark respiration was increased (reduced net photosynthesis could be caused by increased respiration); and transpiration was increased. Results were similar in short- and long-term exposures. Plants generally showed rapid recovery of these processes after termination of exposures lasting up to several days. In sunflower, the rate of reduction of net photosynthesis was related to leaf age *(127)*. Taylor and Selvidge *(128)* reported a slight increase in both net photosynthesis and transpiration at low SO_2 concentrations for short time periods, followed by a decrease in both processes. Higher SO_2 concentrations induced immediate decreases in both processes.

The explosion of research on SO_2 effects has led to many investigations of the effects on enzyme systems and metabolic processes. Some scientists have utilized SO_3^{2-} in solution instead of SO_2; although SO_2 probably rapidly hydrates in solution to H_2SO_3, too little is known about the mechanism of plant response to SO_2 versus SO_3^{2-} to accept results from such experiments. Therefore, we have omitted papers on experiments which use SO_3^{2-} in solution. Effects of SO_2 on enzyme systems have been examined in many studies (e.g., *129–135*). Generally, these studies show changes in the activities of many enzymes and the extent to which these changes are affected by SO_2 concentration, plant species, environment, and plant age. The enzymes studied include ribulose-bisphosphate carboxylase, glutamate–pyruvate transaminase, glutamate–oxaloacetate transaminase, peroxidases, alanine aminotransferase, aspartate aminotransferase, glutamate dehydrogenase, glucose-6-phosphate dehydrogenase, isocitrate dehydrogenase, malate dehydrogenase, adenosine 5′-phosphosulfate sulfotransferase, glycolate oxidase, glyceraldehyde-3-phosphate dehydrogenase, ribulose-5-phosphate kinase, and fructose-1,6-bisphosphatase. In some cases, enzyme activity is increased by exposure of the plants to low levels of SO_2 and decreased by higher concentrations.

Plant metabolism is affected by SO_2 in a variety of ways (e.g., *136–139*). Sulfur dioxide has stimulated phosphorus metabolism *(139)* and reduced foliar chlorophyll concentration *(137, 138)*. Carbohydrate levels were increased by low concentrations of SO_2 and decreased by higher concentrations *(136)*. Barakhtenova and Nikolaevskii *(140)* reported a suppression of photosynthesis and an inhibition in photophosphorylation after exposure of C_3 and C_4 plants to high concentrations of SO_2 (2 or 4 ppm) for short time periods (1 minute); however, the significance of these findings is unclear because the methodology was inadequately described.

The mechanism of action of SO_2 on plants has been studied by comparing sensitive and resistant cultivars of the same species, or by comparing sensitive and resistant species. Differences in sensitivity are often related to SO_2 uptake and thus to stomatal activity. Bressan *et al. (141)* showed this in four cultivars of cucumber, but not for leaves of different sensitivity on the same plant. They concluded that differences in a biochemical or developmental resistance control may exist between leaves. Later, they reported *(142)* that the differential leaf sensitivity was explained by the formation and loss of hydrogen sulfide (H_2S); resistant leaves lost H_2S more rapidly than did sensitive leaves. Olszyk and Tingey *(143)* reported evidence for the photodetoxification of SO_2, but not of O_3. They found greater injury to plants on which stomata remained open during dark exposures to SO_2 and presented a strong case for sulfite as the primary phytotoxicant for acute plant injury. They developed an acceptable hypothesis based on a number of excellent papers (e.g., *142*) for photodetoxification of sulfite.

Plant physiological and biochemical processes are probably more important controllers of plant resistance to SO_2 (tolerance) than is control of gas entry via the stomata (avoidance).

2. Forest Species

The effects of O_3 and SO_2 on the physiology, biochemistry, and morphology of trees have received much less study than those of crops. Most of the research used gas-exchange methodology (i.e., stomatal conductance, net photosynthesis, respiration, and transpiration). Some research has focused on metabolic changes in leaves or the effects on leaf cuticular waxes. Results generally support the thesis that O_3 and SO_2 induce changes in tree physiology that hasten senescence and reduce tree growth; in pine, this may show as early needle abscission and the retention of fewer years of needles. We have highlighted a few of these studies.

a. OZONE. Yang *et al. (144)* found that short-term O_3 exposure of three white pine clones inhibited net photosynthesis and transpiration in two of the more sensitive clones, with recovery an hour after the exposure ended; O_3 increased dark respiration in the most sensitive clone. In long-term exposures, they reported *(145)* a reduction in net photosynthesis for all three clones and a decrease in dark respiration for the sensitive clone; reduction in needle length and biomass correlated well with visible injury of needles. A high O_3 concentration inhibited net photosynthesis during the growing season in Scots pine *(146);* the degree of inhibition was related to clonal sensitivity to O_3.

Coyne and Bingham *(147, 148)*, in an excellent field study in the San

Bernardino National Forest of southern California, used a minicuvette system to study differences in net photosynthesis and stomatal conductance in three chronic injury sensitivity classes of an O_3 stressed Ponderosa pine stand. They reported that O_3 accelerated reductions in net photosynthesis and stomatal function, both normally associated with aging, that correlated well with sensitivity to O_3. Needle abscission occurred when net photosynthesis decreased to about 10% of maximum. In the sensitive trees, stomatal conductance was highest in new needles when they were expanding and O_3 sensitivity might have been greatest. Stomatal conductance decreased as injury became more pronounced. Reduction in photosynthetic rate was greater than would be expected from change in stomatal conductance, leading the authors to suggest that loss of photosynthetic capability was related to loss of chlorophyll. Typically, the O_3-sensitive trees retained only 1 year of needles.

Chronic exposures of hybrid poplar leaves to 0.125 or 0.085 ppm O_3 decreased net photosynthesis and leaf chlorophyll and increased dark respiration *(149)*. These changes were associated with an accelerated aging (senescence) of leaves. Four-year-old Ponderosa pine seedlings showed decreased chlorophyll in all four needle ages after exposure to 0.5-ppm O_3 for 9 hours per day for 18 days *(150)*.

Trimble *et al. (151)* found an increase in the thickness of epicuticular wax of tolerant white pine clones when exposed to 0.3-ppm O_3 for 6 hours per day for 7 days. They also reported higher concentrations of alkane components within the waxes of tolerant than of sensitive white pine clones. These results may or may not be related to O_3 sensitivity of the clones.

In long-term O_3 exposures of Ponderosa pine seedlings, Tingey *et al. (152)* reported increased amino acids in the roots but not in the shoots. Soluble sugars, starch, and phenols increased in the shoots but decreased in the roots; this finding may reflect a stronger sink requirement for photosynthate in the shoots of stressed plants that leaves less photosynthate available for use by the roots.

b. Sulfur Dioxide. Leaf conductance rates in seedlings of two birch (*Betula*) species were correlated with relative susceptibility to SO_2, but not with sulfur accumulation *(153)*. Leaf conductance of hybrid poplar was stimulated by 0.25-ppm SO_2, but not by 0.12-ppm, and 0.12-ppm SO_2 caused changes in foliar water potential and osmotic potential *(154)* that were affected by leaf position and weeks of exposure. Hallgren *et al. (155)* reported higher daytime net SO_2 uptake in Scots pine; stomatal opening was not the primary factor controlling deposition. A lower rate of uptake could be partly explained by a light-dependent process that resulted in the re-emission of reduced sulfur compounds. Stomatal conductance was an important factor in the SO_2 susceptibility of five clones of trembling aspen *(156);* other factors at least partially determined the SO_2 sensitivity of two shrub species from the California chapparal *(157).* Keller *(158)* reported uptake of SO_2 in four deciduous species exposed to SO_2 in the winter. Foliar concentrations of sulfur in the spring growth of two of four deciduous tree species, exposed to SO_2 during the winter, were greater than those of the controls *(159).* Keller also reported *(158)* smaller current annual rings for spruce grafts exposed continuously to SO_2 (<0.2 ppm) for 10 weeks; SO_2 uptake correlated well with CO_2 uptake.

Huttenen and Laine *(160)* reported deterioration of epicuticular waxes of Scots pine that was two to five times as rapid on needles from areas containing high levels of air pollution (primarily SO_2) as on needles from clean areas. Percy and Riding *(161)* exposed white pine seedlings for 11 weeks to 0.05-ppm SO_2 and found that mesophyll parenchyma was injured, and needle ontogeny was slowed.

Changes in metabolic processes or chemical reactions have been demonstrated for trees after single or multiple exposures to SO_2 for various time periods—chlorophyll breakdown [seedling black locust *(162)*], decreased ribulose-bisphosphate carboxylase activity [Scots pine seedlings *(163)*], inhibited adenosine triphosphate formation [pine *(164)*], increased free amino acids [Jack pine *(165)*], increased peroxidase activity [Jack pine and white birch *(166)*], inhibited protein synthesis [Jack pine seedlings *(167)*], increased ethylene evolution [larch, Scots pine, Norway spruce *(168)*], increased hydrogen sulfide emission [Scots pine *(169)*], increased total protein SH [White spruce *(170)*], increased glutathione [spruce *(171)*], inhibition of both photosynthesis and photorespiration [sensitive and resistant Scots pine *(172)*]—and found higher peroxidase in sensitive clones of Eastern white pine not exposed to SO_2 *(173).*

Winner *et al.* *(174)* reported that conductance values of 10 broadleafed trees and shrubs of southern California, in SO_2-free air, were useful in estimating the amount of SO_2 absorbed. The absorbed SO_2 was directly related to SO_2-induced decreases in net photosynthesis, demonstrating a relationship between the conductance values and SO_2 effects on net photosynthesis. This technique would not be useful, without standardization, under other conditions.

B. Factors Affecting Plant Response

Considerable research has studied the effects of various stress factors on the response of plants to O_3 and SO_2. Factors of known importance include genetic (cultivar and species differences, effects on reproductive structures, inheritance of sensitivity); biological (i.e., plant diseases, insects); chemical (i.e., herbicides, insecticides, special additives); environmental [i.e., climatic (temperature, light, humidity); and edaphic (nutrition, soil moisture)] factors.

1. Genetic

Early research concentrated on the determination of relative sensitivities of species, genotypes within a species, and effects on pollen and pollen germination. Little attention was given to understanding the mechanism of inheritance, although some information is provided by Reinert *et al.* *(41)*. Data exist on the influence of genetic factors or response of plants to O_3: (1) effects of O_3 on pollen, meiosis, and adaptation; (2) inheritance of resistance; and (3) cultivar screens.

a. OZONE. Feder *(175)* found that sensitivity to O_3 of germinating pollen *in vitro* was correlated with the O_3 sensitivity of the pollen parent. He identified O_3-resistant and O_3-sensitive pollen in tomato, petunia, and tobacco cultivars and developed a technique for atmospheric O_3 bioassay using the pollen. Significant meiotic damage to chromosomes during meiosis was found when pollen of *Vicia faba* (field bean) was exposed to 2 ppm of O_3 for 4 hours *(176)*. Chromosomal damage may not be found under field conditions, although some physiological injury to pollen may occur.

Walmsley *et al.* *(177)* suggest an adaptive response of radish to continuous exposure to 0.17 ppm of O_3. Although final epicotyl yield was lower than that of controls, new leaves were produced more rapidly and were less sensitive to O_3 than the first leaves, and relative leaf growth rates were about the same at the end of the experiment for both the control and O_3 treatments.

Benoit *et al.* *(178)* found that white pine branchlets, from trees of three different O_3-sensitivity groupings, exposed to 0.10-ppm O_3 (4 or 8 hours per day till anthesis), showed no change in pollen production or viability. Germination of pollen from the more tolerant pine selection was inhibited when grains were exposed on a wet petri dish to O_3 in one experiment but not in a second experiment. Sensitivity of pine pollen does not correlate with the foliar sensitivity of the pollen parent, as shown by Feder *(175)* for several crops. These reports identify issues that need further exploration.

A brief summary of selected studies on the heritability of O_3 resistance is shown in Table III *(179–186)*. These studies suggest that resistance to O_3 is heritable. Heritability may involve several major genes, with partial control by a number of others. Heritability of

Table III Genetic Control of Ozone Sensitivity Selected Studies

Test plant	*Plant selections*	*Ozone exposure characteristics*	*Response*	*Ref.*
Tobacco	7 parents, 21 hybrids	0.50 ppm, 4 hours (greenhouse)	Heterotic response of hybrids was to increase O_3 susceptibility	*179*
Bean, snap	2 sensitive, 2 resistant; F_1, F_2 generations	1.33 ppm, 1 hour (controlled)	Heritability of resistance to O_3 was 0.83; O_3 resistance was recessive and regulated by a few major genes	*180*
Bean, snap	2 sensitive, 2 resistant, F_1, F_2 generations	0.28–0.32 ppm, 8 hour (controlled)	Ozone resistance was dominant; two interacting genes control trait; many genes have minor control	*181*
Bean, snap	1 sensitive, 3 resistant; F_1, F_2 generations	0.41 ppm; 6 hours/day, 2 days (controlled)	Heritability of resistance was 0.66–0.88 (control chambers) and 0.16–0.21 (in the field)	*182*
Potato	14 cultivars, F_1 hybrids	0.40 ppm, 3 hours (controlled)	Dominance for resistance fit model best, general (70%) and specific (20%); combining abilities were significant	*183*
Tall fescue	6 parents, F_1 hybrids	0.50 ppm, 3 hours (controlled)	General combining ability and reciprocal effects were significant; additive genetic variance suggests that breeding for O_3 resistance is possible	*184*
Petunias	7 inbred lines, F_1 generation	Ambient at max of 0.3 ppm, 4 hours; controlled at 0.25–0.50 ppm, 11 hours	Additive gene action for ambient, partial dominance for sensitivity	*185*
Sweet corn	Inbred lines; F_1, F_2 generations	Field exposures	High degree of inheritance for resistance (r) and susceptibility (s); s × r gave mostly susceptible F_1; r × r gave r F_1 and F_2; incomplete dominance for susceptibility	*186*

resistance has been labeled both dominant and recessive, even in the same species *(180, 181)*, suggesting that heritability of O_3 is poorly understood. Hucl and Beversdorf *(182)* suggest that the effect of O_3 on heritability in bean may be affected by the environmental factors present during growth and exposure of the plants. Kress *et al. (187)* identified four sensitivity categories among 18 full sib families of loblolly pine based on foliar injury of primary or secondary needles. Such separations suggest the heritability of O_3 resistance. These categories may be further developed to use loblolly pine as a sensitive indicator for air pollutants. Ozone-sensitivity separations, based on severity of injury, were also reported for three white pine clones *(188)*. The heritability of O_3 resistance is a critical area of study and one that will receive increasing attention as the agricultural community becomes more aware of the extent of crop and forest losses associated with O_3.

Reinert *et al. (41)* have compiled an excellent summary of cultivar screening efforts through 1980. Table IV summarizes several screening efforts since 1975 that suggest a continuing validity of such screens *(189–201)*. Sensitivity extremes, separated in rapid screens, seem to maintain the same relative separations when longer-term screens are performed. Several species were tested both in the field and under controlled conditions *(190, 199)*. Four soybean cultivars were screened in acute and chronic studies *(197)* and 10 soybean cultivars in a field sensitivity study in which four cultivars were grown to yield *(198)*. The results indicate that generalizations based on a single screening effort (single exposure to O_3) either in the field or under controlled conditions, without verification of the relationship between the screening tests and seasonal exposure to chronic O_3 doses using yield as the response measure, may be invalid. In general, however, the extremes in sensitivity appear to keep their relative ranking for crop species.

Considerable research has gone into screening bean germplasm for sensitivity to O_3 (Table IV) for the following reasons: (1) bean, as a species (*Phaseolus vulgaris*), is an important food source throughout the world; (2) bean is a good experimental plant unit and can be grown under a variety of conditions; (3) although the gene pool contains a wide range of sensitivity to acute O_3 exposures, resistant selections do exist; (4) since bean is a legume, nitrogen fixation can be studied; and (5) commercial bean cultivars may be as sensitive to O_3 as the cultivars of any crop species.

Thompson *et al. (202)* exposed 47 annual species from the Mojave Desert to 0- to 0.4-ppm O_3 for 8 hours a day for 4 to 5 days to determine their relative sensitivity. Two species were severely injured at 0.1-

Table IV Relative Cultivar Sensitivity to Ozone: Selected Studies

Test plant	Number of cultivars (cv)	Ozone exposure characteristics	Results	Ref.
Bean, snap	18	1.4 ppm, 1 hour (controlled)	'Sanilac' and 4 other cv were most sensitive; 'Black Turtle Soup' and 'French's Horticultural' were most resistant; visible injury and percentage loss of chlorophyll were highly correlated	*189*
Bean, snap	387	Field ambient	270 resistant, 31 sensitive	*190*
	127	0.35 ppm, 3 hours (controlled)	Good separation of cv was obtained [injury index of 3.8–8.2 (of 10.0) for a 20-cv sample]; correlation with field results was low ($R^2 = 0.20$) but significant	*190*
Bean, snap	24	0.16 ppm, 3 hours (controlled)	Good separation (injury index of 1.5–11.9 of possible 12); use of 2 other O_3 concentrations gave similar results	*191*
Bean, snap	>2,000 Plant introductions	0.60 ppm, 2 hours (controlled)	Identified 54 resistant and 67 highly sensitive introductions	*192*
Begonia	12, Elatior	0.25 ppm (twice) or 0.5 ppm (4 times), 4 hour (controlled)	Two concentrations gave similar cv separations, five cv were most sensitive	*193*
Coleus	15	0.6, 0.8, and 1.3 ppm; 4 hour (controlled)	Nine cv were sensitive; found wide range in tolerance	*194*
Turfgrass species	8 species	0.10 ppm; 3.5 hours/day, 5 days (controlled)	Warm season (2 species) were most tolerant; cool season (6 species) most sensitive	*195*
Potato	59	Field, 2 locations in Ohio	Broad range of resistance: later-maturing were more resistant; relative resistance was similar across tests	*196*
Soybean	4	*Acute:* 0.15, 0.30, 0.45, 0.60, 0.75 ppm; 1.5 hour *Chronic:* 0.03, 0.06, 0.09, 0.12, 0.15 ppm; 6 hours/day, 10 days (greenhouse)	Relative resistance affected by concentration, duration, and measure of response	*197*

Table IV ***(continued)***

Test plant	*Number of cultivars (cv)*	*Ozone exposure characteristics*	*Results*	*Ref.*
Soybean	10	0.038 and 0.106 ppm O_3; 7 hours/day, 28 days (field, open-top chambers)	At young growth stage, foliar injury and growth reductions were not well correlated	*198*
	4	0.025 and 0.101 ppm O_3; 7 hours/day, 3 months (field, open-top chambers)	'Ransom' was not predictable but 'Bragg,' 'Forrest,' and 'Davis' kept same relative sensitivity	*198*
Tobacco	7	0.10, 0.15, 0.20, 0.25 ppm; 2 hours (controlled)	Relative sensitivities were similar in controlled exposures and field studies	*199*
Celery	8	0.30 ppm; 6 hours (controlled) —ambient environment (field)	Tendercrisp was most sensitive in both environments; other cv had comparable injury, but relative values differed in two environments	*200*
Tomato	6	0.40 pm, 1 or 2 hours, 3 ages, 2 exposures (greenhouse)	Relative sensitivity changed with age; plants were most sensitive at 4 weeks; relative rankings were different, depending upon whether response was growth or injury	*201*

ppm O_3 and five were not injured at 0.3-ppm. The remaining species showed slight to intermediate injury at 0.1- to 0.3-ppm O_3.

Townsend and Dochinger *(203)* exposed 1-year-old seedlings of nine pine species to 0.10-, 0.20-, or 0.30-ppm O_3 for 8 hours a day, 5 days a week for about 10 weeks. Japanese white pine, Jack pine, Ponderosa pine, and Scots pine were most sensitive to 0.20- and 0.30-ppm O_3, and only Scots pine showed much injury at 0.10-ppm O_3. In a similar experiment using 11 species and two pine hybrids *(204)*, seedlings from *Pinus jeffreyi* × *P. coulteri* hybrid and *P. monticola* were the most sensitive. Two studies *(205, 206)* with two species of ash suggest that sufficient genetic variability exists for use in breeding resistant ash. However, the extrapolation of seedling results to mature trees has not been verified.

b. Sulfur Dioxide. Pollen germination and germ tube growth on artificial media were inhibited by SO_2 in four plant species *(207)*; moist pollen was more sensitive than dry. DuBay and Murdy *(208)* found inhibition of pollen germination in wild geranium exposed to 0.6-ppm SO_2 at high humidity; inhibition of pollen germination resulted in reduced seed production, but pollen tube growth in the style was not affected. In a similar study with *Lepidium virginicum*, they reported a 50% decrease in pollen germination but no effect on seed set *(209)*. Sulfur dioxide reduced pollen germination in a fir and four pine species when exposed on agar *(210)*.

Several papers have reported the relative sensitivity of plant cultivars or species to SO_2 (Table V) *(211–216)*. These studies used relatively high SO_2 exposures, and the resultant cultivar separations may not be representative of results from chronic exposures. Markowski *et al. (215)* studied six species, using chronic SO_2 exposures over several days, and reported that barley was generally the most sensitive of the six; no consistency was found across species for relative

Table V Relative Susceptibility of Plant Species and Cultivars to Sulfur Dioxide: Selected Studies

Test plant	*Number of cultivars (cv) or species*	*Sulfur dioxide exposure characteristics*	*Results*	*Ref.*
Poinsettia	17	3.0 ppm, 3 hours (controlled)	Bracts were generally less sensitive than true leaves; all cv were injured, most about 50%	*211*
Coleus	15	1.5, 2.3, 3.0 ppm; 4 hours (controlled)	Nine cv were sensitive; found wide range in tolerance	*194*
Turfgrass species	6	2.0 ppm, 6 hours (controlled)	Injury was 4–67% with good separation; relative sensitivity stayed the same in two tests	*212*
Marigold	39	1.0 ppm, 4 hours; 2.0 ppm, 2 hours (greenhouse)	Injury was 0–42% with only one cv showing no injury (1.0 ppm, 4 hours; generally cv were up to 10 times more sensitive at 2 ppm for 2 hours	*213*
Tomato	26	1.0 ppm, 3 hours; 2.0 ppm, 2 hours (greenhouse)	Injury 17–71% with good cv separation	*214*
Selected species	6	0.3 or 0.5 ppm; 5 hours/day, 12 or 16 days	No consistent separation; studied visible injury, biomass, chlorophyll, height	*215*
Barley	12	0.35 ppm, 9 hours (controlled)	Injury 2–12% with cv separation	*216*

sensitivity when using different response parameters. Klein *et al.* *(217)* determined that differential SO_2 uptake was partly responsible for the greater resistance of corn than of pea, but that biochemical mechanisms must be involved. Thompson *et al.* *(202)* exposed 47 annual species from the Mojave Desert to 0- to 1.5-ppm SO_2 for 8 hours a day for 4 or 5 days to determine their relative sensitivity to SO_2. One species was injured at 0.2-ppm SO_2, and three others were occasionally injured at 0.2-ppm SO_2. Two species were slightly injured only at 1.5-ppm SO_2; injury occurred in the remaining species at 0.6 ppm and above.

Ecological responses of plant types to SO_2 have been addressed in several studies. Winner and Mooney *(218)* studied two ecologically matched *Atripex* species and related greater sensitivity to SO_2 in the C_3 species to higher stomatal conductances (in part due to SO_2 stimulation) and to a more sensitive photosynthetic mechanism as compared to that of a C_4 species. Plant species are known to adapt to SO_2 stress. *Lepidium virginicum* plants, growing for over 75 years near a SO_2 source, were more resistant to SO_2 than plants farther from the source *(219)*. On the basis of studies of enzyme activity, broad-leaved dock plants from a SO_2 area probably had adapted to the high pollution levels *(129)*. Horsman *et al.* *(220)* have presented information that supports the hypothesis that perennial ryegrass populations from areas with high SO_2 concentrations have evolved greater tolerances to SO_2 than those from areas with low SO_2. This concept was also shown in five grass species collected from sites differing in SO_2 concentration *(221, 222)* with the suggestion that evolution of tolerance could occur within 17–25 years.

From morphological and physiological studies, Ayazloo *et al.* *(222)* suggested that acute injury differences may be primarily due to avoidance (stomatal control), whereas differences in response to chronic injury may have a biochemical mechanism. In similar work with *Geranium carolinianum* *(223)*, Taylor and Tingey concluded that SO_2 flux does not explain differential sensitivity that is genetically controlled and quantitatively inherited; a physiological–biochemical basis is suggested. In earlier work, Taylor *(224)* analyzed infraspecific variation of SO_2 responses in *G. carolinianum* and concluded that inheritance of SO_2 susceptibility is quantitative and that neither resistance nor susceptibility is dominant.

Sulphur dioxide inhibited rates of photosynthesis and photorespiration more in progeny of susceptible individual Scots pine trees than in progeny of tolerant trees *(225)*. Genys and Heggestad *(226)* reported differences in sensitivity among 16 white pine genotypes to a single acute SO_2 exposure. Relative sensitivities of five pine and three spruce populations to SO_2 varied, depending on whether exposures were acute or chronic *(227);* spruce were generally more sensitive than pine to low-level chronic exposures but more resistant than pine to acute exposures. They concluded *(227)* that acute exposures could not be used to predict the effects of chronic SO_2 exposures on either pine or spruce.

2. Biological

There are many reports on the interrelationships between biological stresses and plant response to air pollutants. These relationships must be understood, as they affect plant growth and productivity. Although mechanisms are poorly understood, available information permits some generalizations (reference *17,* Chapter 16, reference *18,* Chapters 24, 37–39) with regard to effects of O_3 and SO_2 on plant parasites. Heagle (reference *17,* Chapter 16) suggests that obligate fungal parasitism is generally inhibited by O_3 and that some facultative parasites may benefit. Usually effects are indirect through the host. Foliar parasites tend to provide localized protection of the host from O_3 injury.

a. OZONE. The literature on O_3 covers primarily crop species rather than forest species. The studies reported have included interactions of O_3 with insects, bacteria, fungi, and virus on crops and, in some cases, on trees.

Elden *et al. (228)* found that alfalfa resistance to pea aphid was not affected when plants were exposed to O_3; different alfalfa strains were sensitive or resistant to either or both stresses. Benepal *et al. (229)* reported that the Mexican bean beetle fed preferentially on bean leaves exposed to O_3. Numerous observations (reference *18,* Chapters 24, 37) suggest that plants exposed to air pollutants may be attacked more readily by insect species. Lackner and Alexander *(230)* isolated four species of fungi and two species of weevil from several of 25 O_3-sensitive white pine in the Blue Ridge Mountains of Virginia and none from 18 O_3-resistant individuals. They suggested that trees under O_3 stress may be more sensitive to fungal infection or insect attack. Tobacco inoculated with a root-knot nematode was as susceptible to ambient O_3 as the noninoculated plants *(231);* no evidence of increased sensitivity to O_3 by the nematode was presented.

Much of the research on plant–parasite interactions with O_3 is discussed in review articles by Heagle (reference *18,* Chapter 16) and Lawrence *(38).* Highlights of several studies, that are similar to those reported in different review articles, are given in Table VI *(232–250).*

Table VI Interactions of Plant Parasites, Ozone, and Sulfur Dioxide

Test plant	*Parasite*	*Exposure characteristics*	*Results*	*Ref.*
Ozone				
Soybean	*Pseudomonas* sp (bacterium)	0.25, 0.30, 0.35, 0.40 ppm; 2, 3, 4 hours (greenhouse)	Fewer O_3 symptoms if inoculated 24 hours prior; enhanced symptoms if inoculated 4 hours prior; O_3 treated did not show a bacterial response.	*232*
Wild strawberry	*Xanthomonas fragariae* (bacterium)	0.20 ppm; 3 hours (greenhouse)	Ozone inhibited infection of strawberry by the bacterium.	*233*
White bean	Bacterial blight	Ambient O_3; EDU treatment, see II, B, 3, a (field)	Bacterial infection reduced O_3 sensitivity in bean but not when plants were treated with EDU.	*234*
Onion	*Botrytis* sp. (fungus)	Ambient O_3 (field studies)	Increased *Botrytis* infection in O_3 chambers.	*235*
Geranium	*Botrytis cinerea* (fungus)	0.15, 0.30 ppm; 6 hours, 2 times (controlled)	Inhibited sporulation, germination, and the amount of disease.	*236*
Corn	*Helminthosporium maydis* (fungus)	0.06, 0.12, 0.18 ppm; 6 hours/day, various times before or after inoculation (controlled)	Effects on lesion length and sporulation depended on O_3 concentration and timing of exposure.	*237*
Clover	*Rhizoctonia solani* (fungus)	0.3 ppm; 2 hour, 1 or 3 times (field)	Less growth of clover treated with a single acute O_3 concentration was found when the fungus was present; no effect was seen from chronic O_3 exposure.	*238*

(continued)

Table VI ***(continued)***

Test plant	*Parasite*	*Exposure characteristics*	*Results*	*Re*
Tomato	*Glomus fasciculatus* (endomycorrhizal fungus)	0.15, 0.30 ppm; 3 hours, 1 or 2 times/week, 9 weeks (greenhouse)	Infection reduced by 46% and 63% in the 0.15- and 0.30-ppm treatments, respectively.	*23*
Soybean	*Glomus geosporum* (endomycorrhizal fungus)	0.025, 0.049, 0.079 ppm; 7 hours/day, 139 days (field)	Soybean less sensitive to O_3 in presence of fungus; greater P uptake by mycorrhizal plants.	*24*
Fescue	*Glomus mosseae* (mycorrhizal fungus)	0.10 ppm; 3 months, when temperature >27°C and during daylight (controlled)	Reduced intensity of mycorrhizal infection.	*24*
Tobacco	Tobacco streak (virus)	0.30 ppm; 3 hours, 1 or 2 times (greenhouse)	Virus plus O_3 gave a synergistic response for increased injury and decreased biomass.	*24*
Soybean	Tobacco ringspot (virus)	0.35 or 0.40 ppm, 4 hours (greenhouse)	Virus increased soybean resistance to O_3.	*24*
Tomato	Tobacco mosaic virus (TMV) or cucumber mosaic virus (CMV)	0.15, 0.30, 0.45, 0.60, 0.90 ppm; 3 hours (greenhouse)	Amount of injury from O_3 depended upon the virus, tomato cultivar, O_3 concentration, and virus incubation time; TMV plants were more O_3 sensitive than CMV plants; O_3 exposure 21 days after inoculation caused less leaf injury, possibly due to leaf age.	*24*
Pine	*Heterobasidion annosum (Fomes annosus)* (fungus)	Ambient oxidants (field and greenhouse)	Increased infection associated with trees more sensitive to O_3.	*24*

able VI (continued)

Test plant	*Parasite*	*Exposure characteristics*	*Results*	*Ref.*
Fungus	*Heterobasidion annosum (Fomes annosus)* (fungus)	0.05, 0.10, 0.22, 0.45 ppm; 9 hours/day, 3 days (controlled)	Reduced fungus growth rate on agar, slowed growth reduction in conidia germ tubes; O_3 in field too low to affect fungal growth directly.	*246*
ulfur dioxide				
Bean, kidney	*Xanthomonas phaseoli* (bacterium)	0.20 ppm; 6 hours/day, 10 days	Sulfur dioxide had no effects on growth and development of the bacteria.	*247*
Pine, loblolly	*Thelephora terrestris* and *Pisolithus tinctorius* (ectomycorrhizal fungi)	0.05, 0.50 ppm SO_2; 1 hour (also done with 0.05, 0.50 ppm O_3; 1 hour) (greenhouse)	Oxygen uptake in seedlings was reduced by all treatments; nonectomycorrhizal roots showed less uptake, but proportional uptake was similar; SO_2 affected oxygen uptake more than O_3.	*248*
Fungi, 4 species	(Direct effects)	0.004, 0.038, 0.38, 3.8 ppm; 10 days	Mycelial growth was reduced in 3 of 4 species; two species were 50% or less of control; experiment included higher concentrations of SO_2 for time periods from 30 minutes to 20 days.	*249*
Ectomycorrhizal fungi (4 isolates)	(Direct effects on mycelial mats)	0.05, 0.50 ppm; 1 hour (for both SO_2 and O_3)	Reduced mycelial respiration; SO_2 had greater effect than O_3	*250*

The level of interaction between O_3 and parasites usually depends on the growth and exposure conditions, and the concentration and timing of the exposures in relation to inoculation time. Generally, infection of roots by mycorrhizal formation was inhibited by exposure of plants to O_3. The references include details on response measures such as foliar injury, biomass, yield, stomatal effects, and levels of metabolites.

b. SULFUR DIOXIDE. Hughes and associates *(251, 252)* reported that Mexican bean beetle larva fed preferentially on soybean foliage exposed to low levels of SO_2 grew larger and had lower mortality, and that the beetles produced 1.5 times more progeny after one generation than when fed on foliage not exposed to SO_2. More research is needed to determine whether the beetle could be a more destructive pest on soybean in areas of chronic SO_2 pollution. Similarly, a stimulation of growth of the black bean aphid on field bean was shown to be mediated through the host plant after exposure to SO_2 or NO_2 *(253)*.

Sulfur dioxide inhibited infection by the southern corn leaf blight fungus of corn and by the stem rust fungus of wheat *(254)*, two important fungal diseases. Titre of two viruses *(255)* was increased in host plants exposed to chronic doses of SO_2. Several studies are shown in Table VI.

3. Chemical

Several chemicals have been tested to determine whether they can protect plants against O_3 or SO_2 injury. The results of many such studies were summarized by Heck *et al. (9)*. We have highlighted some of the responses which have been reported for several chemical types. Effects of pollutant mixtures will be discussed elsewhere (III,B).

a. OZONE. Czuba and Ormrod *(256)* reported an enhancement of O_3 injury by cadmium (Cd) in plants cultured in increasing concentrations of nutrient Cd; this effect was especially apparent at low concentrations of both O_3 and Cd. Benomyl, a fungicide, suppressed O_3 injury in several plants; benomyl, but not carboxin sprays, reduced O_3 injury in three turfgrass species and gave protection from O_3 to one of the species (creeping bentgrass) in outdoor chambers *(257)*. Sung and Moore *(258)* reported that two of three herbicides tested protected two sensitive tobacco cultivars from O_3 injury when the plants were grown and exposed under controlled conditions (0.15-ppm O_3 for 2.5 hours, on each of 4 consecutive days). However, when the tobacco was exposed in the field to ambient levels of O_3, protection occurred for several weeks after transplant, but not later *(259)*. Hatzios *(260)* found that exposure of sorghum seedlings to O_3 interfered with the herbicide antidote CGA-43089 in protecting the greenhouse-grown seedlings against the herbicide metolachlor. Results suggest that possible interactive effects of O_3 with agricultural chemicals and their effectiveness in an O_3-polluted atmosphere warrant further investigation.

Pretreatment of rice seedlings with abscisic acid (a natural plant hormone) provided protection against O_3 (0.2 ppm for 2 hours) injury as shown by decreased stress-ethylene production and an increased chlorophyll content *(261)*. Natural concentrations of abscisic acid were not related to the severity of O_3 injury in pea *(72)*.

The feasibility of developing an antioxidant spray to protect sensitive plant species from O_3 injury was studied by the E.I. du Pont de Nemours Co. The effectiveness of *N*-[2-(2-oxo-1-imidazolidinyl) ethyl]-*N*-phenylurea (EDU) was tested by both foliar spray and root application *(262)*. The company encouraged independent laboratories to test the effectiveness of EDU. Several studies are highlighted in Table VII *(263–269)* because of our interest in this product as an O_3 protectant and as a possible control treatment for tests to measure the effects of O_3 on crop yields under field conditions. In some cases EDU is effective as an O_3 protectant. Preliminary research in the greenhouse *(267)* showed no effect of EDU on growth of potato and suggested that EDU may have a role in experiments to understand the effects of O_3 on crops grown in the field. However, Hofstra *et al.* *(268)* found that EDU did not protect potato against injury and yield losses from O_3 or SO_2 in the field. Before EDU can be used with confidence in estimating the effects of O_3 on yield in the field, the level of protection and direct effects on yield caused by EDU must be understood.

McClenahen *(270)* showed that EDU sprays on ash and wild cherry seedlings protected them from O_3 injury. Although a significant protection against visible injury was noted, protection against growth inhibition was not tested because the O_3 treatment did not cause measurable growth effects.

b. Sulfur Dioxide. Studies on the interaction of SO_2 with different chemicals in relation to their effects on plants are summarized in Table VIII *(271–275)*. Certain agricultural chemicals tended to protect plants from SO_2 injury, whereas heavy metals, except copper, did not affect plant response to SO_2. Since relatively little research has studied this area, generalizations are premature.

Table VII The Use of EDU as a Protectant against Ozone Effects on Crops

Test plant	*EDU treatment*[a]	*Ozone exposure characteristics*[b]	*Results (Cv = cultivars)*	*Ref.*
Pinto bean	20–500 g/ml (F); 4 mg (S)	0.8 ppm, 12–797 minutes (controlled)	Increasing EDU gave a linear increase in protection against O_3 up to 500 μg/ml; soil applications were equally effective against visible leaf injury.	*262*
Bean, navy 3 cultivars	1.12 kg/ha (F), 2 or 3 times	Ambient (field)	EDU protected against O_3 injury for 7–10 days; yield increase of 37% for 'Sanilac,' 14–30% for 'Seafarer,' and no effect for 'Kentwood.'	*263*
Bean, pinto	500, 1,000 or 5,000 μg/g (F); 1, 3, 7, or 10 days prior to O_3	0.10, 0.25, 0.40, 0.50, 0.75 ppm; 6 hours (greenhouse)	Relatively little protection noted; more occurred at a higher EDU concentration.	*264*
Bean, white	1 kg/ha (F), weekly sprays from flowering	Ambient (field)	EDU gave 38% less O_3 injury and 24% yield increase; defoliation delayed by 3 weeks.	*234*
44 Species	500 ppm as spray (F) or drench (S)	0.30, 0.40, 0.50, 0.60 ppm; 3 hours (greenhouse)	Spray or drench gave similar results; 12 O_3-sensitive species were protected; 4 were not; some species were resistant to the O_3 treatment.	*265*
Alfalfa, 2 cultivars sensitive and resistant (Callus tissue)	1.0 ppm and 0.1 ppm in an agar medium	0.5 ppm, 2 hours (controlled)	1.0 ppm inhibited callus growth; 0.1 ppm did not, but initiated changes modified by O_3.	*266*
Tobacco 'Virginia 115'	1 kg/ha (F), 7 times at 8- to 10-day intervals	Ambient (field)	Yield increase of 17.5%; yield increase of 16% in plants inoculated with nematodes.	*231*

ɔle VII (continued)

Test plant	*EDU treatment*[a]	*Ozone exposure characteristics*[b]	*Results (Cv = cultivars)*	*Ref.*
ato, cultivars; ensitive, and esistant	17.8 kg/ha at Riverside, 8.9 kg/ha at Kern County, Calif.	Ambient at both Riverside and Kern County, Calif.	EDU did not affect potato growth in a greenhouse free of O_3; did not affect yield of O_3-resistant cv but increased yield of O_3-sensitive cv by 208% at Riverside and by 17% at Kern County, Calif.	*267*
ato, cultivars; of ifferent O_3 ensitivity	1.68 kg/ha at 4 field stations	Ambient at each location, 2 years	Four of 24 comparisons gave significant yield increases from EDU treatments; 3 were at one location where SO_2 was also present; maximum increase was 28% on the sensitive 'Norland.'	*268*
unia, a ensitive and esistant ultivar	250–5,000 ppm (F, S)	0.35–0.95 ppm, 3 hours (controlled)	Protection within 4 hours from injury as spray or drench, but unaffected by changed conditions for up to 14 days.	*269*

= foliar spray, S = soil drench.
mbient exposures began when plants emerged.

4. Environmental

Interactions between climatic or edaphic factors and O_3 or SO_2 on plant response are known to occur but are poorly understood. The most extensive discussions of these relationships are found in several reviews of O_3 *(9, 13, 14)* and SO_2 *(15)* effects. Several references on the effects of O_3 and SO_2 have appeared since 1974 *(276–296).*

a. OZONE. Miller and Davis *(276)* reported less injury on Pinto bean primary leaves exposed at 24°C than at 15°C or 32°C; results were not related to stomatal conductance, although leaves exposed to

Table VIII Effects of Chemical Treatment on Plant Response to Sulfur Dioxide

Test plant	*Treatment*	*Sulfur dioxide exposure characteristics*	*Results*	*R*
Petunia, sensitive and resistant cultivars	EDU at 250–5000 ppm (foliar spray, soil drench)	0.75–3.75 ppm, 3 hours (controlled)	No protection against SO_2 concentrations that gave foliar injury.	2
Chrysanthemum, 2 cultivars	Ancymidol, 0.16 and 0.48 mg, soil drench	0.2, 0.5, or 2.5 ppm; 2, 8, or 24 hours (controlled)	Reduced SO_2 injury in both cultivars; related to stomatal conductance.	2
Pea	Cadmium; 5, 10, 15 ppm in soils	0.067 ppm	Cd and SO_2 affect peroxidase; no strong interactions are shown.	2
Species, five	Cadmium, lead, copper, manganese, and silicon, mixture of the oxides in a 1:30:10:15:35 ratio; 1 dusting/week, 4 weeks	0.08 ppm; 13 hours/day, 28 days	Sulfur dioxide did not affect uptake or translocation of heavy metals, did affect yield loss from heavy metals; foliar injury was increased.	2
Barley	Copper, 50 or 100 ppm as soil drench	1 ppm; 6 or 7 hours/day, 2 days	Cu reduced stomatal conductance and plant injury.	2
Cowpea	Ca $(OH)_2$; 0.5% aqueous spray, 4 times at 20-day intervals	About 0.25 ppm; 1.5 hours/day, 40 days	Reductions in chlorophyll and biomass from SO_2 were less for plants receiving the $Ca(OH)_2$ spray.	2
Wheat	Urea spray (3%), 3 times	1 ppm; 2 hours/day, 80 days (field)	Effects were reduced by urea spray.	2

O_3 at 32°C had reduced stomatal diffusion. Dunning and Heck *(277)* reported on the effects of variable temperature, light, and humidity on the response of Pinto bean and Bel-W_3 tobacco to O_3. Each of the three studies included possible interactions between O_3, growth conditions, exposure conditions, and species for each factor. Three-way interactions were found for temperature and light conditions, but only two-way interactions were significant for humidity. The results reinforce the idea that the pollutant–environment–species system is both dynamic and complex. McLaughlin and Taylor *(278)* reported a three- to fourfold increase in O_3 uptake with increasing relative humidity. Their data support the hypothesis that plants in the eastern United States may be affected more by a given O_3 concentration than plants in the west, where relative humidity is lower. They found that O_3 uptake, as affected by humidity, was controlled by internal leaf resistance to O_3 and not by stomatal regulation. Additional investigation is required if we are to understand the environmental influences of O_3 that modify plant growth and productivity.

Generalizations about the modifications of plant response to O_3 by climatic factors, either singly or in combination, are difficult to make. For every generalization, there usually is a known exception. However, evidence indicates that, within limits, lower light intensity during growth, higher light intensity during exposure, higher growth temperatures, and higher growth and exposure humidities increase the sensitivity of many species to O_3. Sufficient information exists to suggest that the increased sensitivity, although related to stomatal conductance, also reflects changed physiological conditions that make the plant more sensitive to O_3. Generally, conditions in the field can affect plant response to O_3 exposure on the following day. A two- to four-day period is usually necessary before plant response to O_3 completely reflects the changed climatic environmental conditions.

Tingey *et al.* *(279)* related plant water status to O_3 sensitivity in bean by using a unique technique to apply a uniform water stress; ethylene release was used to test for an O_3 effect. Ozone sensitivity in bean decreased with increasing plant water stress; protection occurred in 1 day at the highest stress, and recovery from water stress occurred within 6 days. The changed sensitivity was accompanied by only small changes in leaf water potential, which were associated with changed leaf stomatal conductances.

There is considerable interest in the possible preconditioning of plants by treatment with noninjurious concentrations of O_3. Several preliminary reports are included in this section on climatic factors. The effects of low-level O_3 pretreatment(s) for various pretreatment times on the subsequent response of bean to a higher concentration of O_3 were explored by Runeckles and Rosen *(280)*. They found that bean seedlings exposed to 0.02-ppm O_3 for 6 hours per day were more susceptible to subsequent acute O_3 exposures than seedlings not pretreated. Similar results were found for tobacco preexposed to 0.03-ppm O_3 for 12 hours prior to an acute exposure *(281)*. Johnston and Heagle *(282)* found that soybean preexposed to different chronic concentrations of O_3 for 6 hours per day were subsequently more sensitive to an acute O_3 exposure when the chronic exposure alone did not produce visible symptoms; when visible injury was caused by the chronic exposures, subsequent injuries from exposures to acute O_3 episodes caused injury which was either additive or less than additive.

Harkov and Brennan *(283)* reported that hybrid poplar was most sensitive to O_3 (0.10 ppm for 6 hours) at an intermediate foliar nitrogen level (2.69% N) and that a 7–21% reduction in relative leaf water content protected the trees from O_3.

b. Sulfur Dioxide. Environmental factors that influence plant response to SO_2 and are included in this review are light, temperature, humidity, CO_2, freezing, soil moisture, and soil nutrition. Several of these have been studied in combination, but generally the only interaction considered has been with SO_2. A summary of representative studies is shown in Table IX *(284–296)*. Perhaps the most in-depth studies were performed on oat *(285)*. They included interactions between growth temperature, exposure temperature, and SO_2 and between exposure temperature, relative humidity, and SO_2. Root growth was more affected than top growth at all SO_2 concentrations.

Generally plants are more sensitive to SO_2 as light intensity, wind speed, temperature, and humidity increase; elevated CO_2 levels protect plants. Freezing may increase plant sensitivity, and low soil moisture tends to make plants more resistant to SO_2. The differential effect of flooding on two birch species suggests that flooding of some tree species may induce resistance to SO_2 *(296)*.

C. Effects on Growth, Biomass, and Yield

This section highlights greenhouse and controlled-environment research on the effects of O_3 and SO_2 on the biomass and yield of selected plant species. Several field studies are included, but most field research is discussed in Section V of this chapter. Translocation of photosynthate from leaves to other plant compartments, to explain observed biomass or yield changes, has been studied by several investigators. The effects of O_3 and SO_2 on this partitioning of assimilates continues to be a major focus of research.

We have used results from field experiments for assessment activities (Section V) since such experimental results should more accurately reflect effects of ambient pollution levels on plant growth and productivity. This does not detract from the importance of growth chamber, greenhouse, and field pot studies in developing our under-ble IX Effects of Environmental Factors on the Response of Plants to Sulfur Dioxide[a]

Test plant	*Factors*	*Sulfur dioxide exposure characteristics*	*Results*	*Ref.*
mothy	Light: 100 or 400 μE/m²/s; temp.: 30°C day by 19°C night vs. 26°C day by 12°C night	0.11–0.12 ppm, 44 days (controlled)	Lower temp. and light gave greater reduction in stem growth (light, temp., and SO_2 appeared to be interactive); roots were more sensitive to SO_2 regardless of the environmental conditions; net assimilation rate reduced 14.7%.	*284*
t, 2 cultivars	Four growth and exposure temp.; four growth temp. and two exposure RH	0.5, 1.0, 2.0, 4.0 ppm or 0.75, 1.5, 3.0 ppm; 1.5 hours, 2 times (controlled)	Coker 227 more sensitive than Carolee; root dry wt. reduced more than top in all treatments; plants more sensitive at higher growth temp.; exposure temp. did not affect response; plants more sensitive at higher RH; interactions found for SO_2 with growth temp. and exposure RH.	*285*
an, snap	Exposure temp.: 13, 21, 32°C; RH: 40, 60, 80%	0.9 ppm; 1 or 2 hours (controlled)	Higher temp. and RH gave most injury; stomatal conductance increased with temp. and RH.	*286*
an, snap	RH: 35, 78%	0.23, 0.35, 0.47 ppm; 1 hour (controlled)	Two- to threefold increase in SO_2 absorption at higher RH.	*278*
ld geranium	RH: 50, 70, 80, 90%	0.2, 0.4, 0.8, 1.2, 1.5 ppm; 4.7 hours (controlled)	Decreased seed set at RH of 80%; 0.2 ppm at 90% RH gave 32% reduction in seed set; no visible injury.	*287*
ranium	RH and CO_2	1 ppm; 1 hour (controlled)	Injury to leaves greater at high RH and low CO_2; SO_2 in dry air hastened stomatal closure.	*288*

(continued)

Table IX ***(continued)***

Test plant	*Factors*	*Sulfur dioxide exposure characteristics*	*Results*	*Re*
Six species: 3 C_3, 3 C_4	CO_2: 300, 600, or 1,200 ppm	0.25 ppm; 8 hours/day, 5 days week for 2d and 4th weeks of elevated CO_2 (greenhouse)	C_3 plants more sensitive to SO_2 at ambient CO_2 levels; C_4 were more sensitive at elevated CO_2; effect related to effect of CO_2 on stomatal conductance.	28
Soybean	CO_2: 300, 450, 600, and 1,200 ppm	0.18–1.1 ppm, 2 hours (greenhouse)	Photosynthesis decreased with increasing SO_2; effect reduced with increasing CO_2.	29
Ryegrass	Windspeed: 10 or 25 m/minute	1.1 ppm, 4 weeks (controlled)	High wind speeds increased effect of SO_2; reduced leaf area, all dry wts. and root/shoot ratios.	29
Ryegrass	Freezing	0.1 ppm, 3 weeks (controlled)	Greater sensitivity to freezing temp., sensitivity enhanced by added nutrients; effect may be important in field selection of SO_2-resistant genotypes.	29
Plants: 4 species	Soil moisture [adequate (40% FC) and drought (above wilting point)]	0.5 ppm; 5 hour/day, 6 day/week (total length not specified) (field)	Plants in higher soil moisture showed more depression of photosynthesis; corn showed no leaf injury; barley and sunflower did; homozygous inbred corn lines more sensitive than heterozygous lines.	29
Bean, pinto	Soil Moisture at −1/3, −1, −3, −5 atm.	2.2 ppm, 3 hours (controlled)	Severe injury at −1/3 and −1 atm., little at −3 and −5 atm.; injury correlated with soil moisture, stomatal conductance, and water potential; SO_2 decreased stomatal conductance at −1/3 and −1 atm.	29

ble IX (continued)

Test plant	Factors	Sulfur dioxide exposure characteristics	Results	Ref.
assland (native)	Nutrients (S and N)	0.02, 0.05, and 0.10 ppm; growing season (field)	Sulfur and nitrogen addition gave increased biomass; high SO_2 plus nitrogen and sulfur addition decreased aboveground biomass; SO_2 at all levels tended to decrease biomass.	*295*
rch, paper and river	Flooding, 5 weeks	0.35 ppm, 30 hours (exposed after flooding)	Seedlings of river birch affected more by flooding; recovery rapid with faster growth than nonflooded; SO_2 reduced mean relative growth rate of both species in nonflooded conditions but only paper birch in flooded areas.	*296*

bbreviations: μE (microeinstein), temp. (temperature), wt. (weight), RH (relative humidity), FC eld capacity), atm. (atmosphere), m (meter).

standing of the effects of air pollutant stress on plant growth and development. Early assessments using nonfield data have generally underestimated the effects found in field studies, but the estimates have been reasonable. Additionally, the understanding of effects and methodology under nonfield conditions permitted the rapid development of useful field experimental methods and designs.

1. Ozone

The results of greenhouse and growth-chamber research on O_3 effects on growth and yield (Table X) can be summarized in several ways *(297–316)*: (1) Most plants showed O_3 effects on growth and productivity; cultivar differences were usually observed. (2) Where long-term carbon allocation was determined, root growth was affected more than shoot growth. (3) Exposures during early vegetative growth caused changes in growth rate, but when exposure ended,

Table X Growth and Yield of Several Plant Species in Response to Acute or Chronic Exposure t Ozone

Test plant	*Ozone exposure characteristics*	*Results*	*Re*
Bean, white	Ambient O_3; 2,000 ppm EDU (see II, B, 3, a) (flowering to maturity), 3–4 sprays (field)	Yield increases of 0–37% in 3 cultivars across 24 field sites in plants treated with EDU.	*29*
Bean, snap	0.30, 0.60 ppm; 1.5 hour, 2 times; 6 growth stages; harvest 7 days after exposure or at fresh harvest (controlled)	Reduced relative and absolute growth rates, pod production, nodulation, and fixed nitrogen; magnitude varied with O_3 concentration and growth stage; partial recovery from exposure at early growth stage.	*29*
Bean, snap: a sensitive and resistant cultivar	Variable concentrations and durations (controlled)	Short-term changes in allocation to root but no yield effect on resistant cultivar; significant allocation changes persistent for at least 1 week in sensitive cultivar.	*29*
Bean, snap: 2 cultivars	Ambient O_3; weekly benomyl sprays from 5 days after emergence (greenhouse)	'Tempo', 41% increase in yield with benomyl spray (yield in all but first harvest increased); 'Tenderwhite': no effect of spray.	*30*
Tomato, 'Tiny Tim'	0.08–0.10 ppm; 5 hours/day, 5 days/week, 5 weeks (greenhouse)	86% reduction in fruit number, 91% reduction in fruit weight.	*30*
Tomato, 4 cultivars	0.4 ppm; 2 hours, 6 times; seedling stage (greenhouse/field)	Early exposure in greenhouse decreased early marketable yield of plants moved to field, 57% in most sensitive cultivar.	*30*
Tomato, '6718 VF'	Ambient O_3 gradient in southern California (field)	Multiple regression models predicted O_3 was responsible for 85% of reduced fruit size along the gradient; model predicted 50% yield reduction at O_3 dose of 20 ppm-hour for O_3 at concentrations >0.10 ppm.	*30*

Table X *(continued)*

Test plant	*Ozone exposure characteristics*	*Results*	*Ref.*
Tomato, 5 cultivars	Ambient comparisons: Riverside, Calif. (38 ppm-hours) field station (2 ppm-hour), using $O_3 > 0.10$ ppm	Yield reduction fit preceding results for '6718 VF'.	*303*
Tomato	0.15 ppm; 3 hours/day, 2 days/week; 9 weeks; 0.30 ppm; 3 hours/day, 1 day/week, 9 weeks	Mycorrhizal roots were affected by 0.30 ppm O_3 but not roots lacking the fungus; less ^{14}C from photoassimilated CO_2 translocated to roots; greater partitioning to stems.	*239*
Potato, 3 cultivars	Ambient O_3; applied EDU (see II, B, 3, a) soil drench to half of plots (field)	60% leaf injury gave 25% yield loss in 'Norland' and 31% loss in 'Norchip'; 'Norland' had small tubers and 'Norchip' had fewer tubers.	*304*
Potato, 2 cultivars	0.2 ppm; 3 hours at several growth stages (greenhouse)	Decreases in tuber weight and total solids and increases in reducing sugar in O_3-exposed plants.	*305*
Pepper	0.12, 0.20 ppm; 3 hours, 3 times/week (greenhouse)	Decrease in root, stem, and leaf dry matter; 16% and 54% loss in total fruit dry weight, 77% loss in mature fruit (0.2 ppm)	*306*
Carrot, 'Imperator 58'	0.19, 0.25 ppm; 6 hours, 1.5 exposure/week (greenhouse)	32% and 46% decrease in root dry matter, respectively.	*307*
Parsley, "Banquet'	0.20 ppm; 4 hours, 2 times/week, 8 weeks (greenhouse)	23% decrease in plant dry weight, 43% decrease in root dry weight; greatest effects caused by initial exposure.	*308*
Sunflower	0.1, 0.2 ppm; 12 days, from 14 days of age (controlled)	11% and 32% decrease in plant dry weight; roots more affected than shoots.	*309*
Soybean, 4 cultivars	Ambient air, season (field)	No effect on germination of seed produced; seed protein 2% higher; oil, 4% lower.	*310*

(continued)

Table X ***(continued)***

Test plant	*Ozone exposure characteristics*	*Results*	*Ref.*
Soybean, 'Corsoy'	0.022 and 0.112 ppm; 7 hours/day, seasonal (field)	Yield reduced 39% at higher O_3 treatment; no effect on seed protein; 12.6% reduction in seed oil.	31
Cotton, 'Acala SJ-2'	0.20 ppm; 6 hours, 2 times/week, 1 group started at 8 day, 1 at 42 days from seed (greenhouse)	Vegetative biomass and boll production reduced; greatest reduction in boll and root weights; 48% reduction in boll number in both treatments.	31
Clover, 'Ladino'	0.3, 0.6 ppm; 2 hours, 2 times, 7 days apart; 4 stages of development (controlled)	Shoot and root growth reduced; response affected by O_3 concentration and exposure stage, nodulation decreased by exposure at first three stages.	31
Clover, 'Ladino'	0.03, 0.05, 0.08 ppm; 7 hours/day, 6 months (field: pots)	Total forage and forage regrowth reduced for clover and clover–fescue mixture in relation to O_3 concentration; fescue unaffected; O_3 response unaffected by N fertilization.	31
Clover, 'Ladino'	0.03, 0.06, 0.09 pm; 7 hours/day, seasonal for 2 years (field)	Forage regrowth reduced by 14% and 27% in the two higher O_3 treatments during 2d year; 22% and 54% reduction in clover and some increase of fescue as clover was lost; reduction in forage quality.	31
Clover, 'Ladino'	0.05, 0.10, 0.15 ppm; 4 hours/day, 6 days, 32 days after seeding; $^{14}CO_2$ exposures at 48-hours after last O_3 (controlled)	Max. root reduction, 42%; max. shoot reduction, 24%; O_3 reduced net photosynthesis; carbon allocation favored roots at 0.10 ppm O_3 and shoots at 0.15 ppm.	31

normal growth resumed in a short time. (4) Evidence suggests changes in the quality of usable product. (5) Research using ambient O_3 concentrations suggests that growth and yield reductions occur at these concentrations.

The loss of O_3-sensitive species (selections) from plant communities with increasing O_3 stress is important to an understanding of ecosystem dynamics. In agroecosystems, mixtures of clover and grass have been studied with respect to community response to O_3 and the loss of clover from the community. In a crimson clover–annual ryegrass mixture, the more resistant ryegrass increased under O_3 stress at the expense of the clover when exposed to 0.03- or 0.09-ppm O_3 for 8 hours a day for 6 weeks. *(317)*. Similar results were reported for a fescue–ladino clover mixture *(318)* exposed for 2 hours at 0.30-ppm O_3. In the latter study, leaf leachates from O_3-exposed fescue inhibited clover growth and nodulation. In a series of studies *(313–316, 319)* Blum and associates studied the effects of O_3 on the ladino clover–fescue mixture grown in controlled environment, greenhouse, or field chambers. They reported a loss of clover and an increase in fescue in the mixture that correlated with increasing O_3 concentration, although a loss of total biomass occurred at elevated O_3 concentrations.

Stimulation of stem elongation in three of four conifer species was reported during the second year of growth after exposure to O_3 *(320)*; this is not necessarily a healthy effect for the plant, and the response deserves further study. Jensen *(321)* found that O_3 caused decreased stem and leaf dry weight, as well as levels of starch, sucrose, and reducing sugar in the roots of green ash seedlings; the results suggest that photosynthate was used preferentially in the shoot even though a net decrease in shoot growth occurred. Height growth of eight tree species was lower when grown in unfiltered-air containing ambient levels of O_3 in open-top chambers than in charcoal-filtered-air chambers *(322)*. Jensen *(323)* reported a decrease in growth rate of silver maple and eastern cottonwood seedlings as duration and concentration of O_3 exposure increased. McLaughlin *et al. (324)* studied 25-year-old white pine that were sensitive, intermediate, and resistant to O_3 and determined that the decline in sensitive trees resulted from a reduction in needle longevity and increased respiration.

Growth ring analysis *(324)* showed that a 25-year-old, O_3-sensitive white pine had a 70% decrease in ring width over a 15-year period as compared to O_3-resistant trees and had lost the capacity to recover. In similar research *(325)*, using resistant, intermediate, and sensitive Eastern white pine along a 446-km section of the Blue Ridge Parkway, mean annual growth was reduced from 1955 to 1978 for the sensitive and intermediate white pine selections when compared with the resistance selections. As part of the San Bernardino air pollution study *(56–58)*, a tree-ring analysis was developed for the study area *(326)*. Although results are not included here, the concepts of this study should be considered by others doing tree-ring analyses for air pollution research, since fewer than half of the trees in any individual plot could be cross-dated.

2. Sulfur Dioxide

Effects of low SO_2 concentrations on carbon translocation and partitioning and on plant growth and yield have been determined (Table XI) *(327–342)*. Generally, using $^{14}CO_2$ as a tracer, movement of assimilate to developing leaves rather than to roots was observed as a result of SO_2 exposure, although Coughenour *et al.* *(327)* found the reverse. Root growth, in general, was reduced more than shoot growth, and growth was reduced at concentrations well below those usually reported in earlier studies. The results support the contention that plants are more sensitive to low SO_2 concentrations (<0.10 ppm), when exposed continuously, than has previously been supposed. More acceptable experimental designs and facilities are needed to confirm these more recent findings.

Our understanding of effects of low SO_2 concentrations on grasses has been enhanced by both the Colstrip study in the United States *(61, 62)* and by research from the United Kingdom. In the Colstrip study, assimilate partitioning *(327)*, translocation *(328)*, and forage quality *(341)* of native grasses were studied. Although considerable variability in response was shown, the trends appeared consistent, and the experiments were well designed. More interest has been demonstrated in Europe than in the United States in determining effects of low SO_2 concentrations on growth and productivity in forage grasses. The most extensive and controversial studies have been done in the United Kingdom. Cowling and associates reported slight biomass increases or no effect on alfalfa *(335)* and ryegrass *(338)* at low continuous exposures to SO_2, although some changes in assimilates were noted *(343)*. Bell *(344)* presented a summary of research results from the United Kingdom with an emphasis on the different results associated with ryegrass (*Lolium perenne*); he identified some of the reasons for the reported differences. His tabular presentation, showing SO_2 concentration (continuous 24-hour exposures) and days of exposure, indicated that effects on the exposed plants were from $+11\%$ to

Table XI The Effect of Sulfur Dioxide on Aspects of Plant Growth

Response measured	*Test plant*	*Sulfur dioxide exposure characteristics*	*Results*	*Ref.*
Carbon partitioning, translocation (use of $^{14}CO_2$)	Native grasses	0.10 ppm; continuous (field)	Proportion of total assimilated carbon in roots varied with SO_2 treatment, species, and time of year; generally SO_2 caused increased translocation to roots.	*327*
	Agropyron smithii	0.075 ppm; continuous (field)	Stimulation of assimilate translocation to developing leaves.	*328*
	Bean, snap	3 ppm, 2 hours (controlled)	45% reduction in translocation, slow recovery; phloem loading is a major limiting step in translocation.	*329*
	Timothy	0.11 ppm for 3 weeks; 0.06 ppm for 6 weeks; continuous (controlled)	Reduced assimilate movement to roots and increased movement to developing regions of the shoot.	*330*
Yield/biomass	Six species	Constant and stochastic concentration, 0–0.20 ppm	Constant SO_2 concentration underestimated effects compared to the time series treatments; excellent discussion of time series concept; interpretation of results difficult.	*331*
	Bean, snap	0.01, 0.025, 0.10, 0.35 ppm; 12 hours/day, 5 days/week (controlled)	Primary effects above 0.1 ppm include reduced biomass and yield of all parts; reported effects on chlorophyll and other components.	*332*
	Dactylis glomerata	0.11 ppm; 4 weeks, continuous (controlled)	Reduced dry weight of plant components measured, reduced tiller number.	*333*

(continued)

Table XI ***(continued)***

Response measured	*Test plant*	*Sulfur dioxide exposure characteristics*	*Results*	*Ref.*
Yield/biomass	Tobacco and cucumber	0.02 ppm; 4 weeks, continuous (controlled)	Reduced plant dry weight fractions; cucumber more sensitive than tobacco; roots affected more than shoots; no replication.	*334*
	Alfalfa	0.036 ppm, continuous (greenhouse)	Slight increase in biomass with no added sulfate, no effect with added sulfate to growth medium.	*335*
	Sunflower	0.05, 0.1 ppm; 5 weeks, continuous (controlled)	Variable results; higher concentrations gave 20–25% reduction in net assimilation rate; both concentrations accelerated senescence.	*336*
	Arachis hypogaea	0.06–1.0 ppm; 4 hours/day, 6 weeks (controlled)	Above 0.25 ppm found necrosis and reduced net primary production; some beneficial effects at lower concentration.	*337*
	Lolium perenne	0.02, 0.16 ppm; continuous (greenhouse)	No effects in any growth parameters measured; some visual injury at 0.16 ppm and a reduction in specific leaf area.	*338*
	Lolium perenne	0.02 ppm; continuous (greenhouse)	For two nitrogen and sulfate levels with SO_2, found slight increase in growth from added sulfur dioxide with no sulfate; no other effect.	*339*

able XI (continued)

Response measured	*Test plant*	*Sulfur dioxide exposure characteristics*	*Results*	*Ref.*
	Lolium perenne	Variable, means <0.08 ppm for growing season (field)	Growth in filtered and unfiltered air not different in highly polluted area; added SO_2 gave variable results with a maximum growth reduction of 12%.	*340*
ant quality	Western wheat-grass	0.02, 0.038, 0.065 ppm: seasonal means; continuous (field)	Increase in sulfur; no effect on forage digestibility.	*341*
	Tomato	0.12 pm; 72 hours/week, 5 or 10 weeks (controlled)	Slight decrease in ascorbic acid of ripe fruit; no effect on fruit yield, soluble or total solids.	*342*

−68% of the effects on the control plants (the greatest effect was at the lowest concentration). These experiments were complicated, however, by differences in experimental designs (no replication in some instances) and by differences in experimental facilities.

Results of chronic exposures of various tree species to SO_2 are summarized in Table XII *(345–350)*. Most of the research was with seedlings or young grafted trees (same clonal shoot material but different roots). Some research has involved trees grown under different ambient levels of SO_2 for specific time periods.

Several reports, using tree-ring analysis, have been published over the last 10 years; one of these *(349)* considers the historic records well before the start of a large smelter in Nevada (United States), at three distances from the smelter. After start-up of the smelter, changes in width of growth rings at locations close to the smelter are evident, and a probable relationship to the presence of SO_2 is indicated.

D. Dose–Response Functions

Since the Vol. II, 3d ed., Chapter 12 (1977) review, a number of efforts have been made to develop dose–response models to predict

Table XII Effects of Sulfur Dioxide on the Growth of Forest Species

Test plant	*Sulfur dioxide exposure characteristics*	*Results*	*Ref*
Ailanthus	0.1, 0.2 ppm; 1 or 2 week (controlled)	Reduced growth and elongation of 2- or 3-week seedlings; root growth inhibited more than shoot.	*345*
Pine, red or Jack	0.1 ppm (greenhouse)	Lower germination in red pine, effects variable in Jack; growth decreased with visible symptoms only on Jack; symptoms appeared earlier in fall (1 week) than in summer (8 weeks) exposures.	*346*
Five species	Grown at two ambient sites: high seasonal = 0.03 ppm, low seasonal = 0.004 ppm (ave monthly max: high—0.485 ppm, low—0.009 ppm); sites had some O_3 but high had significantly more NO_2; 2-year container plants	Height and rate of shoot elongation reduced for crabapple, sugar maple, white pine; delay in period of max. growth for crabapple and dogwood; flush of growth late in season for sugar maple, white pine, Fraser fir; all related to increased SO_2.	*347*
Four species	0.057 ppm for 15–24 weeks; 0.065 ppm for 14 or 71 weeks	Variable effects, but generally species fairly tolerant to the SO_2 exposures.	*348*
Pine, Pinyon	Smelter emissions, distance	Growth ring analysis indicated decrease in ring width with start-up of smelter.	*349*
Pinus sylvestris	0.055 ppm, 18 months; 3-year plants in 2 soils	Reduced mean relative growth rate, stem diameter, accelerated senescense; no interaction with soil type.	*350*

effects of gaseous pollutants on plant growth and productivity. The most intense effort in this direction is the work by Larsen and coauthors *(351–354)* that uses a log–normal model to predict the response of plants to changing O_3 concentration and duration of exposure. This series of papers presents a useful way to consider specific data sets designed to look at dose–response relationships. The basic model *(352)* is in the following form:

$$c = m_{\mathrm{ghr}} s_{\mathrm{g}}^{z} t^{p} \qquad (1)$$

where c is the pollutant concentration, m_{ghr} is the geometric mean concentration for a 1-hour exposure, s_{g} is the standard geometric deviation, z is plant response (in standard deviations from the median), t is exposure time, and p is the slope of the response line on logarithmic paper. This model worked well with O_3 data sets for different crops and reasonably well with several SO_2 data sets. A large data base of diverse information was reasonably interpreted by the Larsen–Heck model *(351)*. The model was further developed to express the relationship between the effects of gas mixtures (O_3 and SO_2) on plant response *(354)*. The model was again extended *(353)* for use in predicting yield losses based on the use of a weighted mean O_3 concentration (the effective concentration) that gave added emphasis to episodes of higher O_3 concentrations over an extended monitoring period. The data for the yield assessment were obtained from the National Crop Loss Assessment Network (NCLAN) overview publication *(68)*, which is further discussed in Section V of this chapter. Although the log–normal model is not generally recommended for assessment analysis because confidence limits cannot be placed on the predictions, the results *(353)* were very similar to those shown by using the Weibull function *(68)*. The initial injury model *(352)* was tested by Pratt and Krupa *(355)* and was found to fit both visible injury and chlorophyll determinations in short-term exposures of soybean to O_3 concentrations from 0.10 to 0.25 ppm for 0.5–4 hours.

Ormrod *et al.* *(356)* used a dose–response design to study the possible interactions of O_3 and SO_2 on several plants. The results were analyzed by using a response surface technique described by a set of polynomial equations. The technique is similar to that used by Larsen but has added complexity. Ormrod *et al.* *(356)* effectively used a covariate analysis to improve the precision of their analyses. This approach was well developed by the authors and deserves further work with other data sets.

Roberts *(357)* analyzed a large amount of data on the response of plants exposed in chambers to a constant SO_2 addition under a variety of conditions and using different exposure regimes. The best functional form was a linear regression of percentage yield loss on SO_2 concentration. The amount of variability explained was low (10.5%) but significant and was improved with the deletion of several more variable data sets. That a linear regression would best fit a heterogeneous mass of data seems reasonable where a relationship between gas concentration and plant response exists. However, Heck *et al.* *(351)* found the log–normal model could be used for a mass of heterogeneous O_3 data.

We have not attempted to review here the various functional forms that have been used to express relationships between plant yield and productivity. We have addressed some of these efforts, including regression concepts, in Section V of this chapter.

In summary, we believe that research on the effects of air pollutants on plants should use dose–response concepts that are amenable for regression analysis. Some researchers view dose–response approaches as useful only in obtaining yield data. This view reflects the need for information on yield that can be used in assessment activities. However, dose–response concepts are equally important for understanding the mechanisms of plant response. Research must be directed toward understanding the way increasing increments of any gas may change the metabolic patterns within plants before an in-depth modeling of plant response to that gas can be valid.

III. Update of Selected Research Areas

A. Pollutants Other Than Ozone and Sulfur Dioxide

During the 10 years preceding this writing, there was a tendency in the United States to concentrate research efforts on the so-called criteria pollutants. Many other countries have followed this lead. The criteria pollutants include not only O_3, SO_2, and NO_2, which have been the only pollutants discussed so far in this chapter, but also carbon monoxide (which is not toxic to plants at existing levels), total suspended or deposited particulate matter (which are not harmful to plants unless they contain relatively large amounts of specific substances, i.e., toxic metals and, acidic or basic substances), and hydrocarbons (which, except for ethylene, are not toxic at existing atmospheric concentrations). Thus, most research on terrestrial biological systems has concentrated on O_3 and SO_2. Nitrogen dioxide has been studied primarily in combination with O_3 and SO_2 because it is not particularly toxic at existing atmospheric concentrations. This section highlights selective research on NO_2 and several other pollutants that have received some study since 1974.

1. Nitrogen Dioxide

Many investigators are interested in whether NO_2 can be used as a nitrogen source by plants. Durmishidze *et al.* *(358)* used $^{15}NO_2$ with woody plants and showed assimilation of NO_2, but the exposure concentrations were too high (>100 ppm) to predict effects at ambient levels. Uptake of $^{15}NO_2$ and use of the ^{15}N by plants at fairly high concentrations of NO_2 (>1.0 ppm) has been reported *(359, 360)*. Matsumaru *et al.* *(361)* used an indirect method (label of nutrient nitrogen) to show that, at 0.3-ppm NO_2, N was incorporated into the plant system. However, the first published study using ambient levels of NO_2 (as $^{15}NO_2$) was that by Rogers *et al.* *(362)*. They treated bean with 0.097-, 0.152-, or 0.325-ppm NO_2 for 3 hours and reported a linear relation between exposure concentration and uptake. Essentially all the nitrogen was metabolized. Several earlier reports showed greater uptake of NO_2 at night, but Yoneyama *et al.* *(359)* found that night absorption was only about 14% of that absorbed during the day.

A number of interesting studies have documented different responses of plants to NO_2 exposure. Sinn *et al.* *(363)* found that NO_2 did not affect leaf conductance, and that uptake rates correlated well with NO_2 exposure concentrations. Klepper *(364)* reported the release of NO_2 from herbicide-treated soybean plants at night; Tingey *(365)* reported that such releases can be stimulated by sulfite additions. Neither researcher found daylight emissions due to the photosynthetic reduction of nitrite, which appears to be the toxic component resulting from NO_2 exposure. Nitrogen dioxide uptake in bean was negatively correlated with salinity *(366)*; salinity was associated with reduced stomatal conductance.

Mansfield and associates *(367, 368)* reported that both NO and NO_2, associated with production of high levels of greenhouse CO_2, substantially decreased growth and productivity of greenhouse crops. This result should be of concern to greenhouse operators who add CO_2 to the greenhouse by burning a "clean" gas, because such burning can release nitrogen oxides.

2. Fluoride

Most research on the effects of gaseous fluoride on plants, which dates to the 1800s, has used hydrogen fluoride (HF) as the test chemical. Although the research since 1975 has not been extensive, worldwide interest remains because losses occur around point sources.

Many review articles on fluoride effects on vegetation are available; the proceedings of the symposium at Utah State University in 1983 *(1)* are comprehensive and are written by international authorities. We have chosen not to include an extensive review of fluoride exposure in this chapter because of the previously noted symposium volume *(1)*. However, we have selected a sampling of the types of research that have been performed since 1974 for both crop and tree species.

Chamel and Garrec *(369)* determined that the movement of fluoride through the cuticle is not an important mode of entry. MacLean *et al.* *(370)* presented results suggesting that fluoride-induced injury to plants requires light even if the exposure occurs during the dark; the nature of this light activation step was not discussed. Lesion growth of *Xanthomonas campestris* (bacteria) on red kidney bean was decreased by continuous exposure for 5 days to 1 or 3 $\mu g/m^3$ of HF *(371)*. The fluoride content of unpolished rice was proportional to the concentration in the air around a fluoride-emitting source *(372)*. The fluoride content in the thalli of a lichen (*Xanthoria parietina*) was correlated with distance and wind direction from a major emitting source *(373)*; visible injury occurred when the thallus concentration exceeded 68-ppm F. Ghiasseddin *et al.* *(374, 375)* exposed soybean to HF and found reduced yield and seed number at the lowest level of exposure (about 2.2 ppb); fluoride levels in leaves, fruits, and seeds were increased, but seeds contained less fluoride than did leaves and fruits. Continuous exposure of grapevine (64 days) to 0.31-ppb HF resulted in some foliar necrosis, no effect on yield parameters, and no significant fluoride accumulation in the berries *(376)*, but the experimental design suffered from too few replications. Fluoride accumulated in radish leaves and, during long-term exposures, was translocated to the roots, but there was no accumulation in the edible tubercle *(377)*. Facteau *et al.* *(378)* reported a linear relationship between dose and foliar uptake of HF in sweet cherry at HF concentrations below 19.4 ppb.

Garrec *et al.* *(379)* related fluoride concentration in fir needles to calcium, which in turn was related to premature needle aging. Lorenc-Plucinska and Olekayn *(380)* found that 25- and 100-ppb HF reduced net photosynthesis only in susceptible Scots pine trees. Conifer needles showed a correlation of phenol content, without visible needle injury, to distance from a fluoride-emitting source *(381)*. Sidhu *(382)* developed a regression of foliar fluoride concentration in balsam fir to long-term atmospheric fluoride concentrations; relationships were found between foliage exposed for both 1 and 2 years;

levels in foliage stabilized after 2 years. Taylor and Basabe *(383)* correlated patterns of fluoride accumulation and growth reduction in Douglas fir with distance, wind patterns, and air fluoride concentrations near an aluminum smelter.

3. Peroxyacyl Nitrate

Peroxyacetyl nitrate (PAN) is the principal phytotoxic member of the peroxyacyl nitrate family of nitrogenous organic compounds produced in the photochemical oxidant complex. These compounds, especially PAN, received considerable study prior to 1975 because they were components of the Los Angeles, California, photochemical complex and were more phytotoxic than O_3. Temple and Taylor *(384)* collected information on worldwide concentrations of PAN in relation to plant injury and suggested that only in southern California should PAN be considered a major problem. However, PAN, as a component of photochemical oxidant in other areas, may well contribute to plant effects and crop yield reductions.

Several papers about PAN effects on plants are discussed here. Lewis and Brennan *(385)* found that PAN symptoms on tomato may be confused with those resulting from a mixture of O_3 and SO_2. Only lettuce and swiss chard, among eight crops, showed a yield reduction after long-term intermittant exposure to PAN at 0.04 ppm *(386)*. PAN reduced net photosynthesis in two of three lichen species at 0.05 and 0.10 ppm *(387)*. Plant water potential was affected by PAN in a sensitive, but not in a resistant, bean cultivar *(388)*. DeVos *et al.* *(389)* studied the inheritance of PAN resistance in petunia, using the inbred parents of a susceptible and resistant hybrid; they found genotype by environment interactions. Where severe injury occurred, genes for susceptibility showed almost complete dominance; in other cases, resistance appeared to be partially dominant.

We expect PAN to be of continuing interest. Studies using mixtures of PAN with other gases should be continued.

4. Hydrogen Chloride, Hydrogen Sulfide, Analine, and Tetranitromethane

In addition to the aforementioned air pollutants, several pollutants are associated with point sources that have been studied. Many chemicals associated with spills (i.e., chlorine and ammonia) are toxic to plants, but spills are not well documented. We have selected several phytotoxic air pollutants associated with selected processes that will give the reader some idea of the diversity of chemicals that may affect plant life.

Exhaust from solid rocket fuel contains large quantities of hydrogen chloride (HCl). Thus, concern and research were stimulated for the possible local effects of HCl (gas, vapor) on plants near rocket (shuttle) launch sites. Since actual exposures would be short-term but at high concentration, these exposure conditions were used experimentally. For most of the tested species *(4, 5)*, HCl was found to have phytotoxic properties similar to NO_2, with a threshold of 5 – 10 ppm in a 15- to 60-minute exposure period. Swiecki *et al. (390)* reported that bean leaves with their epicuticular wax were more sensitive to HCl. Heath and Endress *(391)* reported permeability changes in bean leaves exposed to HCl. Generally, results suggest that local effects should not be found unless the exhaust cloud contacts plants or is rained out (forming acids).

An interesting series of studies undertaken by Thompson and associates *(392, 393)* involved the continuous exposure of crop and forest species to hydrogen sulfide (H_2S). In a study with seven plant species, they reported moderate to severe injury at both 0.3- and 3.0-ppm H_2S; stimulation of growth was reported in three of the species at 0.03 ppm *(392)*. They reported an increase of glutathione in spinach tissue even at 30-ppb H_2S (393).

Taylor and Selvidge *(128)* determined the relative sensitivity of five sulfur-containing gases on snap bean. Neither carbon disulfide (CS_2) nor methyl mercaptan (CH_3SH) had an effect on photosynthesis or caused foliar injury. The relative sensitivity of the other three compounds was $SO_2 > H_2S >$ COS (carbonyl sulfide) when exposed for 6 hours to 6- to 82-μmol m^{-3} of each gas.

The use of sensitive and resistant clonal material in laboratory and field studies identified aniline as the most likely causal agent of injury and death of loblolly pines growing near two industrial facilities *(6)*. Needle necrosis, short needles, and abscission were the principal symptoms identified.

A study of selected compounds emitted in the production of high explosives identified tetranitromethane (TNM) as the most phytotoxic component *(394)*. The threshold level of response for five plant species was 0.085 – 0.760 ppm in a 120-minute exposure.

B. Pollutant Mixtures

Concern for the effects of pollutant mixtures on terrestrial ecosystems has risen during the past decade; three principal reviews document this concern *(48 – 50)*. Although studies have generally shown increased effects with mixtures of pollutants over effects from individual pollutants, only a few generalizations can be made. When effects have been reported, mixtures at concentrations below those causing visible effects from the pollutants singly tend to give a greater-than-additive (*synergistic*) response; concentrations at or slightly above those causing a threshold effect tend to produce an additive response; concentrations well above threshold tend to cause a less-than-additive (*antagonistic*) response. The term *synergistic* is not necessarily appropriate; it may fit the statistical concept but not be biologically significant since response functions are often not linearly related to pollutant concentration.

The most important two-way pollutant mixture is O_3 with SO_2. The majority of research conducted deals with this combination. The next two major categories involve SO_2 with NO_2 and mixtures involving all three pollutants.

Mixture terminology has not been standardized. Ormrod *et al. (356)* introduced a dichotomy [also used in the United States Environmental Protection Agency review *(48)*] of *no joint action* (when one of the pollutants induces no plant response) and *joint action* (when both pollutants result in some plant response). These terms are ambiguous and unnecessary. First, although *no joint action* implies a possible statistical determination that one pollutant has no effect, such an occurrence is biologically untenable. Second, *joint action* implies that multiple agents join together to produce an effect, but in an additive response the two factors may act independently. The terms *additive* and *interactive* are preferable. An *additive* response to two or more pollutants is equal to the sum of the individual effects (this definition includes the preceding *no joint action* concept). The term *interactive response* includes the concepts of antagonism (less than additive) and synergism (greater than additive). These concepts are clear and simple and can be handled with statistical processes. Except for the concept of synergism (where a statistical synergism may be biologically additive), the statistical and biological interpretations have general agreement. The terms *additive, antagonism,* and *synergism* will be used in the reminder of this section.

Table XIII is a summary of results of pertinent mixture research. We show only a few responses, although in many cases the studies elicited multiple responses. These studies show that our knowledge of the effects of pollutant mixtures on terrestrial ecosystems is fragmentary. Sufficient information is available to suggest that mixtures of pollutants at concentrations similar to those expected in ambient air could be responsible for synergistic responses in a large number of species (cultivars, selections). Where multiple responses have been studied, similar trends are reported.

However, problems exist in our ability to deal adequately with

Table XIII Information on Research Related to Pollutant Mixtures

Mixture	*Test plant*	*Exposure information*	*Plant response to the mixture*	*Ref*
$O_3 + SO_2$	Bean, snap	0.20 ppm each gas; 7 hours/day, 4 days; varied salinity	Stomatal conductance: synergistic (variable); foliar injury: antagonistic; growth: additive; effect changed with salinity.	395
	Bean, snap	0.15 ppm each gas; 6 hours/day, 5 days, consecutive	Foliar injury: antagonistic; stomatal closure: synergistic, variable responses.	396
	Bean, snap	0.25 ppm O_3, 0.8 ppm SO_2; 3 hours; varied temperature	Foliar injury: synergistic or antagonistic; temperature-dependent.	397
	Five species	0.15 ppm each gas; 6 hours/day, 7 days; preexposure with SO_2	Foliar injury: variable results depending on pretreatment, species, and exposure length.	398
	Bean, field	0.05–0.30 ppm O_3; 0.04 ppm SO_2; 4 hours	Net photosynthesis: additive or antagonistic, depending on O_3 concentration.	399
	Pea, garden	0.06–0.27 ppm O_3, 0.11–1.72 pm SO_2; 2, 4, or 8 hours	Foliar injury; reduced leaf area, chlorophyll, leaf weight; data not designed to determine interactive effects.	400
	Pea, 'Alsweet'	0.10 ppm O_3, 0.60 ppm SO_2; 2 hours, 7 exposures spaced over 16-hour photoperiod (controlled)	Not designated to determine interactive effects; injury and chlorophyll decrease correlated; plants less sensitive at first and last exposure periods; stomatal conductance tended to follow injury pattern.	72
	Tomato	0.2 ppm O_3, 0.2 or 0.8 ppm SO_2; 3 or 4 hours, 15 times; infection with nematodes	Growth and yield additive at 0.2–ppm SO_2, antagonistic at 0.8-ppm SO_2.	401
	Lettuce, radish	0.4 ppm O_3, 0.8 ppm SO_2; 6 hours	Use of covariates increased precision for lettuce and radish; lettuce growth and injury effects antagonistic; radish was additive.	402

Table XIII ***(continued)***

Mixture	*Test plant*	*Exposure information*	*Plant response to the mixture*	*Ref.*
$O_3 + SO_2$	Potato	Four O_3 concentrations, filtering of ambient O_3; 0.1 ppm SO_2; for 6 hours/day, 255 hours	Reductions in various growth and yield parameters were additive.	*403*
	Soybean, 2 cultivars	0.25–1.0 ppm O_3, 0.50–1.5 ppm SO_2; 0.75, 1.5, or 3 hours	Foliar injury and reduced shoot fresh weight additive, antagonistic, or synergistic, depending upon concentration and time.	*404*
	Soybean	0.08 or 0.10 ppm O_3, 0.20 or 0.40 ppm SO_2; 2- or 4-hours/day, 1–5 days	Foliar injury and reduced chlorophyll synergistic.	*405*
	Soybean	0.067 ppm O_3, 0.300 ppm SO_2; 7.5 hours/day, 5 days	Synergistic reduction in CO_2 exchange rate (CER), no change in CER with O_3 and only slight with SO_2.	*406*
	Soybean	0.06 or 0.08 ppm O_3, 0.06 or 0.11 ppm SO_2; 5 hours/day, 16 days; in open field facility	Both O_3 and SO_2 caused decreases in a number of yield measures; mixture responses were additive.	*407*
	Grape, 2 cultivars	0.20 and 0.40 ppm O_3, 0.15 and 0.30 ppm SO_2; 4 hours	Foliar injury and reduced shoot length were antagonistic to syngergistic, depending on concentration; leaf abscission was synergistic.	*408*
	Maple, silver	0.05, 0.1, or 0.2 ppm O_3; 0.1 ppm SO_2; 12 hour/day, up to 60 days	Growth analysis and growth parameters; not amenable to mixture interpretation, but $SO_2 + O_3$ did increase growth reductions over O_3 alone.	*409*
	Poplar	0.035 ppm O_3, 12 hours/day; 0.06 ppm SO_2, 24 hours/day; 4 weeks	Reduced dry matter production; interaction not tested but appeared to be additive.	*410*

(continued)

Table XIII ***(continued)***

Mixture	*Test plant*	*Exposure information*	*Plant response to the mixture*	*Ref.*
$O_3 + SO_2$	Aspen, trembling, 11 clones	0.02 ppm O_3, 0.50 ppm SO_2; 0.05 ppm O_3 + 0.35 ppm SO_2; 3 hours	Genetic differences; variation found attributed to genetic differences in response to SO_2 and O_3; not designed to determine mixture effects.	411
	Pine, Scots, (5 seed sources)	0.2 ppm each gas; 6 hours; infection with fungus	Effects on foliar injury were synergistic.	412
	Pine, Scots	1.0 ppm O_3, 66 hours; 1.0 ppm SO_2, 24 hours; 0.5 ppm each gas, 36 hours	Foliar injury; not designed to determine mixture effects but suggestive of antagonistic to synergistic.	413
$SO_2 + NO_2$	Bean, snap	0.1 ppm of each gas; 5 days	Transpiration: increased by individual gases, decreased by combination.	414
	Radish	0.8 ppm of each gas; 2-hour simultaneous or sequential exposures, 1/week, 5 weeks (day exposures); 0.8 ppm or 1.5 ppm each gas; 2-hour simultaneous exposures, 1/week, 5 weeks (night exposures)	Sequential exposures caused no effects; simultaneous exposures reduced growth and relative growth rate; leaf growth favored over edible hypocotyl; night exposures showed similar pattern but a lesser effect.	415
	Tomato	0.11 and 0.05 ppm each gas; continuous, 14 and 28 days	No effects of individual gases; mixture caused decrease in leaf fresh weight and area (14 days), root fresh weight, and dry weight (28 days).	416
	Grass, *Poa pratensis*	0.10 ppm of each gas; 104 hours/week, long-term	Reduced growth of roots and shoots: synergistic in late winter, but not later; shoot recovery during summer; effects on flowering were additive.	417

Table XIII (continued)

Mixture	*Test plant*	*Exposure information*	*Plant response to the mixture*	*Ref.*
$SO_2 + NO_2$	Soybean	0.2, 0.4, 0.6 ppm of each gas; 2 hours	Photosynthesis: synergistic; stomatal conductance: synergistic; respiration: additive.	*418*
	Soybean, 2 cultivars	0.18 or 0.5 ppm SO_2 alone or with 0.06 ppm NO_2; 4-hour/days, 14 days	Yield: no additional effect of NO_2 at low SO_2, but increase in reduction at high SO_2; not designed to determine mixture effects.	*419*
	Soybean	0.13–0.42 ppm SO_2, 0.06–0.40 ppm NO_2; 3-hours/exposure, 10 days during pod fill (2 year study, ambient air with O_3)	Chlorophyll reduction, synergistic; yield reduced from 9 to 25% in 2 years, synergistic.	*420*
	Clover, two populations (from polluted and nonpolluted areas)	0.068 ppm of each pollutant; continuous, summer and winter exposures	Winter exposures: plant growth of population from polluted site not affected by NO_2, more affected by SO_2 than population from unpolluted site, mixture was antagonistic with greater effect on population; summer exposures: plant growth of population from polluted site increased by NO_2 and reduced by SO_2 with a synergistic reduction with both gases; population from unpolluted site increased by NO_2, decreased by SO_2, synergistic reduction by both gases.	*421*
	Trees, 6 species	0.10 ppm of each gas; 104 hours/week, March–August	Reduced growth of seedlings or cuttings: effects on 2 species additive; effects on 4 synergistic.	*422*
$O_3 + PAN$	Petunia; bean, kidney	0.10–0.40 ppm O_3; 0.01–0.10 ppm PAN; 4 hour	Foliar injury, antagonism to synergism depending on concentration of pollutants.	*423*
	Pine, Ponderosa	0.4 ppm O_3, 0.2 ppm PAN	Foliar injury, antagonistic.	*424*

(continued)

Table XIII ***(continued)***

Mixture	*Test plant*	*Exposure information*	*Plant response to the mixture*	*Ref.*
$O_3 + NO_2$	Trees, 7 species	0.10 ppm each gas; 6 hours/day, 28 days	Shoot and root biomass, additive or antagonistic.	*425*
$SO_2 + HF$	Corn, *Pelargonium*,	1.0 ppm SO_2, 0.03 ppm HF; short-term	HF did not affect the stomatal closure response to SO_2 for either species.	*426*
	Corn, sweet	0.09 ppm SO_2, 0.55 ppb HF; 32 days (continuously)	No effect of HF or SO_2, mixture reduced fresh and dry weight of stalk and reduced yield, reduced accumulation of foliar F; interaction not tested.	*427*
$SO_2 + NaF$	Spruce	0.025 or 0.075 ppm SO_2; 5 weeks; 100 ppm F as NaF	Carbon dioxide uptake significantly depressed.	*428*
$NO_2 + HF$	Corn, sweet	0.6 and 1.2 ppm NO_2, 6 hours/day, 4- to 5-days/week; 0.6 and 1.9 ppb HF (continuous)	Foliar injury, additive or antagonistic; interaction shown for increased stomatal resistance.	*429*
O_3 + *Acid Rain*	Alfalfa, 'Saranac'	0.05–0.15 ppm O_3, 7 hours/day, 1 day; 5.6 or 3.0 rain pH, 1 or 2 times/week; together and separate for 2 months or 2 harvests (SO_4^{-2} rain)	Acidity had no effect on yield, but % N decreased by 18–37%; O_3 had no effect on yield or quality; no interactions occurred.	*430*
$O_3 + H_2S$	Bean, snap	0.072 ppm O_3; 0.74, 3.25, 5.03 ppm H_2S; 4 hours/day, 28 days	Changes in net photosynthesis and stomatal conductances appeared additive but were not tested statistically.	*431*
$O_3 + SO_2 + NO_3$	Radish	0.1, 0.2, 0.4 ppm of each gas; 3 hours	Reduced yield: no three-way interaction but an $NO_2 \times O_3$ and an $NO_2 \times SO_2$ interaction; interactions appear synergistic.	*432*
	Radish, Marigold	0.3 ppm each gas; 3 hours, 9 times	Reduced yield: primarily additive responses; $NO_2 \times SO_2$ and $O_3 \times SO_2$ interactions antagonistic.	*433*

Table XIII (continued)

Mixture	*Test plant*	*Exposure information*	*Plant response to the mixture*	*Ref.*
$O_3 + SO_2 + NO_2$	Turfgrass, 6 species, 18 cultivars	0.15 ppm O_3 for 6 hours/day; 0.15 ppm SO_2 and 0.15 ppm NO_2 continuously; 10 days	Foliar injury and reduced leaf area, primarily additive with some antagonism.	*434*
	Sunflower	0.2 ppm each gas, 2 hours	Reduced net photosynthesis: $NO_2 \times O_3$ and $SO_2 \times O_3$, synergistic response; three-way similar to two-way responses.	*435*
	Sycamore, 2 families	0.05 ppm O_3, 0.1 ppm NO_2, 0.14 ppm SO_2; 6 hours/day, 28 days	Reduced growth: various measures appeared additive or synergistic.	*436*

mixture research (i.e., facilities, experimental design, and approaches to interpreting the results). Most laboratories are not equipped to perform research with mixtures (using only one to four chambers at a time), whereas for the most accurate assessment, one chamber per treatment should be used. A number of studies included in Table XIII were not designed to take into account pollutant interactions *(395–436)*. In other reports the results were presented without adequate consideration of possible interactions. Three papers describe statistical means of handling and understanding the results *(354, 356, 432)* and present different but useful approaches to the consideration of pollutant mixtures that could lead to better ways to approach these types of experiments.

Very few studies address the possible biological mechanism(s) for mixture interactions. Most reports address stomatal conductance that affects pollutant entry (avoidance). This is a reasonable approach at the physiological level, but it does not address biochemical mechanisms. Wellburn *et al.* *(437)* have advanced a biochemical explanation for the synergistic action of SO_2 and NO_2 on several grass species. The SO_2 appears to inhibit the ability of the plant to detoxify nitrite; thus, nitrite builds to toxic levels. Continued research of this type is essential if the mechanisms of mixture responses are to be understood.

C. Other Research Areas

We here briefly highlight research on (1) plants as biological indicators, (2) plant breeding programs, (3) effects on plant communities, and (4) economic considerations. The intent is to highlight the importance of these subject areas, not to explore them in depth. These areas are discussed to some extent in other sections of this chapter.

1. Biological Indicators

The material in Vol II, 3rd ed., Chapter 12 on plants as biological indicators of air pollution effects is still pertinent. Research published since that edition adds to our knowledge but neither changes basic concepts nor solves problems presented therein. *Biological indicators* are sensitive plants that respond to low concentrations of a chemical or chemicals in the atmosphere. The best indicators are selective for a given pollutant. Many scientists have wanted to use indicators as monitors that reflect the concentration of a given pollutant in the atmosphere over some time period, with the indicator an integrator of the concentration. Most research supports the thesis that indicators are not good monitors and that such systems can be very expensive. A practical use of indicators would be to calibrate their response to that of different crops of economic importance and then to use the indicator (monitor) to determine crop losses. No efforts have been made to initiate such calibration research. Europeans appear more confident that biological indicators (as monitors) will answer their needs than do North Americans.

The review by Feder *(438)* supports use of bioassay systems for monitoring air quality because of difficulties with physical monitoring systems. Although the use of such systems can be supported, primarily because they will integrate the effects of the ambient pollutant complex, unanswered questions in the development of usable systems remain. Such monitoring systems tend to be subjective, are open to effects from other environmental factors (due to plant response to other stresses), and do not permit the determination of cause-and-effect relationships. Feder suggests that white pine, tobacco, mosses, and lichen should be useful in bioassy systems.

Bel-W_3 tobacco (a sensitive O_3 indicator) was exposed to ambient air in sites throughout the British Isles; injury severity was related to proximity of sources (prevalence of land versus sea wind flow) and to the amount of sunshine *(439)*. This study helped convince Europeans that O_3 should be considered in estimating crop and forest losses from air pollutants. Similar studies were reported around Tel Aviv in Israel *(440)*, and in Germany *(441, 442)*. The tobacco system was developed in the United States in the 1960s and has been used successfully throughout North America and many parts of the world since then.

Fluoride concentration in black locust leaves was used to estimate fluoride levels in forage near an industrial source *(443)*. Injury to bean was used to confirm the presence of fairly high concentrations of NO_2 near a point source *(444)*. Rice *et al.* *(445)* made a baseline study of sulfur and fluoride content in 17 plant species in the American northern Great Plains and recommended that *Juniperus scopulorum* serve as a regional biomonitor. Although foliar elemental concentrations are not generally considered accurate indicators of pollution concentrations, fluoride accumulation is a good indicator of fluoride pollution, and sulfur accumulation may be an indicator of SO_2 pollution.

Interest continues throughout the world in the use of lichens as indicators of air pollution because of their sensitivity. Skye *(446)* presents a good update of research to the late 1970s and makes a reasoned case for employing lichens as indicators. We believe lichens and other cryptogams may be of supplemental use in identifying a potential air pollution problem but are not reliable as primary monitoring tools because their integration requires too long a time period. Lichen transplants *(447)* were used to study the biological impact of air pollution from a smelter. Lichens on sycamore were used to map SO_2 in the United Kingdom as a requirement for environmental planning *(448)*. Winner and Bewley *(449, 450)* made an ecological study comparing the responses of vascular plants and mosses to a SO_2 source in Canada and suggested that mosses would be a good SO_2 bioindicator.

Chemical changes shown to occur within plants exposed to controlled levels of pollutants have been suggested as a means of identifying air pollution injury in the field *(451, 452)*. Although many of these tests are useful when conducted under controlled conditions, they are not useful in the field because the chemical changes caused by pollutants may be duplicated by other stress factors. Jäger *(453)* reviewed biochemical indicators and discussed some of the inherent problems.

Prinz and Scholl *(454)* reviewed an effects-inventory program, which includes biological monitoring, that was carried out from 1972 to 1974. The program included lichen transplants into selected areas and accumulation of various substances by a standardized grass culture. This monitoring program plays a continuing role in air pollution control activities supported by the state of Nordrhein–Westfalen in Germany. Von Rudolph and Halbwachs *(455)* presented the results of a border-crossing network of bioindicators ("Inn–Salzack- area") that was started in Bavaria in 1975 and uses spruce as the indicator species. The system now includes part of Austria. The system was designed to study sulfur and fluoride emissions through needle analysis. The results have been used to map areas of differing levels of fluoride and sulfur.

The most extensive monitoring system and the rationale for its use is presented by Posthumus (reference *18*, Chapter 2). This program represents a major and continuing effort in the Netherlands. It is documented by a series of technical publications that detail the approach and the results obtained. Only their tenth report *(456)* is referenced here because the publications are in Dutch.

2. Breeding Programs

There is interest in adding air pollutant stress (especially O_3) to plant breeding programs as a means of alleviating crop and forestry air pollution problems, but no coordinated research has been started. In early efforts by growers with certain sensitive crops, resistant individuals were selected; the growers then collected and used seed from these plants in subsequent years. Inadvertent selection has occurred: (1) where breeding is done in areas of high O_3, the resultant cultivar releases tend to be more resistant; and (2) where breeding is done in areas of low O_3, the resistant cultivar releases tend to be more sensitive.

The information presented in Section II, B, 1 and Tables III to V clearly indicates that heritability for resistance to O_3 and SO_2 is present in germ plasm that has been tested. Likewise, in species and cultivar screens, the range of response suggests the presence of resistance in gene pools. Thus, the ingredients for successful breeding programs do exist. This does not mean that all species have resistant cultivars, but increased tolerance to O_3 and SO_2 should become available in commercial cultivars. Although a longer time frame would be required, such programs should be considered in managed forest systems.

The introduction of pollutant stress into managed breeding programs is not easy. The breeder must be convinced that air pollution stress plays a significant role in crop productivity and then work closely with air pollution specialists to develop screening procedures. In addition, the breeders must not lose other traits that have been selected for cultivar introductions.

Concern for crop decline (production losses) in several parts of the world is based on lower than expected yields anticipated from new cultural methods and cultivar introductions. Concerns have been expressed for both soybean and cotton production; breeders have considered the commonly accepted stresses as contributing factors in crop decline but have not factored O_3 or other air pollutants into their thinking. Awareness of air pollutant effects is necessary before breeding programs, effective for development of resistant cultivars, can be initiated.

3. Economic Studies

Economic studies are now attempting to determine economic losses of crops and forests to air pollutants (especially O_3). Economic estimates made prior to 1980 were based on crop or tree yield reductions and multiplying these reductions by the current value of the crop or tree used. These calculations were not economic studies *(66)*, but they did provide some guidance for early policy decisions.

Linzon *et al.* *(457)* estimated economic losses from O_3 on crops in Ontario, Canada, using data from the National Crop Loss Assessment Network (NCLAN) *(66)*. They predicted an increase of $15 million per year in crop production in Ontario if the O_3 standard were set at 0.08 ppm for 1 hour. The analysis was derived by calculating yield losses by using the NCLAN data and adjusting for 1980 crop prices.

Adams and co-workers *(66, 458–460)* have made the first extensive attempts to make an economic analysis (econometric approach) of the effects of O_3 on crop production in California *(458)*, on a regional basis *(66, 460)*, or on a national basis *(459)*. Two of these efforts *(66, 460)* have focused on the NCLAN data sets; one *(460)* used "farmers' actions" to estimate crop loss on a regional basis.

Two primary ways are available to obtain biological data for use in an economic analysis. First, use dose–response functions generated experimentally for different crops and crop cultivars (method used by NCLAN) on an area, regional, or national basis; these functions are an integral part of the economic assessment effort discussed in Section V. Second, use the difference in pollution (i.e., O_3) concentration across a region or nation and correlate these concentrations (using some averaging technique) with crop-cultivar yields. The second approach is epidemiological (statistical) and is not adequate for determining effects of O_3 or other atmosphere pollutants on crop yield. Difficulties in the statistical approach include poorly defined differences (1) between O_3 concentrations across a region, (2) in cultivars used, (3) in environmental (climatic and soil) conditions, and (4) in farm management practices. Thus, although such efforts are being tried, only nondefinitive correlations can be provided.

Data are available to initiate preliminary estimates of crop losses from O_3 (references *65–68,* Section V). However, insufficient data exist for even preliminary estimates of losses from other pollutants or losses associated with forest productivity. For these losses, we must rely on subjective estimates by trained investigators.

4. Population and Community Studies

Most of the studies on the effects of air pollutants on plant populations and communities have been associated with large field studies *(56–62)* in the United States and Canada. These studies are highlighted in Table II and are related to studies on SO_2 and photochemical oxidants (especially O_3). The proceedings of a symposium held in Riverside, California *(44),* presented a good overview of many field studies either started or completed by 1980 from a number of countries. The research on the ladino clover–fescue forage system (Section III, C, 1) addresses a competitive situation in which one species is more sensitive to O_3 than the other; the system may give direction for the study of competition in natural communities.

In addition to the large field programs, the following selection of studies in natural systems illustrates the type of research that can be done. Open-top chambers were used to determine the effects of ambient O_3 on milkweed in the Shenandoah National Park (Virginia) in the United States *(461);* the results were confirmed in controlled laboratory exposures. In a similar study *(462),* changes in biomass production presumably due to O_3 were monitored in a high meadow in the same Virginia park. The effects of O_3 on a mixed planting of goldenrod and mugwort favored the growth of the golden rod after 2 years *(463).* In a winter O_3-flux study above a deciduous forest, woody stems and branches removed a measurable portion of the O_3 *(464).* McClenahan *(465)* studied community changes in a deciduous forest in relation to distance from sources of several pollutants (SO_2, F^-, Cl^-) and found a decrease in richness, evenness, and diversity with decreasing distance from source. Liebold *(466)* proposed methods for determining the extent of damage from SO_2 to a spruce plantation based on crown characteristics and changes in tree diameter. Sima and Tollinger *(467)* determined the forest's capacity to remove SO_2 (loading capacity) without damage to the forest. A study on effects of smelter emissions on a forest–tundra transition ecosystem found severe damage attributable to SO_2 *(468)*, as did a similar study on a desert grassland community *(469)*, although toxic levels of heavy metals were reported in the surface soils at both locations. A similar study around a fluoride source in Newfoundland *(470)* related the intensity of effects with distance from the source. A study around an aluminum smelter in Australia *(471)* related changes in plant community composition to variations in species sensitivity to fluoride.

Valuable information can generally be gathered from field studies. Field studies relating to air pollutants have confirmed direct effects on plants, the importance of competitive pressure, the influence of secondary invaders (e.g., parasites and fungi) on a weakened ecosystem, and the need to understand the soil–plant interrelations. The assessment of tree decline in Europe and in the United States should add to our understanding of interactive effects of air pollutants and other stresses on plants, populations, and communities (see Section IV, F).

IV. Acidic Deposition—How Much of a Problem?

One of the most significant air-pollution-related problems to attract the attention and imagination of environmentally concerned individuals during the past decade has been the issue of acidic deposition. A legacy of pollution control strategies (e.g., tall stacks, >200 m), aimed at reduction of ground-level concentrations of sulfur and nitrogen oxides near fossil fuel combustion sources, is currently thought to be a contributing factor in the long-range transport and distribution of acidic substances in the atmosphere. Sulfur and nitrogen oxides emitted higher in the atmosphere can be transported for long distances. During transport, sulfur and nitrogen oxides are oxidixed to sulfate and nitrate before deposition hundreds of kilometers downwind by the wet processes of washout (precipitation passing through a polluted air mass) or rainout (incorporation of the pollutant in cloud-forming processes). The dry processes of sedimentation and impaction (including gases) are also significant, particularly in the eastern United States, where local sources play an important role in regional air quality. Because the processes of atmospheric transport and deposition follow no political boundaries, downwind receptors have frequently been across state, provincial, and national boundaries from sources of pollution. This factor, more than any other, has served to make acidic deposition a major policy issue on both the European and North American continents. Unfortunately, since the term *acidic deposition* is sufficiently broad to cover multiple environmental stress factors, confusion often results from its use, and the term *acid rain* has frequently been used to describe effects whose causes are more accurately attributable to the sum of atmospheric pollutants impacting a unit of landscape.

Acidic deposition has also been implicated in chronic acidification of aquatic ecosystems, where transport of acidic substances through terrestrial ecosystems is insufficient to buffer acid inputs adequately. Our discussion here does not cover acidic deposition effects on aquatic ecosystems, but Chapter 6 and recent reviews of this subject *(27)* are recommended.

A. History of Concern

Acidic deposition in the form of wet deposition is a phenomenon which, at least in the literature, has been recognized within the span of years since the publication of Vol. II, 3d ed., Chapter 12, (1977). Thousands of journal and popular articles have been devoted to the potential and actual environmental consequences of "acid rain." Several historical accounts of research in this area have been written; the most complete is that by Cowling *(472)*. At least three annotated bibliographies have also been published, the most comprehensive of which lists over 3,200 entries as of late 1983 *(473–475)*.

The phenomenon of rainfall acidification by pollutant emissions was recognized by Hales as early as 1757 in England and its effects first examined by Robert Angus Smith as early as the 1870s *(472)*. However, modern attention to the topic awaited examination of long-term precipitation chemistry records collected by the European precipitation chemistry network, beginning in 1948 *(476)*. Twenty-five-year trends of increasing acidity in precipitation and surface waters led Swedish scientists to raise a concern over fishery resources in acidified waters and potential effects on Sweden's forest resources. This concern, voiced to the United Nations in 1968, served to raise the term *acid rain* to the international scientific community. In 1972 a Norwegian national research program was initiated for the purpose of evaluating the effects of acid rain on forest and fisheries resources *(64)*.

As a result of the European studies, North American scientists began to examine the available historical records of precipitation chemistry carefully and reported trends similar to those observed in Scandinavia *(477, 478)*. Canadian and United States research programs to study the potential impacts of acid precipitation were established in the early 1980s *(472)*.

From a historical perspective, the chronology of publications describing effects on vegetation range from observations associated with point or local sources of air pollution *(479–481)*, to hypotheses of potential long-term ecosystem impacts associated with long-range transport and deposition of these pollutants *(476)*, to experimental testing of these hypotheses in a second generation of research in the early 1970s *(482–484)*. From these beginnings, the current rate of scientific publications keyworded by *acid rain* or *acidic deposition* has exceeded 600 per year *(485)*.

Earliest public concerns about acidic deposition were related to acidification of surface waters and effects on fish populations (see Chapter 6). Investigations into potential effects on terrestrial ecosystems, both forest and agricultural, began at approximately the same time.

Because agricultural practices of fertilization and liming represent a much greater source of alteration of agricultural soils than atmospheric deposition reasonably could, forest soils received emphasis earlier than agricultural soils did. However, early screening studies suggested that herbaceous vegetation was more sensitive to direct injury by simulated acidic rain than was woody vegetation. As a result, early work on vegetation effects in the United States focused largely on agronomic crops.

A "third generation" of research evolved in the early 1980s as a result of field observations of dieback and decline of red spruce in high-elevation forest stands of the eastern United States and similar observations for Norway spruce and beech stands in West Germany. Two key pieces of this most recent thinking about the acidic deposition problem have been (1) an evolving recognition of the probable importance of indirect mechanisms of action and (2) acknowledgment of the probable importance of a broad range of interactive stress factors in the overall response pattern. Differences between the symptomology of decline in North American and European forests and an imperfect understanding of the mechanisms of the decline make evaluation of the importance of acidic deposition to forest decline extremely difficult. Spatial correlation between general patterns of decline and regions with high rates of pollutant deposition suggest some relationship to air quality degradation; however, it is difficult to understand and quantify accurately the severity of the acidic deposition problem.

B. Observed Effects on Terrestrial Ecosystems

The potential problems associated with acidic deposition in terrestrial ecosystems arise from both short- and long-range atmospheric transport of sulfur and nitrogen oxides from natural and man-made sources *(27)*. Atmospheric deposition processes contribute sulfur, nitrogen, and a variety of trace metals to natural and managed ecosystems. Sulfur is deposited primarily as SO_2 and sulfate (SO_4^{2-}), and nitrogen in the form of NO_2, nitric acid vapor (HNO_3), and nitrate (NO_3^-). The fate and effects of these pollutants as gases, as wet precipitation, or as dry particle deposition to terrestrial ecosystems are not well characterized on spatial scales larger than small watersheds (e.g., <100 ha) *(27)*. Understanding the processes involved in ecosystem response to such loading is difficult because of the complexity of the biological systems involved, and because both short- and long-range atmospheric processes are involved *(27)*. The impact is regional, the effects are subtle over annual time scales, and many factors interact, not always linearly, to determine the net ecosystem response.

Determining the net effect of sulfur (S) and nitrogen (N) deposition in terrestrial ecosystems requires a knowledge of which processes are likely to be affected and the sum of their varying responses to the deposition patterns characteristic of the impacted region. Critical issues related to S are the S requirements of vegetation, the form of sulfur entering the system, and the ability of certain soils to sequester or immobilize relatively large quantities of S beyond requirement levels. Deposition rates of N and hydrogen ion (H^+) and the fate and effects of the attendant acidity are also important factors in ecological effects. A final evaluation of the effects of atmospheric acidic deposition must also consider the natural sources of acidity in the system.

The temporal and spatial variability in sensitivity of ecosystems, within the total combination of environmental stress factors to which the systems are exposed, must be better understood. Furthermore, both sensitivity and resiliency of aquatic and terrestrial ecosystems must be characterized to make intelligent policy decisions and to identify potential mitigative measures.

Examples for the purpose of our discussion will focus on effects on ecosystem function and structure at the process, organism, and community levels.

C. Effects at the Process Level

The relative toxicity of many substances in the environment may be influenced by widely variable environmental factors that can exert profound regulatory influence on physiological processes. These regulatory influences, in turn, can be extremely important factors in the level of response that an organism might exhibit to acidic deposition, in both wet and dry forms. Figure 1 presents a conceptual model of the relationship between atmospheric S and N, biotic receptors, and biological response.

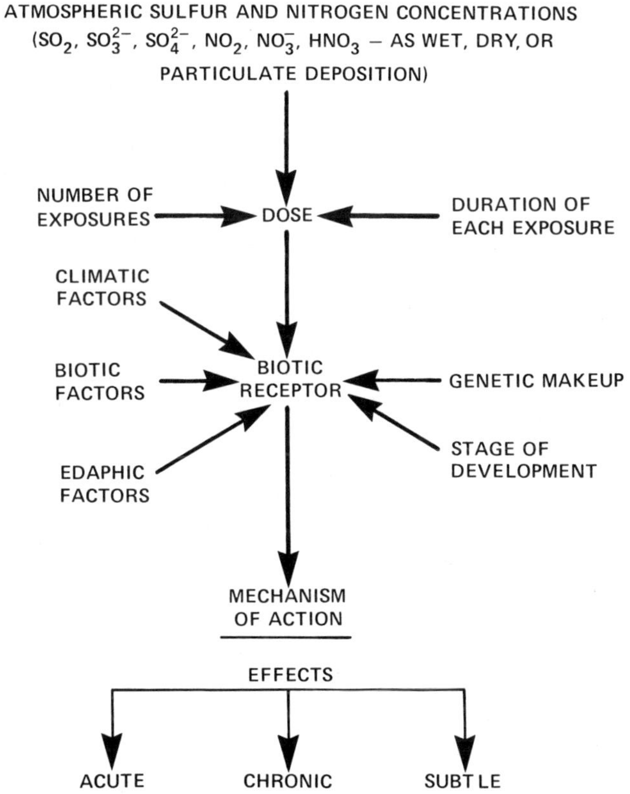

Figure 1. Conceptual model factors involved in biological response to atmospheric sulfur and nitrogen deposition (From Vol. II, 3d ed., Chapter 12, 1977).

Our discussion of effects on process-level responses will focus upon the ways climatic, edaphic, and biotic factors are believed to influence vegetation responses to dry and wet deposition of S and N.

1. Climatic Factors

A broad range of climatic factors, including light intensity, photoperiod, and temperature, may predispose (precondition) plant sensitivity to phytotoxic air pollutants *(486)*. This predisposition may influence plant sensitivity in a complex manner over either short or long time scales. For example, photoperiod, light intensity, and temperature all play regulatory roles in the process of photosynthesis in green plants. The amount of photosynthate stored as sugars in plant tissues can influence SO_2 and NO_2 uptake by regulating osmotic pressure in guard cells of stomata, and hence, gas exchange. Changes in sugar content apparently alter plant sensitivity to SO_2 and NO_2 within only certain sugar concentration ranges, which are dependent not only on light quantity and quality but also on the metabolic rates and translocation of assimilated carbon. On a shorter time scale, low light intensity may directly cause stomatal closure, resulting again in a decreased potential for uptake of SO_2 during periods of unfavorable light quality *(486)*. Similarly, McLaughlin and Taylor *(278)* demonstrated the importance of relative humidity in the uptake of pollutant gases. Regions characterized by conditions of relatively high humidity can be expected to contain vegetation with a higher relative potential for pollutant uptake.

Although similar systematic studies on the role of these climatic factors in altering plant response to wet deposition of acidic substances are limited, observations indicate that they play a less significant role in plant response to wet deposition since stomatal uptake is not considered essential for wet deposition to affect plants *(487)*.

Variables involving the characteristics of the wet deposition event, including duration, intensity, and length of time between events, may be important to plant response *(488)*.

2. Edaphic Factors

Two major soil-related factors affecting vegetation response to atmospheric S and N deposition are soil moisture and mineral nutrition. For dry deposition of SO_2 or NO_2, plant uptake correlates well with soil moisture conditions *(489)*. With adequate soil moisture conditions, stomata tend to remain open under favorable light conditions.

With varying degrees of moisture stress, decrease in plant sensitivity to air pollutants has been correlated with increase in moisture stress. Nutrient content of soils may also affect plant response to gaseous pollutants. For example, an ample supply of mineral nitrogen can increase the resistance of some plant species to SO_2 *(486)*. The mechanism probably involves the plant's ability to utilize available N to increase the incorporation of S into metabolic pathways rather than permitting toxic accumulation of SO_4^{2-} *(490, 491)*. This potential for N inputs in precipitation to influence uptake and accumulation of S in tissues represents a potential interactive effect of acidic deposition components. The interactive potential deserves additional research, especially in consideration of evidence of exceptionally high levels of nitrate deposition in cloud water in high-elevation spruce–fir forests in the eastern United States *(492)*.

3. Biotic Factors

Potential roles for biotic factors as modifiers of vegetation response to S and N deposition are numerous and have been suggested by a broad variety of studies in controlled environments (reference *17*, Chapter 11; *38, 493*). Microorganisms interact with vegetative components of terrestrial ecosystems (1) as pathogens, through changes in relationships among plant growth regulators *(494, 495)* that alter plant response to pollutant uptake (reference *17;* Chapter 11); (2) as pathogens that interfere with stomatal action or membrane integrity *(494);* (3) as symbionts through mycorrhizal or root nodule associations that affect mineral nutrition and water relations *(240, 496);* and (4) as decomposers that influence nutrient turnover and availability *(497)*. Both wet and dry deposition of S or N affect all of these potential pathways, but the potential of any one of them to alter long-term ecosystem function and structure is speculative at the present time. Atmospheric deposition of S may act directly on host–parasite relationships *(498)*.

One of the most frequently cited hypotheses relating acidic deposition effects on vegetation to growth response involves the possible role of mobilized aluminum. This aluminum is mobilized as a result of acidic-deposition-induced soil acidification and may be toxic to fine roots of species such as Norway spruce *(499);* such toxicity may directly affect nutrient uptake. Potential interactions with mycorrhizal associations may represent an additional form of stress contributing to observed forest dieback and decline. Stress may also predispose the roots to a secondary infection by root decay fungi that even further stresses the tree. Effects of atmospheric deposition (e.g., increased leaching of nutrients or decreased photosynthesis) could also be important in tree decline. It should be emphasized that the role of secondary insects and pathogens in the overall decline of forest trees associated with patterns of high S and N deposition is admittedly conjectural but represents an area of needed research.

a. Possible Mechanisms of Plant Response. Ziegler *(500)* has summarized a mechanism by which wet deposition could affect gas exchange rates. In order to reach cells, protons in the water layer must overcome cuticular, stomatal, and plasmalemma resistances. Protons are expected to have an effect on membrane integrity which could inhibit the photosynthetic capacities of chloroplasts. The sulfate counter ion, a competitive inhibitor of the photosynthetic enzyme ribulose-bisphosphate carboxylase, with respect to bicarbonate, may result in increased or decreased CO_2 fixation rates, depending on sulfite concentrations. The preceding mechanisms could result in altered photosynthetic rate and chlorosis of affected tissues. Observations of foliar chlorosis, premature senescence and abscission *(488, 501–503)*, altered photosynthetic rates *(504, 505)*, and photosynthate allocation *(506)* would seem to support the operation of hypothetical mechanisms of response such as those proposed by Ziegler *(500)*.

Other mechanisms for wet-deposition effects on gas exchange rates and nutrient leaching have been hypothesized *(507)*, and some experimental evidence supports these hypotheses. Shriner *(484, 508, 509)* and Crang *(510)* have reported evidence of enhanced weathering of epicuticular waxes of foliar surfaces by simulated acidic rain. Shriner *(484)* observed this to occur after prolonged, chronic exposure of foliage over the course of the growing season (10 minutes per day, 5 days a week, for 8–30 weeks). Paparozzi and Tukey *(511)* were unable to duplicate significant weathering of epicuticular waxes after acute exposures (2 hours/day) for 4 consecutive days, further suggesting that repeated exposure over time is important. These observations of acidity effects have numerous implications for the plant, but most relevant to plant response are the observations that differential or selective penetration of ions occur at the cuticular membrane, that rates of penetration are pH-dependent, and that the ability of the cuticle to act as an ion exchanger may be altered *(487)*. Both cuticular and stomatal transpiration may also be altered by simulated acidic deposition. Measurements of rates of gas exchange were much greater in foliage exposed to simulated rainfall at pH <3.4 when compared to controls *(512)*. These results suggest that foliage exposed to acidic rain is more subject to water stress, as well as being capable of more rapid uptake of gaseous pollutants such as O_3 or SO_2.

b. Nutrient Leaching. Enrichment of throughfall by nutrient elements is a natural component of nutrient cycling processes and has been well documented *(513)*. Evidence of increased leaching of cation nutrients from foliage as simulated precipitation acidity increased has been reported *(514, 515)*. Increases in leaching ranged up to 50% for K^+ when pH 4.0 treatments were compared to pH 5.0 controls *(515)*.

The exchange of H^+ in precipitation for cations on the cuticle exchange matrix can result in significant scavenging of H^+ by a plant canopy. Eaton *et al.* *(516)* found the forest canopy to retain 90% of the incident H^+ from pH 4.0 rain (growing season average), resulting in less acidic (pH 5.0) solutions' reaching the forest floor. The removal of H^+ by exchange processes in the forest canopy does not eliminate the effects of H^+ deposition on the forest ecosystem, however. Cations leached from the foliage may eventually be leached from the ecosystem if the anion associated with H^+ inputs (primarily SO_4^{2-} or NO_3^-) is mobile. Nitrate, although highly mobile, is frequently tied up by biota in the forest floor. The mobility of SO_4^{2-} in forest soils, however, is highly dependent on soil chemical properties such as SO_4^{2-} adsorption capacity *(517)*. Plant response to cation leaching may be (1) accelerated uptake of cations to compensate for foliar cation losses or (2) reduced foliar cation concentrations *(518)*, if H^+ inputs and foliar exchange are of significant magnitude and duration. In either event, the introduction of H^+ with a mobile anion will cause a net loss of cations from the ecosystem, whether the H^+ cation exchange occurs in the forest canopy or the soil.

An example of reduced foliar cation concentrations has been reported by Rehfeuss *et al.* *(519)* for Norway spruce in high-elevation forests of eastern Bavaria. Trees exhibiting symptoms of decline and dieback were paired with nonsymptomatic trees in the same stands and site conditions. Large differences were noted in foliar Ca and Mg content, particularly in older leaves with less Ca and Mg, in declining trees than healthy trees. The Mg content was characterized by the authors as in "extreme deficiency," with calcium in "poor supply." The authors further speculated that since these nutrient deficiencies occurred on soils varying considerably in content of both elements, soil depletion was probably not the dominant contributing factor; rather, the deficiency was mainly a consequence of enhanced leaching of Ca and Mg from the foliage as a result of the deposition of strong acids. The authors *(519)* speculated further that Ca and Mg uptake from soil pools were inadequate to replace the foliar leachates. Such nutritional disorders have been reported to make foliage more susceptible to additional leaching *(520)*. Further evidence to support the hypothesis of Rehfeuss *et al.* *(519)* comes from the observation *(521)* that blossom end rot of tomato, a calcium deficiency symptom, can be induced by simulated acidic rain treatments (pH 3.2) even when soil pools of calcium are not limiting. An alternative explanation could be direct inhibition of Ca uptake and translocation independent of Ca leaching rates *(522, 523)*.

D. Effects at the Organism Level

Although effects discussed at the process level are numerous and complex, the net response by the entire plant is usually manifest in an effect on growth. The S effect may be positive, through S or N nutrition in S or N limited systems *(524)*. However, most forests of the eastern United States receive S deposition in excess of S requirements, suggesting that atmospheric sulfur deposition may not be of major benefit in those systems *(525)*. Where significant N deposition accompanies S deposition, the potential for a positive growth response to the combined S and N inputs would appear plausible. Deposition of S may also represent a significant contribution to crop S requirements *(526, 527)*. Counteracting the potentially beneficial effects are the adverse effects of S as SO_2 or SO_4^{2-} on photosynthetic rate, nutrient leaching from foliage, nutrient uptake, and susceptibility to secondary stress factors such as insects, disease, or drought. These are negative factors capable of reducing growth and competitive fitness of individual plants whether within a crop or forest stand.

Genetic variability in sensitivity to atmospheric deposition represents yet another factor to be considered in an analysis of effects of acidic deposition on populations. Even within populations composed of plants as genetically similar as cultivars of an agricultural crop species, the H^+ concentration required to cause a growth response may vary as much as tenfold *(27)*. Species-to-species variability in sensitivity may be as great as two orders of magnitude in H^+ concentration for similar responses *(27)*. Clearly, understanding the factors controlling such variability will be necessary before conclusions can be drawn about the relative competitive advantage or disadvantage of a species within a plant community. The site nutrient status and duration and rate of inputs must also be superimposed on the inherent biological sensitivity of the system before conclusions about expected changes in forest and crop productivity can be drawn.

E. Effects at the Community Level

The uncertainty with which we can speak of atmospheric deposition effects upon terrestrial ecosystems becomes increasingly greater as we progress from the process to community levels of organization. Factors which influence the competitive ability of an individual or a species within a community will produce, over the long term, changes in species composition, diversity, and system stability. When the principal exposure characteristics that limit growth or yield are identified and when adequate air quality and deposition data are available to quantify exposures, it may be possible to use dose–response equations to estimate effects across theoretically homogenous regions of a country *(528)*.

One of the most promising means to predict community response is the use of mathematical models that integrate data from a wide range of sources to describe the responses of complex ecological systems. The complex structure and perennial growth habit of forest communities make measurement and prediction of response a difficult task *(528)*. This is especially true for the uneven-aged mixed deciduous forest stands characteristic of those regions of the United States currently receiving the highest levels of acidic deposition. Several forest-stand growth models have been developed to illustrate forest community dynamics *(529, 530)*. These models have been used to examine the potential for chronic growth inhibition by wet *(531)* and dry *(532)* deposition processes on individual species and stand-level biomass responses. Such approaches have focused on a few of the components that are potentially operative in ecosystem response, largely because more comprehensive models require data that are currently not available. Inferences on unmeasured ecosystem-level responses at a larger scale must of necessity contain numerous assumptions to compensate for data limitations *(528)*. However, models of this type have identified competition and stand age as important modifiers of forest response.

Although simulation models represent excellent tools for data integration, they are not yet capable of providing quantitative estimates of community response. However, models do indicate the clear need for better data at the process level to achieve a better understanding of relative species' sensitivity to chronic acidic deposition in terrestrial ecosystems.

F. Forest Dieback and Decline

During the past five years, numerous reports have detailed a widespread and apparently synchronous decline in forest growth on the European continent. The rapidity of this decline has been unprecedented. The decline is noteworthy because it involves many species of both softwood and hardwood trees growing across a wide range of soil, site, and climatic conditions *(533)*.

Although acidic deposition has been proposed to explain the phenomenon, it is apparent that no single causal agent can satisfactorily account for the range of effects observed. General hypotheses to explain the decline include (1) toxic elements such as heavy metals *(534)*, (2) O_3 *(535)*, (3) nutrients *(526, 533)*, and (4) growth-altering substances *(533)*.

Many of the symptoms observed in Germany are consistent with an alteration of growth-regulating substances, possibly induced by a combination of excess N deposition and altered carbon allocation. Potential interactions between N nutrition and decreased tolerance to climatic extremes (e.g., drought tolerance and cold hardiness) are also consistent with observed effects. The possible causes of forest decline may be categorized as follows:

1. Acidification of soils, with subsequent mobilization of aluminum, and toxicity of aluminum to feeder roots
2. Gaseous pollutant effects with O_3, SO_2, and nitrogen oxides as primary candidates
3. Magnesium deficiency, resulting from either increased leaching from foliage or decreased uptake by feeder roots
4. Disruption of physiological function, as a result of decreased photosynthetic rate, alteration of carbon allocation, or changes in N dynamics
5. Excess nutrient deposition, especially of N

Although the causes of widespread forest decline are the subject of much speculation, the message for air-pollution effects research is clear: Research on the impact of a single pollutant isolated from the complex of interacting factors impacting vegetation in the field is inadequate.

G. The Net Effect on Terrestrial Ecosystems

The extrapolation of atmospheric deposition impacts to a regional or national scale is critical for an assessment of effects. To make such extrapolations, we must understand the mechanisms which account for the sum of effects on critical processes, including compensatory mechanisms, and consider at least those most important modifying environmental and biological factors likely to vary within and between regions. To apply this knowledge in a meaningful assessment, large-scale data bases that characterize the range and distribution of sensitive resources and the most significant modifying variables are needed. Some of these necessary data bases are available *(536)*, but significant efforts at the national level in each country must be initiated before such assessments can be made to support major policy decisions.

H. Summary

Research on the effects of acidic precipitation on vegetation has focused on forest ecosystems because effects on crop species have been regarded as minimal *(537)*, on the basis of a lack of evidence for measurable effects on more than a few cultivars of major economic crop species. However, if long-term trends of increased demand for United States agricultural production and increased prices of energy and agricultural technology should place an increased pressure on the nation's land and water resources, these trends could shift agricultural production in the United States from chronic surplus to recurring, if not chronic, scarcity *(538)*. If such a shift were to occur, trends in fertilizer, tillage, and pest management practice could make "marginal" effects of acidic precipitation on agricultural productivity a significant factor in agricultural land use.

Our understanding of the effects of acidic atmospheric deposition on terrestrial ecosystems can be characterized by the lack of comprehensive knowledge of effects on ecosystem processes. Additionally, extrapolating available data beyond the site-specific conditions that have been studied to date remains an area of uncertainty. Only a commitment to continued research, of the type identified in this review, will utimately increase the knowledge base and lower this level of uncertainty.

V. Assessment of Effects of Ozone and Sulfur Dioxide on Vegetation

This section addresses assessment methodology to accomplish regional and national assessments of impacts of air pollutants on crop and forest productivity. Ozone is used as an example because most of our available data pertain to O_3. We have discussed available information for SO_2. Although much research reported in earlier sections can be supportive of the assessment effort, it would normally not be used directly, unless field data were not available.

A. Overview of Problem

Primary goals of air quality assessment research are to measure the effects of pollutants on productivity of natural and agricultural ecosystems and to develop principles and techniques for exposure facilities and assessment approaches. Early assessments utilized data from growth chambers, greenhouse studies, closed-top field chamber studies, and field surveys that attempted to relate loss of productivity to foliar injury. Greenhouse and closed-top chamber studies used plants grown under conditions which differed in various ways from field conditions. The degree to which these conditions affected plant response to pollutants is not known. Results of field surveys using foliar injury to estimate yield losses were questionable because identification of the cause of injury symptoms was difficult and because relationships of pollutant dose, injury, and yield were not well defined.

Pollutant dose–plant productivity response information must be obtained from plants exposed to a range of pollutant doses under the same climatic conditions for a given experiment. This condition is necessary to prevent confounding effects of variable environmental factors which affect growth and yield. Ambient pollution gradients can be used to obtain different doses in some instances, provided careful measurement of pollutant concentrations and other factors which can affect plant growth are included. This gradient approach is valid where SO_2 or NO_2 doses vary within short distances (e.g., point sources with low stacks); it is not valid for SO_2 from tall stacks or for NO_2 or O_3 because the distances required to obtain different doses impose environmental variations that will confound the results. Thus, some method of controlling pollutant concentrations while providing relatively normal environmental conditions for plant growth is the preferred procedure for developing dose–response relationships.

Several systems are capable of controlling pollutant concentrations around plants growing under conditions that approximate those in the field (the open-air system). One technique involves the release of pollutants into the open air from holes in pipes or plastic duct systems. The first such system released SO_2 from holes in horizontal pipes at canopy height simultaneously from the four sides of a field plot area *(539)*. Improvement in controlling SO_2 concentrations and an ability to compensate for variable windspeeds were accomplished by using computer control of SO_2 release from the upwind side of the experimental field *(540)*. Similar systems have used (1) horizontal release pipes at intervals within a field *(541)*, (2) a single horizontal release pipe at one side of the field *(542)*, (3) vertical release pipes at regular intervals throughout an experimental area *(543, 544)*, or (4) release of different concentrations from plastic ducts to form a linear concentration gradient along plant rows in a field *(545)*. Detailed descriptions and functions of existing open-air systems have been provided by McLeod and Fackrell *(546)*. All open-air systems have the advantage of growing plants under normal practices. However, control of pollutant levels is difficult because of changes in wind direction and velocity, the experimental plot must have several continuous monitors, and a control treatment cannot be provided for ubiquitous pollutants such as O_3.

Another technique uses cylindrical or hexagonal open-top field chambers *(547–550)*. The ability of these chambers to provide uniform and controllable pollutant concentrations depends on the height of the chambers and the nature of the top attachments that decrease ingress of ambient air. The most widely used open-top chamber is 3 m diameter by 2.4 m tall without a top attachment. There are several reports on the effectiveness of these chambers and on the chamber effects on microclimate *(548, 551–555)*. The primary concern in the use of such systems is the attendant decrease in wind velocity and a slight decrease in light intensity. However, the chambers have not significantly affected yield of most species tested *(65, 556)* and are now in use at many field sites in the United States and other countries.

B. Effects of Ozone on Crop Production

Yield loss due to O_3 is caused by the accumulative effects of daily exposures over the entire growth period. For most crops, symptoms of foliar injury from chronic O_3 exposure are difficult to distinguish from normal foliar senescence that occurs as leaves age and crops mature. The syndrome of early senescence has been observed for all crops tested in open-top field chamber experiments. Early senescence decreases the normal photosynthetic period of all plant leaves, thereby decreasing plant growth rate, total photosynthate production, and final yield.

In a typical season, periods of relatively high O_3 concentrations (0.10–0.15 ppm) are interspersed with periods of lower concentration. The effects of high concentrations are greater than those of low concentrations, but the effects of exposure for 1 or 2 days at 0.10–0.15 ppm would probably not affect the yield of most plants without further daily exposures and additional peaks. The accumulative nature of plant response to daily O_3 exposure was shown for Bel-W_3 tobacco, a sensitive O_3 indicator plant *(557)*. Over eight different 7-day periods, plants exposed to ambient O_3 for seven consecutive days developed more than twice as much foliar injury as the sum of the injury from seven groups of plants, exposed for 1 day each for 7 days.

1. Experimental Design

Open-top field chambers were developed to test the accumulative effects of daily exposures to ambient levels of O_3 *(551)*. In a typical experiment, the variable amount of O_3 in ambient air was used as a baseline exposure level in nonfiltered-air chambers. Different O_3 doses were obtained by adding different but constant amounts of O_3 to the O_3 in the nonfiltered-air chambers for 7 hours a day from 1000–1700 hours. This period of the day was chosen because at this time O_3 concentration in ambient air is usually greatest and plants are likely to be most physiologically active. Chambers with charcoal-filtered air were used as controls. In most experiments, O_3 was added for 7 hours/day from 2–3 weeks after plant emergence until maturity. Additional information on experimental protocol and design can be found in a number of publications *(65, 66, 551, 558–561)*. The statistic chosen to describe the O_3 exposure dose was the mean of the daily 7 hour means over the entire period when different O_3 treatments were applied. In most crop-growing areas of the United States, the seasonal (May–September) 7-hour-per-day mean of O_3 in ambient air is between 0.04 and 0.06 ppm *(559)*. Natural (nonanthropogenic) O_3 is estimated to contribute about 0.025 ppm to this seasonal value, but this estimate is widely debated *(67)*.

2. The National Crop Loss Assessment Network

In 1980, the United States Environmental Protection Agency provided funding for a National Crop Loss Assessment Network (NCLAN) that consists of research groups in different regions of the United States cooperating in field dose–response research for the assessment of immediate and long-term economic effects of air pollution on crop production. The NCLAN field studies provided O_3 dose–yield response relationships for many economically important crops. Publications include several types of mathematical models that describe these relationships *(65–69, 558–561)*.

Field dose–response studies have shown wide differences in sensitivity among crop species. Figure 2 *(66)* was developed by using the Weibull probability distribution of sensitivities because the Weibull

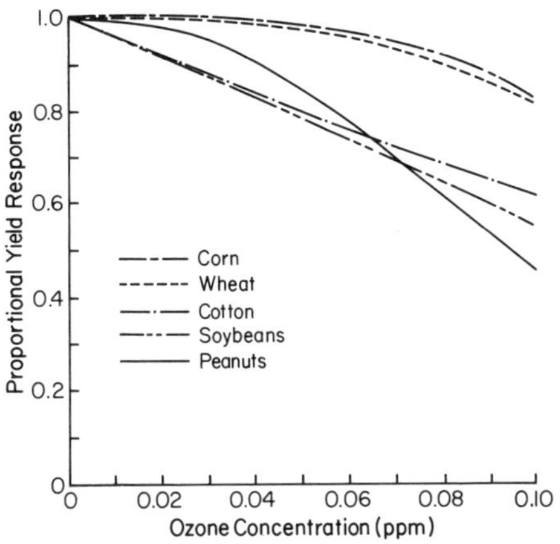

Figure 2. Effects of chronic O_3 exposure on proportional yield loss of five crop species as predicted by the Weibull model, using experimental results from open-top chamber studies. The O_3 concentration is the seasonal 7 hour/day mean. To obtain the proportional response, the value for maximum yield predicted at 0 ppm O_3 of the Weibull model is set at a value of 1.0. The soybean, wheat, and corn responses are based on combined data for five, four, and two cultivars, respectively. The cotton and peanut responses are based on data from one cultivar each. The parameters of the Weibull model used for this figure have been published *(66)*.

function is flexible and allows development of models from combined data sets *(562)*. The Weibull model is given as

$$y = \alpha \exp[-(x/\sigma^{c}] + \epsilon \qquad (2)$$

where y is the yield and x is the O_3 dose (concentration over some period of time). The three parameters to be estimated are α, the hypothetical maximum yield at zero O_3 concentration; σ, the O_3 concentration when yield is 0.37α; c, a dimensionless shape parameter ($c = 1$ gives the exponential loss function, whereas an increasing c gives a threshold-type response); ϵ, the random error associated with each observation. Figure 2 shows the relative response of five crop species at different seasonal O_3 doses as proportions of the maximum yield which would be expected in the absence of O_3. If 0.025 ppm is considered a background O_3 level, the losses predicted for cotton, soybeans, and peanuts at a seasonal 7-hour-per-day mean of 0.05 ppm are about 10%, 12%, and 14%, respectively. However, field corn and winter wheat losses at 0.05 ppm would be less than 3%.

Soybean has been studied more than other crops because it is second to corn in economic importance and because it is sensitive to O_3. Dose–response relationships for different soybean cultivars tested in different years at different sites, and for one cultivar tested at one site in different years, are shown in Figure 3. In spite of the differ-

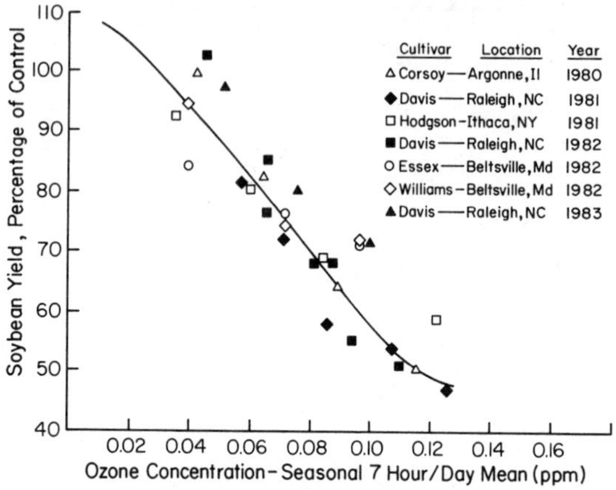

Figure 3. Effects of chronic O_3 exposures in open-top field chambers on yield of soybeans. The seasonal O_3 concentrations for the control treatment ranged from 0.02 to 0.03 ppm, depending on the year and location. References: Corsoy *(561)*, Davis 1981 *(558)*, Hodgson *(559)*, Davis 1982 *(559)*, and Essex and Williams *(559)*.

ences in cultivar, season, and site, the measured losses at a given seasonal 7-hour-per-day mean O_3 concentration were generally within 10% of each other.

3. Yield-Loss Assessment

Any yield assessment, whether local, regional, or national, requires three basic types of information *(563)*: (1) a knowledge of crops and their yields within some type of geographic unit such as a state, a province, or a county (a crop census); (2) an air quality data base from which exposure estimates can be developed for the same geographic unit as the crop census; and (3) a crop response function that relates crop yield to O_3 concentration over some time period.

A crop census, the first basic type of information required for assessment, is provided by the United States Department of Agriculture on a county level every 5 years. In addition, state crop production and yield data are reported each year. These crop yield and acreage figures are averaged across all cultivars for a given species. Thus, it is not possible to base a national assessment on crop cultivars, only on species.

The second critical component of a national assessment is an adequate air quality data base. Such a data base is available from the United States Environmental Protection Agency (EPA) Storage and

Retrieval of Aerometric Data (SAROAD) program. This data base includes hourly O_3 values from many sites throughout the country. The interpolation of SAROAD O_3 data using a statistical technique called *Kriging* (estimating gridded values from irregularly spaced, generally clustered monitoring data) provides an estimate of county-level O_3 concentrations *(559)*. The acceptance of an exposure statistic (concentration and averaging time) that can be used as an air quality standard and in crop response models would be useful. Alternatively, different exposure statistics could be used for the standard and the model, if the distribution of O_3 concentrations in the ambient air would permit using one statistic to estimate the other.

The third critical element of a national assessment is the development of O_3 dose–response model(s). The functional form for the model should be one that best describes the observed biological response (e.g., the Weibull function). The model should use an exposure statistic that is not site-specific. It has been shown that a model developed by using a seasonal mean (1 hour/day or 7 hour/day) statistic from one site can be used to estimate yield loss at a second site, or for a different year at the same site, using the seasonal mean statistic from the second site or different year *(68)*. This cannot be done by using seasonal peak statistics.

In summary, the NCLAN approach to assessing the national impact of O_3 on agricultural crops includes the following steps: (1) developing functional relationships between O_3 doses and crop response for important crops in different regions of the United States; (2) developing estimates of air quality (O_3 statistics) in rural areas; and (3) merging these pieces of information with United States crop yield and acreage information. The result is an estimate of the consequences of exposure of crops to O_3, expressed in physical (yield) units. As the result of these research efforts, knowledge of O_3 dose–crop yield response relationships increased greatly between 1974 and 1984.

4. Economic Assessment

Translation of loss estimates into economic terms involves entry of the dose–response and air quality data into a national economic model of agriculture that incorporates producer and consumer behavior *(564)*. The economic analysis then generates production and consumption responses for those commodities that are keyed to O_3 levels intended to portray alternative O_3 exposure scenarios. Estimates of annual losses to producers from O_3 are from \$1 billion to \$5 billion. These estimates must be considered preliminary for several reasons:

(1) only one or several cultivars of the most important species have been studied; (2) effects of soil moisture on dose–response relationships are poorly understood: most NCLAN studies were conducted with adequate soil moisture, and the effects of drought stress on dose–response relationships were not quantified; and (3) the effects of other factors (e.g., climate, exposure dynamics) on plant response to O_3 in the dose–response relationships developed from open-top chamber studies are unquantified.

C. Effects of Sulfur Dioxide on Crop Production

Sulfur dioxide concentrations in most agricultural areas rarely exceed 0.01 ppm for extended periods of time *(15)*. Daily 4- or 7-hour exposures of cotton and tomato *(559)* or soybeans *(558)* in open-top field chambers, throughout the growing season, have demonstrated that SO_2 concentrations likely to occur regionally in the United States do not cause decreased yield. A similar conclusion was reached by Roberts *(565)*, who reviewed the effects of SO_2 on plant productivity, including results of open-air studies in England. However, emissions of SO_2 near point sources can cause decreased yield in some crop species. Extensive dose–response studies with an open-air exposure system have simulated exposures of soybeans to SO_2 near point sources. Sulfur dioxide released from pipes at canopy height produced a SO_2 concentration gradient within a short distance downwind *(541)*. Soybean yield was consistently decreased by periodic exposures of plants after flowering to total doses of approximately 10–15 ppm-hours of SO_2 (Figure 4) *(542, 566–568)*. The 10- to 15-ppm-

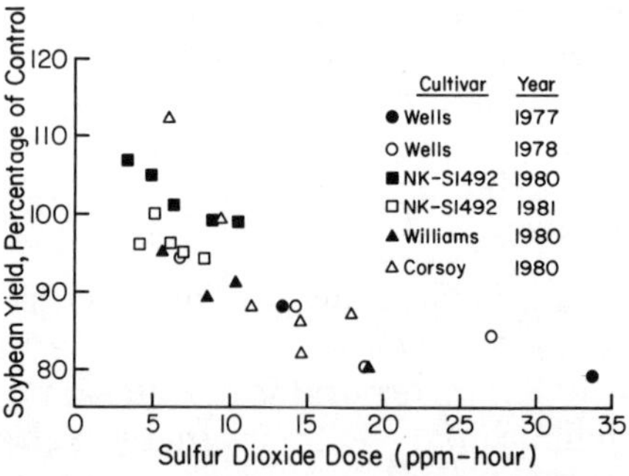

Figure 4. Effects of periodic SO_2 exposures in open air on yield of soybeans. The SO_2 ppm-hour doses are the product of the mean SO_2 concentration, the mean exposure duration, and the number of exposures, which ranged from 0.12 to 0.38 ppm, 2.5 to 4.2 hours, and 10 to 24 exposures, respectively. References for experimental details: Wells 1977 *(567)*, Wells 1978 *(567)*, NK-S1492 1980 *(542)*, NK-S1492 *(568)*, and Williams and Corsoy *(542)*.

hour dose statistics were products of mean exposure durations of 2.5 to 4.2 hours, mean concentrations of 0.12–0.31 ppm, and 19–25 exposures. Doses in the 5-ppm-hour range were either stimulatory or inhibitory. Maximum peak to mean SO_2 concentration ratios were about 2.5 *(541)*. Two field corn cultivars were more resistant to acute SO_2 exposures than were soybeans *(569)*.

D. Effects on Forest Production

Field studies that have described relationships between ambient concentrations of pollution and tree growth have provided important information relating air pollutants to forest production. Controlled dose–response studies have been possible only for seedlings and saplings because few exposure systems that can accommodate larger trees have been developed. The system developed by DeCormis *et al.* *(543)*, which consists of vertical release pipes for open-air exposures, might be useful for forestry research. However, descriptions of effects of pollutants on trees exposed in this system had not been published at the time of this writing. Field studies have been performed around point sources where tree growth was inversely related to decreasing doses of SO_2 at increasing distances from point sources. Several studies were extensively reviewed *(15)* and discussed in Vol. II, 3d ed., Chapter 12 (1977).

Field studies have established relationships between air quality and tree growth in several parts of the United States. Tree core analyses are often included to measure annual growth increments. One study in the San Bernandino mountains at the western edge of the Los Angeles, California, basin compared annual growth rates of Ponderosa pine trees over different 30-year periods. The average growth rate of trees from 1941 to 1971 was 40% less than that from 1910 to 1940 *(58, 570)*. The only environmental difference between the two periods was believed to be ambient concentrations of atmospheric O_3 (and other oxidants), which increased greatly after 1940. Other tree species in the same area (e.g., Jeffrey pine, White fir, California black oak) have also been affected but are apparently less sensitive than is Ponderosa pine *(58)*.

Strong evidence that O_3 inhibited growth of white pine in the eastern United States exists. Eastern white pine trees in the Appalachian mountains that showed symptoms typical of O_3 injury grew up to 15% less than trees rated tolerant to O_3 *(571)*. White pine in Tennessee showed a steady decline in radial growth over the period from 1960 to 1979 *(572)*. Annual radial growth increments were smaller for pine classified as sensitive than for those classified as tolerant.

There are reports of decline of red spruce and other tree species near the top of mountains in the eastern United States (*573,* Section IV,B and personal communication with R. I. Bruck, North Carolina State University). Similar reports exist for trees in Germany's Black Forest region. Preliminary evidence suggests that both problems began in the early 1960s, but cause-and-effect relationships have not yet been determined.

Causes for decreased growth rates of trees have not been positively identified because adequate experimental controls have not been possible. Ozone, SO_2, acidic precipitation, and heavy metal deposition, singly or in combination, have all been implicated. However, climatic extremes, insects, or infectious diseases may also be involved. One thing is certain: the growth of forest species has declined in many areas of the world over the last 30 years. More definitive dose–response studies are needed to identify cause-and-effect relationships between anthropogenic substances and the growth of forest species.

VI. Future Research Directions

The primary goals of research on terrestrial vegetation today are to assess the impact of air quality on growth and productivity and to understand the mechanism(s) of plant responses. Effects on terrestrial systems are probably predictable, at some level of confidence, once response mechanisms are understood.

Research emphasis on air pollutant effects on vegetation has covered a spectrum from studies of the mechanisms and effects of gaseous pollutants (primarily ozone, sulfur dioxide, and nitrogen dioxide), to descriptions of effects of wet deposition of sulfur and nitrogen and associated acidity, to descriptions of effects of combinations of gaseous pollutants, and finally to an appreciation of the probable role of all components of atmospheric deposition on the net response of vegetation. Future research must continue to evolve to a more holistic viewpoint, with studies aimed at understanding the importance of the individual components of the system to the total pollutant–environment–plant complex.

Much of the technical methodology has been designed and tested in the United States and Canada *(19).* However, innovative changes and design ideas are being initiated in research laboratories in many countries. Experimental designs and ways to analyze data have continued to evolve (as indicated in several sections of this chapter). The use of experimental designs that permit regression analysis and the establishment of confidence limits is gaining wide acceptance, although many experimental facilities are not designed for this type of experimental approach. We anticipate major advances in research methodology in the future.

Further refinement in the assessment of air pollution effects on vegetation is needed. Too little standardization of experimental design and technique has been one of the greatest obstacles to the utilization of the vast wealth of information reviewed in this chapter in an assessment of vegetation damage meaningful in policy analysis. The notable attempt to break from this precedent, the National Crop Loss Assessment Network (NCLAN), in the United States represents a milestone in air pollution effects assessment. However, the NCLAN is only a beginning, and it or similar programs must be included in future research plans.

Few areas of environmental research offer as many opportunities or as many challenges to build on the current state of the science as does the future of air pollution effects research, building on the current state of the science.

REFERENCES

1. J. L. Shupe, H. B. Peterson, and N. C. Leone, eds., "Fluorides: Effects on Vegetation, Animals, and Humans," Proc. Int. Fluoride Symp. Paragon Press, Inc., Salt Lake City, Utah, 1983.
2. P. E. Waggoner, *Am. Sci.* **72,** 179 (1984).
3. E. R. Lemon, ed., "Carbon Dioxide and Plants: The Response of Plants to Rising Levels of Atmospheric Carbon Dioxide," Am. Assoc. Adv. Sci. Sel. Sympo. No. 84. Westview Press, Inc., Boulder, Colorado, 1983.
4. S. Lerman, O. C. Taylor, and E. F. Darley, *Atmos. Environ.* **10,** 873 (1976).
5. W. W. Heck, W. M. Knott, E. P. Stahel, J. T. Ambrose, J. N. McCrimmon, M. Engle, L. A. Romanow, A. G. Sawyer, and J. D. Tyson, "Response of Selected Plants and Insect Species to Simulated Solid Rocket Exhaust Mixtures and to Exhaust Components from Solid Rocket Fuels," Tech. Memo. 74109, KSC TR 51-1. Natl. Aeron. Space Admin., John F. Kennedy Space Center, Cape Canaveral, Florida, 1980.
6. J. M. Cheeseman, T. O. Perry, and W. W. Heck, *Environ. Pollut.* **21,** 9 (1980).
7. L. S. Dochinger and T. A. Seliga, eds., "Acid Precipitation and the Forest Ecosystem," Proc. 1st Int. Symp., Gen. Tech. Rep. NE-23. U.S. Department of Agriculture, Forest Service, Upper Darby, Pennsylvania, 1976.
8. U.S. Environmental Protection Agency, "Diagnosing Vegetation Injury Caused By Air Pollution." Applied Science Associates, Inc., Air Pollution Training Institute, Research Triangle Park, North Carolina, 1976.
9. W. W. Heck, J. B. Mudd, and P. R. Miller, *in* "Ozone and Other Photochemical Oxidants," Chapter 11, p. 437. Natl. Acad. Sci., Washington, D.C., 1977.
10. P. R. Miller, *in* "Ozone and Other Photochemical Oxidants," Chapter 12, p. 586. Natl. Acad. Sci., Washington, D.C., 1977.

11. D. C. McCune and T. C. Weidensaul, *in* "Sulfur Oxides," Chapter 6, p. 80. Natl. Acad. Sci., Washington, D.C., 1978.
12. National Academy of Sciences, "Nitrogen Oxides," Chapter 9. Natl. Acad. Sci., Washington, D.C., 1977.
13. U.S. Environmental Protection Agency, "Air Quality Criteria for Ozone and Other Photochemical Oxidants," Chapter 10/11, EPA-60018-78-004. USEPA, Washington, D.C., 1978.
14. U.S. Environmental Protection Agency, "Air Quality Criteria for Ozone and Other Photochemical Oxidants," Chapters 7 and 8, EPA-600/8-84-020A (external review draft). Research Triangle Park, North Carolina, 1984.
15. U.S. Environmental Protection Agency, "Air Quality Criteria for Particulate Matter and Sulfur Oxides," Vol. III, Chapter 8, EPA-600/8-82-029C. Research Triangle Park, North Carolina, 1982.
16. U.S. Environmental Protection Agency, "Air Quality Criteria for Oxides of Nitrogen," Chapter 12, EPA-600/8-82-026. Office of Research and Development, Research Triangle Park, North Carolina, 1982.
17. M. H. Unsworth and D. P. Ormrod, eds., "Effects of Gaseous Air Pollution in Agriculture and Horticulture." Butterworth, London, 1982.
18. M. J. Koziol and F. R. Whatley, eds., "Gaseous Air Pollutants and Plant Metabolism." Butterworth, London, 1984.
19. W. W. Heck, S. V. Krupa, and S. N. Linzon, eds., "Handbook of Methodology for the Assessment of Air Pollution Effects on Vegetation," Proc. Spec. Conf. Air Pollut. Control Assoc., Pittsburg, Pennsylvania, 1979.
20. D. S. Shriner, C. R. Richmond, and S. E. Lindberg, eds., "Atmospheric Sulfur Deposition, Environmental Impact and Health Effects." Ann Arbor Sci. Publ., Ann Arbor, Michigan, 1980.
21. D. P. Ormrod, "Pollution In Horticulture." Elsevier, Amsterdam, 1978.
22. R. Guderian, "Air Pollution, Phytotoxicity of Acidic Gases and Its Significance in Air Pollution Control," Ecol. Stud. No. 22. Springer-Verlag, Berlin and New York, 1977.
23. W. H. Smith, "Air Pollution and Forests: Interactions Between Air Contaminants and Forest Ecosystems." Springer-Verlag, Berlin and New York, 1981.
24. T. A. Mansfield, ed., "Effects of Air Pollutants on Plants," Soc. Exp. Biol.—Semin. Ser. No. I. Cambridge Univ. Press, London and New York, 1976.
25. M. Treshow, ed., "Air Pollution and Plant Life." Wiley, Chichester, England, 1984.
26. W. E. Winner, H. A. Mooney, and R. A. Goldstein, eds., "Sulfur Dioxide and Vegetation: Ecology, Physiology and Policy Issues." Stanford Univ. Press, Palo Alto, California, 1986 (in press).
27. A. P. Altshuller and R. A. Linthurst, eds., "The Acidic Deposition Phenomenon And Its Effects: Critical Assessment Review Papers," Vol. II, EPA-60018-83-016B. U.S. Environ. Prot. Agency, Washington, D.C., 1984.
28. R. Guderian, ed., "Photochemical Oxidants-Formation, Distribution, Control and Effects on Vegetation." Springer-Verlag, Berlin and New York, 1985.
29. M. Verkroost, *Meded. Landbouwhogesch. Wageningen* **74** (1974).
30. L. Skarby and G. Sellden, *Ambio* **13,** 68 (1984).
31. S. N. Linzon, *in* "Sulfur In The Environment: Part III. Ecological Impacts" (J. R. Nriagu, ed.), p. 109. Wiley, New York, 1978.
32. M. Ianculescu, *Rev. Padurilor* **88,** 491 (Translation by SAAD Publications, Karachi, Pakistan, TT77-53275) (1973).

33. J. E. Hallgren, *in* "Sulfur in the Environment" (J. O. Nriagu, ed.), p. 163. Wiley, New York, 1978.
34. T. M. Roberts, *Atmos. Environ.* **18,** 629 (1984).
35. H. J. Jager and H. Klein, *Angew. Bot.* **54,** 337 (1980).
36. M. H. Unsworth and V. J. Black, *in* "Stomatal Physiology" (P. G. Jarvis and T. A. Mansfield, eds.), p. 187. Cambridge Univ. Press, London and New York, 1981.
37. D. N. Alstad, G. F. Edmunds, and L. H. Weinstein, *Annu. Rev. Entomol.* **27,** 369 (1982).
38. J. A. Lawrence, *Z. Pflanzenkr. Pflanzenschutz* **88,** 156 (1981).
39. D. L. Rist and J. W. Lorbeer, *in* "Microbial Ecology of the Phylloplane" (J. P. Blakeman, ed.), p. 305. Academic Press, London, 1981.
40. W. W. Heck, U. Blum, R. A. Reinert, and A. S. Heagle, *in* "Strategies of Plant Reproduction" (W. J. Meudt, ed.), Beltsville Agric. Res. Cent. Symp. No. 6, p. 333. Allanheld, Osmun & Co. Publisher, Inc., Totowa, New Jersey, 1983.
41. R. A. Reinert, H. E. Heggestad, and W. W. Heck, *in* "Breeding Plants for Less Favorable Environments" (M. N. Christiansen, ed.), p. 259. Wiley, New York, 1982.
42. T. W. Tibbitts and J. M. Kobriger, *HortScience* **18,** 675 (1983).
43. K. Kvist and P. Barklund, *Forstwiss. Centralbl.* **103,** 74 (1984).
44. P. Miller (Technical Coordinator), "The Effects of Air Pollutants on Mediterranean and Temperate Forest Ecosystems," Gen. Tech. Rep. PSW-43. Pacific Southwest For. Range Exp. Stn., Berkeley, California, 1980.
45. S. Smidt, *Z. Pflanzenkr. Pflanzenschutz* **85,** 689 (1978).
46. B. Ulrich, *Atmos. Environ.* **18,** 621 (1984).
47. National Research Council, "Acid Deposition Atmospheric Processes in Eastern North America." Natl. Acad. Press, Washington, D.C., 1983.
48. A. S. Lefohn and D. P. Ormrod, eds., "A Review and Assessment of the Effects of Pollutant Mixtures on Vegetation-Research Recommendations," EPA-600/3-84-037. U.S. Environ. Prot. Agency, Corvallis, Oregon, 1984.
49. National Institute for Environmental Studies, "Studies on Effects of Air Pollutant Mixtures on Plants," Parts 1 and 2, No. 65. Ibaraki, Japan, 1984.
50. R. A. Reinert, *Annu. Rev. Phytophath.* **22,** 421 (1984).
51. C. L. Mulchi, C. R. Curtis, H. E. Heggestad, and L. D. Moore, eds., "Air Pollution Impacts on Agriculture and Forestry," NE-121. Agric. Exp. Stn., College Park, Maryland, 1983.
52. W. W. Heck, U. Blum, W. F. Boss, A. S. Heagle, R. A. Linthurst, R. A. Reinert, J. F. Reynolds, and H. H. Rogers, *in* "Review in Environmental Toxicology" (E. Hodgson, ed.), Vol. 1, p. 173. Elsevier/North-Holland, New York, 1984.
53. R. L. Heath, *Annu. Rev. Plant Physiol.* **31,** 395 (1980).
54. T. A. Mansfield and P. H. Freer-Smith, *Biol. Rev. Cambridge Philos. Soc.* **56,** 343 (1981).
55. E. J. Pell, *in* "Plant Disease: An Advanced Treatise" (J. G. Horsfall and E. B. Cowling, eds.), Vol. 4, p. 273. Academic Press, New York, 1979.
56. O. C. Taylor (Principal Investigator), "Oxidant Air Pollution on a Western Coniferous Forest Ecosystem, Task B: Historical Background and Proposed Systems Study of the San Bernardino Mountain Area." Statewide Air Pollut. Res. Cent., University of California, Riverside, 1973.
57. O. C. Taylor, ed., "Photochemical Oxidant Air Pollution Effects On a Mixed Conifer Forest Ecosystem," EPA-600/3-80-002. U.S. Environ. Prot. Agency, Corvallis, Oregon, 1980.
58. P. R. Miller, O. C. Taylor, and R. G. Wilhour, "Oxidant Air Pollution Effects on a

Western Coniferous Forest Ecosystem," EPA-600/D-82-276. Environ. Res. Brief, U.S. Environ. Prot. Agency, Corvallis, Oregon, 1982.
59. A. H. Legge, D. R. Jaques, G. W. Harvey, H. R. Krouse, H. M. Brown, E. C. Rhodes, M. Nosal, H. U. Schellhase, J. Mayo, A. P. Hartgerink, P. F. Lester, R. G. Amundson, and R. B. Walker, *Water, Air, Soil Pollut.* **15,** 77 (1981).
60. H. R. Krouse, A. H. Legge, and H. M. Brown, *Water, Air, Soil Pollut.* **22,** 321 (1984).
61. E. M. Preston and R. A. Lewis, "The Bioenvironmental Impact of a Coal-Fired Power Plant, Sixth Interim Report, 1979–1980 Studies, Colstrip, Montana," EPA-600/3-81-007. U.S. Environ. Prot. Agency, Corvallis, Oregon, 1981.
62. W. K. Lauenroth and E. M. Preston, eds., "The Effects of SO_2 on a Grassland: A Case Study in the Northern Great Plains of the United States," Ecol. Stud. Ser., Vol. 45. Springer-Verlag, Berlin and New York, 1984.
63. D. Drablos, "Reference Key for Research Activities and Field Investigations, 1972–1979," SNSF Proj. Royal Norwegian Meteorological Service, Oslo, 1980.
64. L. N. Overrein, H. M. Seip, and A. Tollan, "Acid Precipitation-Effects on Forest and Fish," Final Report of the SNSF Project 1972–1980, FR 19–80, SNSF Proj. Royal Norwegian Meteorological Service, Oslo, 1980.
65. W. W. Heck, O. C. Taylor, R. Adams, G. Bingham, J. Miller, E. Preston, and L. Weinstein, *J. Air Pollut. Control Assoc.* **32,** 353 (1982).
66. W. W. Heck, R. M. Adams, W. W. Cure, A. S. Heagle, H. E. Heggestad, R. J. Kohut, L. W. Kress, J. O. Rawlings, and O. C. Taylor, *Environ. Sci. Technol.* **17,** 573A (1983).
67. W. W. Heck, W. W. Cure, J. O. Rawlings, L. J. Zaragoza, A. S. Heagle, H. E. Heggestad, R. J. Kohut, L. W. Kress, and P. J. Temple, *J. Air Pollut. Control Assoc.* **34,** 729 (1984).
68. W. W. Heck, W. W. Cure, J. O. Rawlings, L. J. Zaragoza, A. S. Heagle, H. E. Heggestad, R. J. Kohut, L. W. Kress, and P. J. Temple, *J. Air Pollut. Control Assoc.* **34,** 810 (1984).
69. Interagency Task Force on Acid Precipitation, "National Acid Precipitation Assessment Plan, June 1982." Council on Environmental Quality, Washington, D.C., 1982.
70. Interagency Task Force on Acid Precipitation, "National Acid Precipitation Assessment Program," Annual Report to the President and Congress. Council on Environmental Quality, Washington, D.C., 1983.
71. L. L. Knudson, T. W. Tibbitts, and G. E. Edwards, *Plant Physiol.* **60,** 606 (1977).
72. J. M. Kobriger, T. W. Tibbitts, and M. L. Brenner, *Plant Physiol.* **76,** 823 (1984).
73. D. T. Tingey, C. Standly, and R. W. Fields, *Atmos. Environ.* **10,** 969 (1976).
74. H. J. Stan, S. Schicker, and H. Kassner, *Atmos. Environ.* **15,** 391 (1981).
75. R. Leuning, M. H. Unsworth, H. N. Newmann, and K. M. King, *Atmos. Environ.* **13,** 1155 (1979).
76. L. K. Butler and T. W. Tibbitts, *J. Am. Soc. Hortic. Sci.* **104,** 213 (1979).
77. A. Faensen-Thiebes, *Angew. Bot.* **57,** 181 (1983).
78. C. M. Gesalman and D. D. Davis, *J. Am. Soc. Hortic. Sci.* **103,** 489 (1978).
79. M. J. Harris and R. L. Heath, *Plant Physiol,* **68,** 885 (1981).
80. G. E. Taylor, D. T. Tingey, and H. C. Ratsch, *Oecologia* **53,** 179 (1982).
81. M. Dijak and D. P. Ormrod, *Environ. Exp. Bot.* **22,** 395 (1982).
82. K. Onasa, Y. Hashimoto, and I. Aiga, *Environ. Control Biol.* **19,** 85 (1981).
83. J. B. Mudd, S. K. Banerjee, M. M. Pooley, and K. L. Knight, *in* "Gaseous Air Pollutants and Plant Metabolism" (M. J. Koziol and F. R. Whatley, eds.), Chapter 8. Butterworth, London, 1984.
84. H. D. Grimes, K. K. Perkins, and W. F. Boss, *Plant Physiol.* **72,** 1016 (1983).

85. A. Keitel and U. Arndt, *Angew. Bot.* **57,** 193 (1983).
86. R. L. Heath and P. E. Frederick, *Plant Physiol.* **64,** 455 (1979).
87. R. Athanassious, *Z. Pflanzenphysiol.* **97,** 227 (1980).
88. R. Sutton and I. P. Ting, *Am. J. Bot.* **64,** 404 (1977).
89. K. P. Pauls and J. E. Thompson, *Physiol. Plant.* **53,** 255 (1981).
90. A. Rhoads and E. Brennan, *Phytopathology* **68,** 883 (1978).
91. E. J. Pell and N. S. Pearson, *Plant Physiol.* **73,** 185 (1983).
92. J. J. Speroni, E. J. Pell, and W. C. Weissberger, *Am. Potato J.* **58,** 407 (1981).
93. B. Rubin, D. Penner, and A. W. Saettler, *Environ. Toxicol. Chem.* **2,** 295 (1983).
94. U. Blum, G. R. Smith, and R. C. Fites, *Environ. Exp. Bot.* **22,** 143 (1982).
95. A. C. Purvis, *Can. J. Bot.* **56,** 1540 (1978).
96. F. Fong and R. L. Heath, *Z. Pflanzenphysiol.* **104,** 109 (1981).
97. T. Sakaki, N. Kondo, and K. Sugahara, *Physiol. Plant.* **59,** 28 (1983).
98. F. J. Castillo, C. Penel, and H. Greppin, *Plant Physiol.* **74,** 846 (1984).
99. J. F. Petolino, C. L. Mulchi, and M. K. Aycock, Jr., *Crop Sci.* **23,** 1102 (1983).
100. E. V. Pidleckis, C. R. Curtis, and H. E. Heggestad, *Phytopathology* **74,** 572 (1984).
101. R. Athanassious, *Z. Pflanzenphysiol.* **97,** 183 (1980).
102. S. Satoh, K. Kaneka, and T. Unezawa, "Relationship between Intraspecific Difference in Ozone Sensitivity and Leaf Constituent among *Phaseolus vulgaris* L. Cultivars." Denryoku Chuo Kenkyujo Hokoku, Central Research Institute, Electric Power Industry, Japan, 1981.
103. G. Hofstra, A. Ali, R. T. Wukasch, and R. A. Fletcher, *Atmos. Environ.* **15,** 483 (1981).
104. A. S. A. F. Guri, *Can. J. Plant Sci.* **63,** 733 (1983).
105. C. P. Chanway and V. C. Runeckles, *Can. J. Bot.* **62,** 236 (1984).
106. E. H. Lee and J. H. Bennett, *Plant Physiol.* **69,** 1444 (1982).
107. C. R. Black and V. J. Black, *Plant, Cell Environ.* **2,** 329 (1979).
108. V. J. Black and M. H. Unsworth, *J. Exp. Bot.* **31,** 667 (1980).
109. W. E. Winner and H. A. Mooney, *Science* **210,** 789 (1980).
110. J. Bonte, L. De Cormis, and P. Lougvet, *Environ. Pollut.* **12,** 125 (1977).
111. I. M. Rao, R. G. Amundson, R. Alscher-Herman, and L. E. Anderson, *Plant Physiol.* **72,** 573 (1983).
112. J. R. Barton, S. B. McLaughlin, and R. K. McConathy, *Environ. Pollut.* **21,** 255 (1980).
113. V. J. Black and M. H. Unsworth, *Nature (London)* **282,** 68 (1979).
114. Z. Miszalski and H. Ziegler, *Phytopathology* **97,** 144 (1980).
115. D. G. Sprugel and J. E. Miller, *Water, Air, Soil Pollut.* **12,** 233 (1979).
116. G. E. Taylor, Jr., S. B. McLaughlin, D. S. Shriner, and W. J. Selvidge, *Atmos. Environ.* **17,** 789 (1983).
117. G. E. Taylor, Jr. and D. T. Tingey, *Plant Physiol.* **72,** 237 (1983).
118. M. L. Wesley and B. B. Hicks, *J. Air Pollut. Control Assoc.* **27,** 1110 (1977).
119. S. G. Garsed and D. J. Read, *New Phytol.* **78,** 111 (1977).
120. S. G. Garsed and D. J. Read, *Environ. Pollut.* **13,** 173 (1977).
121. M. Plesnicar, *Plant Sci. Lett.* **10,** 205 (1977).
122. V. J. Black and M. H. Unsworth, *J. Exp. Bot.* **30,** 473 (1979).
123. M. Katase, T. Ushijima, and T. Tazaki, *Bot. Mag.* **96,** 1 (1983).
124. S. B. McLaughlin, D. S. Shriner, R. K. McConathy, and L. K. Mann, *Environ. Exp. Bot.* **19,** 179 (1979).
125. H. Saxe, *Physiol. Plant.* **57,** 101 (1983).
126. B. K. Takemoto and R. D. Noble, *Environ. Pollut., Ser. A* **28,** 67 (1982).

127. A. Furukawa and T. Totsuka, *Environ. Control Biol.* **21,** 43 (1983).
128. G. E. Taylor, Jr. and W. J. Selvidge, *J. Environ. Qual.* **13,** 224 (1984).
129. D. C. Horsman and A. R. Wellburn, *Environ. Pollut.* **13,** 33 (1977).
130. Z. Miszalski, *Bull. Acad. Pol. Sci., Ser. Sci. Biol.* **29,** 483 (1981).
131. M. Pierre and O. Queiroz, *Environ. Pollut., Ser. A* **28,** 209 (1982).
132. R. Rabe and K. R. Kreek, *Z. Pflanzenphysiol.* **97,** 215 (1980).
133. G. F. Soldatini and I. Ziegler, *Phytochemistry* **18,** 21 (1979).
134. K. Tanaka, T. Otsuko, and N. Kondo, *Plant Cell Physiol.* **23,** 1009 (1982).
135. H. R. Wyss and C. Brunold, *Physiol. Plant.* **50,** 161 (1980).
136. M. J. Koziol and C. F. Jordon, *J. Exp. Bot.* **29,** 1037 (1978).
137. W. K. Lauenroth and J. L. Dodd, *Water, Air, Soil Pollut.* **15,** 309 (1981).
138. S. N. Pandey and D. N. Rao, *Trop. Ecol.* **19,** 155 (1978).
139. M. Plesnicar, *Int. J. Appl. Radiat. Isot.* **34,** 833 (1983).
140. L. A. Barakhtenova and V. S. Nikolaevskii, *Izv. Akad. Nauk SSSR, Ser. Biol.* **1,** 90 (1983) (Translated: 0098-2164/83/1001-0065807.50. Plenum, New York, 1983).
141. R. A. Bressan, L. G. Wilson, and P. Filner, *Plant Physiol.* **61,** 761 (1978).
142. J. Sekiya, L. G. Wilson, and P. Filner, *Plant Physiol.* **70,** 437 (1982).
143. D. M. Olszyk and D. T. Tingey, *Plant Physiol.* **74,** 999 (1984).
144. Y. S. Yang, J. M. Skelly, B. I. Chevone, and J. B. Birch, *Environ. Int.* **9,** 265 (1983).
145. Y. S. Yang, J. M. Skelly, B. I. Chevone, and J. B. Birch, *Environ. Sci. Technol.* **17,** 371 (1983).
146. G. Lorenc-Plucinska, *Arbor. Kornickie* **24,** 329 (1979).
147. P. I. Coyne and G. E. Bingham, *J. Air Pollut. Control Assoc.* **31,** 38 (1981).
148. P. I. Coyne and G. E. Bingham, *For. Sci.* **28,** 257 (1982).
149. P. B. Reich, *Plant Physiol.* **73,** 291 (1983).
150. W. M. Aitken, W. R. Jacobi, and J. M. Staley, *Plant Dis.* **68,** 398 (1984).
151. J. L. Trimble, J. M. Skelly, S. A. Tolin, and D. M. Orcutt, *Phytopathology* **72,** 652 (1982).
152. D. T. Tingey, R. G. Wilhour, and C. Standley, *For. Sci.* **22,** 2334 (1976).
153. A. R. Biggs and D. D. Davis, *Can. J. For. Res.* **11,** 69 (1981).
154. A. R. Biggs and D. D. Davis, *Can. J. For. Res.* **12,** 612 (1982).
155. J. E. Hallgren, S. Linder, A. Richter, E. Troeng, and L. Granat, *Plant, Cell Environ.* **5,** 75 (1982).
156. T. W. Kimmerer and T. T. Kozlowski, *Plant Physiol.* **67,** 990 (1981).
157. W. W. Winner and H. A. Mooney, *Oecologia* **44,** 296 (1980).
158. T. Keller, *Can. J. For. Res.* **10,** 1 (1980).
159. T. H. Keller, *Environ. Pollut.* **26,** 313 (1981).
160. S. Huttenen and K. Laine, *Ann. Bot. Fenn.* **20,** 79 (1983).
161. K. E. Percy and R. T. Riding, *Can. J. Bot.* **59,** 2558 (1981).
162. W. Suwannapinunt and T. T. Kozlowski, *Can. J. For. Res.* **10,** 78 (1980).
163. J. E. Hallgren and K. Gezelius, *Physiol. Plant.* **54,** 153 (1982).
164. G. W. Harvey and A. H. Legge, *Can. J. Bot.* **57,** 759 (1979).
165. S. S. Malhotra and S. K. Sarkar, *Physiol. Plant.* **47,** 223 (1979).
166. A. A. Khan and S. S. Malhotra, *Biochem. Physiol. Pflanz.* **177,** 643 (1982).
167. A. A. Khan and S. S. Malhotra, *Phytochemistry* **22,** 1325 (1983).
168. J. B. Bucher, *Eur. J. For. Pathol.* **11,** 369 (1981).
169. J.-E. Hallgren and S.-A. Fredriksson, *Plant Physiol.* **70,** 456 (1982).
170. D. Grill, H. Esterbauer, M. Scharner, and C. Felgitsch, *Eur. J. For. Pathol.* **10,** 263 (1980).
171. D. Grill, H. Esterbauer, and K. Hellig, *Phytopathol. Z.* **104,** 264 (1982).

172. G. Lorenc-Plucinska, *Photosynthetica* **17,** 20 (1983).
173. R. T. Eckert and D. B. Houston, *For. Sci.* **28,** 661 (1982).
174. W. E. Winner, G. W. Koch, and H. A. Mooney, *Oecologia* **52,** 16 (1982).
175. W. A. Feder, *Environ. Health Perspect.* **37,** 117 (1981).
176. R. Janakiraman and P. M. Harvey, *Can. J. Genet. Cytol.* **18,** 727 (1976).
177. L. Walmsley, M. R. Ashmore, and J. N. B. Bell, *Environ. Pollut., Ser. A* **23,** 165 (1980).
178. L. F. Benoit, J. M. Skelly, L. D. Moore, and L. S. Dochinger, *Can. J. For. Res.* **13,** 184 (1983).
179. T. R. Huang, M. K. Aycock, Jr., and C. L. Muchi, *Crop Sci.* **15,** 785 (1975).
180. L. K. Butler, T. W. Tibbitts, and F. A. Bliss, *J. Am. Soc. Hortic. Sci.* **104,** 211 (1979).
181. A. S. A. F. Guri, *Can. J. Plant Sci.* **63,** 727 (1983).
182. P. Hucl and W. D. Beversdorf, *Can. J. Plant Sci.* **62,** 861 (1982).
183. N. E. DeVos, R. R. Hill, E. J. Pell, and R. H. Cole, *Crop Sci.* **22,** 992 (1982).
184. W. J. Johnston, R. Dickens, and R. L. Haaland, *Crop Sci.* **23,** 235 (1983).
185. G. P. Hanson, D. H. Addis, and L. Thorne, *Can. J. Genet. Cytol.* **18,** 579 (1976).
186. J. W. Cameron, *J. Am. Soc. Hortic. Sci.* **100,** 577 (1975).
187. L. W. Kress, J. M. Skelly, and K. H. Hinkelmann, *Can. J. For. Res.* **12,** 203 (1982).
188. Y.-S. Yang, J. M. Skelly, and B. I. Chevone, *Phytopathology* **73,** 1234 (1983).
189. L. K. Butler and T. W. Tibbitts, *J. Am. Soc. Hortic. Sci.* **104,** 208 (1979).
190. J. P. Meiners and H. E. Heggestad, *Plant Dis. Rep.* **63,** 273 (1979).
191. A. E. G. Tonneijck, *Neth. J. Plant Pathol.* **89,** 99 (1983).
192. R. A. Reinert, J. A. Dunning, W. W. Heck, P. S. Benepal, and M. Rangappa, *HortScience* **19,** 86 (1984).
193. R. A. Reinert and P. V. Nelson, *HortScience* **14,** 747 (1979).
194. P. Semeniuk and H. E. Heggestad, *J. Hered.* **72,** 459 (1981).
195. G. A. Richards, C. L. Mulchi, and J. R. Hall, *J. Environ. Qual.* **9,** 49 (1980).
196. A. R. Mosley, R. C. Rowe, and T. C. Weidensaul, *Am. Potato J.* **55,** 147 (1978).
197. A. S. Heagle, *Environ. Pollut.* **19,** 1 (1979).
198. A. S. Heagle and M. B. Letchworth, *J. Environ. Qual.* **11,** 690 (1982).
199. H. A. Menser and J. F. Chaplin, *Tob. Int.* **182,** 69 (1980).
200. J. T. A. Proctor and D. P. Ormrod, *HortScience* **12,** 321 (1977).
201. R. A. Reinert and W. R. Henderson, *J. Am. Soc. Hortic. Sci.* **105,** 322 (1980).
202. C. A. Thompson, D. M. Olszyk, G. Kats, A. Bytnerowlez, P. J. Dawson, and J. W. Wolf, *J. Air Pollut. Control Assoc.* **34,** 1017 (1984).
203. A. M. Townsend and L. S. Dochinger, *J. Arboric.* **8,** 186 (1982).
204. P. R. Miller, G. J. Longbotham, and C. R. Longbotham, *Plant Dis.* **67,** 1113 (1983).
205. K. C. Steiner and D. D. Davis, *Can. J. For. Res.* **9,** 106 (1979).
206. D. F. Karnosky and K. C. Steiner, *Phytopathology* **71,** 804 (1981).
207. S. R. K. Varshney and C. K. Varshney, *Environ. Pollut., Ser. A* **24,** 87 (1981).
208. D. T. DuBay and W. H. Murdy, *Bot. Gaz. (Chicago)* **144,** 376 (1983).
209. D. T. DuBay and W. H. Murdy, *J. Environ. Qual.* **12,** 147 (1983).
210. T. Keller and H. Bedan, *Environ. Pollut., Ser. A* **33,** 237 (1984).
211. D. T. Krizek and P. Semenink, *J. Am. Soc. Hortic. Sci.* **108,** 47 (1983).
212. M. Ayazloo, J. N. B. Bell, and R. W. Laycock, *J. Sports Turp Res. Inst.* **6,** 85 (1980).
213. T. K. Howe and S. S. Woltz, *HortScience* **17,** 596 (1982).
214. T. K. Howe and S. S. Woltz, *HortScience* **17,** 249 (1982).
215. A. Markowski, S. Grezesiak, and M. Schramelo, *Bull. Acad. Pol. Sci., Ser. Sci. Biol.* **23,** 637 (1975).
216. P. M. Toironen and G. Hofstra, *Can. J. Plant Sci.* **59,** 475 (1979).

217. H. Klein, H. J. Jager, W. Domes, and C. H. Wong, *Oecologia* **33,** 203 (1978).
218. W. E. Winner and H. A. Mooney, *Oecologia* **46,** 49 (1980).
219. W. H. Murdy, *Bot. Gaz. (Chicago)* **140,** 299 (1979).
220. D. C. Horsman, T. M. Roberts, and A. D. Bradshaw, *J. Exp. Bot.* **30,** 495 (1979).
221. M. Ayazloo and J. N. B. Bell, *New Phytol.* **88,** 203 (1981).
222. M. Ayazloo, S. G. Garsed, and J. N. B. Bell, *New Phytol.* **90,** 109 (1982).
223. G. E. Taylor and D. T. Tingey, *Oecologia* **49,** 76 (1981).
224. G. E. Taylor, *Bot. Gaz. (Chicago)* **139,** 362 (1978).
225. G. Lorenc-Plucinska, *Photosynthetica* **16,** 140 (1982).
226. J. B. Genys and H. E. Heggestad, *Can. J. For. Res.* **13,** 1262 (1983).
227. S. G. Garsed and A. J. Rutter, *New Phytol.* **92,** 349 (1982).
228. T. C. Elden, R. K. Howell, and R. E. Webb, *J. Econ. Entomol.* **71,** 283 (1978).
229. P. S. Benepal, M. Rangappa, and J. A. Dunning, *Bean Improve. Coop. Annu. Rep.* **22,** 50 (1979).
230. A. L. Lackner and S. A. Alexander, *Plant Dis.* **67,** 679 (1983).
231. S. Bisessar and K. T. Palmer, *Atmos. Environ.* **18,** 1025 (1984).
232. E. J. Pell, F. L. Lukezic, R. G. Levine, and W. C. Weissberger, *Phytopathology* **67,** 1342 (1977).
233. J. A. Laurence and F. A. Wood, *Phytopathology* **68,** 689 (1978).
234. P. J. Temple and S. Bisessar, *Phytopathology* **69,** 101 (1979).
235. R. I. Wukasch and G. Hofstra, *Phytopathology* **67,** 1080 (1977).
236. C. R. Krause and C. Weidensaul, *Phytopathology* **68,** 195 (1978).
237. A. S. Heagle, *Phytopathology* **66,** 616 (1977).
238. M. Kochhar, R. A. Reinert, and U. Blum, *Environ. Pollut., Ser. A* **28,** 255 (1982).
239. P. M. McCool and J. A. Menge, *New Phytol.* **94,** 241 (1983).
240. P. F. Brewer and A. S. Heagle, *Phytopathology* **73,** 1035 (1983).
241. I. Ho and J. M. Trappe, *Environ. Exp. Bot.* **24,** 71 (1984).
242. R. A. Reinert and G. V. Gooding, Jr., *Phytopathology* **68,** 15 (1978).
243. R. H. Vargo, E. J. Pell, and S. H. Smith, *Phytopathology* **68,** 715 (1978).
244. D. P. Ormrod and W. G. Kemp, *Can. J. Plant Sci.* **59,** 1077 (1979).
245. R. L. James, F. W. Cobb, P. R. Miller, and J. R. Parmeter, *Phytopathology* **70,** 560 (1980).
246. R. L. James, F. W. Cobb, Jr., and J. R. Parmeter, Jr., *Phytopathology* **72,** 1205 (1982).
247. J. A. Lawrence and K. L. Reynolds, *Environ. Pollut., Ser. A* **33,** 379 (1984).
248. J. L. Carney, H. E. Garrett, and H. G. Hedrick, *Phytopathology* **68,** 1160 (1978).
249. A. Grzywacz, *Folia For. Pol., Ser. A* **22,** 165 (1976).
250. H. E. Garrett, J. L. Carney, and H. G. Hedrick, *Can. J. For. Res.* **12,** 141 (1982).
251. P. R. Hughes, J. E. Potter, and L. H. Weinstein, *Environ. Entomol.* **11,** 173 (1982).
252. P. R. Hughes, A. I. Dickie, and M. A. Penton, *J. Environ. Qual.* **12,** 565 (1983).
253. G. P. Dohmen, S. McNeill, and J. N. B. Bell, *Nature (London)* **307,** 52 (1984).
254. J. A. Laurence, L. H. Weinstein, D. C. McCune, and A. L. Alusio, *Plant Dis. Rep.* **63,** 975 (1979).
255. J. A. Laurence, A. L. Aluisio, L. H. Weinstein, and D. C. McCune, *Environ. Pollut., Ser. A* **24,** 185 (1981).
256. M. Czuba and D. P. Ormrod, *Environ. Pollut., Ser. A* **25,** 67 (1981).
257. D. J. Papple and D. P. Ormrod, *J. Am. Soc. Hortic. Sci.* **102,** 792 (1977).
258. S. J. S. Sung and L. D. Moore, *Weed Sci.,* **27,** 167 (1979).
259. J. J. Reilly and L. D. Moore, *Weed Sci.* **30,** 260 (1982).
260. K. K. Hatzios, *Weed Sci.* **32,** 280 (1983).

261. Y.-H. Jeong, H. Nakamura, and L. Ota, *Jpn. J. Crop Sci.* **50,** 560 (1981).
262. J. E. Carnahan, E. L. Jenner, and E. K. W. Wat, *Phytopathology* **68,** 1225 (1978).
263. G. Hofstra, D. A. Littlejohns, and R. T. Wukasch, *Plant Dis. Rep.* **62,** 350 (1978).
264. T. C. Weidensaul, *Phytopathology* **70,** 42 (1980).
265. H. M. Cathey and H. E. Heggestad, *J. Am. Soc. Hortic. Sci.* **107,** 1035 (1982).
266. J. P. Rier, Jr., V. K. Sood, A. Whitaker, and C. Watson, *Phytopathology* **73,** 1262 (1983).
267. K. W. Foster, J. P. Guerard, R. J. Oshima, J. C. Bishop, and H. Timm, *Am. Potato J.* **60,** 127 (1983).
268. G. Hofstra, R. T. Wukasch, and D. M. Dressler, *Can. J. Plant Pathol.* **5,** 115 (1983).
269. H. M. Cathey and H. E. Heggestad, *J. Am. Soc. Hortic. Sci.* **107,** 1028 (1982).
270. J. R. McClenahen, *Plant Dis. Rep.* **63,** 320 (1979).
271. C. R. Johnson, J. N. Joiner, and D. N. Gower, *HortScience* **13,** 35 (1978).
272. L. Gruenhage and H. J. Jaeger, *Angew. Bot.* **56,** 167 (1982).
273. G. H. M. Krause and H. Kaiser, *Environ. Pollut.* **12,** 63 (1977).
274. P. K. Nandi, M. Agrawal, and D. N. Rao, *Sci. Hortic. (Amsterdam)* **22,** 47 (1984).
275. S. N. Pandey, *Environ. Pollut., Ser. A* **27,** 73 (1982).
276. C. A. Miller and D. D. Davis, *Plant Dis.* **65,** 750 (1981).
277. J. A. Dunning and W. W. Heck, *J. Air Pollut. Control Assoc.* **27,** 882 (1977).
278. S. B. McLaughlin and G. E. Taylor, *Science* **211,** 167 (1981).
279. D. T. Tingey, G. L. Thutt, M. L. Gumpertz, and W. E. Hogsett, *Agric. Environ.* **7,** 243 (1982).
280. V. C. Runeckles and P. M. Rosen, *Can. J. Bot.* **55,** 193 (1977).
281. E. H. Steinberger and S. Naveh, *Agric. Environ.* **7,** 255 (1982).
282. J. W. Johnston, Jr. and A. S. Heagle, *Phytopathology* **72,** 387 (1982).
283. R. Harkov and E. Brennan, *Phytopathology* **70,** 991 (1980).
284. T. Jones and T. A. Mansfield, *Environ. Pollut., Ser. A* **27,** 57 (1982).
285. W. W. Heck and J. A. Dunning, *J. Air Pollut. Control Assoc.* **28,** 241 (1978).
286. D. L. Rist and D. D. Davis, *Phytopathology* **69,** 231 (1979).
287. W. H. Murdy and H. L. Ragsdale, *J. Environ. Anal.* **9,** 493 (1980).
288. J. Bonte and P. Louguet, *Physiol. Veg.* **13,** 527 (1975).
289. R. W. Carlson and F. A. Bazzaz, *Oecologia* **54,** 50 (1982).
290. R. W. Carlson, *Environ. Pollut., Ser. A* **39,** 309 (1983).
291. T. W. Ashenden and T. A. Mansfield, *J. Exp. Bot.* **28,** 729 (1977).
292. A. W. Davison and I. F. Bailey, *Nature (London)* **297,** 400 (1982).
293. S. Grzesiak, *Bull. Acad. Pol. Sci., Ser. Sci. Biol.* **27,** 309 (1979).
294. J. A. Davids, D. D. Davis, and S. P. Pennypacker, *Phytopathology* **71,** 1208 (1981).
295. W. K. Lauenroth, D. G. Milchunas, and J. L. Dodd, *Environ. Exp. Bot.* **23,** 339 (1983).
296. R. J. Norby and T. T. Kozlowski, *For. Sci.* **29,** 739 (1983).
297. P. M. A. Toivonen, G. Hofstra, and R. T. Wukasch, *Can. J. Plant Pathol.* **4,** 381 (1982).
298. U. Blum and W. W. Heck, *Environ. Exp. Bot.* **20,** 73 (1980).
299. S. B. McLaughlin and R. K. McConathy, *Plant Physiol.* **73,** 630 (1983).
300. W. J. Manning and W. A. Feder, *in* "Effects of Air Pollutants on Plants" (T. A. Mansfield, ed.), Soc. Exp. Biol., Semin. Ser. No. I, pp. 47–60. Cambridge Univ. Press, London and New York, 1976.
301. W. R. Henderson and R. A. Reinert, *J. Am. Soc. Hortic. Sci.* **104,** 754 (1979).
302. R. J. Oshima, P. K. Braegelman, D. W. Baldwin, V. van Way, and O. C. Taylor, *J. Am. Soc. Hortic. Sci.* **102,** 289 (1977).

303. R. J. Oshima, P. K. Braegelmann, D. W. Baldwin, V. van Way, and O. C. Taylor, *J. Am. Soc. Hortic. Sci.* **102,** 286 (1977).
304. B. B. Clarke, M. R. Henninger, and E. Brennan, *Phytopathology* **73,** 104 (1983).
305. E. J. Pell, W. C. Weissberger, and J. J. Sperioni, *Environ. Sci. Technol.* **14,** 568 (1980).
306. J. P. Bennett, R. J. Oshima, and L. F. Lippert, *Environ. Exp. Bot.* **19,** 33 (1979).
307. J. P. Bennett and R. J. Oshima, *J. Am. Soc. Hortic. Sci.* **101,** 638 (1976).
308. R. J. Oshima, J. P. Bennett, and P. K. Braegelmann, *J. Am. Soc. Hortic. Sci.* **103,** 348 (1978).
309. H. Shimizu, S. Motohashi, H. Iwaki, A. Furukawa, and T. Totsuka, *Environ. Control Biol.* **19,** 137 (1981).
310. R. K. Howell and L. P. Rose, *Plant Dis.* **64,** 385 (1980).
311. C. Greenwald and A. G. Endress, *J. Agric. Chem.* **32,** 50 (1984).
312. R. J. Oshima, P. K. Braegelmann, R. B. Flagler, and R. R. Teso, *J. Environ. Qual.* **8,** 474 (1979).
313. M. B. Letchworth and U. Blum, *Environ. Pollut.* **14,** 303 (1977).
314. R. A. Montes, U. Blum, and A. S. Heagle, *Can. J. Bot.* **60,** 2745 (1982).
315. U. Blum, A. S. Heagle, J. C. Burns, and R. A. Linthurst, *Environ. Exp. Bot.* **23,** 121 (1983).
316. U. Blum, E. Mrozek, Jr., and E. Johnson, *Environ. Exp. Bot.* **23,** 369 (1983).
317. G. P. Lumis and D. P. Ormrod, *Can. J. Plant Sci.* **58,** 769 (1978).
318. J. P. Bennett and V. C. Runeckles, *Crop Sci.* **17,** 443 (1977).
319. M. Kochhar, U. Blum, and R. A. Reinert, *Can. J. Bot.* **58,** 241 (1980).
320. R. A. Montes, U. Blum, A. S. Heagle, and R. J. Volk, *Can. J. Bot.* **61,** 2159 (1983).
321. K. F. Jensen, *Environ. Pollut., Ser. A* **26,** 147 (1981).
322. S. F. Duchelle, J. M. Skelly, and B. I. Chevone, *Water, Air, Soil Pollut.* **18,** 363 (1982).
323. K. F. Jensen, *Can. J. For. Res.* **12,** 420 (1982).
324. S. B. McLaughlin, R. K. McConathy, D. Duvick, and L. K. Mann, *For. Sci.* **28,** 60 (1982).
325. L. F. Benoit, J. M. Skelly, L. D. Moore, and L. S. Dochinger, *Can. J. For. Res.* **12,** 673 (1982).
326. B. Gemmill, J. R. McBride, and R. D. Laven, *Tree-Ring Bull.* **42,** 23 (1982).
327. M. B. Coughenour, J. L. Dodd, D. C. Coleman, and W. K. Lauenroth, *Oecologia* **42,** 229 (1979).
328. D. G. Milchunas, W. R. Laurenroth, and J. L. Dodd, *Environ. Exp. Bot.* **22,** 81 (1982).
329. K. H. Teh and C. A. Swanson, *Plant Physiol.* **69,** 88 (1982).
330. T. Jones and T. A. Mansfield, *Environ. Pollut., Ser. A* **28,** 199 (1982).
331. L. Male, E. Preston, and G. Neely, *Atmos. Environ.* **17,** 1589 (1983).
332. H. Saxe, *Physiol. Plant.* **57,** 108 (1983).
333. T. W. Ashenden, *Environ. Pollut.* **15,** 161 (1978).
334. V. Mejstrik, *Environ. Pollut., Ser. A* **21,** 73 (1980).
335. D. R. Lockyer and D. W. Cowling, *J. Exp. Bot.* **32,** 1333 (1981).
336. H. Shimizu, A. Furukawa, and T. Totsuka, *Seibutsu Kankyo Chosetsu* **18,** 39 (1980).
337. L. C. Mishra, *Environ. Exp. Bot.* **20,** 397 (1980).
338. D. W. Cowling and M. J. Koziol, *J. Exp. Bot.* **29,** 1 (1978).
339. A. M. Chamberland and J. M. Gauthier, *Atmos. Environ.* **11,** 257 (1977).
340. K. E. Colvill, R. M. Bell, T. M. Roberts, and A. D. Bradshaw, *Environ. Pollut.* **31,** 33 (1983).

341. D. G. Milchunas, W. K. Lauenroth, and J. L. Dodd, *J. Range Manage.* **34,** 282 (1981).
342. R. J. Lotstein, D. D. Davis, and E. J. Pell, *HortScience* **18,** 72 (1983).
343. M. J. Koziol and D. W. Cowling, *J. Exp. Bot.* **31,** 1687 (1980).
344. J. N. B. Bell, *in* "Effects of Gaseous Air Pollution in Agriculture and Horticulture" (M. H. Unsworth and D. P. Ormrod, eds.), Chapter 11. Butterworth, London, 1982.
345. P. E. Marshall and G. R. Furnier, *Environ. Pollut.* **25,** 149 (1981).
346. R. T. Riding and K. A. Boyer, *Environ. Pollut., Ser. A* **30,** 245 (1983).
347. B. R. Roberts and V. M. Schnipke, *J. Environ. Hortic.* **1,** 72 (1983).
348. S. G. Garsed, J. F. Farrar, and A. J. Rutter, *J. Appl. Ecol.* **16,** 217 (1979).
349. M. A. Thompson, *Environ. Pollut., Ser. A* **26,** 251 (1981).
350. S. G. Garsed, A. J. Rutter, and J. Relton, *Environ. Pollut., Ser. A* **24,** 219 (1981).
351. W. W. Heck, R. I. Larsen, and A. S. Heagle, *in* "Crop Loss Assessment," (P. S. Teng and S. V. Krupa, eds.). Misc. Publ. No. 7, pp. 32–49. Agric. Exp. Stn., University of Minnesota, St. Paul, 1980.
352. R. I. Larsen and W. W. Heck, *J. Air Pollut. Control Assoc.* **16,** 325 (1976).
353. R. I. Larsen and W. W. Heck, *J. Air Pollut. Control Assoc.* **34,** 1023 (1984).
354. R. I. Larsen, A. S. Heagle, and W. W. Heck, *J. Air Pollut. Control Assoc.* **33,** 198 (1983).
355. G. C. Pratt and S. V. Krupa, *Phytopathology* **71,** 1129 (1981).
356. D. P. Ormrod, D. T. Tingey, M. L. Gumpertz, and D. M. Olszyk, *Plant Physiol.* **75,** 43 (1984).
357. T. M. Roberts, *Philos. Trans. R. Soc. London, Ser. B* **305,** 299 (1984).
358. S. V. Durmishidze, N. N. Nutsubidze, and N. D. Abashidze, *Izv. Akad. Nauk Gruz SSR, Ser. Biol.* **2,** 410 (1976).
359. T. Yoneyama, H. Sasakawa, S. Ishizuka, and T. Tatsuka, *Soil Sci. Plant Nutr.* **25,** 267 (1979).
360. B. J. Prasad and D. N. Rao, *Indian J. Exp. Biol.* **18,** 879 (1980).
361. T. Matsumaru, T. Yoneyama, T. Totsuka, and K. Shiratori, *Soil Sci. Plant Nutr.* **25,** 255 (1979).
362. H. H. Rogers, J. C. Campbell, and R. J. Volk, *Science* **206,** 333 (1979).
363. J. P. Sinn, E. J. Pell, and R. L. Kabel, *J. Air Pollut. Control Assoc.* **34,** 668 (1984).
364. L. Klepper, *Atmos. Environ.* **13,** 537 (1979).
365. D. T. Tingey, *Atmos. Environ.* **13,** 1475 (1979).
366. J. Fuhrer and K. H. Erismann, *Experientia* **36,** 409 (1980).
367. T. M. Capron and T. A. Mansfield, *J. Exp. Bot.* **28,** 112 (1977).
368. T. A. Mansfield, M. E. Whitmore, and R. M. Law, *in* "Air Pollution by Nitrogen Oxides" (T. Schneider and L. Grant, eds.), pp. 511–520. Am. Elsevier, New York, 1982.
369. A. Chamel and J. P. Garrec, *Environ. Pollut.* **12,** 307 (1977).
370. D. C. MacLean, R. E. Schneider, and L. H. Weinstein, *Environ. Pollut. Ser. A* **29,** 27 (1982).
371. J. A. Lawrence and K. L. Reynolds, *Phytopathology* **74,** 578 (1984).
372. S. Sakuri, K. Stai, and H. Tsunoda, *Fluoride* **16,** 175 (1983).
373. F. B. M. Davies, *Environ. Pollut., Ser. A* **29,** 189 (1982).
374. M. Ghiasseddin, J. M. Hughes, J. E. Diem, and J. W. Mason, *J. Plant Nutr.* **3,** 429 (1981).
375. M. Ghiasseddin, J. M. Hughes, J. E. Diem, and J. W. Mason, *J. Plant Nutr.* **3,** 441 (1981).
376. F. Murray, *J. Am. Soc. Hortic. Sci.* **108,** 526 (1983).
377. S. Kisman, C. Charlot, S. Brun, and J. C. Cabanis, *Plant Soil* **74,** 417 (1983).

378. T. J. Facteau, S. Y. Wang, and K. E. Rowe, *J. Am. Soc. Hortic. Sci.* **103,** 115 (1978).
379. J. P. Garrec, P. Abdulaziz, E. Lavielle, L. Vandeveide, and R. Plebin, *Fluoride* **11,** 186 (1978).
380. G. Lorenc-Plucinska and J. Olekayn, *Fluoride* **15,** 149 (1982).
381. Y. D. Yee-Meiler, *Mitt. Eidg. Anst. Forstl. Versuchswes.* **53,** 201 (1977).
382. S. S. Sidhu, *Stud. Environ. Sci.* **8,** 425 (1980).
383. R. J. Taylor and F. A. Basabe, *Environ. Pollut., Ser. A* **33,** 221 (1984).
384. P. J. Temple and O. C. Taylor, *Atmos. Environ.* **17,** 1583 (1983).
385. T. Lewis and E. Brennan, *Environ. Pollut.* **32,** 101 (1983).
386. L. L. Sigal and O. C. Taylor, *Bryologist* **82,** 564 (1979).
387. O. C. Taylor, P. J. Temple, and A. J. Thill, *HortScience* **18,** 861 (1983).
388. T. E. Starkey, D. D. Davis, E. J. Pell, and W. Merrill, *HortScience* **16,** 547 (1981).
389. N. E. DeVos, R. R. Hill, R. W. Hepler, E. J. Pell, and R. Craig, *J. Am. Soc. Hortic. Sci.* **105,** 157 (1980).
390. T. J. Swiecki, A. G. Endress, and O. C. Taylor, *Can. J. Bot.* **60,** 316 (1982).
391. R. L. Heath and A. G. Endress, *Z. Pflanzenphysiol.* **92,** 271 (1979).
392. C. R. Thompson and G. Kats, *Environ. Sci. Technol.* **12,** 550 (1978).
393. L. J. deKok, C. R. Thompson, J. B. Mudd, and G. Kats, *Z. Planzenphysiol.* **111,** 85 (1983).
394. C. R. Thompson, G. Kats, and R. W. Lennos, *Environ. Sci. Technol.* **13,** 1263 (1979).
395. A. Bytnerowicz and O. C. Taylor, *J. Environ. Qual.* **12,** 397 (1983).
396. D. W. Beckerson and G. Hofstra, *Atmos. Environ.* **13,** 533 (1979).
397. C. A. Miller and D. D. Davis, *HortScience* **16,** 548 (1981).
398. G. Hofstra and D. W. Beckerson, *Atmos. Environ.* **15,** 383 (1981).
399. V. J. Black, D. P. Ormrod, and M. H. Unsworth, *J. Exp. Bot.* **33,** 1302 (1982).
400. D. M. Olszyk and T. W. Tibbitts, *J. Am. Soc. Hortic. Sci.* **107,** 266 (1982).
401. B. B. Shew, R. A. Reinert, and K. B. Barker, *Phytopathology* **72,** 822 (1982).
402. D. P. Ormrod, D. T. Tingey, and M. Gumpertz, *HortScience* **18,** 896 (1983).
403. K. W. Foster, H. Timm, C. K. Labanauskas, and R. J. Oshima, *J. Environ. Qual.* **12,** 75 (1983).
404. A. S. Heagle and J. W. Johnston, *J. Air Pollut. Control Assoc.* **29,** 729 (1979).
405. G. C. Pratt, K. W. Kromroy, and S. V. Krupa, *Environ. Pollut. Ser. A* **32,** 1 (1983).
406. N. M. Le Sueur-Brymer and D. P. Ormrod, *Can. J. Plant Sci.* **64,** 69 (1984).
407. P. B. Reich and R. G. Amundson, *Environ. Pollut., Ser. A* **34,** 345 (1984).
408. R. D. Shertz, W. J. Kender, and R. C. Musselman, *Sci. Hortic. (Amsterdam)* **13,** 37 (1980).
409. K. F. Jensen, *Can. J. For. Res.* **13,** 298 (1983).
410. J. Mooi, *Mitt. Forstl. Bundesversuchsans. Wien* **137,** 47 (1981).
411. D. F. Karnosky, *Can. J. For. Res.* **7,** 437 (1977).
412. T. C. Weidensaul and S. L. Darling, *Phytopathology* **69,** 939 (1979).
413. S. Bialobok, P. Karolewski, and J. Oleksyn, *Arbor. Kornickie* **25,** 289 (1980).
414. T. W. Ashenden, *Environ. Pollut.* **18,** 45 (1979).
415. W. E. Hogsett, S. R. Holman, M. L. Gumpertz, and D. T. Tingey, *Environ. Pollut., Ser. A* **33,** 303 (1984).
416. B. A. Marie and D. P. Ormrod, *Environ. Pollut., Ser. A* **33,** 257 (1984).
417. M. E. Whitmore and T. A. Mansfield, *Environ. Pollut., Ser. A* **31,** 217 (1983).
418. R. W. Carlson, *Environ. Pollut., Ser. A* **32,** 11 (1983).
419. R. G. Amundson, *J. Environ. Qual.* **12,** 454 (1983).
420. P. M. Irving and J. E. Miller, *Can. J. Bot.* **62,** 840 (1984).

421. M. Zafar Iqbal, *Pak. J. Sci. Ind. Res.* **26,** 225 (1983).
422. M. E. Whitmore and P. H. Freer-Smith, *Nature (London)* **300,** 55 (1982).
423. I. Nouchi, H. Mayumi, and F. Yamazoe, *Atmos. Environ.* **18,** 453 (1984).
424. D. D. Davis, *Plant Dis. Rep.* **61,** 640 (1977).
425. L. W. Kress and J. M. Skelly, *Plant Dis.* **66,** 1149 (1982).
426. J. Bonte, C. Bonte, and L. De Cormis, *Fluoride* **16,** 220 (1983).
427. R. H. Mandl, L. H. Weinstein, M. Dean, and M. Wheeler, *Environ. Exp. Bot.* **20,** 359 (1980).
428. T. Keller, *Oecologia* **44,** 283 (1980).
429. R. G. Amundson, L. H. Weinstein, P. van Leuken, and L. J. Colavito, *Environ. Exp. Bot.* **22,** 49 (1982).
430. J. Rebbeck and E. Brennan, *Environ. Pollut., Ser. A* **36,** 7 (1984).
431. P. I. Coyne and G. E. Bingham, *J. Air Pollut. Control Assoc.* **28,** 1119 (1978).
432. R. A. Reinert, D. S. Shriner, and J. O. Rawlings, *J. Environ. Qual.* **11,** 52 (1982).
433. R. A. Reinert and J. S. Sanders, *Plant Dis.* **66,** 122 (1982).
434. T. Elkiey and D. P. Ormrod, *J. Am. Soc. Hortic. Sci.* **105,** 664 (1980).
435. A. Furukawa and T. Totsuka, *Environ. Control Biol.* **17,** 161 (1979).
436. L. W. Kress, J. M. Skelly, and K. H. Hinkelmann, *Agric. Environ.* **7,** 265 (1982).
437. A. R. Wellburn, C. Higginson, D. Robinson, and C. Walmsley, *New Phytol.* **88,** 223 (1981).
438. W. A. Feder, *Environ. Health Perspect.* **27,** 139 (1978).
439. M. R. Ashmore, J. N. B. Bell, and C. L. Reily, *Environ. Pollut., Ser. B.* **1,** 195 (1980).
440. A. I. Goren and A. E. Donagi, *Oecologia* **44,** 418 (1980).
441. G. Braun, *Forstwiss. Centralbl.* **96,** 289 (1977).
442. L. Steubing, *Angew. Bot.* **56,** 1 (1982).
443. J. R. McClenahen and T. C. Weidensaul, *J. Environ. Qual.* **6,** 169 (1977).
444. A. E. Donagi and A. I. Goren, *Environ. Sci. Technol.* **13,** 986 (1979).
445. P. M. Rice, P. C. Tourangeau, C. Johns, and C. C. Gordon, *Environ. Pollut. Ser. B* **7,** 233 (1984).
446. E. Skye, *Annu. Rev. Phytopathol.* **17,** 325 (1979).
447. F. Le Blanc, G. Robitaille, and D. N. Rao, *J. Hattori Bot. Lab.* **40,** 27 (1976).
448. C. M. Vick and R. Bevan, *Environ. Pollut.* **11,** 203 (1976).
449. W. E. Winner and J. D. Bewley, *Oecologia* **33,** 311 (1978).
450. W. E. Winner and J. D. Bewley, *Oecologia* **33,** 221 (1978).
451. U. Schreiber, W. Vidaver, V. C. Runeckles, and P. Rosen, *Plant Physiol.* **61,** 80 (1976).
452. R. Rabe and K. H. Kreeb, *Environ. Pollut.* **19,** 119 (1979).
453. H.-J. Jäger, *in* "Monitoring of Air Pollutants by Plants" (L. Steubing and H.-J. Jäger, eds.), p. 99. Dr. W. Junk Publishers, The Hague 1982.
454. B. Prinz and G. Scholl, *Schriftenr. LIB, Landesanst. Immissions- Bodennutzungsschutz Landes Nordrhein-Westfalem* **36,** 62 (1975).
455. E. von Rudolph and A. Halbwachs, *Eur. J. For. Pathol.* **13,** 334 (1983).
456. Anonymous, "National Meetret Voor Luchtverontreiniging. Verslag over de periode 1 Oktober 1976 – 1 Oktober 1977," Publ. NML-RIV, No. 9. Drukkerijelinkwijk BV-Utrecht, The Netherlands, 1978.
457. S. N. Linzon, R. G. Pearson, J. A. Donnan, and F. N. Durham, "Ozone Effects on Crops in Ontario and Related Monetary Values," ARB-13-84 Phyto. Ontario Ministry of the Environment, Toronto, Ontario, Canada, 1984.
458. R. M. Adams, T. D. Crocker, and N. Thanavibulchai, *J. Environ. Econ. Manage.* **9,** 42 (1982).

459. R. M. Adams, S. A. Hamilton, and B. A. McCarl, "The Economic Effects of Ozone on Agriculture," EPA-600/3-84-090. U.S. Environ. Prot. Agency, Corvallis, Oregon, 1984.
460. J. W. Mjelde, R. M. Adams, B. L. Dixon, and P. Garcia, *J. Air Pollut. Control Assoc.* **34,** 360 (1984).
461. S. F. Duchelle, *Plant Dis.* **65,** 661 (1981).
462. S. F. Duchelle, J. M. Skelly, T. L. Sharick, B. I. Chevone, Y.-S. Yang, and J. E. Nellessen, *J. Environ. Manage.* **17,** 299 (1983).
463. R. Cornelius, *Angew. Bot.* **56,** 243 (1982).
464. M. L. Wesely, D. R. Cook, and R. L. Hart, *Boundary-Layer Meteorol.* **27,** 237 (1983).
465. J. R. McClenahen, *Can. J. For. Res.* **8,** 432 (1978).
466. H. E. Liebold, *Wiss. Z. Tech. Univ., Dresden* **26,** 507 (1977).
467. M. Sima and V. Tollinger, *Ekológia* **1,** 77 (1982).
468. D. Hocking, P. Kuchar, J. A. Plambeck, and R. A. Smith, *J. Air Pollut. Control Assoc.* **28,** 133 (1978).
469. J. L. Dawson and T. H. Nash, III, *Environ. Exp. Bot.* **20,** 61 (1980).
470. L. K. Thompson, S. S. Sidhu, and B. A. Roberts, *Environ. Pollut.* **18,** 221 (1979).
471. F. Murray, *Environ. Pollut., Ser. A* **24,** 45 (1981).
472. E. B. Cowling, *Environ. Sci. Technol.* **16,** 110A (1982).
473. D. A. Wilshire and M. L. Evans, "Acid Precipitation—An Annotated Bibliography," Circ. No. 923. U.S. Geol. Surv., Alexandria, Virginia, 1984.
474. D. L. Rambo, "An Acid Rain Bibliography," CERL-053. U.S. Environ. Prot. Agency, Corvallis, Oregon, 1980.
475. Technical Information Service, "Acid Precipitation: A Bibliography," DOE/TIC-3399. U.S. Dep. Energy, Washington, D.C., 1983.
476. S. Oden, "The Acidification of Air and Precipitation and Its Consequences in the Natural Environment," Bull. No. 1. Ecology Committee, Swedish National Science Research Council, Stockholm, 1968 (Translation Consultants, Ltd., Arlington, Virginia, TR-1172).
477. G. E. Likens, F. H. Bormann, and N. M. Johnson, *Environment* **14,** 33 (1972).
478. C. V. Cogbill and G. E. Likens, *Water Resour. Res.* **10,** 1133 (1974).
479. R. A. Smith, "Air and Rain: The Beginnings of a Chemical Climatology." Longmans, Green, London, 1872.
480. C. Crowther and H. G. Ruston, Jr., *Agric. Sci.* **4,** 25 (1911).
481. A. G. Gordon and E. Gorham, *Can. J. Bot.* **41,** 1063 (1963).
482. B. Jonsson and R. Sundberg, "Has the Acidification by Atmospheric Pollution Caused a Growth Reduction in Swedish Forests?" Res. Note No. 20. Department of Forest Yield Research, Royal College of Forestry, Stockholm, Sweden, 1972.
483. L. Wiklander, *Grundforbaettring* **26,** 155 (1973).
484. D. S. Shriner, Ph.D. Thesis, North Carolina State University, Raleigh, 1974.
485. N. S. Dailey and S. G. Winslow, "Health and Environmental Aspects of Acid Rain An Abstracted Literature Collection," NLM/TIRC-80/1. National Library of Medicine/Federation of America Societies for Experimental Biology, Rockville, Maryland, 1980.
486. I. Ziegler, *Environ. Qual. Saf.* **2,** 182 (1973).
487. L. S. Evans, *Environ. Exp. Bot.* **22,** 155 (1982).
488. J. S. Jacobson and P. van Leuken, *Proc. Int. Clean Air Cong. 4th, 1977,* pp. 124–127 (1977).
489. T. T. Kozlowski, *BioScience* **30,** 88 (1980).

490. M. Katz, *Ind. Eng. Chem.* **41,** 2450 (1949).
491. S. N. Linzon, P. J. Temple, and R. G. Pearson, *J. Air Pollut. Control Assoc.* **29,** 520 (1979).
492. G. M. Lovett, W. A. Reiners, and R. K. Olson, *Science* **218,** 1303 (1982).
493. D. S. Shriner, *Phytopathology* **68,** 213 (1978).
494. R. N. Goodman, Z. Kiraly, and M. Zaitlin, "The Biochemistry and Physiology of Infectious Plant Disease," pp. 232–280. Van Nostrand-Reinhold, Princeton, New Jersey, 1967.
495. S. B. McLaughlin and D. S. Shriner, *in* "Plant Disease: An Advanced Treatise" (J. G. Horsfall and E. B. Cowling, eds.), Vol. 5, pp. 407–431. Academic Press, New York, 1980.
496. D. S. Shriner and J. W. Johnston, *Environ. Exp. Bot.* **21,** 199 (1981).
497. G. Abrahamsen, J. Hovland, and S. Hagvar, *in* "Effects of Acid Precipitation on Terrestrial Ecosytems" (T. C. Hutchinson and M. Havas, eds.), pp. 341–362. Plenum, New York, 1980.
498. D. S. Shriner and E. B. Cowling, "NATO Advanced Research Institute on the Effects on Acid Rain on Terrestrial Ecosystems," Eco-Sci. Ser., pp. 435–442. Plenum, New York, 1980.
499. B. Ulrich, R. Mayer, and P. K. Khanna, *Soil Sci.* **130,** 193 (1980).
500. I. Ziegler, *in* "Proceedings of the International Workshop on Effects of Acid Precipitation on Vegetation, Soils, and Terrestrial Ecosystems" (L. S. Evans and G. R. Hendrey, eds.), pp. 19–20, BNL-51195. Brookhaven Natl. Lab., Upton, New York, 1979.
501. L. S. Evans, N. F. Gmur, and J. J. Kelsch, *Environ. Exp. Bot.* **22,** 155 (1977).
502. J. W. Johnston, D. S. Shriner, C. I. Klarer, and D. M. Lodge, *Environ. Exp. Bot.* **22,** 329 (1982).
503. D. S. Shriner, M. E. Decot, and E. B. Cowling, *Proc. Am. Phytopathol. Soc.* **1,** 112 (1974).
504. R. W. Ferenbaugh, *Ame. J. Bot.* **63,** 293 (1976).
505. P. M. Irving and J. E. Miller, *J. Environ. Qual.* **10,** 473 (1981).
506. J. S. Jacobson, J. Troiano, L. J. Colvaito, L. I. Heller, and D. C. McCune, *in* "Polluted Rain" (T. Y. Toribara, M. W. Miller, and P. E. Morrow, eds.), pp. 291–305. Plenum, New York, 1980.
507. C. O. Tamm and E. B. Cowling, *USDA For. Serv. Gen. Tech. NE* NE-23, 854-855 (1976).
508. D. S. Shriner, *in* "Proceedings of the 12th University of Rochester Symposium on Environmental Toxicity: Polluted Rain," Environ. Sci. Res. Ser., pp. 259–272. Plenum, New York, 1980.
509. D. S. Shriner, *in* "Air Pollutants and Their Effects on Terrestrial Ecosystems" (A. H. Legge and S. V. Krupa, eds.), pp. 365–388. Wiley, New York, 1986 (in press).
510. R. Crang, University of Illinois, Urbana (personal communication).
511. E. T. Paparozzi and H. B. Tukey, Jr., *in* "Acid Precipitation Series" (R. A. Linthurst, ed.), Vol. 5, pp. 13–18. Butterworth, Boston, Massachusetts, 1984.
512. L. S. Evans and G. R. Hendrey, eds., "Proceedings of the International Workshop on Effects of Acid Precipitation on Vegetation, Soils, and Terrestrial Ecosystems," BNL-51195. Brookhaven Natl. Lab., Upton, New York, 1979.
513. H. B. Tukey, Jr., *in* "Effects of Acid Precipitation on Terrestrial Ecosystems" (T. C. Hutchinson and M. Havas, eds.), pp. 141–150. Plenum, New York, 1980.
514. J. A. W. Fairfax and N. W. Lepp, *Nature (London)* **255,** 324 (1975).
515. T. Wood and F. H. Bormann, *Ambio* **4,** 169 (1975).

516. J. S. Eaton, G. E. Likens, and F. H. Bormann, *J. Ecol.* **61,** 495 (1973).
517. D. W. Johnson and G. S. Henderson, *Soil Sci.* **128,** 34 (1979).
518. H. W. Zottl and E. Mies, *Mitt. Dtsch. Bodenkd. Ges.* **38,** 423 (1983) (Translation: Oak Ridge Natl. Lab. tr-5168. Oak Ridge, Tennessee).
519. K. E. Rehfuess, C. Bosch, and E. Pfannkuck, "International Workshop on Growth Disturbances of Forest Trees." Int. Union For. Resh. Organ./Jyvaskyla, Finland, 1982.
520. H. B. Tukey, Jr., *Annu. Rev. Plant Physiol.* **21,** 305 (1970).
521. D. S. Shriner and J. W. Johnston, "EPA Acidic Deposition Ecological Effects Research Review," Pap. AG-21., North Carolina State University, Raleigh, 1982.
522. P. I. Arnon and C. M. Johnson, *Plant Physiol.* **17,** 525 (1942).
523. C. D. Foy and J. C. Brown, *Proc. Soil Sci. Soc. Am.* **27,** 403 (1963).
524. M. A. Tabatabia, *Biogeochemistry* **1,** (1984).
525. D. W. Johnson, *Biogeochemistry* **1,** 29 (1984).
526. J. C. Noggle, *in* "Atmospheric Sulfur Deposition—Environmental Impact and Health Effects" (D. S. Shriner, C. R. Richmond, and S. E. Lindberg, eds.), pp. 289–297. Ann Arbor Sci., Publ., Ann Arbor, Michigan, 1980.
527. U. S. Jones and E. L. Suarez, *in* "Atmospheric Sulfur Deposition—Environmental Impact and Health Effects" (D. S. Shriner, C. R. Richmond, and S. E. Lindberg, eds.), pp. 377–396. Ann Arbor Sci. Publ., Ann Arbor, Michigan, 1980.
528. S. E. Lindberg and S. B. McLaughlin, *in* "Air Pollutants and Their Effects on Terrestrial Ecosystems" (A. H. Legge and S. V. Krupa, eds.), pp. 449–503. Wiley, New York, 1986 (in press).
529. D. B. Botkin, J. F. Janak, and J. R. Wallis, *J. Ecol.* **60,** 849 (1973).
530. H. H. Shugart and D. C. West, *J. Environ. Manage.* **5,** 161 (1977).
531. D. B. Botkin and J. D. Aber, "Some Potential Effects of Acid Rain on Forest Ecosystems: Implications of a Computer Simulation," BNL-50889. Brookhaven Natl. Lab., Upton, New York, 1979.
532. D. C. West, S. B. McLaughlin, and H. H. Shugart, *J. Environ. Qual.* **9,** 43 (1980).
533. P. Schutt and E. B. Cowling, *Plant Dis.* **69,** 548 (1985).
534. G. Tyler, *Ambio* **1,** 52 (1972).
535. G. H. M. Krause, K. D. Jung, and B. Prinz, *in* Proceedings of the International Colloquium of the VDI on Acid Precipitation: Causes and Effects." Lindau, West Germany, 1983 (Transaction: Oak Ridge Natl. Lab., tr-5070. Oak Ridge, Tennessee).
536. R. J. Olson, C. J. Emerson, and M. K. Nungesser, "Geoecology: A County-Level Environmental Data Base for the Conterminous United States," ORNL/TM-7351. Oak Ridge Natl. Lab., Oak Ridge, Tennessee, 1980.
537. P. M. Irving, *in* "Acid Rain: Origin and Effects," VDI-Ber., No. 500, pp. 215–221. VDI-Verlag GmbH, Dusseldorf, West Germany, 1983.
538. P. R. Crosson and S. Brubaker, "Resource and Environmental Effects of U.S. Agriculture." John Hopkins Univ. Press, Baltimore, Maryland, 1982.
539. J. J. Lee, E. M. Preston, and R. A. Lewis, *Jt. Conf. Sens. Environ. [Conf. Proc.], Pollut. 4th, 1977,* pp. 49–53 (1978).
540. P. Greenwood, A. Greenhalgh, C. H. Baker, and M. H. Unsworth, *Atmos. Environ.* **16,** 2261 (1982).
541. J. E. Miller, D. G. Sprugel, R. N. Muller, H. J. Smith, and P. B. Xerikos, *Phytopathology* **70,** 1124 (1980).
542. J. E. Miller, H. E. Smith, and P. B. Xerikos, "Variability of The Yield Response of Field-Grown Soybean Cultivars to SO_2," Rep. No. ANL-80-115, pp. 15–18. Argonne Natl. Lab., Argonne, Illinois, 1980.

543. L. De Cormis, J. Bonte, and A. Tisne, *Pollut. Atmos.* **17,** 103 (1975).
544. A. R. McLeod, K. Alexander, and P. Hatcher, "A Prototype System for Open-Air Fumigation of Agricultural Crops: 2. Construction and Description," CERL Note No. TPRD/L/2474/N83. Cent. Elect. Res. Lab., Leatherhead, England, 1983.
545. J. H. Shinn, B. R. Clegg, and M. L. Stuart, "A Linear-Gradient Chamber for Exposing Field Plants to Controlled Levels of Air Pollutants," UCRL Rep. No. 80411. Lawrence Livermore Lab., Livermore, California, 1977.
546. A. R. McLeod and J. Fackrell, "A Prototype System for Open-Air Fumigation of Agricultural Crops: 1. Theoretical Design," CERL Note No. TPRD/L/2474/N83. Cent. Electr. Res. Lab., Leatherhead, England, 1983.
547. A. H. Buckenham, M. A. Parry, C. P. Whittingham, and A. T. Young, *Environ. Pollut., Ser. B* **2,** 475 (1981).
548. A. S. Heagle, D. E. Body, and W. W. Heck, *J. Environ. Qual.* **2,** 365 (1973).
549. R. H. Mandl, L. H. Weinstein, D. C. McCune, and M. Keveny, *J. Environ. Qual.* **2,** 371 (1973).
550. T. M. Roberts, R. M. Bell, D. C. Horoman, and K. E. Colvill, *Environ. Pollut., Ser. A* **31,** 9 (1983).
551. A. S. Heagle, R. B. Philbeck, H. H. Rogers, and M. B. Letchworth, *Phytopathology* **69,** 15 (1979).
552. D. M. Olszyk, T. W. Tibbits, and W. M. Hertzberg, *J. Environ. Qual.* **9,** 610 (1980).
553. M. H. Unsworth, A. S. Heagle, and W. W. Heck, *Atmos. Environ.* **18,** 373 (1984).
554. M. H. Unsworth, A. S. Heagle, and W. W. Heck, *Atmos. Environ.* **18,** 381 (1984).
555. L. Weinstock, W. J. Kender, and R. C. Musselman, *J. Am. Soc. Hortic. Sci.* **107,** 923 (1982).
556. A. S. Heagle and W. W. Heck, *in* "Crop Loss Assessments" (P. S. Teng and S. V. Krupa, eds.), E. C. Stakman Commem. Symp., Misc. Publ. No. 7, pp. 296–305. Agric. Exp. Stn., University of Minnesota, St. Paul, 1980.
557. A. S. Heagle and W. W. Heck, *Environ. Pollut.* **7,** 247 (1974).
558. A. S. Heagle, W. W. Heck, J. O. Rawlings, and R. B. Philbeck, *Crop Sci.* **23,** 1184 (1983).
559. W. W. Heck, O. C. Taylor, R. M. Adams, G. Bingham, J. E. Miller, E. M. Preston, L. H. Weinstein; R. G. Amundson, R. J. Kohut, J. A. Laurence, W. C. Cure, A. S. Heagle, J. T. Gish, H. E. Heggestad, L. W. Kress, G. E. Neely, J. O. Rawlings, and P. Temple, "National Crop Loss Assessment Network (NCLAN) 1982 Annual Report," EPA-600/3-84-049. U.S. Environ. Prot. Agency, Corvallis, Oregon, 1983.
560. R. Kohut and J. A. Laurence, *Environ. Pollut., Ser. A* **32,** 233 (1983).
561. L. W. Kress and J. E. Miller, *J. Environ. Qual.* **12,** 276 (1983).
562. J. O. Rawlings and W. W. Cure, *Crop Sci.* **25,** 807 (1985).
563. D. S. Shriner, W. W. Cure, A. S. Heagle, W. W. Heck, D. W. Johnson, R. J. Olson, and J. M. Skelly, "An Analysis of Potential Agriculture and Forest Impacts of Long Range Transport Air Pollutants," ORNL-5910. Oak Ridge Natl. Lab., Oak Ridge, Tennessee, 1984.
564. R. M. Adams and B. A. McCarl, *J. Environ. Econ. Manage.* **12,** 264 (1985).
565. T. M. Roberts, *Atmos. Environ.* **18,** 629 (1984).
566. J. E. Miller, H. E. Smith, P. B. Xerikos, and P. M. Irving, "Sulfur Dioxide Effects on Soybean Yield: Dose-Response Relationships from 1979 Field Studies," Rep. No. ANL-79-65. Argonne Natl. Lab., Argonne, Illinois, 1979.
567. D. G. Sprugel, J. E. Miller, R. N. Muller, H. J. Smith, and P. B. Xerikos, *Phytopathology* **70,** 1129 (1980).
568. P. M. Irving, P. B. Xerikos, and J. E. Miller, "The Combined Effects of Sulfur Dioxide and Nitrogen Dioxide on Tube Growth and Productivity of Soybean,"

Rep. No. ANL-81-85, Part III, pp. 17–22. Argonne Natl. Lab., Argonne, Illinois, 1981.
569. J. E. Miller, H. J. Smith, and W. Prepejchal, "Evidence for Extreme Resistance of Field Corn to Intermittent Sulfur Dioxide Stress," Rep. No. ANL-81-85, Part III, pp. 27–29. Argonne Natl. Lab., Argonne, Illinois, 1981.
570. P. R. Miller and M. J. Elderman, eds., "Photochemical Oxidant Air Pollutant Effects on a Mixed Conifer Forest Ecosystem: A Progress Report," Rep. No. EPA-600/3-77-104. U.S. Environ. Prot. Agency, Corvallis, Oregon, 1976.
571. L. F. Benoit, J. M. Skelly, L. D. Moore, and L. S. Dochinger, *Can. J. For. Res.* **12,** 673 (1982).
572. S. B. McLaughlin, R. K. McConathy, D. Duvick, and L. K. Mann, *For. Sci.* **28,** 60 (1982).
573. A. H. Johnson and T. G. Siccama, *Environ. Sci. Technol.* **17,** 294A (1983).

Lists of books relevant to this chapter published since the lists in Chapter 8, Volume II, 3d edition (1977) can be found in Chapter 1, Table IX, Volume VII.

6

Effects of Acidic Deposition on Aquatic Ecosystems

Magda Havas

Institute for Environmental Studies
University of Toronto
Toronto, Ontario, Canada

I. Introduction

The term *acid rain* was first used by Robert Angus Smith in 1872 *(1)* in his description of precipitation contaminated with "sulphuric acid" or "acid sulphate" as a consequence of coal combustion near industrial centers in England, Scotland, and Germany. The damaging effects of acidic deposition on plants and materials were also noted at that time *(2, 3)*. By the early 1920s the adverse effects of acidity on trout populations were recognized and well documented in Norway *(4–6)*. Despite continued studies since the 1920s, it was not until the publications of Svante Oden *(7–9)* that acidic deposition was linked with lake water acidification and received worldwide attention. Oden *(8)* found that both precipitation and surface water in Sweden were becoming more acidic. He attributed this acidity to oxides of sulfur and nitrogen emitted as by-products of coal combustion, smelting of sulfide-bearing ores, and automobile exhaust originating in industrial centers in England and Germany. Oden predicted adverse effects to aquatic biota.

At the First International Symposium on Acid Precipitation and the Forest Ecosystem, in 1975 *(10)*, European and North American scientists were first brought together to discuss their findings. Several conferences dealing with aquatic ecosystems have been held since 1975. The most notable include one sponsored by the SNSF Project (Norwegian Interdisciplinary Research Programme "Acid Precipitation—Effects on Forests and Fish") *(11)*; a conference on the "Effects of Acid Precipitation on Ecological Systems" *(12)*; and one related to "Acid Rain/Fisheries" *(13)*.

Acidification of aquatic ecosystems in both Europe and North America has been reviewed by Haines *(14)*, and "Acidification Today and Tomorrow," *(15)* prepared for the 1982 Stockholm Conference on the "Acidification of the Environment," summarized the Swedish results. A National Research Council of Canada *(16)* 1981 report summarized Canadian results, and the final report of the SNSF Project *(17)* summarized the Norwegian data from 1972 to 1980.

This chapter presents some of the major findings on the effects of acidic deposition on surface water and ground water chemistry, sediment chemistry, and aquatic biota. A discussion of remedial measures is also included. Controversial topics are presented but are not discussed in detail because of space limitations. For a detailed review of some of these issues, refer to Havas *et al.* *(18)*.

II. Effects of Acidic Deposition

A. Water Chemistry

1. Lake Water

a. ACIDIFICATION. Acid-sensitive lakes are soft-water, nutrient-poor lakes, with low concentrations of dissolved salts and limited buffering capacity. These lakes are located on relatively hard, impermeable quartz-bearing bedrock, such as granites and gneisses, which are characteristic of the Precambrian Shield in both North America and Northern Europe (Figures 1 and 2) *(18a)*. Soft-water lakes downwind from sulfur dioxide and nitrogen oxide sources are the most susceptible to acidification.

Only a small fraction of the precipitation falls directly onto a lake. Most of the precipitation which falls within the drainage basin can be chemically altered by the soil and vegetation *(17)*. However, high-altitude headwater lakes above the treeline, where the soils are shallow or nonexistent, are strongly influenced by precipitation chemistry. Also, during spring snow-melt, while the soils are still frozen and the vegetation is dormant, the melt water, which contains wet and dry deposition of acidic substances accumulated during the entire winter, can significantly acidify the surface water in lakes and depress the

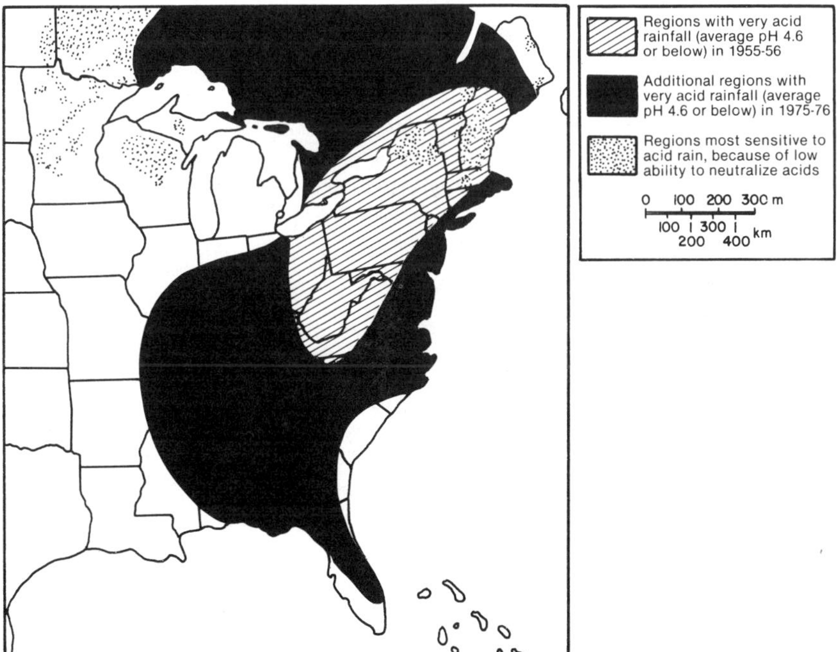

Figure 1. Map of eastern North America showing regions most sensitive to acidic deposition (stippled). The areas which received acidic rainfall (average pH 4.6 or lower) are much smaller in 1955–1956 (hatched) than in 1975–1976 (black) (From G. E. Likens, "The 1981 Yearbook of Science and the Future," pp. 212–227. Encyclopaedia Britannica, 1981. Used with permission).

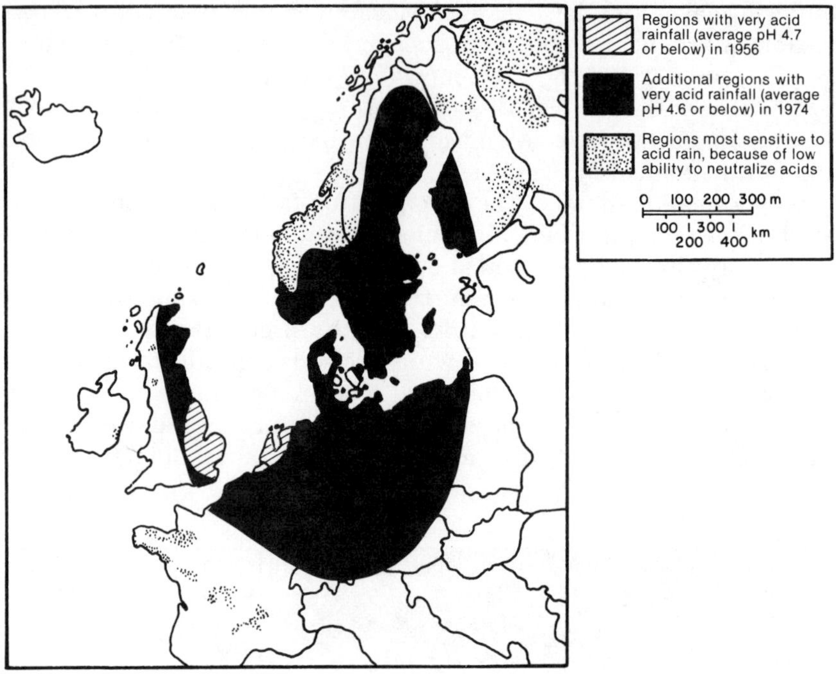

Figure 2. Map of Europe showing regions most sensitive to acidic deposition (stippled). The areas which received acidic rainfall (average pH 4.6 or lower) are much smaller in 1956 (hatched) than in 1974 (black) *(18a)*. (Data was unavailable for Eastern Europe.)

pH in streams. In contrast, deep glacial till deposits, as well as pockets of base-rich minerals such as limestone or dolomite, can impart local immunity to the acidifying effects of the precipitation.

Evidence from a number of regions, including Sweden, Norway, Jutland, Scotland; in Canada from Ontario, Quebec, and Nova Scotia; and in the United States from New York, the New England states, North Carolina, and Florida, indicates that lakes are becoming more acidic *(14, 18)*. This conclusion is based not only on pH measurements of the water, which can be variable seasonally and therefore misleading, but also on increased acidification of surface sediments *(19)* and changes from alkaline-loving to acid-loving diatoms deposited in the surface sediments of lakes *(20)*. Data from lakes which once supported an active sports fishery but are now too acidic to support even the more acid-tolerant species also suggest that these lakes have acidified *(21)*. Regional surveys indicate that the lakes with the low-

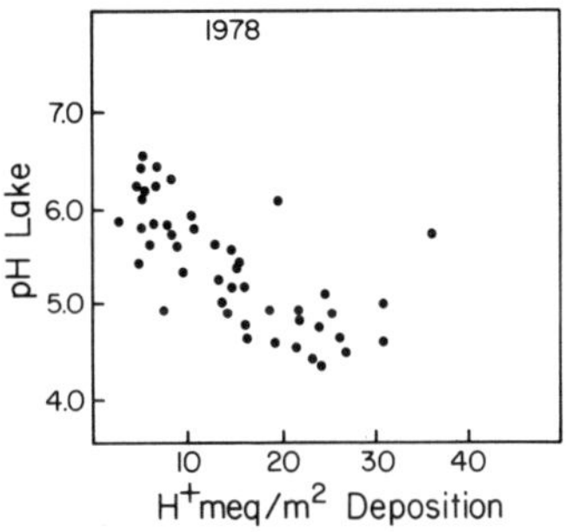

Figure 3. Scatter diagram for pH in lake water versus deposition of hydrogen ions via precipitation *(22)*.

est pH are found in areas with the highest deposition of hydrogen ions in the precipitation (Figure 3) *(17, 22)*.

Although the rate of acidification is difficult to estimate largely as a result of its nonlinearity (Figure 4) *(23)*, limited historical data, and seasonal fluctuations in lake water pH (i.e., spring pH depression as mentioned previously, Figure 5) *(24a)*, some of the estimates from Norway, Sweden, and North America are quite similar. Some lakes in

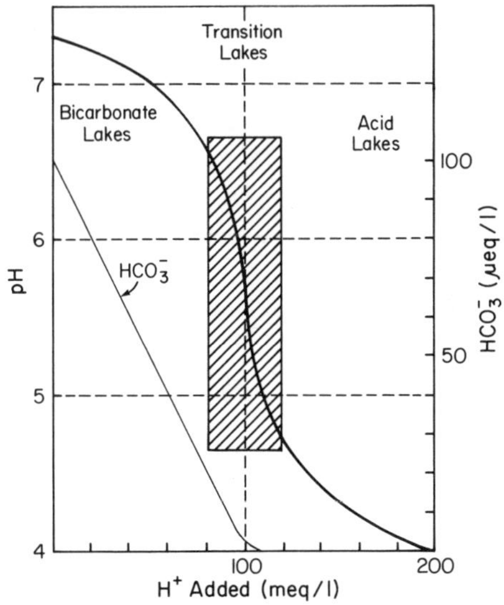

Figure 4. Titration curve for a bicarbonate solution with an initial concentration of 100 μEq/liter (Reprinted with permission from *Water Res.*, **14**, A. Henriksen, Copyright 1980, Pergamon Press, Ltd.).

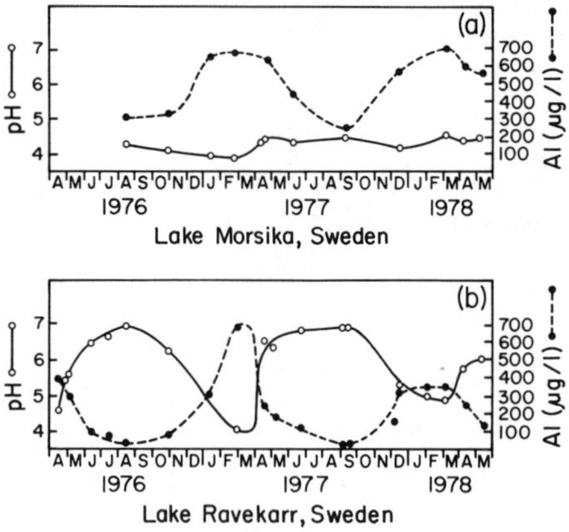

Figure 5. Changes in aluminum concentrations with pH in Lakes Morsika (a) and Ravekarr (b), Sweden [From W. Dickson *in* "Acid Precipitation" (G. R. Hendry, ed.), Proc. Int. Workshop, Sagamore Lake Conf. Cent., Sagamore Lake, New York, Brookhaven Natl. Lab., Uptown, New York, 1978. Used with permission].

Sweden have acidified by 1.4–1.7 pH units within the past 30–40 years according to Hendrey *et al.* *(25)*. Almer *et al.* *(26)* estimated a maximum acidification rate of 1.8 pH units within 30 years. In a group of Norwegian lakes the pH has decreased as much as 1.25 pH units within 30–40 years *(27)*. A frequency distribution of the change in

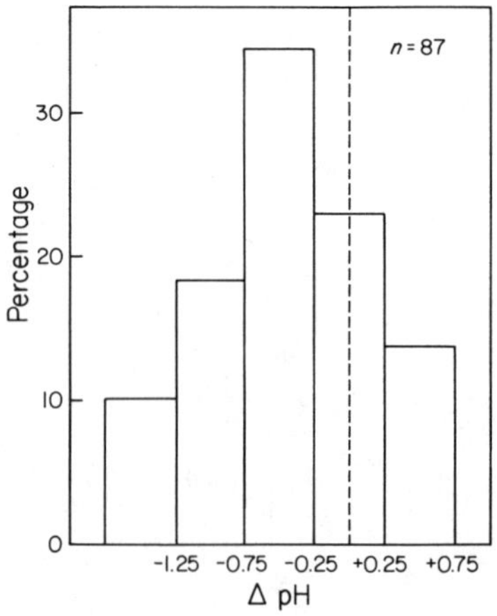

Figure 6. Frequency distribution for the changes in pH (Δ pH) between results obtained in 1920–1949 and results from the 1970s for 87 lakes in southern Norway *(27)*.

pH for lakes in southern Norway (Figure 6) shows that a decrease in pH is more common and more extreme than an increase in pH, which may occur as a consequence of natural weathering. In North America, Beamish and Harvey *(21)* estimated the average decline in pH, based on 46 lakes, to be 0.16 pH units per year. In the Adirondacks, the median pH of high-elevation lakes has dropped by approximately 1.5 units in 45 years *(28)*.

As a consequence of this rapid acidification, many lakes have been chemically altered along the western coast of Sweden, in southwestern Norway, in the northeastern United States, and in southeastern Canada. Approximately 93% of the lakes surveyed along the western coast of Sweden in the 1970s had pH values below 5.5, a pH which appears to be critical for the survival and successful reproduction of many species *(14, 29, 30)*.

b. Major Elements: Ca, Mg, HCO_3, NO_3. The water chemistry changes in several ways in addition to a loss of buffering capacity and a decrease in pH. Bicarbonate, which is normally the dominant anion in these lakes, is replaced by sulfate *(30, 31)*. "Excess" sulfate (corrected for marine sources) in lake water appears to be fairly well correlated with sulfate deposition. Although some of the sulfate is taken up by the drainage basin, a fair amount enters the water, as well

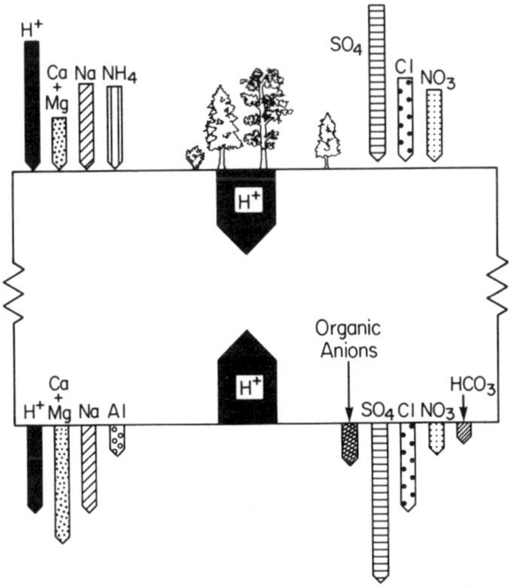

Figure 7. Schematic illustration of the inputs to and outputs from a forested catchment *(31a)*.

(Figure 7) *(31a)*. Nitrogen oxides are also elevated in the precipitation. The terrestrial ecosystem, however, is frequently nitrogen-limited and readily assimilates the incoming nitrogen before it reaches the lakes and streams. Only during the spring can a significant nitrogen pulse be observed in run-off. Concentrations of calcium and magnesium may increase slightly as a result of acid leaching from soils, although this increase is marginal and is unlikely to affect the biota significantly.

The relationship between alkalinity and the concentrations of calcium and magnesium is well defined for nonacidified soft-water lakes. When these lakes are exposed to acid precipitation, the alkalinity decreases, although the concentrations of calcium and magnesium remain constant or increase slightly. Henriksen *(32)* has applied these observations and developed an empirical model to identify lakes which have recently acidified. His model in combination with the "Schofield sausage" *(33)* can be used to predict lake water acidification in Norway, in the Adirondacks, and near Sudbury, Ontario, accurately, on the basis of calcium, magnesium, and sulfate levels.

c. METALS. Metals such as aluminum, iron, manganese, zinc, cadmium, mercury, and lead, which may be present in the precipitation, leached from the soils (Figure 7) and sediments, or mobilized by microbial activity, tend to increase in concentration in acidified lakes

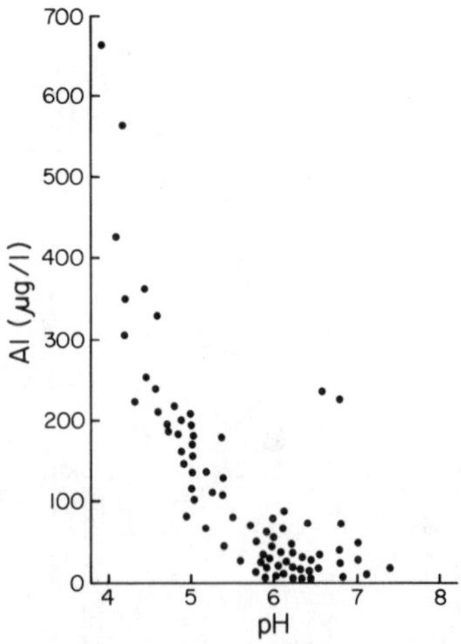

Figure 8. The pH and concentrations of aluminum (total) in Swedish lakes during summer and autumn 1976 *(19)*. (The two "outliers" are very humic waters.)

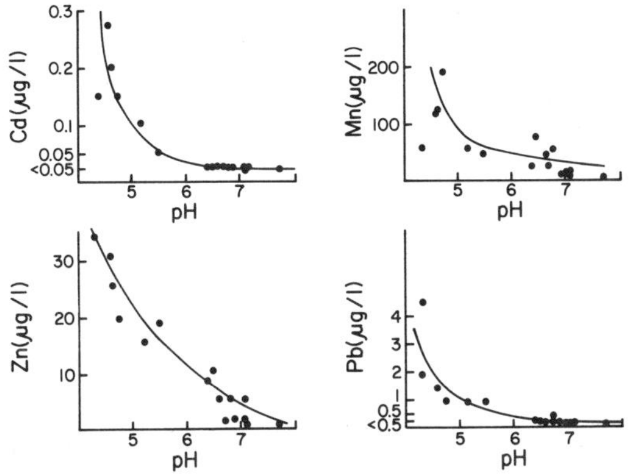

Figure 9. Metals in 16 lakes on the west coast of Sweden with similar metal deposition but with different pH, December 1978 *(19)*.

(Figures 8 and 9). Some of the metals, such as mercury, can become elevated in the biota as a result of increased bioavailability *(34)*. Other metals, such as aluminum, may be toxic to aquatic organisms *(35)*. Background levels of aluminum are approximately 0.05 mg/liter in nonacidified lakes. Concentrations approaching 1.0 mg/liter have been reported in acidic surface water in both North America and Scandinavia *(19, 36)*. Laboratory studies have shown that 0.1 mg/liter may be fatal to some species of fish, particularly in soft water low in calcium and organic ligands *(35)*.

Aluminum can also complex with phosphate which is already limiting primary productivity in these waters. There is some evidence that phosphate levels are lower in some acidic lakes containing elevated levels of aluminum *(19)*. The effects of acidification have been called *oligotrophication* (i.e., a decrease in lake productivity) *(31)*, in contrast with *eutrophication* (i.e., an increase in lake productivity, characterized by algal blooms, fish kills, and foul-smelling water). As a consequence of oligotrophication, acidified lakes are generally clear, highly transparent, pristine-looking lakes ideal for many recreational activities except fishing.

d. STAGES OF ACIDIFICATION. Poorly buffered oligotrophic lakes exposed to acid precipitation pass through three stages during their acidification which can be represented by a bicarbonate titration curve (Figure 4). The first stage corresponds to bicarbonate lakes and is characterized by a steady decrease in the acid-neutralizing, or buffering, capacity. The pH is usually above 6 with temporary dips in pH following heavy autumn showers and spring snow melt [Figure 5(b)]. Acid-sensitive organisms, unable to tolerate even short periods of acidity, are often eliminated during this stage.

Once the acid-neutralizing capacity is exhausted, the pH starts to decrease more rapidly. This second stage (transition lakes, Figure 4) can occur quite rapidly, with as much as a 1 unit drop in pH within a decade. Concentrations of bicarbonate ion continue to decrease, and concentrations of sulfate and metals begin to increase in the water column. Many of the moderately sensitive plant and animal species are eliminated or reduced in number during this stage.

During the third and final stage (acid lakes, Figure 4), the pH begins to stabilize. This process is partly a function of atmospheric acidity (mean annual pH tends to be between 4.0 and 4.5) and partly of the presence of aluminum and iron, both of which act as acid buffers and help maintain a pH in the mid to high 4's [Figure 5(a)], in the water column. By this stage, acid-sensitive species have been eliminated, and only the most acid-tolerant species remain. Some of the remaining species can be found at or near the sediment surface, where the pH tends to be slightly higher than in the water column.

2. Stream Water

Streams reflect the chemistry of drainage water more readily than do lakes because of their relatively short retention times. Fluctuations in the pH, buffering capacity, and metals therefore tend to be more extreme in streams, especially in shallow, low-order streams (unbranched or having one to two branches).

In one study of an acid brook in the Adirondack Mountains, New York, Schofield and Trojnar *(36)* reported a range of pH from 4.6 in March 1977 to 6.5 in July of the same year. The low pHs were invariably associated with elevated levels of aluminum. Levels of aluminum varied from less than 0.1 to 1.0 mg/liter.

Similar results were obtained during experimental acidification of a small stream (Norris Brook) in New Hampshire *(37)*. As the pH was lowered from 5.7–6.4 to 3.9–4.5, concentrations of aluminum increased from less than 0.05 to peak values in excess of 0.2 mg/liter. At the same time, the drift of macroinvertebrates increased dramatically. It is believed that stream invertebrates can avoid acid and metal stress by releasing their hold on sediments and stones and drifting downstream.

In addition to seasonal variations, concentrations of hydrogen ions and aluminum ions also vary along the length of a stream. Provided that the geology does not change substantially, hydrogen ion concentrations tend to decrease downstream *(38)*. Aluminum, which is soil-derived, will be elevated only if the acidic drainage water comes into contact with soils and sediments (Figure 10). Drainage of vast areas of exposed bedrock is characterized by low pH and low levels of aluminum in stream water.

One of the major concerns regarding stream acidification is that episodic fluctuations in pH and toxic metals can have a devastating effect on stream biota. Fish which spawn in the upper reaches of cool, clear streams during the spring and fall, where there is little buffering provided by the gravel beds, are at particular risk.

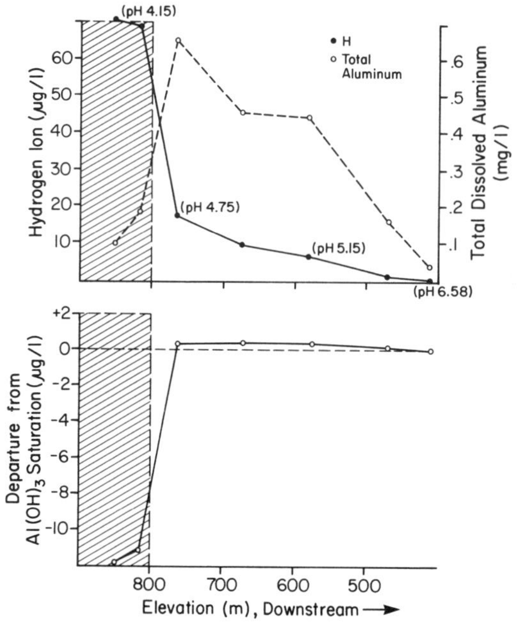

Figure 10. Dissolved aluminum and acidity of stream water draining the summit of Mount Cube, New Hampshire, April 4, 1980. Cross-hatched area represents metamorphic rocks. Aluminum saturation was based on pH, aluminum hydroxide complexes, and a theoretical solubility product of 10^{-9} for $Al(OH)_3$. Fluoride and organic complexes were not considered. (Reprinted with permission from *Geochim. Cosmochim. Acta,* **45,** N. M. Johnson, C. T. Driscoll, J. S. Eaton, and G. E. Likens, Copyright 1981, Pergamon Press, Ltd.).

3. Groundwater

There is good evidence that groundwater has also been affected by acidic deposition. This effect is of particular concern in areas where private wells provide the major source of drinking water. Shallow wells less than 10 m deep with sandy or gravelly soil are most susceptible to acidification *(40, 41)*.

One study compared the pH of groundwater in shallow wells from three soil types. The median pHs of the well water in sandy soil, moraine soil, and clay-rich soil were 5.10, 5.45, and 6.10, respectively. Many of the acidic wells contained high levels of metals such as aluminum, manganese, and iron from the soil, as well as copper, lead, and zinc from the pipes in the home. Although levels of zinc in "raw" water may be less than 0.002 mg/liter, in acidic tap water values as high as 4.9 mg/liter have been reported *(15)*. High aluminum concentrations *(40)* as well as copper concentrations up to 40 mg/liter have been found in groundwater in Sweden. The World Health Organization in 1971 *(41a)* recommended that the maximum permissible copper content in tap water not exceed 1.5 mg/liter.

Some of the problems associated with high acidity and metal levels include corrosion of pipes and metallic-tasting water. Some of the more serious side effects include diarrhea in children and excessive aluminum accumulation by people having kidney dialysis machine treatment. Although there have been no clinical studies on the adverse effects of increased aluminum consumption, several have found a relationship between elevated levels of aluminum in brain tissue of patients with Alzheimer's disease, which is a form of premature senility *(42)*. Clearly both the extent and rate of groundwater acidification in addition to the possible effects on people require more study.

B. Sediment Chemistry

Sediment profiles, which are laid down gradually with time, can reflect changes in the chemical composition of the overlying water. A number of studies have examined the effect of acidic deposition on lake water acidification, but relatively few have looked at the effect on sediment chemistry. In one study Dickson *(19)* showed that surface sediments were more acidic than deeper sediments in Lake Horsikan in Sweden (Figure 11). The increase in sediment acidity was attributed to contact with acidic lake water (pH 4.28–4.54) and was associated with elevated concentrations of metals such as zinc, cadmium, and lead in the surface sediments. This pattern of metal accu-

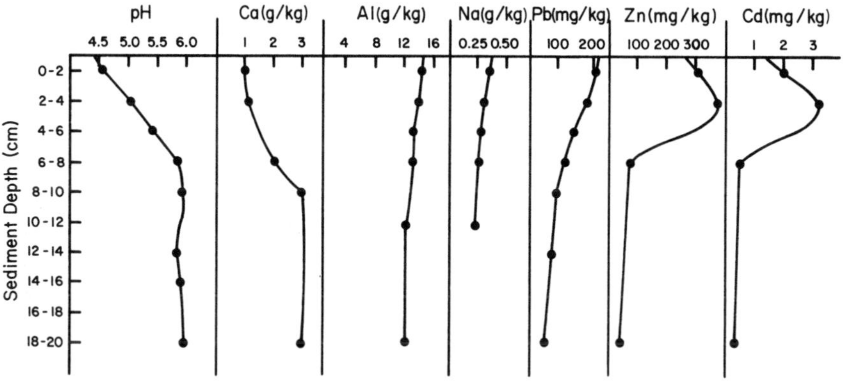

Figure 11. The pH and metals in the sediments of Lake Horsikan, Swedish West Coast, 1977–1979 *(19)*.

mulation is characteristic of areas exposed to aerial metal deposition *(43)*. Since sediments are better buffered than the overlying water, changes in sediment pH provide a conservative estimate of the onset of lake water acidification. Furthermore, temporary drops in the pH of surface water which may have a profound impact on the biota are unlikely to affect bottom sediments, except near the shore.

Although a wealth of information can be gleaned from sediments, some of the metal data must be approached with caution. Strongly reducing sediments, for example, can generate alkalinity and can also alter the concentration and speciation of metals such as iron and manganese at the sediment–water interface *(44)*. Altered redox potential, however, cannot explain the elevated levels of zinc, cadmium, and lead in the surface sediments of Lake Horsikan, Sweden.

C. Biota

1. Decomposers

Nutrients in a lake are derived partly from the drainage basin and partly from internal nutrient cycling initiated by decomposers. Some evidence suggests that the rate of decomposition decreases in acidified lakes, particularly once the pH drops below 5.0. Bacteria, which are the most common decomposers in circumneutral lakes, appear to be particularly sensitive to acidity, although certain groups, such as *Thiobacillus*, thrive in acidic environments. They are replaced by fungi, which on occasion have been observed to form dense fungal mats on the sediment surface *(45)*. Whether these mats are composed of fungal mycelium or filamentous algae remains a matter of debate *(46)*. Nevertheless, these mats in combination with coarse organic detritus, which accumulates on the sediment surface, further impede nutrient cycling from the sediments to the water column and contribute to the process of oligotrophication mentioned earlier.

Some of the earlier experiments conducted both in the laboratory under controlled conditions and in the field under more natural conditions indicated that oxygen consumption and microbial decomposition of peptone are reduced, the rate of phosphorus cycling (based on ^{32}P studies) becomes limited, and oxidation of ammonium is inhibited at low pH *(25)*. Some of the later studies provided conflicting data. Gahnstron *et al.* *(47)* found no significant differences in decomposition rates (based on oxygen production and glucose turnover) in microcosms, using sediments from acidic lakes (pH < 5.2) and circumneutral lakes (pH 6.5–6.8). Similarly, during the experimental acidification of Lake 223 at the Experimental Lakes Area (ELA) in northwestern Ontario, there was no evidence of a reduced rate of decomposition in the lake *(48)* or in microcosms *(49)*, although the pH of the lake was reduced only 1 pH unit from an average initial pH of 6.6.

At least part of this discrepancy may be explained by the populations of decomposers compared. For example, the bacteria within the water column of an acidic lake may be quite different from those found in a circumneutral lake. In the sediments, which have not yet acidified, the populations of bacteria may be the same *(50)*. Furthermore, one may expect an immediate effect following rapid acidification, which may not necessarily occur if the rate of acidification is gradual, as it was in the Experimental Lakes Area (ELA) study in Ontario. A third factor which may be important is the pH at which the experiment is performed. Whereas pH 5.0 appears to be critical for the inhibition of decomposition and changes in the populations of decomposers, above this pH there may be little or no effect.

2. Primary Producers

a. *ALGAE.* In lakes and streams, the three major groups of primary producers include algae, mosses, and macrophytes. The algae are either free-floating (planktonic), attached to other plants (epiphytic–periphytic) or to rocks (epilithic–lithophytic), or simply growing on the sediment surface (benthic). They vary in size from single-celled individuals less than 20 μm in diameter to filaments several centimeters long. Phytoplankton provide the major source of food for the filter-feeding crustaceans and rotifers; the periphyton are grazed by aquatic insects and snails that have specialized mouth parts for scraping the surface of rocks and plants.

In circumneutral lakes the algal flora may consist of 40 or more species at any one time (Figure 12). Below pH 6, there are seldom more than 20 species and below pH 5 seldom more than 10 species of phytoplankton. Among the different taxa, blue-greens (Cyanophyta) are the most acid-sensitive. They are rare below pH 5 and absent below pH 4 *(51)*. In only one study were blue-green algae reported to increase with decreasing pH *(52)*. Blue-green algae, like bacteria, are prokaryotic and lack a protective nuclear membrane. Likewise their chlorophyll is dispersed throughout the cell and is not enclosed in a membrane as in all other eukaryotic algae. This may be one reason for their apparent acid sensitivity.

Diatoms have both acid-sensitive and acid-tolerant species. These different groups have been classified as *acidobiontic* (optimum below pH 5.5), *acidophilous* (below pH 7), *indifferent* (around pH 7), and *alkaliphilous* (mainly above pH 7). Since the silica frustule, or "shell," is relatively resistant to decay, and since species can be identified by the shape, size, and markings on their frustules, accumulation of dead diatom shells provides a historical record of acidification. With a knowledge of the pH limits of individual species and groups, it is possible to reconstruct, with some degree of certainty, the changes in lake water pH. These diatom profiles have been constructed for a

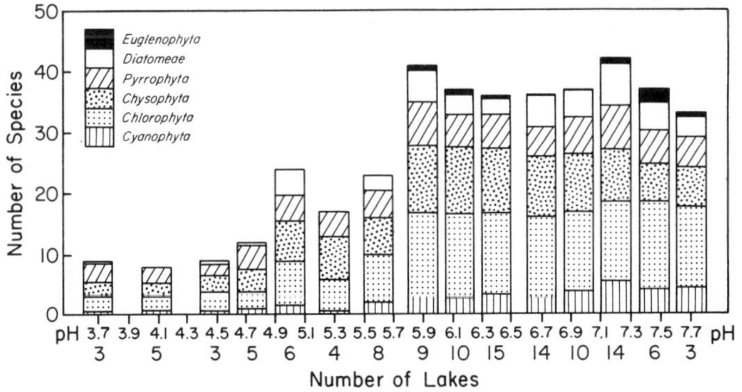

Figure 12. Number of phytoplankton species in lakes with different pH values. A striking decrease is found at below pH 5.8. (From B. Almer, W. Dickson, C. Ekström, E. Hornstrom, and U. Miller, *Ambio* **3,** 303–336, 1974.).

number of recently acidified lakes and indicate a shift from acid-sensitive to acid-loving species *(20)*.

The decrease in species diversity and the shift from acid-sensitive to acid-tolerant species have also been reported for other taxa, including the Chlorophyta (green algae), Chrysophyta (golden-brown algae), and Pyrrophyta (dinoflagelates) (Figure 12). Among the dinoflagelates, *Peridinium inconspicum* and *Gymnodinium* species appear to be acid-tolerant and are commonly found in acidic lakes. Euglenophyta are rare in oligotrophic lakes, being limited as much by the lack of nutrients as by any other chemical parameter. Interestingly, a few species, such as *Euglena mutabilis*, are uncommon in these waters but become abundant in extremely acidic (<pH 4), metal-contaminated environments *(53)*.

Although there is a definite decrease in species diversity, there appears to be no significant change in community biomass. The few species that are able to tolerate the low pH constitute a larger percentage of the total biomass in acid-stressed lakes than in similar unstressed lakes *(26, 54)*. As a consequence, changes in algal biomass or in productivity may be undetected as a lake acidifies and are not considered good early-warning indicators of acidification *(55)*.

Although the community biomass may not change during the early stages of acidification, there is a definite shift from planktonic to attached and benthic forms of algae. Species such as *Mougeotia, Batrachospermum, Lyngbya, Oscillatoria,* and *Pseudoanabaena,* which are not normally abundant in circumneutral lakes, can form dense colonies in acidified lakes and streams. Experimental acidification of streams both in Norway (Tovdal River) and in North America (Norris Brook, New Hampshire) resulted in an increase in attached algae *(56)*. Similarly, during the experimental acidification of a lake (Lake 223, Canada), growth of filamentous algae was noted. Why the shift from a planktonic-dominated community to one dominated by benthos occurs is poorly understood. It may relate to reduced grazing pressure by invertebrates, increased light penetration, or nutrient limitations in the water column.

b. MOSSES AND MACROPHYTES. Aquatic mosses and macrophytes include both vascular and nonvascular plants, submerged and emergent plants, those rooted into the sediment, as well as those which are free-floating on the water surface. They are found predominantly in the littoral zone along the edge of lakes, where the water is shallow and where there is enough light for photosynthesis. In clear lakes macrophytes may also be seen several meters below the surface of the water. Rooted macrophytes stabilize the sediments and reduce soil erosion by wind and wave. They also provide shelter and support for invertebrates, a substrate for algae, and protection for fish.

In Norway and Sweden acidification of lakes has been accompanied by a shift from communities dominated by *Isoetes*, *Litorella*, and *Lobelia* to ones dominated by *Sphagnum* and *Utricularia*. The presence of *Sphagnum* in lakes is not highly unusual; what is unusual is its rapid rate of encroachment. In Lake Orvattnet, Sweden, the area covered by *Sphagnum* increased from 8% in 1967 to 63% by 1974 in shallow areas. During the same period the pH of the water decreased from 5.6 to 4.8 *(6)*. The success of *Sphagnum* has been attributed to its ability to use carbon dioxide from solution, the predominant form of inorganic carbon in acidic lakes. For photosynthesis, most other species use bicarbonate, which is converted into carbon dioxide during acidification.

Sphagnum, unlike the other aquatic plants, provides an unfavorable habitat for aquatic organisms. It has a strong cation exchange capacity and readily removes divalent cations such as calcium and magnesium from the surrounding water. Availability of these ions is already low and decreases further when *Sphagnum* is present. *Sphagnum* releases hydrogen ions in exchange for calcium and magnesium and therefore promotes the acidification of its own environment. Estimates indicate that as much as 1.3 Equ H^+ may be released per square meter of *Sphagnum (46)*.

This invasion by *Sphagnum* has been reported in Scandinavia but has not been found to any great extent in North America. The only possible exception is a study of Lake Colden in the Adirondacks by Hendrey and Vertucci *(46)*. The flora of Lake Colden was first described in 1932, at which time the pH of the lake water was assumed to be above 5.4. This lake was resurveyed in 1979, when the lake water pH varied between 4.8 and 5.1. Several changes in species composition and abundance had occurred during the intervening 47 years. *Eleocharis acicularis*, *Nuphar leuteum*, and *Lobelia dortmanna* had decreased; *Eriocaulon septangular* and *Utricularia* had increased in abundance. *Sphagnum pylaesii*, which was not mentioned in the early survey, was abundant in 1979.

There have been few studies on the specific effects of low pH on aquatic macrophytes. In one study, growth of *Lobelia dortmanna* was retarded and the flowering delayed at pH 4.0 *(6)*. Unfortunately, this pH is about 10 times more acidic than that of many of the lakes in which changes in aquatic plants have already been described. The concentrations at which hydrogen ions become harmful to aquatic plants have not been established. Nor do we know whether these plants are affected by hydrogen ions directly or indirectly, by the effect of hydrogen ions on nutrient availability and metal toxicity. This is one area that needs more study.

The effect of aluminum, which is mobilized at low pH, on aquatic plants is also poorly understood. Aluminum is extremely toxic to land plants growing in acidic soils and has been known to interfere with the uptake of calcium and phosphorus *(57)*. Similar studies are needed for aquatic plants.

3. Invertebrates

The invertebrate fauna in lakes and streams consists of a diverse assemblage of organisms, among which are the protozoa, rotifers, crustaceans, and insect larvae, as well as those animals which spend only a portion of their life-cycle in water, such as salamanders and frogs. These organisms range in size from fractions of a millimeter to several centimeters and can be found free-floating in the water column (zooplankton) or crawling among the rocks and weeds near the sediment surface (benthos). Their eating preferences are highly variable. Some feed on algae (herbivors), others on bacteria and decomposing material (detritivors); a third group are predacious on other invertebrates (carnivors). These invertebrates in turn may be consumed by fish predators or by predatory birds and thus provide a key link in the energy flow within the environment.

a. *Planktonic Species.* The effects of acidification on the distribution of lake-dwelling and stream-dwelling invertebrates have been studied in Norway *(58–61)*, Sweden *(26)*, England *(62)*, Canada *(63–65)*, and the United States *(66, 67)*. These studies indicate that a decrease in species diversity, a decrease in community biomass, a shift from planktonic to benthic forms during lake acidification, and greater invertebrate drift during stream acidification occur. Many of these effects begin to appear between pH 5 and 6. In Norway, for example, circumneutral clearwater lakes have an average of 16 species of zooplankton, whereas clearwater lakes below pH 5 have only 7 *(17)*. Figure 13 *(68)* shows that there are fewer species of rotifers, cladocera, and copedpoda in acid clearwater lakes than in circumneutral lakes. Among these taxa, cladocera show the greatest reduction in species diversity.

The distribution of some rotifers (*Keratella cochlearis* and *Kellicottia longispina*) appears to be fairly independent of pH; others are common only at high pH *(Polyarthra vulgaris)* or at low pH *(Polyarthra rematia, Keratella serrulata) (26, 60)*. Although rotifers

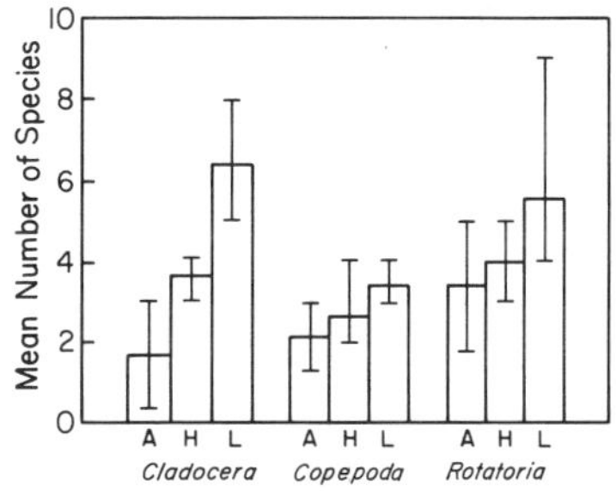

Figure 13. Mean number of zooplankton species in acid lakes (A), humic acid lakes (H), and less acid lakes (L) in western Norway (From A. Hobaekand G. G. Raddum, "Zooplankton Communities in Acidified Regions of Southern Norway," SNSF-Proj. IR75/80. NISK. 1432 Aas-NHL, Norway, 1980. Used with permission.).

are generally rare in acidified lakes, several species are acid-tolerant. These include *Lecane luna* (which has been found at pH 4.62), *Asplanchna priodonta* (pH 4.55), *Notholca squanula* (pH 4.45), *Polyarthra dolichoptera* (pH 4.45), *P. euryptra* (pH 4.1), *P. vulgaris* (pH 4.1), *Synchaeta pectinata* (pH 4.1), *Keratella quadrata* (pH 4.1), *K. taurocephala* (pH 4.05), *K. hiemalis* (pH 4.05), *Monostyla lunaris* (pH 4.05), *Kellicottia bostoniensis* (pH 4.05), *Trichocerca cylindrica* (pH 4.05), *T. multicrinis* (pH 3.96), *Keratella cochlearis* (pH 3.98), *Kellicottia longispina* (pH 3.98), *Lecane closterocerca* (pH 2.8), and *Brachionus urceolaris* (pH 2.8) *(64, 69, 70)*. The environments in which the latter two species are found are more characteristic of acid mine drainage ponds than lakes acidified by acid precipitation.

In North America the typical planktonic crustaceans found in acidified lakes include *Bosmina longispina, Diaptomus minutus,* and *Holopedium gibberum.* In several lakes the diversity has been reduced to only one species: *D. minutus (63).* Although *Daphnia* are considered to be acid-sensitive, *Daphnia ambigua, D. catawba, D. galeata mendotae, D. retrocurva, D. pulicaria,* and *D. longiremis* have all been found at pHs below 5.0 *(71).* Laboratory experiments have shown that *D. catawba* and *Holopedium gibberum* are considerably more acid-tolerant than *D. magna,* which exists only in alkaline ponds and lakes *(71).*

Experiments have shown that sodium and chloride regulation is impaired at low pH *(70, 72, 72a).* Sensitive individuals are unable to control their salt balance, have problems with osmoregulation, and quickly die. Elevated aluminum concentrations can also be toxic to these species in acidic lakes. Maximum aluminum toxicity seems to occur around pH 5.5 in these soft-water lakes at 0.3 mg Al/liter. Higher levels of calcium (>3 mg/liter) can provide some protection against both hydrogen ion and aluminum toxicity. Humic substances can completely mask the toxic effects of aluminum, presumably as a result of competitive binding *(24)*; brown water lakes and ponds are often productive despite their low pH, possibly as a result of such masking.

b. Benthic Species. The distribution and biomass of benthic invertebrates are also affected at low pH. Snails are absent below pH 5.2 and are rare at pH 5.2–6.6 in Norwegian lakes *(58)*. Freshwater mussels are also uncommon in acidifed lakes *(59)*. Molluscs have a high calcium requirement and cannot survive in softwater lakes low in dissolved calcium (Figure 14); they need calcium for their shells. In one study *(73)* two species of adult mussels were transplanted from a neutral lake to an acid lake (pH 4.4) that was devoid of mussels. At the end of the 46-day experiment, there was evidence of shell corrosion. The authors concluded that although adult mussels may be able to survive at pH 4.4, young mussels may have difficulty secreting material for their shells.

Crayfish are increasingly rare in acidified lakes *(26)*. Their last refuge seems to be at the sediment surface, where the pH is generally higher than in the water column *(74)*. Experiments have shown that crayfish are particularly acid-sensitive immediately after molt and have problems with calcium uptake below pH 5.75 *(75)*.

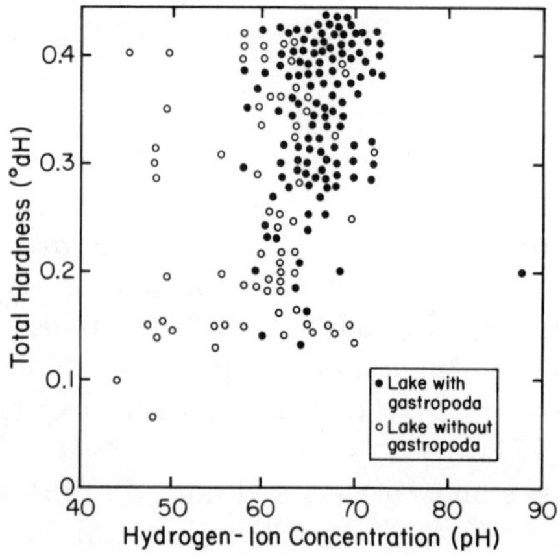

Figure 14. The pH and total hardness in low-calcium lakes in southeastern Norway, showing presence or absence of snails. °dH = European scale for total hardness 1 °dH = 10 mg CaO/liter ≃ 2.86 mg Ca/liter *(58)*.

Aquatic insects may be either acid-sensitive or acid-tolerant. Mayfly larvae (Ephemeroptera) and stonefly larvae (Plecoptera) are more acid-sensitive than either caddisflies (Trichoptera) or true flies (Diptera) and disappear as the pH decreases in both lakes and streams. This phenomenon has been demonstrated in both laboratory studies *(76)* and stream acidification experiments *(77)*. The 30-d TL_{50} for *Ephemerella subvaria*, a mayfly, occurred at pH 5.38; of the nine species tested, it was the most acid-sensitive. The 50% successful emergence occurred at pH 5.9, and only *Isogenus fronalis*, a stonefly, was more acid-sensitive. In contrast, the caddisflies *Hydropsyche betteni* and *Brachycentrus americanus* were both extremely acid-tolerant and could survive (30-d TL_{50}) at a pH as low as 3.38 and 2.45, respectively *(76)*. In the field, chironomids, chaoborus, corixids, and dytisid beetles appear to be particularly acid-tolerant *(78, 79)*. Dense populations of *Chironomus riparius* have been found at as low as pH 2.8 in tundra ponds at the Smoking Hills *(53)*.

In lakes, insects can avoid the acid stress by burrowing into the surface sediments or staying near the sediment–water interface, where the pH is generally higher than in the water column. In streams, insects drift after an acid pulse and thus avoid the acidity by being washed into the lower reaches of a stream, where the water is usually more alkaline *(37, 38)*. Despite the fact that some insects can flourish at low pH, biomass of even this group is reduced in acidified lakes.

Although the general pattern shows a decrease in the number of invertebrate species in acidic lakes, it is not clear to what degree this reduction is due to direct hydrogen ion toxicity, metal toxicity, low levels of calcium, or complex biotic interactions (i.e., shifts in predation, competition, parasitism, and so on). Nor is it known to what degree shifts in the invertebrates affect higher and lower trophic levels.

4. Fish

The concern about the effects of acidification of fresh waters in the late 1970s was due largely to concern about the loss of fish populations. Fish were rapidly disappearing from clear, pristine-looking lakes which were many miles away from industrial sources of pollution, both in Scandinavia and in North America.

a. *FISH SURVEYS: FIELD STUDIES.* Some of the earliest reports on the effects of acidification on fish population came from Norway. Annual catch of Atlantic salmon (*Salmo salar*) decreased rapidly in the early 1900s from seven rivers in the southernmost part of the country in comparison to rivers in the rest of the country (Figure 15). Although this decline was due primarily to recruitment failure, massive fish kills were also reported in 1911 and have continued sporadically until the present. In all cases, these fish kills occurred after heavy autumn showers or after early June snow melt, which we now know can be extremely acidic *(6)*.

During one fish-kill episode in March 1975, thousands of dead brown trout (*Salmo trutta*) were found along the Tovdal River. There were no obvious signs of fish disease. Laboratory experiments later showed that the pH of the water (pH 5.0) was toxic enough to kill adult fish in several days. This low pH was caused by melt of the snow pack with its accumulation of acidic radicals.

Surveys of water quality and fish populations indicate that trout have become extinct in thousands of lakes in Norway and that the remaining fish populations are affected by low pH in 20% of the region south of 63°N *(6)*.

Along the western coast of Sweden many of the lakes are now too acidic to support viable fisheries. It has been estimated that 50% of the lakes in this area and approximately 10,000 lakes in all of Sweden have pH values below 6. Arctic char (*Salvelinus alpinus*) are particularly sensitive and are now extinct in many of the acidified lakes.

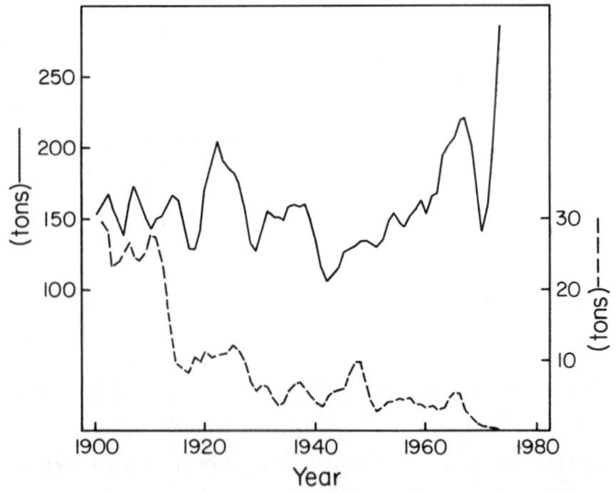

Figure 15. Salmon fisheries statistics. Yearly yield for Atlantic salmon (*Salmo salar* L.) fisheries in seven rivers from the southernmost part of Norway (bottom curve) compared with 68 rivers from the rest of the country (top curve). Catch data are based on official statistics and are for fish caught in the river during spawning migration *(6)*. The seven rivers are Audna, Kvina, Lygna, Mandelsena, Nidelva, Sogna, and Tovdalselva.

Populations of roach (*Leuciscus rutilus*), minnow (*Phoxinus phoxinus*), trout, cisco, perch (*Perca fluviatilis*), and pike (*Esox lucinus*) have also been affected *(26)*. In contrast, the eel (*Anguilla vulgaris*) appears to be the least affected by acidification, perhaps because it spawns in the sea.

Fish kills have been reported in several Swedish lakes in the province of Varmland. In one study *(80)* high levels of aluminum were found on the gills of the dead fish, and it was believed that aluminum leached from the watershed contributed to the mortality of these animals.

In the United States some of the oldest data on the effects of acidity on fish populations come from the Adirondack Mountains of New York state. A survey of several hundred high-elevation lakes conducted between 1929 and 1932 and repeated in 1975 indicated that whereas only 4% of the lakes had pH values below 5 and were devoid of fish in the 1930s, by 1975 over 50% had pH values less than 5 and 90% of these lakes were devoid of fish life *(28)*.

The loss of fish populations from these lakes was attributed not only to low pH but to elevated levels of aluminum. In one study of Little Moose Lake, levels of aluminum increased from a prethaw level of less than 20 μg/liter to a high of 630 μg/liter during the spring thaw *(81)*. In an acid brook flowing into this lake, concentrations of aluminum varied from 100 to 1000 μg/liter from March until August 1977. Laboratory studies have shown that although aluminum toxicity is strongly influenced by pH and the presence of complexing ligands, concentrations as low as 100 μg/liter can be toxic to white sucker fry (*Catostomus commersoni*) *(35)*.

In Canada, the loss of fish populations as a consequence of acidification has been documented in Ontario and Nova Scotia *(82)*. In the Clyde River, Nova Scotia, the annual catch of Atlantic salmon has decreased sharply since the 1940s and is reminiscent of the decline in Norway which occurred at the turn of the century. In the La Cloche region of Ontario, among the important sports fish there has been a steady disappearance of lake trout (*Salvelinus fontinalis*), smallmouth bass (*Micropterus dolomieui*), largemouth bass (*Micropterus salmoides*), and walleye (*Stizostedion vitreum*). Near Sudbury, Ontario, 25% of the lakes with pH less than 5.5 have lost brooktrout (*Salvelinus fontinalis*) and lake trout populations *(83)*. In Lumsden Lake, Ontario, the loss of fish populations was particularly rapid. Yellow perch (*Perca flavescens*) and burbot (*Lota lota*) were last reported in 1960. Slimy sculpin (*Cottus cognatus*) and lake trout were last captured in 1967. In 1968 white suckers suddenly became rare and, by 1969, along with trout, perch, and lake herring, had disappeared completely. In 1970 lake chub was added to the list of disappearing species *(21)*. Therefore, within 10 years, eight species had disappeared from Lumsden Lake.

b. Physiological Responses: Laboratory Studies. In addition to fish surveys, experiments in both the field and laboratory, as well as detailed physiological studies, have contributed greatly to our understanding of the mechanisms involved in the extinction of fish populations and to the identification of acid-sensitive life-cycle or seasonal stages. The loss of fish populations from recently acidified lakes has been attributed primarily to reproductive failure, which is characterized by missing age classes (Figure 16) *(84)* and a shift to large older individuals. Since reproduction is a sensitive stage within the life cycle, species such as white sucker, which spawn in the spring and deposit their eggs on the gravel beds in the shallow streams, are most likely to be affected adversely by acidification.

The physiological and biochemical effects of hydrogen ions and aluminum on fish have been intensively studied *(35, 85–87)*. One of the most sensitive stages to low pH is reproduction. Problems with reproduction seem to relate to problems with calcium regulation. Lockhart and Lutz *(88)* found that reproductively mature female white suckers in an acidic lake had lower levels of serum calcium than

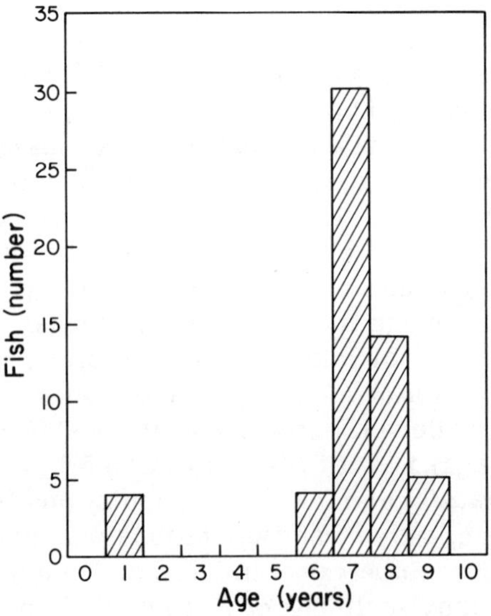

Figure 16. Age composition of a sample of yellow perch taken from Patten Lake, Ontario *(84)*.

those in a nearby circumneutral lake. The females with the lower levels of calcium produced fewer eggs or resorbed their eggs.

Laboratory experiments with the fathead minnow (*Pimephales promelas*), which is extremely acid-sensitive, have shown similar results *(89)*. The adults could survive even at pH 4.5, but egg production per female was significantly reduced at pH 6.6 and completely inhibited at pH 5.2 compared with the control at pH 7.5.

In addition to the effects on reproduction, problems with calcium regulation have also been linked to deformed spines in white suckers *(21)*, to smaller heads in fathead minnows *(89)*, and to bands devoid of calcium in the otoliths, or auditory organs, of fish *(90)*.

At lower pHs (≤pH 5), fish are unable to maintain high internal concentrations of sodium and chloride. These two ions are maintained at a constant level irrespective of external concentrations and are essential not only for water balance, but for the activation of essential enzymes and conduction of nerve impulses. In acidic water, particu-

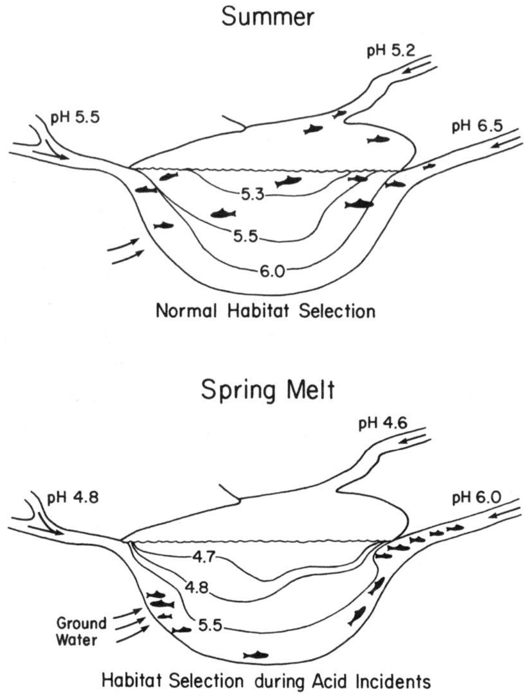

Figure 17. Habitat selection in freshwater fish related to spatial and temporal gradients in water chemistry *(91a)*.

larly if aluminum is present, fish are unable to regulate their tissue chloride levels *(91)*. Some of the distress symptoms associated with either hydrogen ion or aluminum stress include lack of appetite, listlessness, loss of fright reaction, loss of negative phototaxis, loss of equilibrium, darkening coloration, gill necrosis, irregular operculum movement, mucus on gills, and coughing. Once the fish begin to exhibit these symptoms, if they are not allowed to avoid the stress, they die within a few hours or days. In acidified lakes fish avoid areas of low pH (Figure 17) *(91a)*.

At even lower pH values (<pH 4) death has been attributed to suffocation and acidification of the body fluids. At low pH, hydrogen ions penetrate the body fluids. If the rate of penetration is slow, the hydrogen ions can be buffered. However, once the buffering capacity of the body fluids is exhausted, or when the rate of penetration is more rapid than the rate of buffering, the body fluids acidify. Even a small decrease in serum pH can be lethal *(86)*. The ability of the blood to transport oxygen to the cells is reduced. At the same time, mucus accumulates on the gill membrane, further impeding oxygen uptake, and the animal suffocates to death.

III. Remedial Measures

It is apparent that acidic deposition has damaged poorly buffered aquatic systems in both Europe and North America. Unless steps are taken to protect the environment, many more lakes and many more species will be adversely affected. But before we rush into a remedial program we must know the ecological and economic costs. We must also acknowledge the difference between short-term "stop-gap" measures which may be more economical in the short run and long-term solutions which may be more economical in the long-run. Finally, the solution should be individualized for a particular region. Reducing sulfur dioxide emissions may be effective in Europe, where sulfur dioxide is the major acidifying agent but may be less effective in North America, where nitrogen oxide emissions contribute substantially to precipitation acidity.

There are basically three methods by which we can protect the environment. These include the following:

1. Reducing acidic emissions
2. Buffering lakes to decrease their susceptibility to acidification
3. Breeding acid-tolerant and metal-tolerant species

A. Reducing Acidic Emissions

Irrespective of the program used in a particular region, at least part of the solution will have to include reducing acidic emissions *(17, 92)*. The extent to which acidic emissions must be reduced is still highly speculative. Estimates by Overrein and coworkers *(17)* for southern Norway based on the Henrikson titration curve (Figure 4) *(32)* suggest a 60–70% reduction in hydrogen and sulfate (no distinction made) concentrations to give a yearly weighted average precipitation pH of 4.7.

Dickson *(92)* based his acceptable acid load on soil neutralizing capacity. He divided Sweden into regions based on soil type. In the most sensitive regions, the base production potential of soils and bedrock is 0.1–0.2 keq H^+/ha/year, corresponding to 0.17–0.33 g S/m^2/year (wet plus dry deposition). In the least-acid-sensitive regions, where there is a higher weathering potential, lakes acidify if the sulfur in wet deposition exceeds 1.0 g S/m^2/year. Dickson suggests that lakes should remain healthy if the acid load is less than 0.5 g S/m^2/year (0.3 keq H^+/ha/year). European levels are currently 10 times higher than this value.

A target loading of 20 kg SO_4/ha/year of wet deposition was estimated to be the threshold level to protect most sensitive areas in North America in the "U.S.–Canada Memorandum of Intent on Transboundary Air Pollution" *(93)*.

Reducing sulfur emissions by 50% would cost approximately $780 per ton of sulfur in terms of the 1980 dollar for northwestern and southern Europe (OECD's calculations) *(94)*. Applied to the whole of Europe, excluding the Soviet Union, the cost of a 50% reduction in sulfur dioxide would be $8 billion per year. This would increase the cost of energy by 3%.

The Office of Technology Assessment made similar calculations for the United States *(95)*. A reduction of sulfur dioxide emissions by 10 million tons/year would cost $3–4.5 billion (1980 dollar) and would increase electricity rates by 2–3% across the board and by 10–15% for some states. The accuracy of these estimates and the validity of comparing the European costs directly with the American costs are difficult to determine. The bottom line, however, is whether a reduction in acidic emissions will prevent further deterioration of freshwater systems and allow acidified lakes to recover. In other words, is acidification reversable?

There are only two studies which indicate that a reduction in sulfur dioxide emissions will indeed result in lake recovery. In one study

(96), a small oligotrophic lake (Baby Lake) near a copper–nickel smelter in the town of Coniston, Ontario, was sampled yearly from 1968 until 1983. In 1972, when the nearby smelter closed, local air quality improved dramatically. While the smelter was operating, the pH of Baby Lake varied from 4.0 to 4.2 with one value as high as 4.4 in 1970 (Figure 18). After the smelter closed in 1972, the pH increased steadily to 5.4 by 1983. At the same time concentrations of sulfate, copper, and nickel—which were emitted by the smelter—decreased in the water (Figure 18). Even concentrations of the nonsmelted metals iron, manganese, and zinc decreased, presumably as a result of reduced metal solubility at higher pHs *(96)*. Calcium concentrations did not change with time. Within a few years this lake should be able to support fish populations. Currently it is devoid of fish.

In another study *(97)* near Sudbury, Ontario, Clearwater Lake was sampled from 1973 until 1983. Prior to 1978, emission of sulfur dioxide from Sudbury were in excess of 1.3 million metric tons per year. From 1978 until 1982 these emissions decreased substantially (0.39–0.94 million metric tons/year). After 1978 hydrogen ion and sulfate concentrations in Clearwater Lake began to decrease. By 1983 the sulphate concentration in the lake was 380 μeq/liter compared to greater than 500 μEq/liter prior to 1978 (Table I). The pH increased from less than 4.33. to 4.6 during the same period.

These two studies indicate that lakes can recover quite rapidly after reduced emissions of sulfur dioxide. The period the biota will require in order to respond is not yet known.

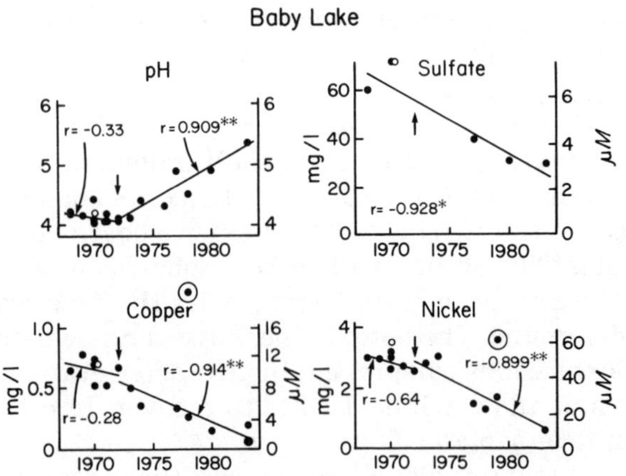

Figure 18. Changes in the chemical composition of Baby Lake, Ontario, from 1968 until 1983. In 1972 the nearby copper–nickel smelter at Coniston closed permanently (arrow) and dramatically improved local air quality *(96)*.

Table I Emissions of Sulfur (10^6 mt/year) at Sudbury, Ontario, Average pH, H^+ and Sulfate Concentrations ($\bar{x} \pm$ s.e.; μeq/liter) in Clearwater Lake, as Well as the Number of Samples Collected per Year (*n*) *(97)*.

Year	SO_2	*n*	$[H^+]$	*pH*	$[SO_4^{2-}]$
1973	1.46	16	51.3 ± 3.2	4.29	579 ± 7
1974	1.47	23	57.5 ± 2.4	4.24	533 ± 9
1975	1.39	13	46.8 ± 1.6	4.33	519 ± 14
1976	1.41	20	58.9 ± 2.5	4.23	542 ± 7
1977	1.34	11	79.4 ± 1.8	4.10	550 ± 7
1978	0.68	12	39.8 ± 1.8	4.40	498 ± 4
1979	0.50	6	38.9 ± 4.5	4.41	458 ± 4
1980	0.94	12	33.1 ± 2.4	4.48	457 ± 5
1981	0.84	7	33.1 ± 2.9	4.48	408 ± 6
1982	0.39	6	33.1 ± 1.6	4.48	396 ± 3
1983	—[a]	6	25.1 ± 2.2	4.60	379 ± 5

[a] Data currently unavailable.

B. Chemical Alterations

Shortly after acidic deposition was recognized as affecting poorly buffered soft-water lakes, liming was introduced as a solution to the problem. Unfortunately, addition of limestone to lakes or to their drainage basins, at best, is a temporary measure and, at worst, can be detrimental to fish and other forms of aquatic life.

Major liming programs were initiated in Sweden *(15, 98)*, in Canada *(99)*, and in the United States *(100)*. Unfortunately in some of these programs, limestone was added to remote lakes, but follow-up studies on the effects on water quality and on the biota were not conducted. Therefore, despite the fact that thousands of lakes have been limed, conclusive data are available for only a few.

Several different products, including agricultural limestone ($CaCO_3$), dolomitic limestone ($CaCO_3 \cdot MgCO_3$), quicklime (CaO), and calcium hydroxide [$Ca(OH)_2$], have been tested to raise the pH of lakes. These compounds vary in their relative solubility, cost, and ability to impart acid-neutralizing capacity to lakes. Calcium hydroxide is twice as expensive as agricultural limestone, but only one-fifth the quantity is needed to neutralize a lake. On the basis of these criteria, calcium carbonate alone or in combination with calcium hydroxide seems to be the preferred neutralizing agent.

The amount of limestone added to lakes depends on a number of factors, including lake water pH, lake volume, and flushing time. Treatments of 1–2 tons per acre have been recommended *(100)*.

The cost of these liming programs is highly variable. Lakes having no road access are more expensive to lime—the process requires helicopter time. Blake *(100)* estimated an average cost of \$40/ha or \$100/acre for pond liming. The cost of neutralizing Bowland Lake, a small head-water lake in Ontario, was \$30,000; the cost for Trout lake (four times larger) was \$82,000 *(101)*. It is expected that these two lakes may need to be neutralized again in 4–5 years.

Not all lakes need to be limed every 4–5 years. Studies in Ontario *(16, 102)* showed that Middle Lake, which was neutralized in 1973 with $Ca(OH)_2$ and $CaCO_3$ to pH 7.01 still had a biologically safe pH (6.54) by 1979 (Figure 19). Only phosphorus was added during the intervening years. In contrast, Lohi Lake was neutralized in 1973, 1974, and 1975. The pH dropped from 6.1 in 1975 to 5.27 by 1977, and to 4.79 by 1978 *(99)*.

Considering the number of lakes affected by acidic precipitation and the temporary effect of liming, it is obvious that whole-lake liming is not the solution. Liming selected streams or small bays within a larger lake ecosystem may be a less expensive alternative to liming the entire lake, although more studies are needed to evaluate the biological effectiveness of such a scheme.

Shortly after limestone is added to a lake, the pH, the acid-neutralizing capacity, and calcium concentrations increase, and metals de-

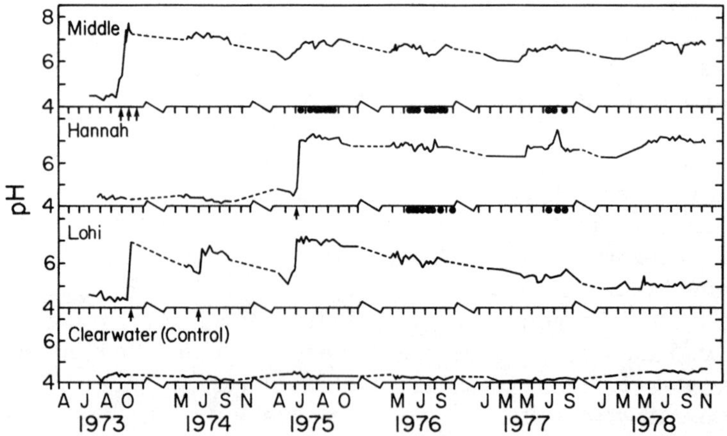

Figure 19. The pH values of Middle, Hannah, Lohi, and Clearwater lakes, Ontario. Treatments with neutralizing agents [$CaCO_3$, $Ca(OH)_2$] are denoted by arrow, additions of phosphorus by solid circle (From P. J. Dillon, N. D. Yan, W. A. Scheider, and N. Conroy, *Arch. Hydrobiol. Beih. Ergebn. Limnol.* **13,** 317–336, 1979. Used with permission.).

Table II Chemical Changes Associated with Liming in Lake Nelson, Canada *(99)*

Lime addition	*Before*	*After*	*% Difference*
Alkalinity (mg/liter)	2.5	3.8	+49
pH	5.7	6.4	−80
Al (μg/liter)	81	13	−84
Mn (μg/liter)	63	18	−71
Fe (μg/liter)	61	19	−69
Ca (mg/liter)	4.06	5.41	+33

crease. This effect is clearly demonstrated in Nelson Lake, Ontario (Table II).

The beneficial effects of liming on biota are not clear. Liming of Lohi Lake showed only a marginal increase in the number of algal genera (Table III). Total biomass of the algae decreased after liming. The

Table III Response of Plankton to Neutralization of Lohi Lake, Ontario *(16)*.

Parameter		*1973*	*1974*	*1975*	*1976*
Phytoplankton data					
Number of genera per collection					
Cyanophyta		0	1.3	1.0	0.5
Dinophyceae		1.0	0.5	1.1	1.1
Cryptophyceae		1.6	0.6	0.7	0.8
Chrysophyceae		1.1	2.5	3.2	3.2
Chlorophyta		3.1	2.9	4.0	2.7
Bacillariophyceae		0.2	1.0	0.8	0.8
	Total	7.0	8.8	10.8	9.1
Biomass (mg wet weight/liter)	Total	0.65	0.32	0.25	0.24
Percentage Chrysophyceae		1	63	37	43
Percentage Dinophyceae		32	11	42	35
Percentage Cryptophyceae		24	3	3	15
Percentage Chlorophyta		42	15	16	5
Percentage Others		1	8	2	2
Zooplankton data					
Number of species per collection		3.6	3.1	3.0	3.2
Biomass (mg dry weight/square meter)		24.0	1.3	0.7	8.1
Percentage Cladocera		81	77	45	79
Percentage Copepoda		19	23	55	21
Percentage Dominants[a]		B. long	B. long	C. vern and B. long	B. long

[a] B. long, *Bosmina longirostris;* C. vern, *Cyclops vernalis.*

relative biomass of Chrysophyceae increased from 1% to 63% after lime additions, whereas Dinophyceae, Cryptophyceae, and Chorophyta decreased dramatically.

Similarly for the zooplankton, there was no significant change in species diversity, and biomass actually decreased from 24 to 1.3 mg dry weight/m^3 as a consequence of liming. In other words, the population crashed: a most undesirable effect.

Introduction of rainbow trout in submerged enclosures into both Lohi Lake and Middle Lake resulted in rapid mortality of the fish despite the addition of base *(99)*. It was believed that copper levels were still too high to allow for a successful fish-stocking program.

The liming program in Sweden *(15)* was considerably more successful than that described in Ontario, Canada. Phytoplankton diversity increased dramatically from less than 20 to greater than 50 species when the pH was raised from 4.7 to 7.5. Summer populations of planktonic crustaceans were more abundant after liming. Crayfish stocking programs and trout stocking programs were very successful. In a non-copper-contaminated Swedish lake liming actually allowed reestablishment of reproductive populations of perch (Figure 20).

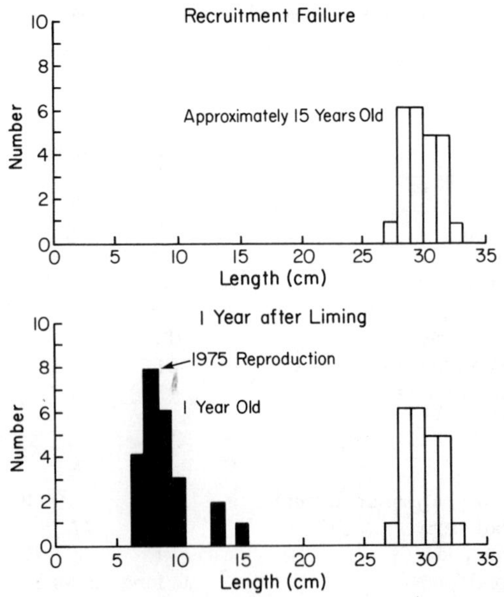

Figure 20. Liming of Lake St. Skarsjon, Sweden, in 1975 reestablished the perch population *(91a)*.

If levels of aluminum are elevated in the water, as they sometimes are in acidified lakes, liming may actually contribute to the death of the resident fish population by increasing aluminum toxicity *(39)*. Fish-stocking programs are likely to fail if fish are introduced immediately after liming. Sufficient time for the aluminum to precipitate out of the water column into the sediments must be allowed before fish are introduced.

C. Fish Stocking and Breeding Programs

Since both recreational and commercial fishing are important industries, attempts have been made to restock selected acidified lakes as well as neutralized or limed lakes with sports fish. Some species of fish may be able to survive for one or more seasons, although, if the lake is too acidic, they may not be able to reproduce successfully. Stocking programs have had variable success in the past. Despite safe pH values, elevated levels of metals often resulted in poor recapture statistics.

There are at least two methods by which survival of fish in acidic lakes can be prolonged. One involves short-term acclimation to both low pH and elevated aluminum; the other involves selection of genetically tolerant strains. Studies by Guthrie *(103)* showed that brook trout exposed to gradually increasing levels of aluminum and hydrogen ions became acclimated and were able to survive longer than nonacclimated individuals (Figure 21). Within 21 days at pH 5.0 and 0.4 mg Al/liter, all of the nonacclimated brook trout died, whereas 50% of the acclimated trout were still alive.

There is now some evidence that acid tolerance may be heritable. The eggs and fry of wild brown trout populations from acidic environments have been reported to be more acid-tolerant than those from nonacidic environments. Experiments for the breeding of acid tolerance, however, have met with limited success and remain at an experimental stage of development *(104)*.

Another factor that must be considered is the metal load of fish in acidic lakes. There is evidence that mercury and cadmium levels may be elevated in these fish. Although this does not appear to be a universal phenomenon, fish metal levels should be monitored during the evaluation of the relative success of fish-stocking programs. Stocking acid lakes or recently limed lakes may simply facilitate metal transfer from fish to man.

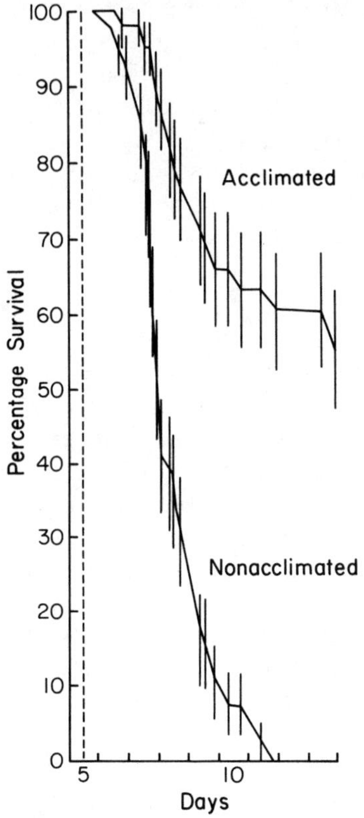

Figure 21. Survival of acclimated and nonacclimated yearling brook trout to nominal pH of 5.0 and aluminum 0.4 mg/liter [From W. A. Flick, C. L. Schofield, and D. A. Webster, *in* "Acid Rain/Fisheries" (R. E. Johnson, ed.) Am. Fish. Soc., Bethesda, pp. 287–306. 1982. Used with permission.].

IV. Conclusion

Acid deposition has had a significant impact on poorly buffered, soft-water lakes and streams in Europe and North America. In addition to the decrease in pH, metals—leached from the soils and sediments or deposited into the precipitation—have increased in concentration in the water column. Some of these metals are toxic to aquatic biota, and others are accumulated by aquatic biota. They can then enter the human food chain, either through drinking water or consuming fish.

Acidification affects both plants and animals and is frequently reflected by a decrease in species diversity beginning at around pH 5.5. Changes in biomass, productivity, and decomposition are less sensitive parameters and are generally not noticed until the pH drops below 5.0. In acidified lakes there tends to be a shift from planktonic to benthic production with occasional nuisance growths of mosses and filamentous algae. Fish may be scarce or absent in these waters.

Many lakes and rivers have already been drastically changed by acidic deposition. If we wish to prevent further degradation of our aquatic ecosystems and to allow acidified lakes to recover, a program of reduced acidic emission is necessary. Until such a program can be legislated, liming of selected lakes and streams, combined with fish stocking and fish breeding programs, may provide an interim solution.

REFERENCES

1. R. A. Smith, "Air and Rain: The Beginnings of Chemical Climatology." Longmans, Green, London, 1872.
2. E. Gorham, "A Historical Review of Interactions between the Atmosphere and the Biosphere," Document submitted to the Committee on the Atmosphere and Biosphere, Commission on Natural Resources. Natl. Acad. Sci., Washington, D.C., 1980.
3. E. B. Cowling, *In* "Acid Rain/Fisheries" (R. E. Johnson, ed.), pp. 15–31. Am. Fish. Soc., Bethesda, Maryland, 1982.
4. K. Dahl, *Salmon Trout Mag.* **46,** 35–43 (1927).
5. S. E. Sunde, Annual Report of the Fishery Inspector. Department of Agriculture, Oslo, Norway, 1926.
6. H. Leivestad, G. Hendrey, I. P. Muniz, and E. Snekvik, *In* "Impact of Acid Precipitation on Forest and Freshwater Ecosystems in Norway" (F. H. Braekke, ed.), Res. Rep. 6/76, pp. 86–111 (ISBN 82-90/53-27-9). Oslo, Norway, 1976.
7. S. Oden, Dagens Nyheter, Stockholm, 1967.
8. S. Oden, Ecology Committee Bulletin No 1, 68. Swedish Nat. Res. Council, Stockholm, 1968.
9. S. Oden, *Water, Air, Soil Pollut.* **6,** 137–166 (1976).
10. T. Dochinger and T. A. Seliga, eds., "Proceedings of the First International Symposium on Acid Precipitation and the Forest Ecosystem, 1975." Ohio State Univ. Press, Columbus, 1976.
11. D. Drablos and A. Tollan, eds., "Proceedings of the International Conference on Ecological Impact of Acid Precipitation" (ISBN 82-90376-07-3). Oslo, Norway, 1980.
12. F. M. D'Itri, "Acid Precipitation Effects on Ecological Systems." Ann Arbor Sci. Publ., Ann Arbor, Michigan, 1982.
13. R. E. Johnson, ed., "Acid Rain/Fisheries." Am. Fish. Soc., Bethesda, Maryland, 1982.
14. T. A. Haines, *Trans. Am. Fish. Soc.* **110,** 669–707 (1981).
15. Swedish Ministry of Agriculture, "Acidification Today and Tomorrow", Stock-

holm Conference on the Acidification of the Environment (Translated by S. Harper). Swed. Minist. Agric., Stockholm, Sweden, 1982.
16. National Research Council of Canada, "Acidification in the Canadian Aquatic Environment: Scientific Criteria for Assessing the Effects of Acidic Deposition on Aquatic Ecosystems," NRCC No. 18475. NRCC, Ottawa, 1981.
17. L. N. Overrein, H. M. Seip, and A. Tollan, "Acid Precipitation—Effects on Forest and Fish," Final Report of the SNSF-Project 1972–1980 (ISBN 82-90376-07-3). Oslo, Norway, 1980.
18. M. Havas, T. C. Hutchinson, and G. E. Likens, *Environ. Sci. Technol.* **18,** 176A–186A (1984).
18a. G. E. Likens, "The 1981 Yearbook of Science and the Future," pp. 212–227. Encyclopaedia Britannica, 1981.
19. W. Dickson, *In* "Proceedings of the International Conference on Ecological Impact Acid Precipitation" (D. Drablos and A. Tollan, eds.), pp. 75–83. Oslo, Norway, 1980.
20. R. B. Davis and F. Berge, *In* "Proceedings of the International Conference on Ecological Impact of Acid Precipitation" (D. Drablos and A. Tollan, eds.), pp. 270–271 (ISBN 82-90376-07-3). Oslo, Norway, 1980.
21. R. J. Beamish and H. H. Harvey, *J. Fish. Res. Board Can.* **29,** 1131–1143 (1972).
22. E. Mohn, E. Joranger, and R. F. Wright, "Regional Lake Surveys 1975–1978, A Statistical Data Analysis," SNSF-Proj. IR 62/80 (ISBN 82-90376-07-3). Oslo, Norway, 1980.
23. A. Henriksen, *Water Res.* **14,** 809–813 (1980).
24. M. Havas, *Can. J. Fish. Aquat. Sci.,* **42,** 1741–1748 (1985).
24a. W. Dickson, *In* "Limnological Aspects of Acid Precipitation" (G. R. Hendry, ed.), pp. 37–41, Proc. Int. Workshop, Sagamore Lake Conf. Cent., Sagamore Lake, New York Brookhaven Natl. Lab., Uptown, New York, 1978.
25. G. R. Hendrey, K. Baalscrud, T. S. Traaen, M. Laake, and G. Raddum, *Ambio* **5,** 224–227 (1976).
26. B. Almer, W. Dickson, C. Ekström, E. Hornstrom, and U. Miller, *Ambio* **3,** 303–336 (1974).
27. R. F. Wright, "Historical Changes in the pH of 128 Lakes in Southern Norway and 130 Lakes in Southern Sweden over the Period 1923–1976," SNSF-Proj. TN34/77. (NISK) 1432 Aas-NLH, Norway, 1977.
28. C. L. Schofield, *Ambio* **5,** 228–230 (1976).
29. W. Dickson, Rep. No. 54, pp. 8–20. Institute of Freshwater Research, Drottingholm, Sweden, 1975.
30. R. F. Wright and E. T. Gjessing, *Ambio* **5,** 219–223 (1976).
31. O. Grahn, H. Hultberg, and L. Landner, *Ambio* **3,** 93–94 (1974).
31a. H. M. Seip, *in* "Proceedings of the International Conference on Ecological Impacts of Acid Precipitation" (D. Drablos and A. Tollan, eds.), pp. 358–365 (ISBN 82-90376-07-3). Oslo, Norway, 1980.
32. A. Henriksen, *in* "Acid Rain/Fisheries" (R. E. Johnson, ed.), pp. 103–121. Am. Fish. Soc., Bethesda, Maryland, 1982.
33. R. F. Wright, N. Conroy, W. T. Dickson, R. Harriman, A. Henriksen, and C. L. Schofield, *in* "Proceedings of the International Conference on Ecological Impact of Acid Precipitation" (D. Drablos and A. Tollan, eds.), pp. 377–379 (ISBN 82-90376-07-3). Oslo, Norway, 1980.
34. G. H. Tomlinson, R. J. P. Brouzes, R. A. N. Mclean, and J. Kadlecek, *in* "Proceedings of the International Conference on Ecological Impact of Acid Precipitation"

(D. Drablos, and A. Tollan, eds.), pp. 134–135 (ISBN 82-90376-07-3). Oslo, Norway, 1980.

35. J. P. Baker, *in* "Acid Rain/Fisheries," (R. E. Johnson, ed.), pp. 165–176. Am. Fish. Soc., Bethesda, Maryland, 1982.
36. C. Schofield and J. Trojnar, *in* "Polluted Rain" (T. Toribara, M. Miller, and P. Morrow, eds.), pp. 341–366. Plenum, New York, 1980.
37. R. J. Hall, G. E. Likens, and S. B. Fiance, *Ecology* **61,** 976–989 (1980).
38. N. M. Johnson, C. T. Driscoll, J. S. Eaton, and G. E. Likens, *Geochim. Cosmochim. Acta* **45,** 1421–1437 (1981).
39. W. Dickson, *Vatten* **39,** 400–404 (1983).
40. H. Hultberg and A. Wenblad, *in* "Proceedings of the International Conference on Ecological Impact of Acid Precipitation" (D. Drablos and A. Tollan, eds.), pp. 220–221 (ISBN 82-90376-07-3). Oslo, Norway, 1980.
41. U. Sibul and L. Reynolds, "Asynoptic Survey of the Acidity of Ground Waters in the Sudbury Area of Ontario, 1981," APIOS Rep. No. 005/82. Ontario Ministry of Environment, Water Resources Branch, Toronto, 1982.

41a. World Health Organization, "International Standards for Drinking Water." WHO, Geneva, 1971.

42. D. R. Crapper-McLachlan and B. J. Farnell, *In* "Effects of Aluminum in the Canadian Environment" (M. Havas and J. F. Jaworski, eds.). Natl. Res. Coun. Can., Ottawa, 1986 (in press).
43. T. C. Hutchinson, A. Fedorenko, J. Kitchko, A. Kuja, J. van Loon, and J. Lichwa, *in* "Environmental Biogeochemistry" (J. O. Nriagu, ed.), pp. 565–585. Ann Arbor Sci. Publ., Ann Arbor, Michigan, 1976.
44. R. G. Wetzel, "Limnology." Saunders, Philadelphia, Pennsylvania, 1975.
45. H. Hultberg and O. Grahn, *Great Lakes Res.* **2,** 208–217 (1976).
46. G. R. Hendrey and F. Vertucci, *in* "Proceedings of the International Conference on Ecological Impact of Acid Precipitation" (D. Drablos and A. Tollan, eds.), pp. 314–315 (ISBN 82-90376-07-3). Oslo, Norway, 1980.
47. G. Gahnstron, G. Andersson, and S. Fleischer, *in* "Proceedings of the International Conference on Ecological Impact of Acid Precipitation" (D. Drablos and A. Tollan, eds.), pp. 306–307 (ISBN 82-90376-07-3). Oslo, Norway, 1980.
48. D. W. Schindler, R. Wagemann, R. B. Cook, T. Ruszczynski, and J. Prokopowich, *Can. J. Fish. Aquat. Sci.* **37,** 342–354 (1980).
49. P. Muller, *Can. J. Fish. Aquat. Sci.* **37,** 355–363 (1980).
50. W. A. Scheider, J. Adamski, and M. Paylor, "Reclamation of Acidified Lakes near Sudbury, Ontario." *Ontario Ministry of the Environment,* Rexdale, 1975.
51. T. D. Brock, *Science* **179,** 480–483 (1973).
52. R. E. Kwiatkowski and J. C. Roff, *Can. J. Bot.* **54,** 2546–2561 (1976).
53. M. Havas and T. C. Hutchinson, *Nature (London)* **301,** 23–27 (1982).
54. J. W. Hargreaves, E. J. H. Lloyd, and B. A. Whitton, *Freshwater Biol.* **5,** 563–576 (1975).
55. D. W. Schindler, personal communication 1984.
56. G. R. Hendrey, *in* "Acid Rain/Fisheries" (R. E. Johnson ed.), pp. 125–135. Am. Fish. Soc., Bethesda, Maryland, 1982.
57. C. D. Foy, R. L. Chaney, and M. C. White, *Annu. Rev. Plant Physiol.* **29,** 511–566 (1978).
58. J. Okland, *in* "Proceedings of the International Conference on Ecological Impact of Acid Precipitation" (D. Drablos and A. Tollan, eds.), pp. 322–323 (ISBN 82-90376-07-3). Oslo, Norway, 1980.

59. K. A. Okland, *in* "Proceedings of the International Conference on Ecological Impact of Acid Precipitation" (D. Drablos and A. Tollan, eds.), pp. 324–325 (ISBN 82-90376-07-3). Oslo, Norway, 1980.
60. G. G. Raddum, *in* "Proceedings of the International Conference on Ecological Impact of Acid Precipitation" (D. Drablos and A. Tollan, eds.), pp. 330–331 (ISBN 82-90376-07-3). Oslo, Norway, 1980.
61. J. P. Nilssen, *Int. Rev. Gesomten Hydrobiol.* **65,** 177–207 (1980).
62. D. W. Sutcliff and T. R. Carrick, *Freshwater Biol.* **3,** 437–462 (1973).
63. W. G. Sprules, *J. Fish. Res. Board Can.* **32,** 389–395 (1975).
64. J. C. Roff and R. E. Kwiatkowsky *Can. J. Zool.* **55,** 899–911 (1977).
65. W. Keller, "Planktonic Crustacea in Lakes of the Greater Sudbury Area," Technol. Rep. Ontario Ministry of the Environment, Rexdale, 1981.
66. T. L. Chrisman, R. L. Schulze, P. L. Brezonik, and S. A. Bloom, *in* "Proceedings of the International Conference on Ecological Impact of Acid Precipitation" (D. Drablos and A. Tollan, eds.), pp. 296–297 (ISBN 82-90376-07-3). Oslo, Norway, 1980.
67. J. L. Confer, T. Kaaret, and G. E. Likens, *Can. J. Fish. Aquat. Sci.* **40,** 36–42 (1983).
68. A. Hobaek and G. G. Raddum, "Zooplankton Communities in Acidified Regions of Southern Norway," SNSF-Proj. IR75/80. (NISK) 1432 Aas-NHL, Norway, 1980.
69. Ministry of Environment, Ontario, "Studies of Lakes and Watersheds near Sudbury Ontario: Final Limnological Report," Supplementary volume to SES 009/82. Ontario Ministry of Environment, Rexdale, 1982.
70. M. Havas and T. C. Hutchinson, *Can. J. Zool.* **61,** 241–249 (1983).
71. M. Havas and G. E. Likens, *Can. J. Zool.* **63,** 1114–1119 (1985).
72. W. T. W. Potts and G. Fryer, *J. Comp. Physiol.* **129,** 289–284 (1979).
72a. M. Havas, T. C. Hutchinson, and G. E. Likens, *Can. J. Zool.* **10,** 1965–1970 (1984).
73. H. W. Brown and M. E. Jewell, *Trans. Am. Microsc. Soc.* **45,** 20–34 (1926).
74. N. C. Collins, A. P. Zimmerman, and R. Knoechel, *in* "Effects of Acidic Precipitation on Benthos" (R. Singer, ed.), pp. 35–48. North Am. Benthol. Soc., Springfield, Illinois, 1981.
75. D. F. Malley, *Can. J. Fish. Aquat. Sci.* **37,** 364–372 (1980).
76. H. L. Bell, *Water Res.* **5,** 113–119 (1971).
77. R. J. Hall and G. E. Likens, *Nature (London)* **292,** 329–331 (1981).
78. J. H. D. Vangenechten and O. L. J. Vanderborght, *in* "Proceedings of the International Conference on Ecological Impact of Acid Precipitation" (D. Drablos and A. Tollan, eds.), pp. 342–343 (ISBN 82-90376-07-3). Oslo, Norway, 1980.
79. T. Wiederholm and L. Eriksson, *Oikos* **15,** 235–247 (1977).
80. O. Grahn, *in* "Proceedings of the International Conference on Ecological Impact of Acid Precipitation" (D. Drablos and A. Tollan, eds.), pp. 310–311 (ISBN 82-90376-07-3). Oslo, Norway, 1980.
81. C. L. Schofield, Res. Proj. Technol. Compl. Rep. A-072-NY. Office of Water Research and Technology, Department of the Interior, Washington, D.C., 1977.
82. H. H. Harvey and C. Lee, *in* "Acid Rain/Fisheries" (R. E. Johnson, ed.), pp. 45–55. Am. Fish. Soc., Bethesda, Maryland, 1982.
83. N. I. Conroy, K. Hawley, and W. Keller, "Extensive Monitoring of Lakes in the Greater Sudbury Area 1974–1976," Sudbury Environ. Study Rep. Ontario Ministry of the Environment, Rexdale, 1978.

84. P. M. Ryan and H. H. Harvey, *J. Fish. Res. Board Can.* **34,** 2079–2080 (1978).
85. P. O. Fromm, *Environ. Biol. Fish.* **5,** 79–93 (1980).
86. M. Havas, *in* "Effects of Acidic Precipitation on Benthos" (R. Singer, ed.), pp. 49–65. North Am. Benthol. Soc., Springfield, Illinois, 1981.
87. D. J. Spry, C. M. Wood, and P. V. Hodson, "Canadian Technical Report of Fisheries and Aquatic Sciences," No. 999. Dept. of Fisheries and Oceans, Ottawa, Ontario, 1981.
88. W. L. Lockhart and A. Lutz, *Water, Air, Soil Pollut.* **7,** 317–332 (1977).
89. D. I. Mount, *Water Res.* **7,** 987–993 (1973).
90. H. Hultberg, *Water, Air, Soil Pollut.* **7,** 279–294 (1977).
91. I. P. Muniz and H. Leivestad, *in* "Proceedings of the International Conference on Ecological Impact of Acid Precipitation" (D. Drablos and A. Tollan, eds.), pp. 320– 321 (ISBN 82-90376-07-3). Oslo, Norway, 1980.
91a. I. P. Muniz and H. Leivestad, *in* "Proceedings of the International Conference on Ecological Impact of Acid Precipitation" (D. Drablos and A. Tollan, eds.), pp. 84–92 (ISBN 82-90376-07-3). Oslo, Norway, 1980.
92. W. Dickson, *VDI-Ber.* **500,** 387–391 (1983b).
93. U.S.-Canada Memorandum of Intent, "Emissions, Costs, and Engineering Assessment," Final Report of Work Group 3B of the US-Canada Memorandum of Intent on Transboundary Air Pollution Study, June. 1982.
94. Organization for Economic Co-operation and Development, "The Costs and Benefits of Sulphur Oxide Control. A Methodogical Study. OECD, Paris, 1977.
95. Office of Technology Assessment, "Acid Rain and Transported Air Pollutants: Implications for Public Policy," Summary. U.S. Congress, Washington, D.C., 1984.
96. T. C. Hutchinson and M. Havas, *Water, Air, Soil Pollut,* 1986 (in press).
97. B. D. LaZerte and P. J. Dillon, *Can. J. Fish. Aquat. Sci.* **41,** 1664–1677.
98. B. Bengtsson, W. Dickson, and P. Nyberg, *Ambio* **9,** 34–36 (1980).
99. N. D. Yan and P. J. Dillon, *in* "Studies of Lakes and Watersheds near Sudbury Ontario: Final Limnological Report," SES 009/82. Ontario Ministry of Environment, Rexdale, 1982.
100. L. M. Blake, *in* "Acid Precipitation Effects on Ecological Systems" (F. M. D'Itri, ed.), pp. 251–260. Ann Arbor Sci. Publ., Ann Arbor, Michigan, 1982.
101. H. Mason, *Can. Res.*, September, pp. 28–29 (1984).
102. P. J. Dillon, N. D. Yan, W. A. Scheider, and N. Conroy, *Arch. Hydrobiol. Beih. Ergebn. Limnol.* **13,** 317–336 (1979).
103. C. A. Guthrie, M. S. Thesis, Cornell University, Ithaca, New York, 1981.
104. W. A. Flick, C. L. Schofield, and D. A. Webster, *in* "Acid Rain/Fisheries" (R. E. Johnson, ed.), pp. 287–306. Am. Fish. Soc., Bethesda, Maryland 1982.

Lists of books relevant to this chapter published since the lists in Chapter 8, Volume II, 3d edition (1977) can be found in Chapter 1, Table IX, Volume VII.

7

Effects on Human Health

John R. Goldsmith

Ben Gurion University of the Negev
Beer Sheva, Israel

I. Introduction

The body of knowledge concerning the health effects of atmospheric pollutants has two principal uses: for deciding which constraints or regulations are indicated in order to protect the health of exposed populations and to advance biomedical and environmental science. Since our primary concern is for its health protection uses, it is proper to start with Ferris's *(1)* excellent 1978 review, which summarized knowledge of health effects of exposures to low levels of the six then-regulated pollutants in the United States. He said, "The present primary standards for the six regulated pollutants seem adequate to protect the health of the public. . . . . Until more data are available they should not be changed up or down." For the most part, this is still true.

The purpose of this chapter is to present and interpret studies which have been published since Vol. II was finished in 1977. In some cases earlier references are provided for the reader's convenience.

A. Impact of Emission Control Programs

Driven largely by evidence of human health effects, most developed countries have implemented controls on stationary sources, and some mobile sources as well. This has resulted, in most cases, in a decrease, first in particulate matter emissions and exposures, and later in gaseous emissions and exposures. The reduction in particulate emissions as a result of the Clean Air Act of Great Britain is given much credit for the substantial reduction of morbidity and mortality from chronic bronchitis in that country *(2)*.

In the United States reduction in emissions has not always resulted in a reduction of exposures estimated from mean ambient air monitoring levels *(3)*. (See Chapter 1, this volume, for air quality trend data).

Although no convincing health-monitoring data demonstrate unequivocally the health benefits of the reductions which have occurred, it is widely believed that health benefits have been the principal benefits from fixed-source pollution control in the United States *(4)*.

Control programs in developed countries have made it more attractive to locate pollution-generating activities, such as smelters, refineries, pulp mills, and aluminum reduction plants, in less-developed countries (LDCs), which usually have less restrictive pollution emission regulations.

As a result of the "export of polluting sources," future health effects research will be necessary in such countries to a greater extent than in the past. In addition, LDCs will need to have assistance in devising regulatory principles which will both protect the health of their populations and sustain their legitimate developmental goals.

B. Energy Conservation and Its Impact

The abrupt increase in fuel oil and gasoline prices in late 1973 led to a reduction in the maximum automobile speed limit, the introduction of more fuel-efficient automobiles, and more careful attention to home insulation. There was an abrupt drop in United States motor vehicle crash fatalities, estimated to represent a saving of over 10,000 lives a year, as well as a sharp drop in gasoline consumption. However, there has been no documented favorable health impact as a result of decreased pollution from motor vehicles.

Brown *et al.* *(5)* examined deaths from cardiovascular diseases, chronic lung diseases, and all causes except motor vehicle accidents in the first quarter of 1974 comparing them with those for the preceding 3 years for San Francisco and Alameda counties, California. They noted that there seemed to be a sharp drop in 1974, which they attributed to the reduced pollutants. However, no confirmation of this report has appeared, and no general reduction of ambient pollution levels was noted.

Home insulation has produced both a specific problem related to off-gassing of formaldehyde from urea-formaldehyde foam insulation *(6)* and a general problem of sealing homes and buildings which aggravates the effects of indoor pollution from cooking and heating fires. This also apparently has produced a new problem related to discomfort associated with being in a sealed environment *(7)*. (See also Section VII).

The use of larger amounts of wood for home cooking and heating may very well be at the expense of increased emissions, as is shown in Chapter 8, Vol. VII. The problem may, in some dwellings, be even more severe than these data suggest, since a substantial proportion of the emissions from wood burning may be emitted within the dwelling and thus lead to high levels of indoor pollution *(8, 9)*.

C. Reemergence of Coal as a Preferred Energy Source

The abrupt price increase of crude oil, starting in 1973, has made the use of coal rather than fuel oil economically attractive for large-scale energy sources such as power plants. The public image of coal as a fuel which is dirty to handle and a source of health-damaging pollution has been an impediment to its more widespread use. Although it is true that the use of coal for household heating in an open grate *was* the major source of the soot (black suspended particulate matter) and SO_2 which contributed to the excess mortality episodes and to the high chronic bronchitis rates of British and other population groups, emissions from a modern, coal-fired power plant bear little relationship to the emissions from the thousands of chimney pots which, in an historic sense, they replace. This difference is particularly great for particulate matter from a modern efficiently run coal-fired power plant, which contains very little carbon, is not black, and, being emitted from a stack usually several hundreds of meters in height, is well dispersed before it enters the breathing zone of a human population. Most such plants are equipped with a baghouse or other particulate matter control system, and increasingly they are being equipped with flue gas scrubbers to remove SO_2. The properly designed and operated coal-burning power plant need no longer produce any health risks due to emissions of black suspended particulate matter or SO_2. However, the collective impact of many such plants together may increase atmospheric acid droplet formation and deposition with possibly serious effects on vegetation and freshwater ecosystems.

In places where such plants are a dominant source and community concern is great, epidemiological monitoring strategies have been designed and put into operation in order to assure that any possible effects are detected in time for effective preventive action *(10)*. Such strategies have yet to be evaluated.

D. Responsibilities of Scientists for Risk Assessment

In past decades, public health agencies had the dual task of determining what the health risks of pollution were and which groups were affected, as well as deciding which regulations were appropriate for health protection. These two tasks of risk assessment and determination of the acceptable level of risk to be used for regulatory purposes required quite different types of skills and often lead to confusion as to which professional group was responsible for which portion of the dual task. The Committee on Science in the Promotion of Human Welfare of the American Association for the Advancement of Science first pointed out *(11)* that the scientific task of determining the risks could and should be separated from the value-laden task of deciding which risks were acceptable to a given community and which protective actions were therefore appropriate. Although the committee took this position initially with respect to radiation risks, it quickly recognized that the principles were applicable to air pollution as well, and it established the "Air Conservation Commission," whose report *(12)* became one of the bases for the preventive approach to air pollution, often designated *air quality management*.

Lowrance *(13)* provided the most general treatment of the problem for environmental scientists in his report "Of Acceptable Risk," prepared for the National Academy of Sciences. Risk Assessment has become a professional activity in its own right with books and journals dedicated to it *(14–16)*. What this means for air pollution health effects is that the research to be undertaken may have a more precise focus on the accurate and valid assessment of human health risks. The necessity of considering whether the risks will or will not require certain standards or regulations becomes the task of another professional group trained to undertake this task, informed and guided by the results and evaluations of health and environmental scientists.

Related to this process is the emergence of "decision analysis" and "cost-benefit" analysis. Of particular interest to the field of health effects research is the dominant role attributed by some analysts in the United States to health benefits from stationary-source air pollution control. After reviewing the economic assessments of pollution impacts and their reduction, for example, Freeman *(4)* gives as the most reasonable estimate of the realized benefits from prevention of morbidity and mortality the figure of $17 billion out of a total of benefits of $21.7 billion. The range given for the estimated health benefits is from $3.1 to $40.6 billion! Possibly if more health scientists participate in the development of such estimates, the validity would be greater and the range of uncertainty narrower.

1. Framework for Scientific Input to Regulatory Decisions

For most pollutants there exist several possible health effects and for each there is usually some potential input from both experimental and epidemiological studies. Each such valid study contributes one or more points on a dose–response relationship. Such relationships when graphed indicate the cumulative proportion of a population that is likely to be affected at a given exposure or dose level. For each effect there is expected to be an S-shaped relationship of the proportion of that population exposed who manifest such an effect at each exposure or dose level. The populations studied may consist of all those who are exposed, a subset of them, for instance, in a given age group or a given health status, or a population of animals or plants. If this set of hypothetical cumulative dose–response relationships is plotted with a probability scale on the vertical axis and a logarithmic scale on the horizontal axis, a diagram such as Figure 1 will be produced. Because of the probability scale, the S-shaped dose–response relationship becomes linear if and when the response relationship is symmetrically normal (i.e., Gaussian).

We now can identify four regions or zones: to the left a "zone of indifference" within which variations in exposures are not thought to have unfavorable effects of any consequence on health or well-being. With increasing exposures we enter the "zone of incremental harm," in which it is reasonable to seek decreases in exposures in order to reduce risks of damage, but the urgency of such reductions may be modified by the possible costs and the degree to which such effects are acceptable to the community in question. At the higher doses there is a "zone of imperative protective action" for which no responsible scientist can escape the obligation to urge prevention or reduction in exposures to prevent serious consequences. These zonal boundaries are in essence vertical strips on such a graph. It may happen that a type of relatively mild effect could occur to a large fraction of the population, and at the same dose a more serious effect could occur to a tiny fraction of the exposed population. Such very small numbers of hypersensitive individuals in a large population

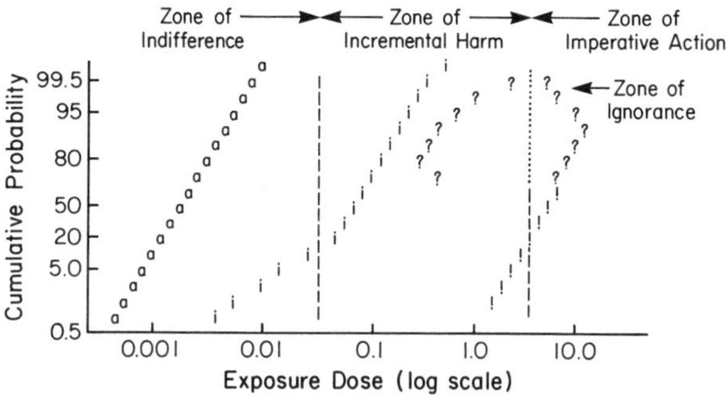

Figure 1. Schematic framework for the input of health scientists to the regulatory decision system.

This is a log-probability graph of what is known for a given pollutant about dose-response information, with the ordinate representing on a probability scale the cumulative proportion of the population affected at a given dose.

Four "zones" are identified, three approximately separated by vertical lines, and the fourth a zone of ignorance (surrounded by question marks). The curve to the left (*a*), represents, for example, the cumulative percentage of persons affected by soiling or by visibility impairment. The second from the left (*i*) represents, for example, the impairment of lung function during physical activity. The inflection (bend) in the curve reflects a type of distribution which can be produced by observations on a population, including an unusually reactive subgroup. It might be redrawn if possible to reflect the relative sensitive population and the relative insensitive one, which would have different slopes. The third curve (*!*) might reflect the proportion of a population whose chronic disease is affected by the given doses of air pollution. Much of the third curve is in the area of ignorance where we lack useful dose-response information.

may more reasonably be protected by procedures prescribed by their physician rather than by emission controls.

Finally, there exists in most real dose-response data systems a "zone of ignorance," which may occur within any portion of the graphical system; such zones should be designated by specific questions as to effect and dose.

When using the cumulative probability scale, one sometimes finds, as in Figure 1, that the available data plot as two linear segments rather than a single line. This indicates the likelihood that the data reflect responses of more than one group in the population. It also indicates the desirability of trying to divide the population into two or more groups according to criteria of their responsiveness. When it is possible to identify unusually sensitive groups, their responses should, if possible, be plotted separately, as though they represented two different types of effect.

An important advantage of such a plot is that it identifies a band of dose–response relationships for which it is expected and reasonable for scientists to participate in cost–benefit trade-offs in the "zone of incremental harm." This is in contrast to the attitude to be expected of health scientists with respect to responses in the adjacent zones of medical indifference and medical urgency.

2. Time–Concentration–Effect Relationships

In the previous section the notion "dose" is used to imply both a concentration and a sufficient duration to lead to a specific effect. In order to be effective input to regulatory processes, data concerning effects must be expressed in terms of both concentration and duration of exposure; to be most useful in relation to air quality data, the duration of exposure may be expressed in terms of the averaging times often used with air quality monitoring data.

Table I, for example, indicates some different ranges of averaging times for selected effects. A logarithmic plot has been shown to yield a linear boundary for the relationship of averaging time and maximal observed concentration for a large variety of pollutants and monitoring sites, based on a year's complete data sets *(16a)*. In Figure 2 are plotted a number of concentration–duration averaging times for se-

Table I Some Characteristic Response Times for Various Effects

	Characteristic response times		
Type of effect	*Short-term (seconds to minutes)*	*Medium-term (hours to days)*	*Long-term (months to years)*
Human health	Odor, visibility, eye irritation	Acute respiratory effects, increased Pb body burden	Chronic respiratory disease
Animals, vegetation	Field crop and ornamental plant damage	Field crop and ornamental plant damage	Fluorosis of livestock, impaired yield of fruit and forest crops
Materials	Acid droplet pitting, nylon hose and film destruction	Rubber cracking, tarnishing, blackening of paint	Corrosion, soiling, building materials deterioration

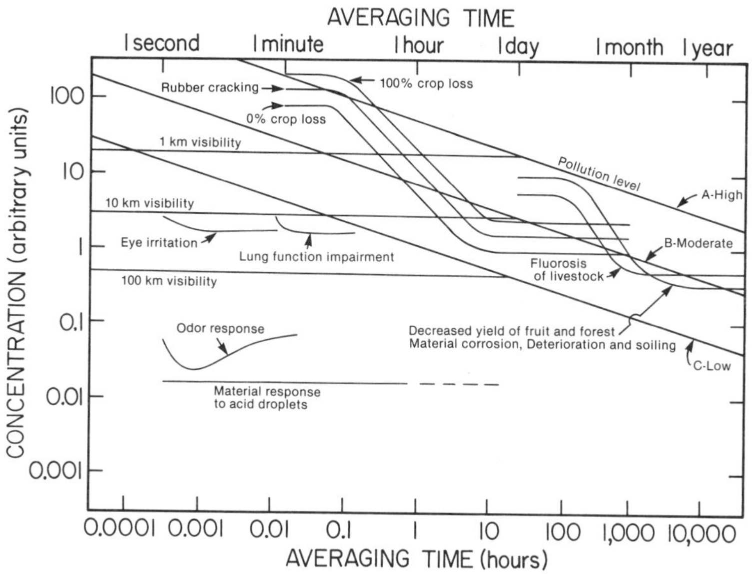

Figure 2. Schematic framework for reflecting the time–concentration–effects relationships. The scales in this figure are logarithmic on both vertical and horizontal axes and are such that they produce a linear relationship for the maximum value for a given averaging time when a long term set of monitoring data are plotted *(16a).* Such a plot can also be used to relate different concentrations and averaging times which reflect the same level of air quality—the parallel sloping lines in the figure *(16b).* In addition the framework permits visualization of the time–concentration relationships of different types of effects. Shown as examples are horizontal lines reflecting visibility impairment; sloping lines reflecting brief exposure to odorants, eye irritants, and acid droplets; and sigmoid curves reflecting longer-term exposures to pollutants that corrode, deteriorate, and soil materials, crack rubber, and cause fluorosis of livestock, damage to crops and forests, and impairment of lung function.

lected effects on the same scale. Lines *A*, *B*, and *C* drawn parallel to the commonly observed boundary for maximal values provide a basis for relating concentration–averaging times and possible effects *(16b).*

The relationship of these response curves to ambient air quality is shown by lines *A*, *B*, and *C*, which represent the maximum or any other chosen percentile line from a display which shows actual air quality. Where the air quality is poor (line *A*), essentially all the adverse effects displayed will occur. Where the air quality is good (line *C*), most of the intermediate and long-term adverse effects displayed will not occur. Where the air quality is between good and poor, some of the adverse intermediate and long-term effects will occur, but in an attenuated form, in comparison with those of poor air quality.

II. Sulfur Oxides and Black Suspended Particulate Matter

New findings and analyses have led to important changes in our perception of the health impact of this major group of pollutants. It was once thought that the two pollutants interacted and that both had to be elevated in order to observe health effects; sulfate aerosol, the most plausible product of their interaction, now seems not to be producing attributable effects at realistic concentrations. It was thought that the black suspended particulate matter was the most important member of the combination; effects of SO_2 at realistic concentrations now seem to be more relevant.

Adult chronic bronchitis was formerly considered the most likely condition to be related to such exposures; chronic respiratory diseases in children and adults with airway hyperreactivity are now thought to be the most relevant in the United States. Total suspended particulate matter (TSP) was considered the most appropriate pollutant to measure for estimating the health impact of particulate exposures; there is now general agreement that respiratory tract penetration of smaller-sized particulate matter is required for health effects to be manifest. Therefore, the measurement of particulate matter smaller than 10–15 μm and its chemical and physical characterization is now the objective.

A. Health Effects of Sulfate Aerosols

Emitted sulfur dioxide is generally thought to be oxidized slowly by atmospheric oxygen to sulfur trioxide, SO_3, a hygroscopic gas, readily combining with water to form sulfuric acid aerosol (H_2SO_4). Ultimately the aerosol reacts with atmospheric particles or surfaces to form sulfates *(17)*. Among these reactive surfaces are the particulate matter mats held in the high-volume sampler. Sulfates vary in their physical properties and biological reactivities. Sulfuric acid and calcium sulfate are at the extremes of such reactivity and irritancy. A widely used experimental system, the measurement of increased pulmonary flow resistance in the guinea pig, for example, illustrates the effect of oxidation state, particle size, and cations for the various sulfur compounds (see also discussion in Chapter 6, Vol. II, 3rd ed.). We can see from such data *(18)* that if all of the atmospheric sulfur were present as sulfuric acid of a particle size of 2.5 μm, one-thirtieth as many sulfur atoms would be required to produce a 50% increase in flow resistance, compared to the number required if all the sulfur were present as SO_2. When the particle size is as low as 0.3 μm and all the sulfur is present as zinc ammonium sulfate, the augmentation is nearly a thousandfold! At equal particle sizes, sulfuric acid is the most reactive of the "sulfates," with acid sulfates such as ammonium acid sulfate or zinc ammonium sulfate next in order of irritancy.

Only a small fraction of the sulfur compounds present are in the form of such irritants as sulfuric acid mists, and these reactive materials are transient states in a complex process. Further, under realistic circumstances particle sizes and cation mixtures are variable. Nevertheless in the late 1960s and early 1970s, when sulfur dioxide was thought likely to have no effect on health at concentrations lower than 3.0 ppm, the sulfate aerosol as the active ingredient in the sulfur oxides–black suspended particulate matter complex was the leading hypothesis.

For example, Amdur *(19)* in 1969 wrote: "the toxicological literature combined with the literature on atmospheric chemistry suggests that sulfur dioxide levels be controlled in terms of the potential formation of irritant particles."

The possible importance of sulfate aerosols for human health was augmented by the evidence of their importance in visibility impairment *(20)*. The so-called acidic deposition (acid rain) phenomenon gives evidence of persisting atmospheric acidity over substantial distances from major sources *(21, 22)*.

1. Experimental Evidence

Ammonium sulfate is much less irritating than is zinc ammonium sulfate or sulfuric acid. Fortunately, humans appear to have a mechanism for neutralizing inhaled acid aerosols based on the reaction with ammonia produced in the oronasal passages *(23)*. At equivalent particle sizes, ammonium sulfate so produced is only a tenth as potent as sulfuric acid in increasing the pulmonary flow resistance of the guinea pig.

However, the results of exposing human subjects, both "normal" and "asthmatic," to sulfuric acid aerosols or to zinc ammonium sulfate at levels likely to be found in polluted areas has not yet provided strong support for the notion that such particulate matter has a substantial health hazard for humans.

Avol *et al. (24)*, for example, considered that the "worst case" level of pollution might be as much as 100 μg of sulfate per cubic meter, with a particulate size of 0.3 μm. They used such doses of ammonium sulfate, ammonium acid sulfate (also known as *ammonium bisulfate*), and sulfuric acid particulates in a blind design with both normal and asthmatic subjects for 2-hour exposures, of which 1/2 hour was spent on a bicycle ergometer (a device for standardizing the exercise level of a subject) in order to raise the respiratory volume to approximately twice its resting level. The six asthmatic subjects, whose average age was 48.7 years, had a 10% increase in respiratory resistance when exposed to ammonium bisulfate and some increase in respiratory symptoms as well. No effects were found in the normal subjects. (See also Section II,B,2).

Sackner *et al. (25)* had previously found no change in symptoms or in lung function of asthmatic subjects with exposures to sulfuric acid aerosol of 0.1 – 0.2 μm at a concentration of 1 mg/m^3, but Sackner's subjects were older, were having steroid therapy, and presumbly had more inherent variability of their tests. Sackner's subjects did not exercise, and this characteristic may have led to delivery of a much lower dose of sulfate particulate matter to the tissues and membranes of the airway.

Utell *et al. (26)* have reported a study which involved 16 normal and 17 asymptomatic asthmatic subjects. The mean ages were 27 and 26 years, respectively, and all were nonsmokers. None of the asthmatic subjects was receiving corticosteroid therapy. All subjects had a provocative test with carbachol aerosol, which in sufficient concentrations can cause bronchoconstriction and an increase in airway resistance. All subjects were exposed by mouthpiece so that the nasal defense mechanisms would not be operative. Exposures were to salt (NaCl), ammonium bisulfate, ammonium sulfate, sulfuric acid aerosol and sodium bisulfate, with particle sizes between 0.5 and 1.0 μm and concentrations of 100 μg and 1.0 mg per cubic meter. The subjects were blind as to the nature of the material and the concentrations. For the normal subjects no change in airway conductance occurred, even at high concentrations. (Airway conductance is the reciprocal of airway resistance. Irritants cause a drop in conductance.) The bronchoconstrictor action of carbachol was potentiated by the aerosols more or less in proportion to their acidity. For the asthmatic subjects, 1 mg of sulfuric acid or sodium bisulfate produced a significant fall in conductance; sulfuric acid potentiated the carbachol effect. For the two asthmatic subjects with the greatest reaction to sulfuric acid at high dose levels, there was potentiation also at low doses, but the asthmatic group as a whole did not show a statistically significant effect at low doses. Utell *et al. (27)* later demonstrated a similar reaction to sulfuric acid mist in asthmatics at 450 μg/m^3 but not at 100 μg/m^3.

Lippman *et al. (28)* have shown that at doses of 100 μg per cubic meter there is an increase in the rate of clearance of particles by the mucous layers of the respiratory system, whereas at 1 mg per cubic meter there is a retardation in the rate of clearance. These alterations, like those due to smoking, tend to persist with repeated exposures.

Koenig *et al.* *(29)* have shown that adolescent asthmatics, when exercised enough to increase their resting ventilatory rate by five to six times and exposed to 100 μg each of NaCl and sulfuric acid with a particle size of 0.6 μm, inhaled by mouthpiece, had a significant decrease in the forced expiratory volume as well as a significant increase in respiratory resistance.

Thus in contrast to results that might have been expected from the experiments on guinea pigs, human exposure studies, even with asthmatics who may be assumed to be hyperreactive, do not yet demonstrate significant health risks from levels of sulfates present even in the most polluted communities.

2. Epidemiological Evidence

The epidemiological evidence of health impairment from sulfates is based on analyses of morbidity and mortality, mostly of the type that is called *ecological analysis*, meaning that the exposure is inferred on the basis of location of residence, and not measured or estimated specifically for the persons whose health reactions are being studied. Increasingly priority is being given to studies in which the exposure is specifically measured for the populations at risk. Examples are the studies of lead in dentine (see Section VI) and studies of the so-called six communities (cf. Refs. *84–86, 89*) in which extensive use is made of personal samplers.

Since sulfates are the ultimate oxidation products of emitted sulfur dioxide, their effects are inevitably confounded by the levels of copollutants and the meterological factors which affect these reactions. Finally, although a given measured level of soluble sulfate ions extracted from a high-volume filter must be thought to include all of the sulfuric acid, zinc ammonium sulfate, and all other sulfates, we have only a vague idea as to how much of what is measured was actually an active irritant when it was in the atmosphere. We know that some of the sulfate is formed on the high-volume filter itself *(30)*. For all these reasons, even in the best designed and conducted "ecological" studies of community health impact of the sulfur oxide–black suspended particulate matter complex are not likely to provide strong evidence for a specific effect of sulfates, so long as the measurements are based on high-volume filter samples.

Four community epidemiological studies often cited in evaluations of sulfate health risks will be evaluated. Each of them uses a different endpoint, as well as a different method.

Dohan *(31)*, Dohan and Taylor *(32)*, and later Ipsen *et al.* *(33, 34)* analyzed data for sickness absence in the Pittsburgh area in relation to pollution levels and meterological variables and by time of year for several populations of industrial workers. The pollutant measurements available at the time were suspended sulfates and smoke shade. Ipsen *et al.* *(34)* concluded: "No single air pollution measurement has a notable additional effect to those of the three weather variables, but their sum is positively correlated with the prevalence measurement on the same day and on the seventh day following." No measurements of other sulfur oxides were available. This study cannot, therefore, provide evidence in relation to the specific effects of sulfates as opposed to the effects of sulfur pollutants in general.

Lave and Seskin *(35)* have presented multivariate analyses of mortality among United States Standard Metropolitan Statistical Areas (SMSAs) initially for 1960 and later for the years to 1969. In their 1960 data set for 117 SMSAs, sulfate and particulate levels were the only pollution variables available. Thibodeau *et al.* *(36)* have provided a critique of the work, and in particular of the instability of the regression coefficients in relation to outlying values for some of the SMSAs. They recommended the deletion of data for several of the deviant SMSAs (statistically designated as outliers), and when data for these nine locations are excluded, the coefficients change substantially as shown in Table II *(36)*. The analyses done for 1969 also included a subset of 68 SMSAs for which sulfur dioxide monitoring data were also available. When the SO_2 data were included, the regression coefficients and elasticities for the effect of sulfate were reduced, indicating that the data could be interpreted as reflecting the

Table II Estimates of Regression Coefficients of Mortality with Sulfates and Total Suspended Particulate Matter for United States Standard Metropolitan Statistical Areas (SMSAs), 1960 *(36)* [a]

	Sulfates		*Particulates*	
Choice of pollution indicator	*All 117 SMAs*	*Excluding 9 outlying SMAs*	*All 117 SMAs*	*Excluding 9 outlying SMAs*
Minimum sulfate or TSP	0.47	0.08	0.20	0.57
Mean sulfate or TSP	0.17	0.26	0.30	0.01
Maximum sulfate or TSP	0.03	0.08	−0.02	0.01

[a] Depending on whether all 117 are included or whether a reduced set excluding 9 outlying SMSAs is analyzed.

complex of sulfur-containing emissions, and not necessarily an effect specific for sulfate. Of passing interest is the fact that for these 68 SMSAs there is a negative association or mortality with nitrate levels; that is, with a higher nitrate pollution, the mortality is lower. The fact that *maximal* pollutant values have *lower* regression coefficients than *mean* or *minimum* values (Table II) argues against those associations being causal.

The third data set often cited as relevant to sulfate effects is the CHESS data on respiratory morbidity in relationship to sulfur oxides in Utah in the vicinity of smelters and in New York City *(37)*. In the former data, the levels of sulfate given were, in fact, derived from sulfur dioxide measurements and from emissions from the smelter and therefore cannot be considered to be specific *(38, 38a)*. For the New York City data, after some criticisms had been made of the original report, a follow-up study was done in the same areas by Lan and Shy *(39)*. No effects specific for sulfates were identified.

Goldsmith *et al. (40)*, using a new method, have analyzed the number of emergency room admission (visits) by day for four sites within a 5-mile radius of several hospitals in the Los Angeles basin for a 2-year period in relation to a variety of pollution and meteorological measurements.

They used a procedure called *path analysis* in order to take advantage of what is thought to be known about the pollutant–meterological interactions and the plausible relationships of primary pollutants such as SO_2, CO, and NO to such secondary pollutants as oxidant, NO_2, and sulfates. Although no interesting relationships with sulfate were found in the community with the highest levels, Riverside, sulfate levels had significant correlation coefficients and high path coefficients for a variety of different plausible path models for two of the locations, Long Beach and Lennox. (A *high path coefficient* indicates that sulfate was strongly associated with emergency room visits when all other variables were allowed to exert their effects according to the relations specified by the model or path diagram.) In no community did SO_2 have a strong association with emergency room admissions. Although oxidant levels had substantial correlations with emergency room admissions in three of the communities, when temperature and other pollutants and environmental variables were allowed to enter the plausible models, the path coefficients were positive in only one community, Azusa. The data used were for weekdays. Day-of-week and long-term trends were minimized by computational adjustments. In all locations the dominant effect was of temperature. The authors conclude that the sulfate pollution at Lennox and at Long Beach had an important and possibly causal association with the demand for emergency room admissions.

3. Interpretation of Data Concerning Sulfates

If there is a specific effect of sulfates as distinct from the effects of sulfur dioxide, such methods as those of Goldsmith *et al.* *(40)* seem likely to be able to detect it. However, sampling of atmospheric sulfate through direct means, rather than through a high-volume filter, is more likely to provide convincing and generalizable evidence of such effects.

Taken together, the experimental and epidemiological data do not add up to evidence for a specific effect of sulfate aerosol. Since the impact of any regulations which might be derived from an air quality standard for sulfate would be on emissions of SO_2, from the point of view of health protection, there seems little urgency to add a new set of sulfate regulations, as long as sulfur emissions are kept low; this is especially true in light of the nonspecificity of both monitoring and health effects data. The situation may change quickly, however, as new data accumulate.

Some of the indicated directions are sampling and analysis for acid aerosols of particles in the respirable size ranges and adjustment for the formation of sulfate on the filters. Thurston and Spengler *(41)*, for example, have shown for Steubenville, Ohio, a relatively polluted area and for Watertown, Massachusetts, a relatively clean area, that more of the sulfate on the filter in the heavily polluted area was in the nonrespirable size range, as can be seen in Table III. The table also shows the correction for artifactual sulfate using Coutant's procedure *(30)*.

Table III Mean Pollution Concentrations in Steubenville, Ohio, and Watertown, Massachusetts, May 1976 to December 1978, μg/m³ *(41)*

	N = *327*	*N* = *149*
	Watertown	*Steubenville*
Sulfur dioxide	19.8	57.9
Total suspended particulate	50.6	115.8
Respirable particulate	18.2	44.7
Total sulfate	6.3	17.3
Respirable sulfate	4.7	11.3
Nonrespirable sulfate	1.6	6.0
Artifactual sulfate	2.0	3.2
Corrected total sulfate[a]	4.3	14.0

[a] Corrected on the basis of Coutant's procedure *(30)*.

In any event, studies of effects of sulfur dioxide exposures on respiratory function in physically active persons with airway hyperreactivity now appear to explain the joint effects of this complex of pollutants better.

B. Effects of Sulfur Dioxide Exposures on Physically Active Asthmatic Subjects

Until 1978, our understanding of the reactions of human subjects to inhalation of SO_2 was that no effects on ventilatory function could be expected with exposures below several parts per million for about 10 minutes.

However, epidemiological analysis (cf. Chapter 7, Vol. II, 3rd ed.) of effects of pollution episodes of combined sulfur dioxide and black suspended particulate matter indicated that populations were adversely affected by exposures to sulfur dioxide of about 0.2 ppm on a 24-hour average basis *(42–44)*. It was logical to assume that the black suspended particulate matter was speeding up the oxidization of the SO_2 to SO_3, the anhydride of sulfuric acid, and that the oxidized material, the acid sulfates, were the active agents. The World Health Organization, among others, recommended that the air quality standards reflect the joint presence of these two pollutants.

1. Airway Hyperreactivity

The possibility that some persons were unusually reactive to SO_2 might have explained the apparent discrepancy between experimental and epidemiological findings.

Possible mechanisms of airway hyperreactivity were pointed out by Nadel and Widdicombe and Nadel *et al.* *(45, 46)*, when they showed that humans exposed to 2–5 ppm of SO_2 for 3–10 minutes had increases in airway resistance which could be reversed or prevented by isoproterenol (a drug which on inhalation blocks effects mediated by the vagus nerve). Furthermore, in cats in which the upper airway was anatomically isolated from the lungs and lower airway, exposure of the upper airway to SO_2 led to bronchoconstriction (narrowing) of the lower airway by reflex mechanisms mediated by the vagus nerve. The effect could be reversed by cooling or sectioning of the vagus nerve or by the drug atropine.

In 1962, Frank *et al.* *(47)* reported on the exposures of 11 human subjects to 1, 5, or 13 ppm by mouthpiece and found that pulmonary flow resistance was increased significantly in the group at 5 or 13 ppm and that the increase was less with nasal breathing. One of their subjects had a significant increase in resistance at 1 ppm. Other investigators also observed that an occasional subject, otherwise normal, reacted strongly to concentrations at which most subjects were unreactive. In London, Lawther *et al.* documented the unusual sensitivity of one of their subjects *(48)*, and Jaeger *et al.* noted such an unanticipated reaction in 1 out of their 40 subjects *(49)*.

As early as 1967 Goldsmith *(50)* pointed out that these unusually reactive experimental subjects may represent the populations which were adversely affected during the air pollution episodes and urged that hyperreactivity and its prevalence and mechanisms be assessed.

Orehek *et al. (51)*, using the reactivity of asthmatic subjects to the inhalation of mecholyl (a chemical simulating the natural stimulation of the autonomic nerves), showed that their reactivity was increased after exposures to nitrogen dioxide. Subsequently, the hyperreactivity of asthmatic subjects to similar agents and the augmentation of such activity after air pollution exposures have been shown by a number of studies *(51a)*. About 80% of persons with asthma and from 2% to 10% of "normal" subjects can be shown to be hyperreactive, when changes in airway resistance are measured after exposures to mecholyl or similar substances.

2. Experimental Exposure Studies of Exercising Asthmatics

Exercise augments the responses in inhalational studies by two mechanisms: First, the depth of breathing increases with exercise and therefore the lung is exposed to a greater volume of air, along with whatever pollutants there are in it.

Second, with increasing exercise, the velocity of airflow in the respiratory system increases, and the absorption of the soluble gases in the upper airway is decreased. SO_2 is a relatively soluble gas, most of which never gets deeper than the back of the throat in normal breathing. However, above a certain level of exercise (enough to produce an increase to about three to four times the resting volume per minute of respired gas, called the *minute volume*), open mouth breathing replaces normal nasal breathing; as a result, the proportion of removal of soluble gases is much diminished.

Kleinman *(51b)* has developed a mathematical model reflecting minute ventilation, partitioning between oral and nasal breathing, and differences in pollutant "scrubbing" (removal) in the oral and nasal airway. With this model, apparent discrepancies in data for mouthpiece and oronasal exposures are greatly diminished.

For these reasons, exercising asthmatic subjects have become recognized as one of the "sensitive populations" most likely to be unfavorably affected by exposures to atmospheric irritants, such as SO_2 *(52)*.

Sheppard *et al.* *(53, 54)* showed that asthmatic subjects, while breathing through a mouthpiece, reacted more to 1, 3, and 5 ppm of SO_2 than did nonasthmatic atopic (allergic) or normal subjects. With exercise sufficient approximately to double the resting minute volume, asthmatic subjects regularly had increased airway resistance with 0.5 ppm, and in two of the subjects, significant increases occurred with exposures to 0.1 ppm. Similar results were reported by Koenig *et al.* *(55)* for adolescent asthmatic subjects using exercise sufficient to increase resting ventilation by five to six times.

In a preliminary study, Linn *et al.* *(56)* had similar findings, but under exposure chamber conditions, with normal routes of breathing (i.e., without using a mouthpiece), these effects were not observed. Kirkpatrick *et al.* *(57)* repeated such studies with a face mask, with and without nasal occlusion, and confirmed that responses are greater for mouthpiece breathing than for breathing by normal routes of ventilation, although they found detectable effects at 0.5 ppm in exercising asthmatics even when they breathed with a face mask. Data from these and subsequent studies are shown in Table IV and Figure 3 *(58–62)*. Koenig *(60)* has shown similar reactions with adolescent asthmatics, using a procedure for measuring total respiratory resistance.

It is clear that airway reactivity is variable among subjects; it has been found to be increased after acute respiratory infections *(63)* and after 2-hour ozone exposures *(64)*. Prior to 1978, most of the human exposure studies had not assessed the airway reactivity of the subjects. A typical study was that of Frank *et al.* *(47)*, discussed in Section II,B,1.

Many of the clinical as well as experimental studies share the same weakness, namely averaging of data for all the subjects, rather than seeking to identify the subjects with hyperreactivity. The choice of asthmatic subjects is a step in the right direction. A second important step is the use of exercise to simulate the effects of physical activity. In persons with normally patent nasal passages, the initial effects of exercise are simply to increase the dose of pollutants with which the body must cope. For those with nasal obstruction, or for normal persons at high minute volumes (above 50 liter/minute), the effect of exercise is to shunt an increased amount of the incoming air stream directly to the major airways of the lung, thereby limiting the removal in the nasal passages of soluble gases and smaller particles.

Table IV Dose–Response Data for Experimental Exposures of Exercising Asthmatic Subjects to SO_2 by Different Routes of Inhalation with Resulting Increases in Specific Airway Resistance (SR_{aw}) in Percentage Compared to Sham or Preexposure Levels[a]

Route of inhalation	*Senior investigator*	*Reference no.*	*Concentration ppm*	*Ventilation rate 1/minute*	*Dose rate μl/minute*	*% Increase in SR_{aw}*
Mouthpiece						
	Sheppard	*53*	1.00	10[b]	10	25
	Sheppard	*54*	0.25[c]	42	10.5	30
	Linn	*56*	0.50	27	13.5	40
	Bethel	*59*	0.50	27	13.5	33
	Kirkpatrick	*57*	0.50	41	20.5	99
	Bethel	*59*	0.50	41	20.5	215
	Sheppard	*54*	0.50	42	21	115
	Sheppard	*53*	3.00	10[b]	30	76
	Sheppard	*54*	1.00	30	30	95
	Linn	*58*	0.75	40	30	246
	Bethel	*59*	0.50	62	31	225
	Sheppard	*53*	5.00	10[b]	50	126
Oronasal						
	Linn	*56*	0.25	27	6.75	−7.5
	Kehrl	*61*	0.25	38	9.5	7.3
	Linn	*62*	0.20	48	9.6	1.3
	Linn	*56*	0.50	27	13.5	1.0
	Bethel	*59*	0.50	27	13.5	25
	Kehrl	*61*	0.50	38	19	44
	Linn	*62*	0.40	48	19.2	25
	Bethel	*59*	0.50	41	20.5	32
	Kirkpatrick	*57*	0.50	42	21	51
	Linn	*62*	0.60	48	28.8	104
	Linn	*58*	0.75	40	30	155
	Bethel	*59*	0.50	62	31	74[d]
	Bethel	*59*[d]	0.50	62	31	140[e]
	Linn	*62*	0.75	48	36	122
	Kehrl	*61*	1.00	38	38	112
Nasal						
	Kirkpatrick	*57*	0.50	43	21.5	28

[a] Airway resistance is estimated from measurements of alveolar pressure divided by respiratory flow. It tends to be higher in a given individual at a lower lung volume. SR_{aw}, specific airway resistance, is the product of the airway resistance and the lung volume at which it is measured.
[b] Strictly speaking, these are exposures at rest, with an assumed ventilation rate of 10 1/min
[c] In two of the most reactive of these subjects, 0.1 ppm (or μl/1) at this ventilation rate led to increases in SR_{aw} which were statistically significant.
[d] With a face mask
[e] Freely breathing in a chamber

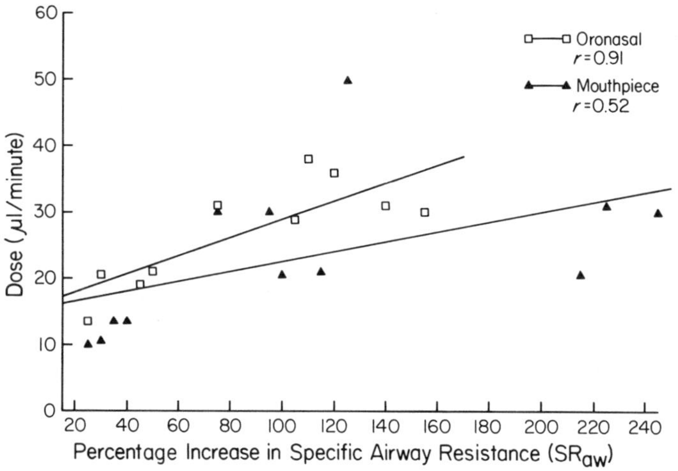

Figure 3. Effects of dose rate and route of delivery of SO_2 for experimental studies of physically active asthmatic subjects by a variety of investigators, based on data in Table IV. Percentage increase in the airway resistance has been derived from the various reports, comparing the effects of the indicated degree of exercise with SO_2 to the same degree of exercise without the gas. Minute volumes of ventilation (liters per minute) are multiplied by the concentration in parts per million (that is microliters per liter) to yield a dose rate of microliters of SO_2 per minute. The oronasal breathing route includes both face mask breathing and exposures in a chamber, but even so there is less scatter as reflected in the higher r, (correlation coefficient) than with mouthpiece breathing. Since it is generally felt that a doubling of airway resistance is needed for healthy subjects to note some distress in breathing, such values are more likely at dose levels above 20 microliters/min.

Sheppard *et al. (53, 54)* confirmed that the rate of airflow was the critical variable in their exercising subjects by using voluntary hyperventilation at the equivalent airflow and showing that equivalent responses occurred. Hazucha and Bates *(65)* in 1975 had used exercise sufficient to double the minute ventilation of their subjects and found a slight decrease in lung function at 0.37 ppm.

It is likely that hyperreactors or asthmatic subjects spend only a small proportion of their time at levels of exertion equivalent to fourfold increases in resting ventilation, but if those times are also times of elevated pollution, such subjects may well have symptoms such as shortness of breath.

C. *Mortality Associated with Sulfur Oxides and Black Suspended Particulate Matter*

The London, England, mortality for the years from 1959 to 1972 was examined in 1982 by Mazumdar *et al. (66)*. They studied associations with black suspended particulate matter and SO_2 for the winter months. During this period a striking drop in pollution occurred; black suspended particulate matter dropped by 89% from 547 $\mu g/m^3$ in the winter of 1958–1959 to 60 $\mu g/m^3$ in the winter of 1971–1972. During the same time interval, sulfur dioxide dropped from 404 to 181 $\mu g/m^3$. The astonishing point is that the mortality associations have not decreased very much, as can be seen in Table V. A linear regression of mortality on smoke did not seem to fit as well as a curvilinear one, based on both smoke and the square of smoke measurements for each day. The quadratic regression of mortality on smoke plus $(\text{smoke})^2$ has decreased only by about half, when expressed as the percentage of daily mortality associated with both terms. This suggests that the observed associations with pollution levels during earlier years reflected, in part, some sort of effect for which the pollutants themselves were a surrogate. Mazumdar *et al.* suggest that this surrogate behavior may be due to some sort of unobserved toxic fraction or associated atmospheric condition which has not diminished. Atmospheric stagnation and a decrease in temperature are often associated with increases in pollution.

Another possibility is that the choice of portions of a year (the winters) during which both mortality and weather variables as well as pollution have a sinusoidal variation could lead to a purely artifactual association. This type of problem was the major subject of Yule's Presidential Address to the Royal Statistical Society in 1926 *(67)!*

An even more astonishing observation is that the same phenomenon seems to have been found in New York City between 1962 and 1972. Buechley *et al. (68)* in 1973 reported the association of excess mortality with SO_2 levels for the years 1962–1966. Then for the follow-up period of 1967–1972, they *(69)* continued to find the same patterns of association between mortality and SO_2 despite a reduction of SO_2 during the period to one-fourth of its initial level!

If there were a causal basis for the associations observed between smoke and mortality in London in 1959, and the smoke were reduced about 90%, one would expect that the association would virtually disappear; similarly, if the associations with SO_2 in New York were causal and the levels decreased by one-fourth, the association should have become much weaker. In neither case did this occur. The most plausible reason is that the measured and presumably controlled pollutant was in part a surrogate for the true cause of the excess. Apart from the possibility that these observations represent a statistical artifact, the most reasonable generalization is that the mortality was associated with atmospheric stagnation and cold weather and not

Table V Associations of Daily Winter Mortality in London with Levels of Smoke and Sulfur Dioxide, μg/m³ *(66)*[a]

	Winter means			Joint regression coefficients (×1000)		Mean effects[b]	
Years	*Deaths*	*SO_2*	*Smoke*	*SO_2*	*Smoke*	*Smoke + smoke²*	*Smoke + SO_2*
58–59	334	404	547	17.99	2.87	11.79	2.65
59–60	269	302	351	−11.55	19.15	5.53	1.20
60–61	315	281	253	−17.97	32.58	5.94	1.03
61–62	316	328	196	5.81	22.97	7.93	2.03
62–63	362	419	206	22.33	−12.28	5.51	1.89
63–64	278	338	190	−11.87	31.12	4.63	0.68
64–65	272	269	135	−27.99	60.22	4.22	0.22
65–66	284	233	106	9.91	17.37	3.15	1.46
66–67	263	243	98	11.68	11.50	2.77	1.51
67–68	310	236	83	−26.32	66.07	2.91	−0.23
68–69	279	219	66	8.84	28.90	3.94	1.38
69–70	291	201	73	16.78	7.47	3.67	1.41
70–71	253	191	67	14.36	6.08	2.82	1.25
71–72	266	181	60	4.29	57.03	5.47	1.62
First 7 winters	307	334	268	−3.32	22.41	6.51	1.59
Second 7 winters	278	215	79	5.65	27.77	3.82	1.23
All 14 winters	292	275	174	1.16	25.09	5.16	1.60

[a] This table compares in the last two columns, the percentage of mean numbers of deaths per day (given in the second column) associated with a quadratic form of dependence on smoke (smoke + smoke²), and a joint linear dependence on smoke levels and sulfur dioxide. In each case the smoke + smoke² "explains" more than the smoke plus SO_2. But while the smoke levels decrease by 1971–1972 to less than a third of their 1961–1962 levels, the proportion of deaths associated does not decrease very much.

[b] As an example, we calculate the mean effects as a percentage for the joint regression of smoke and SO_2 for all 14 winters: the contribution of SO_2 is $1.16 \times 275/1{,}000 = 0.319$; the contribution of smoke is $25.09 \times 174/1{,}000 = 4.366$; the sum of these is 4.685, which is 1.60% of 292, the mean number of deaths per day for all 14 winters.

specifically with smoke in London and SO_2 in New York. This conclusion does not imply that the variation in these pollutants did not make a contribution to increases in mortality during the period of 1950 to 1960, but even in that time period the association may not have been entirely specific.

Certainly the evidence is convincing that the sulfur oxide–black suspended particulate matter complex can produce excess mortality in heavily exposed populations on the basis of studies of the air pollution episodes. These are discussed in Vol. II, 3rd ed. They affected populations in the Meuse Valley in 1930, Donora in 1948, London in 1952 and 1962, New York City in 1953, and Osaka in 1962 *(43)*.

Concerning pollution-related mortality in London, Waller *(70)* writes: "The winter of 1962 to 1963 was in fact the last in which appreciable changes in daily deaths or hospital admissions were seen in relation to pollution . . . and it seems that the effect of the Clean Air Act on the occurrence of periods of high pollution has been even greater than anticipated."

He then discusses the winter of 1975 to 1976, in which daily averages for smoke reached as high as 800 $\mu g/m^3$ in some places and SO_2 reached as high as 1200 μg during a spell of cold, foggy weather. Although some perturbations in the daily mortality data were detectable, Waller feels that these may be as readily associated with the cold weather as with the pollution. Thus, this episode points in the same direction as do the previous analyses: mortality effects of the sulfur oxide–black suspended particulate matter complex are, in New York and London, no longer distinguishable from effects of meteorological conditions under which atmospheric stagnation occurs.

This interpretation is supported by two corollary trends in Great Britain. Since the winter of 1969–1970, it has no longer been possible to detect an association between the aggravation in the health status of panels of chronic bronchitic patients and changes in pollution. There has also been a substantial decline in the mortality from chronic bronchitis, which in 1946–1955 for men aged 60 to 64, was 300 per 100,000 per year; it had dropped by 1973 to 160 per 100,000 *(71, 72)*. It is possible that some of this drop was due to changed diagnostic and coding procedures. Chinn *et al.* *(73)* in an analysis of total and cause-specific mortality in various locations in Great Britain for the years 1969–1973 found no specific associations with pollution *(73)*.

D. Morbidity Studies of Adults

In the final report of an 8-year follow-up of working men in London, Fletcher *et al.* *(74)* conclude that their study provides no valid information concerning the part played by general air pollution in causing irreversible airflow obstruction. In an earlier report *(75)*, this group had speculated that a decrease in the yearly rate of decline in the forced expiratory volume in 1 second, $FEV_{1.0}$, might have been due to the beneficial effects of a decline in pollution. In this final report, they attributed this change to other factors. Their study made a clear distinction between the natural histories of airflow obstruction and mucus hypersecretion and showed that cigarette smoking is a dominating cause for both, although general air pollution was thought to contribute more to the latter.

E. Morbidity Studies of Children and Adults

Because school children are less likely to have their respiratory health affected by their own smoking and because respiratory conditions in childhood may predict the likelihood of respiratory problems in adult life, a number of studies have involved school children. The European Regional Office of the World Health Organization undertook a series of coordinated studies based on a common protocol in Czechoslovakia, Denmark, Greece, the Netherlands, Poland, Romania, Spain, and Yugoslavia *(76)*. Although effects associated with pollution exposures were not found in all countries, and measurements of lung function gave inconsistent results (possibly due to calibration problems), there appeared to be a strong association between air pollution and a variety of indicators of respiratory disease in children. Smoke was reported to have a "far greater effect" than SO_2 (see the following paragraphs, for a conflicting report from France). The data did not allow the establishment of a threshold level, but it was thought that the initial effects of SO_2 and smoke on respiratory indices should be sought from studies done at places with mean annual concentrations of between 50 and 200 $\mu g/m^3$ of either pollutant.

In a large cross-sectional study of school children in a number of communities in Great Britain, Melia *et al.* *(77)* found in 1975 significant associations of smoke and of SO_2 with the frequency of respiratory symptoms for both boys and girls within the winter average range below 100 $\mu g/m^3$ (Tables VI and VII). In a longitudinal analysis of a portion of the multiyear data set for 1973 and 1974 *(78)*, they tested the hypothesis that in the communities in which the pollution decreased most, the symptoms would also decrease the most. This hypothesis was not supported by the data, which led the authors to conclude that the cross-sectional associations might not be due to causal relationships. Their analytical strategy has led to some questions *(79)* which may be resolved when the complete set of data become available.

Boudik *et al.* *(80)* found a higher prevalence of chronic cough and phlegm among smokers aged 20–54 in a polluted area of Bohemia

Table VI Numbers of School Children with Various Levels of Symptom Frequencies, Unadjusted, in Relation to Levels of Smoke, 1975, in British Towns *(77)*

Symptom frequencies	*Smoke levels, winter means μg/m³*		
	<30	*30–60*	*60+*
	Boys		
0	685	607	214
1–2	225	142	72
3+	52	45	25
	$Chi^2 = 12.18$	4df	
	$p = 0.016$		
	Girls		
0	702	628	244
1–2	178	125	79
3+	30	36	27
	$Chi^2 = 20.6$	4df	
	$p = 3.9 \times 10^{-4}$		

Table VII Numbers of School Children with Various Levels of Symptom Frequencies, Unadjusted, in Relation to Levels of SO_2, 1975, in British Towns *(77)*

Symptom frequencies	*SO_2 levels, winter means μg/m³*		
	<30	*30–60*	*60+*
	Boys		
0	703	429	374
1–2	169	148	122
3+	47	38	37
	$Chi^2 = 11.54$	4df	
	$p = 0.021$		
	Girls		
0	686	457	431
1–2	157	103	122
3+	29	19	45
	$Chi^2 = 20.48$	4df	
	$p = 3.9 \times 10^{-4}$		

(Czechoslovakia) than in a nonpolluted area; in the same area, Zapletal *et al.* *(81)* found impaired respiratory function among 10- to 11-year-old children.

In Cracow, Poland *(82)*, Sawicki showed that both smoking and pollution interacted in their associations with chronic bronchitis. Such findings justify study of school children at ages before they begin to smoke.

A French national study was conducted from 1974 to 1976 in 20 locations in six cities and in a newly industrialized zone. Subjects included 7,581 men and 9,083 women between the ages of 25 and 59, as well as 1,291 boys and 1,236 girls between 6 and 10 years of age. Standardized questionnaires and lung function tests were used. Sulfur dioxide was measured by a specific method as well as by level of acidity, and both smoke stain methods and total suspended particulate matter measurements were used to estimate exposures to particulate pollution. Oxides of nitrogen were also measured. Results were adjusted for cigarette smoking, age, sex, and socioeconomic level of the head of the household. Since the sample was restricted to nonmanual workers, occupational exposures were unlikely to have confounded the results. The correlations resulting from the survey are shown in Table VIII, which shows that lower respiratory symptoms were significantly associated with both estimators for SO_2 for men and for the specific measure for women. Among children, upper respiratory symptoms were significantly associated with levels of specific SO_2. Among men, women, and children, there was a significant association of the $FEV_{1.0}$ with the strong acid measurements. Neither black suspended particulate matter nor total suspended particulate matter was associated with symptoms or with lung function impairment. No symptoms were associated with community exposures to oxides of nitrogen *(83)*.

Among the most original and productive community studies in the United States have been the so-called six cities studies by a team from the Harvard University School of Public Health *(84)*. In each of six cities, a sample of about 1,500 adults and 1,500 children were enrolled, examined, and followed. The six cities in approximate ascending order of pollution by sulfur oxides and black suspended particulate matter at the time of the initial selection were Portage, Wisconsin; Topeka, Kansas; Kingston-Harriman, Tennessee; Watertown, Massachusetts; St. Louis, Missouri; and Steubenville, Ohio. The latter two exceeded the then-promulgated ambient air quality standards.

In the initial report there were the expected significant gradients

Table VIII Correlations between Atmospheric Pollution in Six French Cities and Respiratory Findings among Men, Women, and Children, 1974–1976 *(83)*

	Respiratory condition	*Sulfur dioxide, µg/m³*		*Particulate matter, µg/m³*	
Population	*Range of levels, µg/m³*	*Acidometric 13–127*	*Specific 22–85*	*Black smoke 18–152*	*Total suspended particulate 45–243*
Men					
	Lower resp. conditions	0.57[a]	0.53[a]	0.03	0.12
	Upper resp. conditions	0.18	0.35	−0.13	−0.03
	$FEV_{1.0}$	−0.52[a]	−0.34	0.32	0.39
Women					
	Lower resp. conditions	0.31	0.49[a]	−0.19	−0.08
	Upper resp. conditions	0.11	0.22	−0.35	−0.32
	$FEV_{1.0}$	−0.57[b]	−0.42	0.32	0.31
Children					
	Lower resp. conditions	0.19	0.11	0.01	0.13
	Upper resp. conditions	0.32	0.53[a]	0.29	0.20
	$FEV_{1.0}$	−0.70[b]	−0.46	0.39	0.34

[a] $p < 0.05$.
[b] $p < 0.01$. There being 20 locations, there are 19 degrees of freedom.

for adults in both symptoms and lung function with cigarette smoking. Among the heavily smoking women, there was increasing frequency of wheezing with increasing air pollution; among male nonsmokers there was a consistent ordering of frequency of phlegm with increasing air pollution. The prevalence of wheezing among both males and females appeared to be associated with mean levels of "mass respirable particulates" (Figure 4) *(84a)*.

A major contribution of the studies is the use of personal monitors for the estimation of the relative importance of indoor pollution levels to the integrated personal exposures. For example, Dockery and Spengler *(85, 85a)* evaluated the respirable particulate exposures and respirable sulfate exposures for 18 Watertown and 19 Steubenville residents, comparing personal sampler measurements with indoor and outdoor samples. Using a time-weighted exposure model which

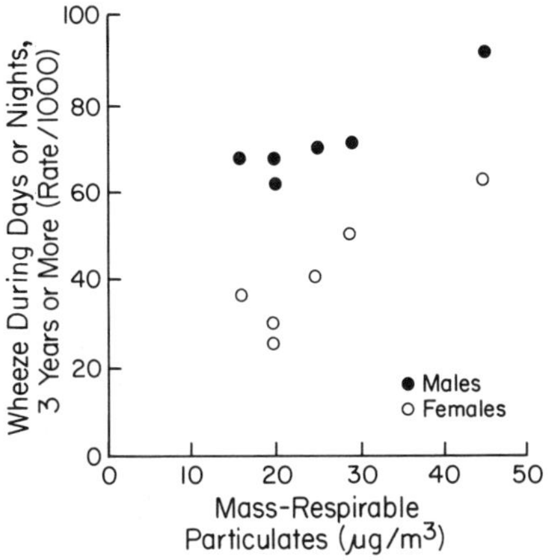

Figure 4. The apparent association between the frequency of positive response (per thousand) to the questions "Do you wheeze during most days or nights?" and "Have you been this way for 3 years or more?" and the mean concentration of mass-respirable particulate matter for men and women respondents in the "Six Cities" study *(84a)*.

included an estimated 20 $\mu g/m^3$ of respirable particulates for each cigarette smoker in the home, the average exposure was 12 $\mu g/m^3$ of respirable-sized sulfates and about 50 $\mu g/m^3$ of respirable particulates.

Samet *et al. (86)* analyzed the relationship between air pollution with SO_2 and particulate matter and the number of emergency room visits in Steubenville, Ohio. They found that the levels of 24-hour means for both pollutants had a statistically significant association with visits for respiratory conditions through a multivariate regression. The overall mean for total suspended particulate matter (TSP) was 156 $\mu g/m^3$ (range 14–696) and for SO_2 90 $\mu g/m^3$ (range 4–369).

1. Responses of Children during Episodes

Stebbings *et al. (87)* studied lung function and symptoms of school children immediately after an episode in Pittsburgh, Pennsylvania, in 1975. For the whole population of children, they found no improvement in lung function or decrease in symptoms in the period of 7 days after the episode. But in a reanalysis of the data studying each child separately, they found that a subgroup of children had a strong upward trend in forced vital capacity over a period of 1 week after the episode *(88)*.

Dockery *et al.* *(89)* studied a group of about 200 Stubenville school children before and immediately after air pollution alerts in the fall of both 1979 and 1980, as well as following sham alerts in the spring of 1980 *(89)*. The combined results for each child were used to calculate a regression coefficient of lung function on the average pollution levels for the preceding 24-hour period ending at 8 A.M. on the day of examination. Such a strategy allows for the detection of effects on a subgroup, as suggested by the second Stebbings study. The distribution of these regression coefficients was centered significantly below zero. The forced vital capacity was about 2% lower during the periods following each alert, with the lowest values occurring 1–2 weeks after the episode. The $FEV_{0.75}$ did not change after the 1978 alert, but was 4% lower after the 1979 alert. No unfavorable effects were observed after the sham episodes of 1980. The maximal TSP values were 422 and 271 $\mu g/m^3$ and for SO_2, 281 and 455 $\mu g/m^3$, for 1978 and 1979, respectively. The response pattern is as shown in Figure 5 *(90)*.

A similar study by Saric *et al.* *(91)* in Yugoslvia found a 5% decrease in $FEV_{1.0}$ in association with SO_2 in the range of 89–235 $\mu g/m^3$ but found no association for smoke.

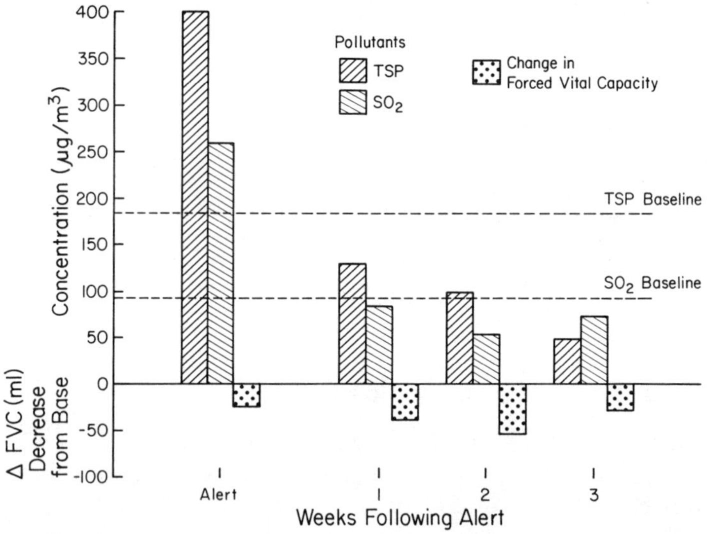

Figure 5. The change in forced vital capacity in children associated with an air pollution "alert" in the Fall of 1978 with about 200 school children *(90)*. Baseline values representing average pollutant concentration were obtained, as shown by the horizontal lines, and then weekly measurements were made. Note that the mean decrease in Forced Expiratory Volume persisted for three weeks after the air pollution peak.

F. Interpretations of These Studies

The studies discussed and their interpretation have led to a substantial reorientation of our understanding of the health risks of exposures to sulfur dioxide and total suspended particulate matter, compared with, for example, the position taken by the World Health Organization in its "Environmental Health Criteria for Sulfur Oxides and Suspended Particulate Matter" *(44)*, which was based on reports up to 1976.

We should now and in the future do the following:

1. Pay more attention to effects on school children both as a result of episodes and of long-term exposures
2. Depend less on regression analyses of mortality
3. Consider that physically active young asthmatic subjects are one of the "most sensitive groups"
4. Look toward sampling of respirable acid sulfate aerosols in the atmosphere as a more valid index than TSP of health risks from particulates
5. Recognize that increased demand for emergency services may be a sensitive and useful guide to the costs and effects of pollution exposures.

SO_2 in the range of 0.05–0.25 ppm for 1 hour is in the "zone of incremental harm" (Figure 1). Above 0.30 ppm, exposures are in the "zone of imperative action."

Although these may seem to be drastic changes, it is necessary to stress that earlier studies and their interpretation provided a strong basis for much-needed control efforts. These efforts had and are continuing to have a highly favorable impact on public health and well-being. Under no circumstances should our reorientation be allowed to justify a laxity in enforcing existing regulations.

It seems likely that the next source of pressure for control of this group of pollutants will derive from concern about acidic deposition (acid rain) and visibility, and that health effects of SO_2 and TSP may be of less concern than they were in previous decades.

III. Ozone and Oxidants

This pollutant complex, the result of photochemical reactions, has primarily affected areas with high levels of incident sunlight and substantial emissions of hydrocarbon vapors and oxides of nitrogen. Health effects have been documented in California for decades and have been troublesome for many months of the year. See Chapter 7, Vol. II, 3rd ed., for a more detailed treatment.

Health reactions have also been documented in Japan, western Pennsylvania, and Connecticut, primarily during the late spring and summer months. New findings as to the duration and pathophysiology of effects of ozone on the airways which conduct gases to the gas-exchanging areas of the lung will be reviewed, as well as evidence for adaptation to ozone or oxidant exposures. A major concern has been that there may be long-term effects of oxidant exposures in the Los Angeles, California, area, and some data on this topic will be reviewed.

Asthma was one of the first conditions suspected of being aggravated by oxidant exposures; new methods of analysis now provide a clearer picture of the relationship of oxidant exposures, copollutants, and asthma. Although ozone is a mutagen for animals, there is as yet no evidence of its being a human mutagen under conditions of community air pollution exposure.

Two thorough reviews have been published and may be consulted for more comprehensive discussion of health effects of ozone and oxidants *(92, 93)*.

A. Pathophysiology of Ozone Effects on the Airways

Golden *et al. (94)* exposed eight healthy nonsmoking subjects for 2 hours to 0.6 ppm of ozone and tested their airway resistance both before and after inhalation of histamine. (Histamine is a substance produced in the body during inflammation, which on inhalation will increase airway resistance. If the increase is very great with a fixed dose of histamine, one may say that the airways are hyperreactive. Carbachol, as mentioned elsewhere, has a similar effect.) They had earlier shown in dogs that hyperreactivity developed during a few hours after exposure. For the group of eight human subjects, hyperreactivity developed during the first hour after exposure, and the average data showed that it disappeared after a day. However for two of the eight subjects, hyperreactivity persisted for up to 3 weeks! Although the authors considered that this variability of duration of ozone hyperreactivity might be due to different breathing patterns of the subjects, they also hypothesized that in those subjects with persisting effects, there may have been damage to the epithelial (or lining) cells of the airways. Other investigators have shown that ozone exposures in animals increase the permeability of the epithelium *(95)*. This hyperreactivity following ozone exposure has been confirmed *(96, 97)*. Holtzman *et al.* *(98)* found that dogs also varied in their hyperreactivity to ozone, and that the hyperreactive dogs had *inflammatory* cells (polymorphonuclear leuckocytes, cells which react to and accompany inflammatory reactions) in the epithelium after exposure, whereas those which failed to show hyperreactivity did not; this finding tends to confirm that the hyperreactivity is an early abnormality.

Boushey and Holtzman *(98a)* review the four physiological and biochemical steps which characterize the progression from ozone-induced inflammation to airway hyperreactivity. They further discuss the ways in which ozone-induced hyperreactivity in humans and animals resembles and differs from the airway hyperreactivity which characterizes asthma.

B. Adaptation to Oxidants and Ozone

It has long been recognized that animals show tolerance to ozone, in the sense that when rodents are exposed to as little as 0.3 ppm, they become "tolerant" to subsequent exposures of several parts per million which would produce massive pulmonary edema in animals exposed for the first time *(99,100)*. This phenomenon is fully discussed in Vol. II, 3rd ed., Chapter 6.

When human subjects are experimentally exposed to such concentrations as 0.3 ppm at intervals of a day or so, some of them show diminished reactivity with later exposures *(101, 102)*. This response is designated as *adaptation*. A difference in the effect of ozone exposure on lung function test results was reported for new arrivals in Los Angeles, California, as compared with the relatively less reactive long-term residents. This was thought to be a manifestation of adaptation *(103)*.

In a report on the persistence of adaptation, Linn *et al.* *(104)* exposed 11 healthy volunteers daily for 4 days to 0.47 ppm of ozone and then reexposed them weekly, studying the forced expiratory lung function tests and symptoms. The initial exposures were for 2 hours with intermittent exercise, and most subjects manifested adaptation. One subject never adapted. Two other subjects showed larger responses on later exposures than they did on initial exposures. In those who adapted, the adaptation was partly lost after 4 days and nearly completely so in 7 days. In an accompanying editorial, Bromberg and Hazucha *(105)* suggest that adaptation is an undesirable effect and link it with the possibility of epithelial damage.

Certainly, there is wide variability in response, and it is difficult to know whether those who adapt or those who do not are to be considered at greater long-term risk. What does seem to be true is that the initially more reactive subjects seem to adapt more slowly; so for acute effects, the most reactive subjects are probably at greatest risk.

Avol *et al.* *(105a)* studied 46 boys and 13 girls between the ages of 12 to 15 years who exercised (mean ventilation 32 L/min) either in an atmosphere of natural photochemical pollution or in purified air in Duarte, California. The mean exposures were 0.144 ppm ozone and 153 $\mu g/m^3$ of TSP, and for purified air 0.001 ppm of ozone and 49 $\mu g/m^3$ TSP. The $FEV_{1.0}$ significantly decreased after the 1 hour exposure to naturally polluted air of both boys and girls, and had only partially recovered during the following hour. Unlike adults with similar exposures and decrements in lung function, the adolescents had no increase in symptoms from such exposures. Presumably having grown up under these conditions they had "adapted" as far as symptoms were concerned. The authors suggest that the lack of symptoms of irritation may put such adolescents at more risk from ambient exposures.

C. Oxidant Pollution and Asthma

Schoettlin and Landau's 1961 report on asthma aggravation *(106)* was one of the earliest and most influential pieces of evidence on health effects of photochemical pollution. Methods of data collection and analysis have improved considerably in the intervening years *(107)*.

Whittemore and Korn *(108)* noted that two problems still handicap the interpretations of such panel studies: the dependence of the attack probability on a given day with experience on previous days and the possible bias due to erratic reporting of panelists with either high or low attack frequencies. By computing a separate multivariate logistic regression for each panelist's attack data, the authors showed the way to overcome these difficulties. They analyzed data for 16 panels collected by the CHESS program in various parts of the Los Angeles (California) basin between 1972 and 1975.

A residual problem of the data was that some subjects were informed that the investigating team was particularly interested in whether asthma attacks occurred during smoggy weather. Hence the possiblity of a response bias cannot be excluded.

The most influential variable was the presence of an attack on the preceding day, and there was a tendency for attacks to be more frequently reported earlier in the study period, which may equally well reflect the tendency for asthma to be episodic. After also adjusting for day of week, there was a significantly increased likelihood of attacks on days with elevated oxidant or total suspended particulate matter, such that there was approximately 60% greater likelihood of an attack on a day with 0.3 ppm or more of oxidant than on a day with less than 0.05 ppm. For days with 0.15 ppm, there was a little more than 20% increased risk of an attack. For particulate matter, there was about 25% increased risk when the level of pollution was 300 $\mu g/m^3$, compared with the risk when values were below 50 $\mu g/m^3$.

In an effort to determine individual attributes that might affect the likelihood of reactivity, researchers assessed each of the 443 panelists response characteristics. Panelists who reported increased attack frequencies in cold weather were significantly less likely to be affected by oxidant. Through examining the residual variation, it became clear that there were increased frequencies in April and September and a deficit in midsummer and midwinter.

Katz and Frezieres *(109)* have subsequently confirmed the responsiveness of some members of an asthma panel to oxidant and have also found significant associations with levels of sulfate and particulate matter.

Goldstein and Hartel *(109a)* have summarized data from studies in New Orleans and New York City, and they have evaluated the usefulness of the methods demonstrated by Whittemore and Korn with additional data sets.

D. Oxidant Health Effects from Non-California Locations

Substantial evidence of health impacts of oxidant pollution has been reported in Japan *(92, 110)*, especially among school children. This is discussed in Chapter 7, Vol. II, 3rd ed.

Imai *et al. (110a)*, reviewing the Japanese studies of oxidant effects during 1970–1980, point out that after reaching a peak pollution level in 1973 of over 300 days when the 1-hour oxidant exceeded 0.12, oxidant pollution declined so that only 86 days had that level of pollution in 1980. All of the positive epidemiological studies in Japan were made before 1976. Although some evidence of nervous system effects was found, it is not clear in what way they are to be interpreted.

Zagraniski *et al. (111)* studied two populations in Connecticut during the summer months, using postcard diaries, in much the same way as asthma panels have been studied. The two populations were employees of the telephone company and patients in a university asthma and allergy clinic. The data were analyzed by correlation analysis with 8-hour and 24-hour oxidant and with sulfate levels. Although researchers reported using a multiple regression analysis, the results are not obvious on careful reading. Some of the correlations are shown in Table IX. Since the criteria for statistical significance of these correlations used by the authors are not clearly stated, significance statements are omitted from the table. The data remain suggestive evidence that there are sensitive individuals reacting in Connecticut to oxidant in ways that resemble the pattern of reactions of populations in California.

Lippmann *et al.* *(112)* measured daily lung function of 83 children in a summer camp in western Pennsylvania over a 2-week period. On 4 of the days, the estimated ozone exposures based on nearby monitoring locations exceeded 0.10 ppm. On these days, there was also an elevation of the acidity of particulate matter obtained by the high-volume sampler using acid-treated quartz fiber filters. For the children tested on each camp day, the highest correlations were with total suspended particulate matter, and the ozone correlations were low. The forced vital capacity test showed a significant regression with ozone (oxi-

Table IX Correlations of Eye and Respiratory Symptoms with 8-Hour Oxidant Levels in Two Groups of Connecticut Subjects *(111)*[a]

		Correlation coefficients			
Population	*Number*	*Eye irritation*	*Nasal irritation*	*Cough*	*Chest discomfort*
Clinic patients	82	0.10	0.13	0.13	0.12
Telephone workers	192	0.35	0.28	0.02	0.22
No respiratory condition	174	0.20	0.23	−0.02	0.05
Hay fever	83	0.15	0.10	0.22	0.04
Asthma	27	0.14	0.17	0.13	0.11
Never smoked	98	0.02	0.09	0.08	0.20
Cigarette smokers	96	0.36	0.32	0.12	0.10
1+ pack/day smokers	19	0.15	0.32	0.24	0.12

[a] The maximal 8-hour ozone level was 0.142 ppm and the mean 0.048. The maximal 1-hour level was 0.235 and the mean of the daily hourly maxima was 0.080.

dant) levels for girls, and one of borderline significance for boys ($p < 0.05$) was used as the criterion of statistical significance. The authors concluded that the effects of ozone "did not appear to be complicated by the effects of other pollutants," primarily on the basis that the TSP levels were low compared to the National Ambient Air Quality Standards, and that the H_2SO_4 was less than 5 $\mu g/m^3$. This argument is not altogether convincing in light of the findings they report. Such studies must be replicated with adequate provision for local measurements of both oxidant and acid aerosols.

Although it was an occupational health study, the report of respiratory symptoms by high-flying aircraft cabin attendants by Reed *et al.* *(113)* is important evidence of the possible health hazards of ozone. A longer-range version of a standard aircraft was flown at somewhat higher altitudes than previously, where the stratospheric ozone levels were high enough that the heated surfaces of the compressors supplying the cabin ventilating air supply did not remove all of the ozone. Cabin attendants began reporting chest soreness and respiratory irritation, and some had persisting cough. By comparing symptoms on various routes and types of aircraft, Reed *et al.* were able to pinpoint the problem as one peculiar to the new type of aircraft flying for long distances at slightly higher altitudes than previously. Although measurements of ozone were not made during the flights studied by Reed *et al.*, it was previously known that ozone levels could reach 0.4–0.6 ppm on such flights. The cabin attendants may have been more affected than the passengers because they are physically active during the flights and therefore expose their respiratory system to larger doses of pollution; it is also less likely that passengers would be able to recognize the occasional aggravation or irritation clearly and to be able to distinguish it as characteristic of certain flights, but it is reasonable to assume that passengers, too, could have been affected. Modifications in the air supply system solved the problem for these aircraft, their passengers, and crew.

E. Evidence of Long-Term Effects

Although aggravation of symptoms of chronic bronchitis and asthma by photochemical pollution has been a recognized effect, a major question existed as to whether long-term exposure might lead to persisting health effects. A team of investigators at the University of California at Los Angeles along with staff and facilities of the Los Angeles Lung Association undertook the long and painstaking effort which is needed to answer this question. Their initial report *(114)* compared respiratory symptoms and lung function from selected census tracts in Burbank, a relatively polluted area, and Lancaster, a relatively unpolluted area northeast of the Los Angeles basin, but a place to which many Los Angeles residents have moved, some to escape the effects of pollution. The census tracts were chosen to be as similar as possible in socioeconomic attributes and demographic characteristics *(114)*. A third community in the Los Angeles, California, basin, Long Beach, was also studied by the same methods *(115)*. Its primary pollutant exposures are to SO_2. Each area was within 2 km of an official air quality monitoring station, and a mobile lung function laboratory was used to collect data on symptoms through a standard questionnaire, electronically monitored forced expiratory volume tests, single breath nitrogen tests (intended to detect relatively poorly ventilated portions of the lung, which may occur in emphysema) and body plethysmography (used to measure airway resistance). The methods were carefully validated by random replications and by comparison of a sample of the subjects' test results with the same tests performed at the pulmonary disease clinic laboratory at the University of California at Los Angeles Medical Center. The initial results are of the cross-sectional study of 3,465 persons of a possible 4,968 in 2,034 households in Burbank; 4,509 person of a possible 5,722 residents in 2,143 households in Lancaster; and 3,786 of a possible 5,007 residents of 2,514 households in Long Beach. Among the male respondents, more reported a change of occupation because of adverse respiratory problems at work in Lancaster than in Long Beach or Burbank. Persons reporting a change in residence or occupation due to respiratory problems were therefore not included in the analysis.

Among those aged 18–59, wheezing was reported more frequently within each smoking category in Long Beach, using age- and sex-standardized prevalence data. Otherwise symptoms were less frequent among adults in Burbank than in the other two areas. But for children 7–17 years of age, cough, wheezing, and breathlessness were reported more frequently in Burbank than in Lancaster.

Predicted values for lung function tests were derived from randomly split samples of Burbank residents who reported never having smoked and who reported no history of symptoms of respiratory disease. The proportion who had $FEV_{1.0}$ tests lower than 75% of that predicted was highest among both nonsmoking and smoking groups in Burbank, and airway resistance was highest in Burbank adults in both smoking categories, whether or not other symptoms were present.

The average daily maximum oxidant level during the years of the study was 9.0 pphm in Burbank, 6.5 pphm in Lancaster, and 3.7 pphm in Long Beach. The authors conclude that "This study suggests that long term exposure to air pollutants in the community is associated with respiratory impairment manifested by dysfunction of the large airways." This result was found after "controlling for" effects of smoking, sex, age, social status, and prior respiratory problems severe enough to provoke change in occupation or residence. Prospective study of these populations is continuing, and the results will help to answer the question of whether exposure to photochemical oxidants can cause chronic respiratory disease. The evidence to date is that it may do so if criteria of large airway functional impairment are sufficient to indicate disease.

The results to date are sufficient to put health risks from frequent oxidant exposures of more than 0.2 ppm as a 1-hour maximum in the "zone of imperative action."

F. Oxidant Exposures and Cancer

There is concern for a possible cancer-causing effect of pollution based on the findings that particulate extracts can lead to skin cancer in experimental animals, and that ozone has mutagenic properties in animals and is associated with the occurrence of chromosomal breaks in samples of white blood cells taken from occupationally exposed men. So far, no evidence indicates that residents of Los Angeles have elevated rates of cancer when compared to other urban populations with similar smoking and occupational exposures. One group of investigators *(116)* reported that in one part of Los Angeles County, men had unusually high rates of lung cancer, although women did not. This was subsequently found to be due largely to smoking and occupational exposures, and not to community air pollution as first suspected. Efforts to identify an increase in chromosome breaks in the lymphocytes of recent migrants to Los Angeles were not successful. (See also Section IX).

There is as yet no convincing evidence of any effect of exposure to photochemical oxidant on human risk of cancer. In a search for additional evidence, priority should be given to occupationally exposed groups, whose exposures can be well characterized. There is a surprising scarcity of prospective cohort studies of workers occupationally exposed to ozone.

IV. Oxides of Nitrogen

Experimental studies with nitrogen dioxide are considered risky because of the evidence that effects of the exposures under occupational circumstances may have delayed effects as long as several days after exposure was terminated. For further discussion, see Chapter 7, Vol. II, 3rd ed.

Two important findings have been reported since Vol. II, 3rd ed., was published in 1977.

Orehek *et al.* *(51)* in a study of asthmatic subjects, found that exposures to 0.1 ppm of NO_2 significantly increased their responsiveness to carbachol, an agent like histamine or mecholyl which simulates the natural bronchoconstrictive response of the smooth muscles of the airway. This demonstration of the role of NO_2 in aggravating the hyperreactivity of the airway has not been confirmed, although the similarity to effects of ozone (reviewed previously) is impressive. Studies in physically active asthmatic subjects would also be of interest. Such exposures may reflect somewhat more sensitive response patterns than are included in the EPA Draft Air Quality Criteria for Oxides of Nitrogen, which cite a level of 1.0 ppm for 15 minutes as being capable of leading to increased airway resistance in bronchitics *(116a)*.

Like carbon monoxide, some of the most interesting new findings are related to exposures to nitrogen dioxide as a component of indoor pollution, which is more fully discussed in Section VII,A). Briefly, evidence suggests that gas cooking and heating of homes, when not vented, can increase the exposures to NO_2, and that such exposures may be associated with increased respiratory problems among young children with weekly mean exposures above 0.02 ppm.

The findings of Ehrlich *(117)* that NO_2 exposures of animals may impair defense mechanisms against respiratory infections, for which a certain amount of support was found by the studies of Shy *(118)*, have stimulated some of the research on indoor pollution. But at present such risks remain in the "zone of ignorance."

V. Carbon Monoxide

Since 1977, a consensus has solidified as to the role of carbon monoxide exposures in cardiovascular disease (see Chapter 7, Vol. II, 3rd ed.). This has been reflected in several reviews *(119–121)*. The importance of indoor exposures has been increasingly recognized (see

Section VII). Carbon monoxide exposure from cigarette smoking, both for smokers and for bystanders (passive smoking), is greater than that from community air pollution. Little work has been done to define the risks of persons whose dwellings are in close proximity of heavy vehicular traffic more clearly, despite the availability of simple and portable apparatus for measuring carbon monoxide in expired air *(121a)*.

Some confusion continues on three questions. These are the background level of carboxyhemoglobin (COHb); the added risks of carbon monoxide exposures at altitude; and the question of whether an unusually sensitive population of children can be defined.

Normal persons in good health produce and excrete some carbon monoxide in their expired air at all times. The concentration of CO in expired air is usually 2–3 ppm, and the background levels of COHb in the blood are in the range of 0.5% of the circulating hemoglobin. An increase of COHb to about 2.5% has been shown to decrease the exercise capacity of persons with peripheral or coronary vascular disease, representing an increase of about 2.0%. However, there is no scientific consensus on this effect's having a threshold. Since one of the possible effects is the impairment of survival of persons suffering an acute vascular episode such as a stroke or myocardial infarction (heart attack), the risk for persons with vascular disease associated with the likelihood of exposures producing more than a 2% increase in COHb is in the "zone of imperative action."

In terms of added risks at altitude, the discussion was initiated by the concern in California about exposures to tourists who went to the Lake Tahoe basin area (above 5,000 ft, or approximately 1,800 m). Both altitude and carbon monoxide exposures interfere with the supply of oxygen to the tissues, and, therefore, together they constitute more of a stress than each alone. The California Air Resources Board, using some physiological assumptions and a set of mathematical equations as the basis, set a more restrictive standard for locations at altitudes greater than 5,000 ft. Collier recognized that the assumptions were erroneous, and Collier and Goldsmith *(122)* published a critique of the basis for the standard, which pointed out the need for some field studies to provide a more suitable basis for the regulation of carbon monoxide risks at altitude. Issues such as this also demonstrate the need to consider whether environmental conditions in resort and recreation areas should be more hygienic and free of risk than those in other locations.

As far as risks to children are concerned, the averaging times used in evaluation of CO risks are based on the expected rate of uptake of adults. This leads to such interpretations as that a resting adult requires about 8 hours to reach a level near the equilibrium concentration of COHb associated with a given fixed level of CO exposure. (Of course, exposures under realistic circumstances fluctuate over a large range.) Using adult uptake rates is quite appropriate so long as the population at greatest risks is a population with vascular disease, nearly all of whom are adults. It is obvious that children, with their smaller body mass, will, even at rest, reach equilibrium more rapidly than will adults; however, children are less often at rest. If there were a health risk of low-level COHb for a population of children, then a shorter averaging time would be appropriate. The questions then become (1) is there a population of high-risk children, and (2) do repeated small interferences with oxygen transport have a deleterious effect on some children? These concerns classify the risk as being in the "zone of ignorance."

VI. Lead

Neurobehavioral effects are among the most important and most controversial results of exposure to airborne lead. They are in addition to the recognized risks from interference with prophyrin metabolism.

A. *Proportional Contribution of Atmospheric Lead to Lead Burdens in Children*

Human populations are exposed to lead in food, water, air, and dust. Some children deliberately or unconsciously ingest nonfood items, such as dirt, paint chips, newspaper, toothpaste, and a wide variety of other items. Children who habitually ingest large amounts of such items are said to have the trait of *pica*.

For populations with high lead burdens as manifested in high blood lead levels, three questions arise: (1) what are the health risks? (2) what is the source of the lead? and (3) how is it taken into the body? The need or urgency for control depends on the answer to the first question, and the means of prevention depends on the answers to the second and third questions.

We have accumulated a great deal of information about lead burdens in children and know that they begin even before birth. It is clear that preschool children are a "most sensitive group" with respect to ambient lead pollution exposures. The most heavily exposed children have either lived in the vicinity of primary or secondary smelters or in urban areas. Table X shows mean blood lead levels for preschool children and the proportion with elevated values from a nationwide survey in the United States.

Regardless of the *source,* there are two principal *routes* for lead exposures for preschool children; by inhalation of airborne lead largely from the exhaust of motor vehicles burning gasoline containing tetra-ethyl lead, and by oral ingestion of lead in dust. The contamination of play areas by sedimented lead particulate matter is a major route of exposure for lead emitted into polluted air. In older urban housing, originally painted with lead-based paint, the dust adjacent to such housing may also be contaminated. These two sources for excess lead ingestion by urban children, paint and motor vehicle exhaust, both have two routes of exposure: a direct one, eating of paint chips or inhalation of polluted air, and an indirect one, the hand-to mouth ingestion of dirt or dust from areas where children play. If urban children have grassy or otherwise uncontaminated play areas, they ingest less lead regardless of the source. Lead ingestion is thus likely to be greater among children from poor families than among those better off financially. As Table X indicates, the lowest mean lead values are found among the comparatively prosperous and rural children, and the highest values among the poor, black, and urban children.

The United States Center for Disease Control has designated a level of 30 μg of lead per deciliter of blood (in combination with 50–250 μg/dl of free erythrocyte protoporphyrin, FEP) as evidence of "undue lead absorption" *(122a).* As Table X shows, a sizable proportion of poor, urban, and black children exceed this value. For the United States population as a whole, 4% of the children were in this category *(122a, 122b).* Such findings create urgency for answers to the questions of the sources of the lead, and its potential harm *(123).*

No doubt, lead-based paint made a large contribution in the past, as did lead in food and beverages. A British Royal Commission on Environmental Pollution *(124)* undertook a modeling effort to try to estimate the proportion of lead uptake by 2-year-old urban and rural children which came from gasoline (petrol) (Table XI). Although the ranges are broad, the midpoints suggest that a substantial part of the uptake, especially in urban children was from leaded gasoline.

Each major lead source has a characteristic distribution of lead isotopes. Yaffe *et al. (125)* studied four cases of unduly high lead absorption in California. The results pointed to paint exposure in two cases and to dust brought home from the workplace in the other two.

Table X **Blood Lead Levels in Children Aged 6 Months Through 5 Years by Race, Annual Family Income, and Degree of Urbanization of Place of Residence, United States, 1976–1980 [*(122a)*, Based on the National Health and Nutrition Survey II (NHANES II)]**

	White children			Black children		
Annual family income	*Number examined*	*Blood lead μg/dl*	*% over 30 μg/dl*	*Number examined*	*Blood lead μg/dl*	*% over 30 μg/dl*
<$6,000	256	18.1 ± 0.6	5.9 ± 1.3	917	22.9 ± 0.9	18.5 ± 3.6
$6,000–14,999	887	15.3 ± 0.5	2.2 ± 0.5	163	20.7 ± 0.6	12.1 ± 1.9
$15,000+	690	13.7 ± 0.4	0.7 ± 0.3	60	17.2 ± 0.8	2.8 ± 1.2
Degree of urbanization						
Urban >1,000,000						
Central city	133	17.4 ± 0.8	4.5 ± 1.9	143	23.1 ± 1.3	18.6 ± 2.8
Noncentral city	224	16.2 ± 0.6	3.8 ± 0.8	29	19.2 ± 0.7	(3.3 ± 1.4)[a]
Urban <1,000,000	699	15.4 ± 0.7	1.6 ± 0.4	205	20.3 ± 0.8	10.2 ± 2.4
Rural	819	13.5 ± 0.6	1.2 ± 0.5	42	18.3 ± 2.6	(10.3 ± 5.3)[a]

[a] Estimate likely to be unstable as it is based on less than 50 persons.

Table XI Comparison of Lead Exposures and Routes of Ingestion for Rural and Urban 2-Year-Old Children, Modeled for British Conditions ***(124)***

	Rural or small town	*Inner city*
Inhalation route		
Ambient air levels μg/m³ Pb	0.15	1.0
Indoor air levels μg/m³ Pb	0.11	0.75
Hours indoors/day	20	20
Volume of respiration m³/day	6.0	6.0
Lead intake from air μg/day	0.7	4.8
Percentage absorption	70	70
Lead uptake by inhalation μg/day	0.5	3.4
Gastrointestinal route		
Dietary intake μg Pb/day	47	47
Beverage tap water intake μg Pb/day	4.4	4.4
Cooking water intake μg Pb/day	0.4	0.4
Lead levels in dust:		
play areas μg Pb/day	800	1,400
Lead intake from dust μg Pb/day	80	140
Total oral intake μg Pb/day	132	192
Percentage absorption	53	53
Lead uptake from gut μg Pb/day	70	102
Percentage attributable to petrol	5–64%	6–76%
Total estimated uptake μg Pb/day	70.5	105
Amount attributable to petrol μg Pb/day	3.3–45	5.9–80

Rabinowitz *et al. (126)*, studying adult volunteers, was able to show that in Los Angeles, California, a high proportion of the lead in the body entered during inhalation. A switch in the source of lead and a corresponding change in the isotope ratio for motor fuel additives in Turin, Italy, permitted an estimate of the proportion in the blood of the population of the city which was due to gasoline. First results indicated that 24–27% of blood lead was from gasoline *(127)*. If the data were restricted to children, it is reasonable to expect that the proportion might have been as high as 50%.

Another means of estimating the source of blood lead in populations of children is observing changes that occur when lead emissions decrease. Between 1976 and 1980, the mean blood lead levels in the United States population dropped from 15.8 to 10.90 μg/dl, paralleling the decrease in the use of leaded gasoline during the same period (Figure 6) *(128, 128a)*. In the Boston area a mean decrease in umbilical cord lead levels of 11% per year occurred between 1979 and 1981 *(129)*.

We may never know and probably never need to know exactly what

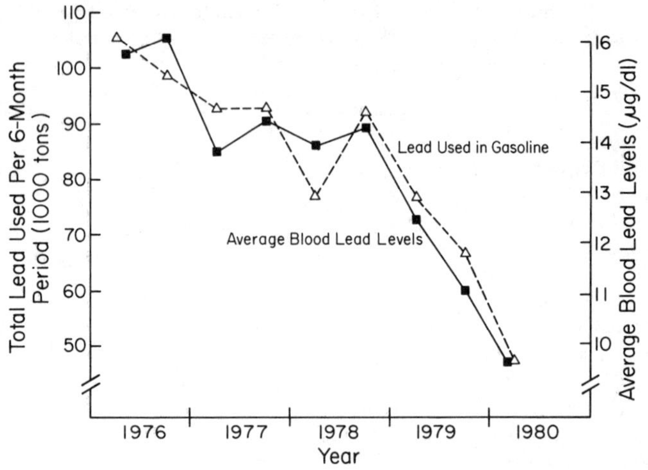

Figure 6. Parallel decrease in the average blood levels obtained in the Health and Nutrition Survey and amount of lead consumed as a motor fuel additive by year in the United States.

fraction of the high levels of lead in the blood of urban preschool children was due to motor vehicles and what fraction was due to paint. We do know how to reduce these levels by reducing industrial and motor vehicle lead emissions and by providing clean play areas and nontoxic paints.

B. Relationship between Atmospheric Lead and Body Burden

The relationships between atmospheric lead levels and lead body burdens have been reviewed by Chamberlain *(130)* and commented on by Goldsmith *(131)* and by Russel-Jones *(132)*. Chamberlain concentrates on the incremental increase in blood lead with each unit increase in atmospheric lead. Goldsmith *(131)* presents data for school children from middle-income census tract homes in the Los Angeles, California, area which show an unusually steep gradient of blood lead with increases in atmospheric exposure. Russel-Jones *(132)* estimates the proportional contribution of gasoline to lead levels in a 2-year-old British urban child as 69% and cites the 70% drop in mean blood levels in Japan since 1967 as reflecting the banning of lead in motor fuel.

Above 30 µg Pb per deciliter of blood, there is interference with the synthesis of hemoglobin and other porphyrin-related biochemical processes. The effect may begin with levels of 15–20 µg/dl *(133)*.

Children with pica may have such high uptakes that convulsions, brain damage and death, can occur *(134–136)*. Studies of minimally brain-damaged children have shown that they have elevated blood lead levels, but one cannot determine from cross-sectional studies whether the lead contributed to the brain damage, or whether brain-damaged children ingest more lead than do normal ones *(137, 138)*.

C. Neurobehavioral Effects of Lead in Children

A new chapter in our understanding of the possible health effects of lead opened with the studies undertaken by Needleman *et al.* *(139)* of intelligence tests and classroom behavior of children whose individual cumulative lead exposures were estimated by lead levels measured in their deciduous teeth. In this study, the investigators systematically reviewed the possible difficulties which could occur on the basis of experience in other studies. They designed their study to avoid these difficulties, where possible:

1. Whereas most previous studies had used blood lead as the marker of recent exposure, the investigators used lead levels in the dentine of deciduous incisor teeth, which is an indicator of exposures during previous years.
2. Whereas most previous studies had obtained subjects from clinics or from volunteers, the investigators used all accessible first- and second-grade children.
3. Confounding variables including perinatal, nutritional, socioeconomic, genetic, and cultural factors can affect intellectual and behavioral capabilities. Confounding with poverty is a particular problem; these investigators used covariance analysis to minimize the effect of confounding.
4. In order to avoid insensitive tests of intellectual performance, well-standardized tests were used; staff were well trained and blind to the lead exposure categories of the children they were testing.
5. A relatively large sample size is needed for such studies in order to have adequate sensitivity, that is, in order to detect as being statistically significant a difference of a modest amount. In addition, comparing the psychological tests for children in the high-PbT (lead in teeth) and low-PbT categories improves the chances for identifying a difference.

Originally the high-lead (PbT) group was chosen on the basis of being in the highest decile (upper 10% of the group), i.e., greater than 24 ppm of PbT. After assuring that the high-lead group all had duplicate analyses greater than 20 ppm, as well as interviews and satisfactory psychological tests, there were 58 children in the high-PbT group. The low-PbT group had 100. All the children, including those with intermediate levels of PbT, had their classroom behavior evaluated by their teachers.

Of the 58 high-lead children, 23 had had their blood lead levels measured 4 years earlier (1973–1974) and had a mean of 35.5 μg/dl. Of the low-lead children, 58 had such tests with a mean of 23.8 μg/dl.

The two communities from which the school samples were drawn were part of the Boston, Massachusetts, Standard Metropolitan Statistical Area (SMSA). Most of the residents live in housing built before 1950 (91.2 in Chelsea and 94.0% in Somerville, compared to 68.9% for the state of Massachusetts as a whole). As "noncentral cities" in a large SMSA, the white children in this study would have been expected to have a mean blood lead level of 16.2 μg/dl according to the survey summarized in Table X. The mean levels for even the low-PbT children were higher than the United States average. Because of the high proportion of old housing and the relatively high average blood lead, it is reasonable to assume that these children had a relatively high exposure to lead originally in paint.

The principal results of the study as shown in Table XIIA were that, after taking into account the possible influence of covariables by covariance analysis, there was a statistically significant association of high PbT and impairment of verbal performance, auditory discrimination, language processing, and attention and their reflection in the

Table XIIA Test Results for First and Second Grade Children of the Possible Relationship of Lead Levels in Deciduous Teeth and Psychological Tests

Test	*Low-lead group mean*	*High-lead group mean*	*Significance*[a]
Full scale IQ	106.6	102.1	0.03
Verbal IQ	103.9	99.3	0.03
Performance IQ	108.7	104.9	0.08
Seashore rhythm	21.6	19.4	0.002
Sentence repetition	12.6	11.3	0.04
Reaction time (12-second delay)	0.41	0.47	0.001

[a] These are two-sided tests, treated as individual results, with covariables of mother's age at the child's birth, mother's educational level, father's socioeconomic status, number of pregnancies, and parental IQ taken into account.

Table XIIB Stratification for Pica in Analysis of Verbal and Performance IQ Tests for First and Second Grade Children in Association with Levels of Dentine Lead

(Means ± Standard Deviations)[a]

Test	*Pica*	*Low lead*	*High lead*	*Significance*
Verbal IQ	+	101.8 (10) ±12.0	99.2 (15) ±13.6	ns
	−	105.1 (87) ±12.8	98.8 (42) ±12.6	$t = 2.62$ $p = 0.0091$
Performance IQ	+	106.9 (10) ±12.3	103.4 (15) ±14.4	ns
	−	110.7 (87) ±13.5	106.0 (42) ±11.7	$t = 2.01$ $p = 0.044$

[a] ns: not significant. *t:* Student's *t* test.

intelligence quotient (IQ). When the high- and low-PbT children were compared, there was a mean difference of 3.5 IQ points. Although this may seem a small mean difference, particularly in light of the variability of IQ testing, it reflects the fact that children with elevated lead levels were three times more likely to have a verbal IQ below 80. Although 5% of the low-lead children had IQ levels in what is called the superior range of >125, no child with an elevated lead level scored in this range. Teachers' ratings of the children showed a dose–response relationship between PbT and such problems as distractability, hyperactivity, daydreaming, inability to follow directions or sequences, and impairment of overall functioning (Figure 7).

A history of pica was present in 10.9% of the low-PbT group and in 28.8% of the high-PbT group. When the data were reanalyzed within strata by the history of pica, there were still significant differences in the nonpica group (Table XIIB).

Similar studies have been reported with similar findings from both Great Britain and Germany *(140–144)*. Ratcliffe *(140)* reported an impairment of mental development and abnormal behavior among British children in association with elevated lead in blood. Yule *et al.* *(141)* studied 166 children living near a British lead smelter in the vicinity of London. Their mean blood level was 13.5 μg/dl, with the highest quartile ranging from 17 to 32. The children were from 6 to 12 years of age. Covariance analysis with social class "partialled out" showed a significant gradient for blood lead with verbal IQ and for full-scale IQ. The estimated proportion of the intelligence associated

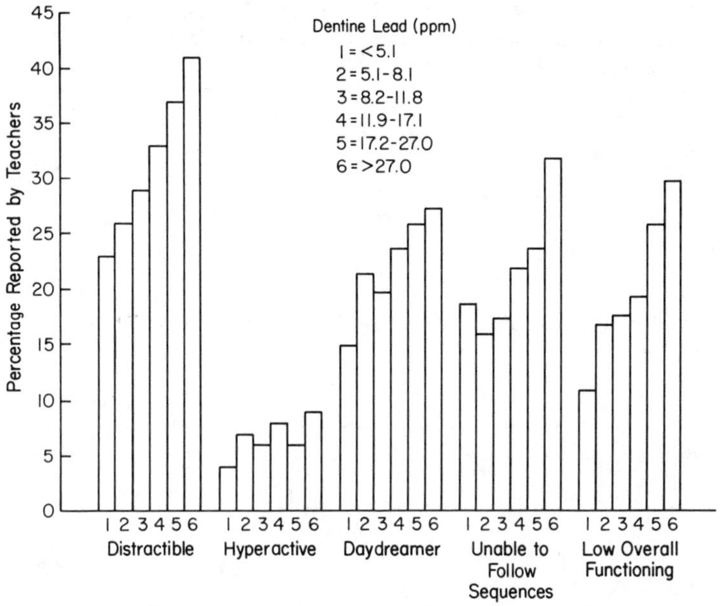

Figure 7. Relationship between teachers rating of School children in the Chelsea and Somerville (Massachusetts) schools and dentine levels of deciduous teeth. There is a general increase in unfavorable ratings as dentine lead concentrations increase. (Reproduced with permission from Ref. *139*)

with blood lead was 2.5–4.5%. Smith *et al.* *(142)* undertook a study in three London boroughs more closely resembling the areas in the Needleman *et al.* *(139)* study. They measured PbT in 6- and 7-year-old children. From a total of 2,357 eligible children, 403 were selected on the basis of their PbT, they were from two social classes: manual and nonmanual workers. After adjustment for different tooth types, three groups were formed: low, medium, and high-PbT, with about 150 children in the two extreme groups and 100 in the intermediate or medium group. The children whose fathers had manual and nonmanual occupations in the low group had mean PbT in ashed samples of 2.8 and 2.7 μg/g of ash, respectively; the intermediate groups had means of 6.0 and 5.9, and the two high groups 11.2 and 10.9 μg/g of ash. Home interviews, psychological tests, and school performance evaluations by teachers were the dependent variables. Lead levels were linked with home cleanliness and with maternal smoking. The teachers' ratings showed consistently lower values in the high-PbT children, but the differences did not reach statistical significance. Without controlling for covariates, there were statistically significant differences between low- and high-PbT groups; but once the social variables were taken into account, the differences were slightly reduced and became statistically nonsignificant, but only slightly diminished in their gradients. The authors cautiously conclude "it is uncertain whether measurable improvements in the children's intelligence, educational attainment, behaviour, etc. would result from a reduction in the body-lead burden of children."

Winnecke *et al.* *(143)* compared results from a battery of psychodiagnostic tests on primary schoolchildren from Duisberg, Germany, having low PbT (<3 ppm) and high PbT (>7 ppm). A deficit of 5–7 IQ points in the high-lead group was of "borderline significance," but significant differences were found for perceptual–motor integration tests and problem-solving capabilities. A larger study was then undertaken *(144)* in the nonferrous metal mining and processing community of Stolberg, Germany, in which emissions controls had reduced the airborne lead levels from 1.3–7.5 $\mu g/m^3$ in 1973 to 0.4–2.0 in 1980. The study examined 115 children whose mean PbT was 6.16 ppm. Analysis was by stepwise regression, including birth and family history, as well as social variables. Significant or nearly significant associations were found between PbT and perceptual–motor integration, reaction–performance, distractability, restlessness, lack of information, and wasting of time. These last four ratings were by mothers of the children. After sociohereditary background variables had been taken into account, the association between IQ and PbT was no longer significant, but, even after adjustment, the high-PbT group scored 4.6 points lower on the mean IQ tests than the low-PbT group.

D. Interpretation of the Studies

Since five different investigators in three countries have had generally similar findings, the biological question of whether lead exposures of children can affect their school performance and psychometric tests seems beyond question. The two remaining questions are interpreting the interaction of poverty or low socioeconomic status with lead exposures and determining the extent to which this effect can be reduced by reducing atmospheric lead levels. Both of these questions have been debated in major scientific journals and in United States Congressional hearings *(145–152)*. A British Royal Commission *(124)* recommended that lead be phased out of gasoline by 1990, within the framework of the decision process of the European Economic Community.

One aspect of the argument in the United States is trivial; Marshall *(151)* quotes EPA consultants as saying that there is a question as to whether Needleman's original study *(139)* reached "statistical significance": that is, a result that would have only a 1 in 20 likelihood ($p = 0.05$) of occurring by chance. First, no such probability level is required by administrative law; second, there are well-established methods for pooling the probabilities of different studies *(153, 154)* when several have results of "borderline" significance.

As to interaction with socioeconomic factors and the finding that in covariance analysis, the inclusion of such factors diminishes the differences by such an amount as to make the results "not statistically significant" [as was reported by Smith *et al. (142)* and Winnecke *et al. (144)*], there is reason for doubting the appropriateness of covariance analysis. Covariance analysis makes the implicit assumption that lead exposures can affect the other variable (socioeconomic status) as readily as socioeconomic status can affect lead exposures. It is more plausible that socioeconomic level affects lead exposure than the reverse, in which case, alternative methods of analysis would be more suitable. For example, stratifying, as Smith *et al. (142)* started to do, would help to determine whether the relationship of lead exposure and performance was restricted to the lowest socioeconomic group. There are more complicated methods for dealing with such problems *(40)*, but, for the present, it suffices to say that the analytical strategy used may have been insensitive. Nevertheless, even if the effect of preschool lead exposures on school performance and psychometric tests is restricted to the lower socioeconomic group or the inclusion of social variables in the analysis slightly weakens the association, neither finding diminishes the urgency for attempting to abate it. The effect remains in the "zone of imperative action" (Figure 1), not because of its widespread prevalence or the level of statistical significance reached, but because of the long-term implications of school performance and the potentially great benefit from abating these effects.

As for the importance of lead from motor vehicles to the lead burden of low-income urban children, the striking decline of 40% in the blood lead of United States children between 1976 and 1980 associated with the decline in motor vehicle lead consumption already provides one answer to the question *(155)*. Figure 8 shows the parallel drop in blood levels of black preschool children in New York City and drops in air lead and lead used in motor vehicle fuel *(128)*. The demonstration of the importance of the hand-to-mouth route of exposure and the realization that much of emitted lead becomes part of the settled dust of streets and play areas provide evidence for this mechanism. This

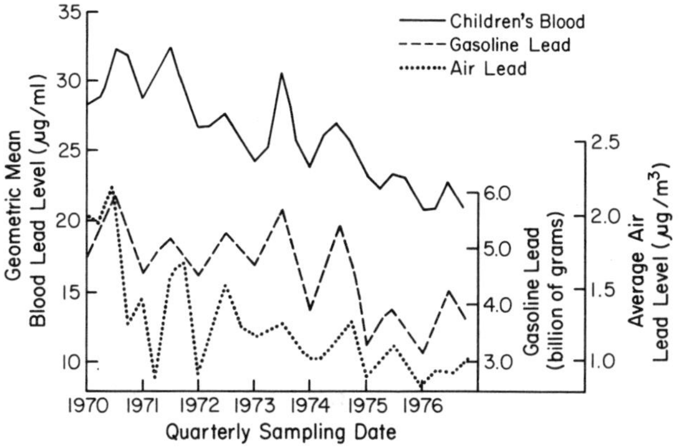

Figure 8. Parallel changes in lead in air, lead used for motor fuel and blood lead levels of black pre-school children in New York City *(128)*.

relationship has been vividly confirmed by a controlled trial of dust control measures among a population of black inner-city children. Charney *et al. (155a)* introduced dust control measures in the homes of 14 children of between 15 and 72 months, who had at the outset confirmed venous blood lead levels of between 30 and 40 μg/dl; these were compared to 35 otherwise similar children who also had had their homes "deleaded," but with no special attention to dust control. The 14 experimental children also had biweekly visits by a dust control team who measured lead levels in dust, removed all dust which could be a source for the children's exposure, and encouraged hand washing. In the experimental group, the average blood lead fell from 37.6 to 31.7 μg/dl, and the average free erythrocyte protoporphyrin (FEP) fell from 234 to 144, whereas the comparison group had no reduction in blood lead. The children with the highest levels had the greatest reduction.

This work has established that exposures of preschool children over a period sufficient to increase blood lead to greater than 30 μg/dl can lead to impairment of attention and language skills (or *perceptual–motor integration*) with associated educational handicaps and poor classroom performance. Such effects may be aggravated by social, economic, genetic, and other environmental factors. The effect can be prevented by reducing all forms of lead pollution. Restricting lead-based paints and eliminating lead from motor fuel are two essential steps. Dust control measures and provision of unpolluted play areas for inner-city children are measures which will help reduce the immediate impact of the problem. The data do not support the existence of a threshold. Further studies to monitor the progress of abatement are urgently needed.

E. Lead and Hypertension

Pirkle *et al. (155b)* have demonstrated for white males aged 40–59 an association between blood lead levels and high blood pressure. The data are from the second National Health and Nutrition Examination Survey carried out between 1976 and 1980. Using multiple regression to "adjust" for effects of age, body mass, nutritional and biochemical variables, there is persisting association of blood pressure and blood lead level. The associations were not altered whether or not hypertensive medication was used, whether the subjects are urban or rural residents, whether the subjects had incomes above or below $10,000, or when subjects' education, the season of the year, and the subject's family history of hypertension were considered.

Since blood lead levels decreased strikingly during this period from an average for adult white males of 16.7–10.5 μg/l, even a very small unrelated decrease in blood pressure could have accounted for the association. It may be premature to predict, as *Pirkle et al.* did, the number of hypertension-related cardiovascular events which would be prevented by lowering the population's body burden of lead.

Among 431 civil servants, age 24–55, in Paris, France, Moreau *et al. (155c)* and Orssaud *et al. (155d)* found a consistent association between hypertension and blood lead when the systolic blood pressures were adjusted for body mass index, age, and alcohol consumption.

VII. Indoor Pollution

There are four principal reasons for concern about the health effects of indoor air pollutants, in addition to the overriding concern that, like the weather, "Nobody is doing anything about it!" It has always been recognized that populations usually spend more time within buildings than they do outside; that the surface properties of building interiors affect introduced pollutants; that there are important indoor sources of pollution; and that personal preferences and behavior, as well as cultural practices, have a great influence on indoor air quality (for reviews and comment in general, see *156, 159*; see also Chapters 1 and 4, this volume).

If experience with occupational and ambient pollution is relevant, the level of awareness of potentially at-risk groups and the seriousness of long-term health effects will be critical in mobilizing an effective interventive effort to prevent some of these possible effects.

This review includes the possible effects of gas cooking and heating equipment and the resulting levels of NO_2 on lower respiratory tract disease, the effects of indoor smoking on nonsmokers, possible hazards from carbon monoxide and organic vapors, and finally the possible relationship of exposures to building materials to the later development of cancer.

A. Gas Heating and Cooking Appliances, Nitrogen Oxides, and Lower Respiratory Tract Conditions

For those of us who early learned that the temperature within a closed combustion chamber is important to the yield of nitrogen oxides, it comes as surprise to learn that there is a substantial amount of nitrogen dioxide produced from the open flame of a gas stove or hot water heater.

Spengler *et al. (160)* as a part of the six cities study, measured and compared SO_2 and NO_2 inside and outside homes and pointed out the public health importance of better defining the indoor NO_2 exposure risk. Sexton *et al. (161)* modeled the exposure risks and showed that long-term NO_2 exposures are more influenced by the type and amount of fuel used for cooking than by the level of outdoor pollution. For example, in the least polluted community of the six community study, Portage, Wisconsin, with an annual mean ambient NO_2 of 15 $\mu g/m^3$, residents with gas cooking stoves were estimated to have weekly average personal exposures of 41 $\mu g/m^3$. Residents of a more polluted community, Steubenville, Ohio, with mean ambient levels of 62 $\mu g/m^3$, if they used electric cooking stoves, would have mean weekly exposures of 39 $\mu g/m^3$, but if they had gas stoves, their exposure would be 70 $\mu g/m^3$.

The importance of these exposures came into focus with a report of a British team, Melia *et al. (162)*, which was carrying out a national study of respiratory symptoms among school children. The investigators recognized that Ehrlich's animal studies *(117)* had indicated that NO_2 might increase the susceptibility to respiratory infections, and that Shy *(118)* had put forward evidence that 7- and 8-year-old children had lower lung function in a community with relatively high NO_2 than comparable children from a relatively unpolluted community. Melia *et al. (162)* found that both boys and girls living in homes with gas cooking had more frequent reports of cough, colds going to the chest, and bronchitis; girls from such homes more frequently reported wheezing. The results were stratified by age and social class. Of 1,240 boys living in homes with gas cooking, 67 had four or more respiratory symptoms, compared to 49.7 who might have been expected to have such symptoms on the basis of rates in homes with electric cooking. Comparably for girls living in homes with gas cooking, 54 had four or more symptoms, whereas only 27.3 were expected on the basis of rates in homes with electric cooking. Both the effect of gas and the overall frequencies of symptoms were increased in the homes of the lower social classes. The effect of latitude, outdoor levels of SO_2 and smoke, family size, and overcrowding did not explain the "cooking fuel" effect.

Keller *et al.* *(163)* failed to confirm such an effect among upper-middle-class United States midwestern families. The more extensive and systematic studies of Speizer *et al.* *(164)*, Melia *et al.* *(165)*, Goldstein *et al.* *(166)*, Florey *et al.* *(167)*, and Ware *et al.* *(167a)* seem to point to a linkage between NO_2 exposures in the home, use of gas for cooking, and childhood respiratory disease or impairment of lung function, independent of the effects of social class and parents' smoking.

For example, in a multivariate analysis, using a log-linear model, of the influence of the relevant variables acting together on the prevalence of "serious respiratory disease" before age 2, Speizer *et al.* found that the low-social-class children had a 13% excess, those with one parent who was a smoker had 32% excess, those whose families used gas cooking had a 12% excess, and all these effects were statistically significant at the 0.05 level.

However, subsequent studies of these cohorts of children failed to confirm the specific effect of exposures from gas stoves on respiratory disease but did suggest that exposure to gas stoves may be associated with reduced pulmonary function *(167a)*. Children of smokers, however, had both impairment of function and more frequent symptoms in this analysis (see Section VII,B).

Melia *et al.* repeated their earlier study, which was based on 1973 data *(162)*, with data for 1977 *(165)*, adjusting as well for smoking, and found an association between gas cooking and respiratory symptom frequencies, independent of other variables, which was highly significant for urban boys and nearly significant for urban girls. In rural areas an association was found for younger girls only. Goldstein *et al.* *(166)*, in one urban community, determined outdoor, kitchen, and bedroom weekly mean concentrations for homes with gas and electric cooking. The outdoor mean NO_2 averaged 0.014–0.024 ppm; in 428 kitchens with gas cooking, the weekly mean was 0.112 (range

0.003–0.317); in 87 electric kitchens the weekly mean was 0.018 (range 0.006–0.188). In the bedrooms of 107 children with gas cooking the mean was 0.031 (range 0.004–0.169), compared to the means of 18 children's bedrooms in houses with electric cooking facilities of 0.014 (range 0.003–0.037).

Thus a gas stove may be estimated to contribute about 0.040 ppm to overall exposure, which in this area was more than twice the outdoor level. The presence of a cigarette smoker in the home contributed a further 0.008 ppm, but as discussed later, NO_2 is not the only relevant product of cigarette smoking to affect health.

Florey *et al.* *(167)* studied not only respiratory symptoms, but also lung function, in children in the same community in which indoor NO_2 determinations were made. Although a clear dose–response relationship was shown for at least one symptom, (Table XIII), no relationship was discovered for lung function. These data for 6- to 7-year-old children tend to reinforce the impression that the possible effect of NO_2, if present, predominates in young children; this is also supported by the follow-up of the children first studied in 1973 by Melia *et al.*, in which the relative risk seemed to decrease with increasing age. Of course, as children grow older, a larger proportion of their time is spent outdoors.

Hosein and Bouhuys *(168)* have pointed out that studies of children over 7 years and of adults in two parts of the United States failed to find an association with gas cooking, but it is precisely the younger children who spend such a large proportion of their time in the house, so the observed predilection for them to respond to NO_2 is plausible and consistent with the reaction patterns to parental cigarette smoking.

Table XIII Crude Prevalence of at Least One Positive Response to a Respiratory Questionnaire among Children for Whom Bedroom Weekly Mean NO_2 (ppb) Was Measured *(167)*

	Weekly mean NO_2 in Bedroom			
	0.0–0.019	*0.020–0.039*	*0.040+*	*Total*
Boys	10/23	11/19	9/13	30/55
	43%	58%	69%	55%
Girls	11/25	9/15	6/8	26/48
	44%	60%	75%	48%
Both sexes	21/48	20/34	15/21	56/103
	44%	59%	71%	54%

A Dutch team, building on the excellent longitudinal studies of van der Lende *et al.* *(168a)*, has looked at the cross-sectional and longitudinal effects of indoor NO_2 and exposures to tobacco smoke on respiratory conditions and lung function among women living in two towns *(168b, 168c, 168d)*. They measured NO_2 levels using diffusion tubes in bedrooms, living rooms, and kitchens of 152 homes, 47% of which were farm houses, and the remainder in small rural villages. Most of the homes used gas appliances. They found among data for 97 nonsmoking women a negative association between NO_2 exposures and several parameters of lung function; by contrast they did not find any association between NO_2 exposures and the rate of decline with age in pulmonary function. The ranges for estimated exposures were 10-391 for kitchens, 8-198 for living rooms and 8-53 for bedrooms (measured in units of micrograms per cubic meter). Despite the relatively greater exposures in the kitchen, the significant associations were with levels in living and bedrooms. Pulmonary function values were adjusted for age, height, passive smoking, and educational level.

In the same population a similar impairment of lung function by passive smoking was observed by cross-sectional analysis, but no association was found between passive smoking and rate of lung function decline with age *(168d)*.

B. *Indoor Pollution with Cigarette Smoke: The Passive Smoking Syndrome*

Being exposed to some other person's cigarette smoke is often unpleasant and may endanger the health of unusually sensitive persons. Since this problem is extensively reviewed elsewhere *(169, 170)*, only the evidence that relates to other air pollution health effects studies will be cited.

The health effects of carbon monoxide are discussed in Chapter 7, Vol. II, 3d ed., and briefly in Section V. One can estimate the relative magnitude of recent exposure to carbon monoxide by determining COHb. Passive smoking and outdoor pollution make a relatively small contribution compared to cigarette smoking and work exposure (for which levels may be high enough to be lethal). Passive smoking exposure to carbon monoxide generally will not lead to high exposure. Its impact is usually in the range of an additional 0.2–4% of an adult's total hemoglobin. The average pack-a-day adult smoker who inhales has about 5% COHb *(119–121)*. An indoor recreational exposure may occur in ice skating rinks, as a consequence of using gasoline-powered equipment for rink maintenance, with mean levels between 23 and 90 ppm and estimated COHb levels in hockey players between 7% and 16% *(171)*.

One of the most serious effects of passive smoking is on respiratory disease in small children. Harlap and Davies *(172)* reported in 1974 that babies in Jerusalem whose mothers smoke had substantially more frequent hospital admissions than did babies of nonsmoking mothers. Colley *et al.* *(173)* followed 2,205 British babies from birth and found that the risk of bronchitis or pneumonia during the first year of life was twice as great if the parents smoked as it was for babies whose parents did not smoke. At later ages the effect was less consistent. Other variables studied could not explain the difference. Fergusson *et al.* *(174)* followed a group of New Zealand children for the first years of life; if the mother smoked, the risk of bronchitis or of pneumonia during the first year was also doubled. There seemed to be a dose–response relationship for the amount of smoking. Effects of father's smoking were small, and adjustments for age of mother, education of mother, family size, and living standards did not explain the effect of mother's smoking. Burchfiel *et al.* *(175)* in Michigan and Goren *et al.* *(176)* in Israel have found similar associations.

Kauffmann *et al.* *(176a)* have shown effects of husband's smoking on respiratory conditions among their wives, using the data from the French national respiratory survey.

Ware *et al.* *(167a)* in two successive examinations of 10,106 white children in six cities found that maternal cigarette smoking was associated with 20–35% increase in the rates of eight respiratory illnesses and symptoms investigated and paternal smoking with smaller but substantial increases. Levels of $FEV_{1.0}$ were significantly lower for children of current cigarette smokers than for children of nonsmokers in both examinations, and these effects were linearly related to daily cigarette consumption.

It is thought that small infants, particularly, are unable to escape being exposed to the smoke their parents produce, which accounts for their higher risk of respiratory conditions.

The risk of respiratory disease in infants and children as a result of the smoking of their parents is in the "zone of imperative action," despite occasional negative reports, some of which are due to choice of relatively resistant populations, small sample size, or too rigid application of statistical criteria.

C. Organic Fumes and Formaldehyde

Urea-formaldehyde foam insulation (UFFI) not only keeps houses from losing heat in winter and gaining it in summer, but its propensity also to give off formaldehyde fumes has generated a great deal of heated discussion *(6, 177)*. In Canada its use has been abandoned. Formaldehyde, also an ingredient in photochemical pollution *(178)*, has been shown to be mutagenic for bacteria and mammals, and, in sufficient doses, can produce nasopharyngeal cancer in rats and mice. The lowest dose which produces eye or respiratory irritation has been given by various authors as from 0.02 to 0.4 ppm. Indoor formaldehyde levels in excess of 0.1 ppm have been reported in a number of countries. Since mobile homes and other low-cost housing are likely to have such insulation, the issue has unusual economic interest.

As more energy-efficient buildings are built and used, occupants often complain of a variety of vague symptoms such as fatigue, dizziness, headache, and eye or respiratory tract irritation. Although often vague, they nevertheless are clearly associated with the time of being present in the building. This has been called the "sick building syndrome" *(7)*, although it is not the building which is sick. Formaldehyde and other organic vapors have been considered as possible causes *(177)*. Weselowski has presented the cumulative frequency distribution of formaldehyde levels for non-mobile home residences. The median value found was 52 ppb for a 7-day average *(178a)*.

A large variety of other organic vapors, including solvents, cooking fumes and odors, spray can diluents, various cosmetic and drug preparations, wood preservatives, and materials used for hobbies and crafts, may be emitted in and lead to health effects in buildings. Most of such effects are of local relevance and are not likely to be identified in general community studies.

D. Building Materials and Cancer Hazards (See also Section VII and Chapter 1)

Asbestos is one of the best-known carcinogens. Occupationally exposed persons may develop mesothelioma, a form of cancer of the lining membranes of the chest and abdominal cavity, which is a relatively rare tumor among those with no asbestos exposure. With sufficient occupational exposure, especially among smokers, high rates of lung cancer and elevated rates for cancer at other sites have been reported *(179–181, 181a,b)*. Asbestos is also much used in construction, as insulating material and as sheets or pipes of asbestos cement. Asbestos cement products should not produce a respiratory hazard to persons living in the buildings in which they are used. Respiratory hazards are therefore principally related to dust produced from remodeling or demolition. In some schools with asbestos insulation, vandalism has added to the respiratory hazard. Insulation materials readily become a source of airborne asbestos and must be installed, repaired, or removed with the utmost care to prevent disseminating the hazardous dust *(181b)*.

Granitic materials contain minute amounts of natural radioactivity and give off radon gas, which decays with the emission of radiation. Building materials of stone, brick, and other granitic minerals continuously emit small amounts of radon. The cancer risks from radon exposures have been quantified primarily for underground miners, and there is general agreement that there is no threshold for such a cancer risk. Although residents of stone houses have much lower exposures and lower excess risks than do miners, it can be estimated that the excess risks from indoor radon exposures may be in the thousands of cases in the United States *(94)*.

Energy-efficient housing poses an especially high risk because of the low rates of air exchange, which can allow radon to accumulate in such places as basements and crawl spaces.

E. Protection Against Risks from Indoor Pollution

Because there is no governmental mechanism for dealing with risks from indoor pollution, knowing the facts does not lead inevitably to an effective course of protective action. The most urgent and beneficial option is to urge mothers of small children not to smoke in order to prevent respiratory illness during infancy. To require that gas stoves be vented is a proposal worthy of discussion; sensible restrictions on the use and handling of asbestos are being imposed. Limiting the use of UFFI to circumstances in which the off-gassing of formaldehyde will not produce either symptoms or detectable increases in background levels seems desirable. For further discussion, see references *158*, *160*, and *170*.

VIII. Cancer and Ambient Air Pollution

An urgent long-term problem for air pollution health effects is the question, Do some air pollutants in some communities under some circumstances cause increases in cancer? In 1975 the evidence that general community pollution had caused cancer was not convincing, despite the realization that carcinogenic activity could be found in particulate samples. Evidence that some increases in cancer risk had occurred in proximity to point sources of asbestos or arsenic was noted *(182)*. The frequently observed excess of lung cancer among urban as compared to rural residents has often been interpreted as though the excess must be caused by exposures to urban air pollution. This assumption would be inappropriate if the following could be shown:

1. There really is no more exposure to polluted air among urban than among rural residents.
2. Other exposures such as occupation and cigarette smoking could account for the apparent urban excess.
3. The rural–urban gradient was least in the areas in which rural–urban pollution gradients were greatest.

There is little likelihood that finding (1) will be substantiated, and as a generalization it seems unlikely. Chapter 7, Vol. II, 3d ed., devotes considerable attention to discussing the findings of type (3).

Between 1978 and 1983 the possibility of (2) has been evaluated by several authors, especially Goldsmith *(183)* and Weinberg *et al.* *(183a)*. Goldsmith *(183)* reviews the data for occupational contributions to the "urban" factor in lung cancer and speculates on the possibility that urban residents began smoking earlier, which might readily account for the urban lung cancer excess. Weinberg *et al.* *(183a)* analyze in detail the differences in cancer risk from smoking in two communities in Allegheny County, Pennsylvania, in which an apparent pollution association with cancer was present. When smoking and occupational differences were taken into account, most of the lung cancer differences were found to be due to cigarette smoking, and effects which could be attributed to air pollution were minimal. This study provides a model of the way the "urban" factor should be adjusted for the differences among populations in the intensity and duration of smoking.

A thorough review of the situation at a conference held at the Karolinska Institut in Stockholm was published in 1978 *(184)*.

At that conference Friberg and Cederlof *(185)* said,

> it is at present not possible to provide epidemiological proof that air pollution per se causes an increase in lung cancer incidence and much less to ascribe an observable increase to a specific substance. At the same time, failure to verify a causal hypothesis in a strict sense does not justify a conclusion that air pollution is causally unrelated to lung cancer.

In the summary, however, it is stated that "Combustion products of fossil fuels in ambient air, probably acting together with cigarette smoke, have been responsible for cases of lung cancer in large urban areas, the numbers being produced being of the order of 5–10 cases per 100,000 males per year" *(186)*.

A committee made up of leading scientists from the American Cancer Society, the National Cancer Institute, other leading organizations, and the International Agency for Research on Cancer concluded that "general air pollution at present has very little effect, if any, on the lung cancer death rate" *(187)*.

A. Use of Mutagenicity Tests

Three widely used laboratory procedures for assessing mutagenicity have been applied to estimation of the risks from air pollution exposures, on the assumption that mutagenicity is a reasonably good predictor of carcinogenic risk. In addition, the assumption that mutagenicity would be reflected in unfavorable reproductive outcomes has led some investigators to assess the possible association of pollutant exposures with reproductive outcome. As yet there are no reports suitable for evaluation.

The situation with respect to coal combustion products and motor vehicle combustion products has been extensively reviewed at a second Karolinska Institut Conference *(188)*. The consensus report (*188*, p. 14). stated:

> Diesel vehicles have higher mutagenic emission rates than gasoline noncatalyst vehicles, which in turn have higher rates than gasoline catalyst vehicles. Residential wood stoves have higher mutagenic emission rates per joule than residential oil heaters. No significant difference in mutagenic emission rates per joule have been reported between coal and oil in similar utility boilers. The mutagenic emissions are dependent on the mode and efficiency of the combustion, and are generally low in well operated units.

In its recommendations on risk assessment, however, the conference concluded:

> In issues involving long-range decisions it will be important to use the combined results of animal bioassasys and short term *in vitro* tests and whenever possible epidemiological evidence. *In vitro* bioassays should be used for screening purposes, whenever appropriate and with due caution, to provide guidance on the comparative carcinogenicity of fossil combustion products, particularly when a series of engineering alternatives for control of emissions are under consideration (*188*, p. 4).

No convincing evidence has been found for the carcinogenicity of diesel emissions in community epidemiological studies *(189)*.

The three basic methods for assessing mutagenicity are bacterial bioassays (using the Ames test), cultured mammalian (including human) cells, and study of chromosomal changes in cultured human lymphocytes, and sister chromatid exchange (SCE) *(190, 191)*.

The Ames test *(191a)* tests bacterial growth in cells of strains which have alterations in their nutritional requirements. Mutagenic agents can cause reversion of these strains and resulting growth. An important advantage of the tests conducted on the human lymphocytes is that exposures may be either *in vivo* or *in vitro*. Cigarettte smokers have increases in SCE, as do workers occupationally exposed to benzene, but miners operating diesel engines did not have any increases when compared to office employees and construction workers *(192)*. Although urinary detection of mutagenic activity has been carried out for industrially exposed workers, as yet no one has reported that community exposures are related to increases in urinary mutagenic activity. Such evidence would be important in placing the uses of these *in vitro* tests in the proper perspective. The possibility that chromosomal changes in circulating lymphocytes could be associated with oxidant pollution was investigated in Los Angeles, California, with negative results *(193)*. Thus, the principal contribution of the rapid developments in testing for mutagenicity of pollutants is its use for screening when alternative control strategies are being assessed.

B. Assessment of the Situation

Extrapolation from the experiences of industrially exposed workers permits a crude estimate of possible risk for asbestos *(193a)* and radon *(158)* as indoor pollutants. Similarly, extrapolation from data on possible exposures to benzo[*a*]pyrene can be used to deal with the possibility of risk from combustion products *(194)*. Speizer concludes that the effect of air pollution on lung cancer is greater than zero but is unlikely to exceed 2% of all lung cancer in urban males. Therefore, there is no basis for altering the 1978 views of Friberg and Cederlof *(185)*.

There is general agreement that there is no threshold for carcinogenesis; hence, there is virtually no way to utilize the abundant laboratory data on carcinogenesis for establishing criteria for complete protection against carcinogenic hazards of pollutants. This limitation gives greater relevance to criteria based on other effects for abating the hazards which may be related to carcinogenesis *(195)*.

Two principles are available for controlling point source emissions of pollutants which may be carcinogens: the in-plant environment must be maintained at a level which minimizes the risk for workers, according to principles of good occupational health practice *(195)*, and estimated community exposures may be evaluated on the basis of extrapolation, as illustrated by Speizer *(194)*.

At present no valid dose–response data are available for carcinogenic effects of community air pollution; we remain at present in the "zone of ignorance."

IX. Toxic Pollutants and Odor

The so-called toxic air pollutants are receiving increasing attention, because of their potency and possible long-term effects. Sound judgments as to general population risk are likely to be based on extrapolation from occupational health studies. Principles for such extrapolation should take into account the different criteria for susceptibility among the general population and among workers. See Chapter 7, Vol. II, 3rd ed., for a discussion of this point.

Reproductive effects are receiving increasing concern in occupational health *(196, 197)*, and its is reasonable that general community studies of possible reproductive effects should be undertaken.

Prevention of odor annoyance by using panels of residents in a type of monitoring system has been demonstrated to be practical *(198)*. There remain some important questions as to whether other health effects are not present at times when odorous materials are emitted, but support for such studies has not materialized *(199)*. Such unresolved issues may best be addressed by a systematic application of epidemiological monitoring *(10)*.

REFERENCES

1. B. G. Ferris, Jr., *J. Air Pollut. Control Assoc.* **35,** 482 (1978).
2. Royal College of Physicians, London, "Air Pollution and Health," Committee on Smoking and Atmospheric Pollution. Pitman Medical, London, 1970.
3. Council on Environmental Quality, "Environmental Quality 1980: The Eleventh Annual Report of the Council on Environmental Quality." U.S. Govt. Printing Office, Washington, D.C., 1981.
4. A. M. Freeman, III, "Air and Water Pollution Control: A Benefit-Cost Assessment." Wiley, New York, 1982.
5. S. M. Brown, M. M. Marmot, S. T. Sacks, and L. W. Kwok, *Nature (London)* **257,** 306 (1975).
6. Committee on Aldehydes, National Research Council, "Formaldehyde and Other Aldehydes." Natl. Acad. Press, Washington, D.C., 1981.

7. T. D. Sterling, *Int. J. Health Serv.* **13,** 277 (1983).
8. J. A. Cooper, "Environmental Quality 1980: The Eleventh Annual Report of the Council on Environmental Quality," p. 277. U.S. Govt. Printing Office, Washington, D.C., 1981.
9. M. M. Waldrup, *Science* **211,** 914 (1981).
10. J. R. Goldsmith, ed., "Epidemiological Monitoring for Environmental Health Hazards," Sci. Total Environ., Spec. Issue, Vol. 32, No. 3. Elsevier, Amsterdam, 1984.
11. Committee on Science and the Promotion of Human Welfare, *Science* **132,** 68 (1960).
12. American Association for the Advancement of Science, "Air Conservation," Commission on Air Conservation, AAAS, Washington, D.C., 1965.
13. W. Lowrance, "Of Acceptable Risk: Science and the Determination of Safety." W. Kaufman, Los Altos, California, 1976.
14. R. C. Schwing and W. A. Albertson, "Societal Risk Assessment." Plenum, New York, 1980.
15. "Risk Analysis." Society for Risk Analysis, McLean Virginia, 1981.
16. L. B. Lave, "Quantitative Risk Assessment in Regulation." Brookings Institution, Washington, D.C., 1982.
16a. R. Larsen, *J Air Pollut. Control Assoc.* **19,** 24 (1969).
16b. A. Stern, *in* 'Fundamentals of Air Pollution" (A. C. Stern, R. W. Boubel, D. B. Turner, and D. L. Fox, eds.), p. 59. Academic Press, New York, 1984.
17. Subcommittee on Airborne Particles, Committee on Medical and Biological Effects of Environmental Pollutants, National Research Council, "Airborne Particles." University Park Press, Baltimore, Maryland, 1979.
18. M. O. Amdur, *J. Air Pollut. Control Assoc.* **19,** 638 (1969).
19. M. O. Amdur, *J. Air Pollut. Control Assoc.* **19,** 644 (1969).
20. R. J. Charlson, A. H. Vanderpol, D. S. Covert, A. P. Waggoner, and N. C. Ahlquist, *Atmos. Environ.* **8,** 1257 (1974).
21. B. Bolin, *in* "The Impact on Environment of Sulfur in Air and Precipitation," Norstadt & Sons, Stockholm, 1972.
22. P. H. Abelson (Editorial), *Science* **221,** 115 (1983).
23. T. V. Larson, D. S. Covert, N. R. Frank, and R. J. Carlson, *Science* **197,** 161 (1977).
24. E. L. Avol, M. P. Jones, R. M. Bailey, M.-N. N. Chang, M. T. Kleinman, W. S. Linn, K. A. Bell, and J. D. Hackney, *Am. Rev. Respir. Dis.* **120,** 319 (1979).
25. M. A. Sackner, D. Ford, R. Fernandez, J. Cipley, D. Perez, M. Kwocka, M. Reinhard, E. D. Michelson, R. Schreck, and A. Wanner, *Am. Rev. Respir. Dis.* **118,** 497 (1978).
26. M. J. Utell, P. E. Morrow, and R. W. Hyde, *Ann. Occup. Hyg.* **26,** 691 (1982).
27. M. J. Utell, P. E. Morrow, D. M. Speers, J. Darling, and R. W. Hyde, *Am. Rev. Respir. Dis.* **128,** 444 (1983).
28. M. Lippman, R. B. Schlesinger, G. Leikauf, D. Spektor, and R. E. Albert, *Ann. Occup. Hyg.* **26,** 677 (1982).
29. J. Q. Koenig, W. E. Pierson, and M. Horike, *Am. Rev. Respir. Dis.* **128,** 221 (1983).
30. R. W. Coutant, *Environ. Sci. Technol.* **11,** 873 (1977).
31. F. C. Dohan, *Arch. Environ. Health* **3,** 387 (1961).
32. F. C. Dohan and E. W. Taylor, *Am. J. Med. Sci.* **240,** 337 (1960).
33. J. Ipsen, F. E. Ingenito, and M. Deane, *Arch. Environ. Health* **18,** 458 (1969).
34. J. Ipsen, M. Deane, and F. M. Ingenito, *Arch. Environ. Health* **18,** 462 (1969).
35. L. B. Lave and E. P. Seskin, "Air Pollution and Human Health." Resources for the Future, Baltimore, Maryland, 1977.

36. L. A. Thibodeau, R. B. Reed, Y. M. M. Bishop, and L. A. Kammerman, *Environ. Health Perspect.* **34,** 165 (1980).
37. J. G. French, G. Lowrimore, W. C. Nelson, J. F. Finklea, T. English, and M. Hertz, *Arch. Environ. Health* **27,** 129 (1973).
38. Subcommittee of the Committee on Science and Technology of the U.S. House of Representatives, "The Environmental Protection Agency's Research Program with Primary Emphasis on the Community Health and Environmental Surveillance System (CHESS): An Investigative Report," Fed. Regist., 12 April, 45FR21702. U.S. Govt. Printing Office, Washington, D.C., 1980.
38a. J. H. Ware, L. A. Thibodeau, F. E. Speizer, S. Colome, and B. G. Ferris, Jr., *Environ. Health Perspect.* **41,** 255 (1981).
39. S.-P. Lan and C. Shy, *Environ. Health Perspect.* **42,** 203 (1981).
40. J. R. Goldsmith, H. L. Griffith, R. Detels, S. Beeser, and L. Neumann, *Am. J. Epidemiol.* **118,** 759 (1983).
41. G. D. Thurston and J. D. Spengler, "Meterological and Pollutant Influences on Total and Respirable Sulfate Levels: A Multivariate Statistical Analysis," Proc. Annu. Meet. Air Pollut. Control Assoc., Pittsburgh, Pennsylvania, 1980.
42. World Health Organization, "Air Quality Criteria and Guides for Urban Air Pollutants," Tech. Rep. Ser. No. 506. WHO, Geneva, 1972.
43. J. R. Goldsmith and L. T. Friberg, *in* "Air Pollution" (A. Stern, ed.), Vol. II, 3rd ed., Chapter 7. Academic Press, New York, 1977.
44. World Health Organization, "WHO Environmental Health Criteria #8: Sulfur Oxides and Suspended Particulate Matter." WHO, Geneva, 1979.
45. J. A. Nadel and J. Widdicombe, *J. Appl. Physiol.* **17,** 861 (1962).
46. J. A. Nadel, H. Salem, B. Tamplin, and Y. Tokiwa, *Arch. Environ. Health* **10,** 175 (1965).
47. N. R. Frank, M. O. Amdur, J. Worcester, and J. L. Whittenberger, *J. Appl. Physiol.* **17,** 252 (1962).
48. P. J. Lawther, A. J. MacFarlane, R. E. Waller, and A. G. F. Brooks, *Environ. Res.* **10,** 355 (1975).
49. M. J. Jaeger, D. Tribble, and H. J. Wittig, *Lung* **156,** 119 (1979).
50. J. R. Goldsmith, *Am. J. Public Health* **57,** 1532 (1967).
51. J. Orehek, J. P. Massari, P. Gaylord, C. Grimaud, and J. Charpin, *J. Clin. Invest.* **57,** 301 (1976).
51a. J. Bahous, A. Cartier, G. Ouimet, L. Pineau, and J.-L. Malo, *Am. Rev. Respir. Dis.* **129,** 216 (1984).
51b. M. T. Kleinman, *J. Air Pollut. Control Assoc.* **34,** 32–37 (1984).
52. J. R. Goldsmith, *in* "Pulmonary Diseases and Disorders" (A. P. Fishman, ed.), Chapter 37. McGraw-Hill, New York, 1980.
53. D. Sheppard, W. S. Wong, C. F. Vehara, J. A. Nadel, and H. A. Boushey, *Am. Rev. Respir. Dis.* **122,** 873 (1980).
54. D. Sheppard, A. Saisho, J. A. Nadel, and H. A. Boushey, *Am. Rev. Respir. Dis.* **123,** 486 (1981).
55. J. Q. Koenig, W. E. Pierson, and N. R. Frank, *Environ. Res.* **25,** 340 (1981).
56. W. S. Linn, R. M. Bailey, D. A. Shamoo, T. G. Venet, L. H. Wightman, and J. D. Hackney, *Environ. Res.* **29,** 220 (1982).
57. M. B. Kirkpatrick, M. Sheppard, J. A. Nadel, and H. A. Boushey, *Am. Rev. Respir. Dis.* **125,** 627 (1982).
58. W. S. Linn, D. A. Shamoo, C. E. Spier, L. M. Valencia, V. T. Anzar, T. G. Venet, and J. D. Hackney, *Environ. Res.* **30,** 340 (1983).

59. R. A. Bethel, D. J. Erle, J. Epstein, D. Sheppard, J. A. Nadel, and H. A. Boushey, *Am. Rev. Respir. Dis.* **128,** 592 (1983).
60. J. Q. Koenig, W. E. Pierson, M. Harike, and N. R. Frank, *J. Toxicol. Environ. Health* **11,** 129 (1983).
61. H. Kehrl, L. J. Roger, M. Hazucha, and D. Horstman, *Am. Rev. Respir. Dis* **127,** 160 (Abstr.) (1983).
62. W. S. Linn, T. G. Venet, D. S. Shamoo, L. M. Valencia, U. T. Anzar, C. E. Spier, and J. D. Hackney, *Am. Rev. Respir. Dis.* **127,** 278 (1983).
63. A. J. Aquila, W. J. Hall, R. G. Douglas, Jr., and M. J. Utell, *Am. Rev. Respir. Dis.* **122,** 3 (1980).
64. M. J. Holtzman, J. H. Cunningham, J. R. Sheller, G. B. Irslager, J. A. Nadel, and H. A. Boushey, *Am. Rev. Respir. Dis.* **120,** 1059 (1979).
65. M. Hazucha and D. V. Bates, *Nature (London)* **257,** 50 (1975).
66. S. Mazumdar, H. Schimmel, and I. T. T. Higgins, *Arch. Environ. Health* **37,** 213 (1982).
67. G. U. Yule, *J. Royal Stat. Soc. Ser. A* **89,** 1 (1926).
68. R. W. Buechley, W. B. Riggan, W. Hasselblad, and J. B. Van Bruggen, *Arch. Environ. Health* **27,** 134 (1973).
69. R. W. Buechley, "SO_2 levels in 1967-1972 and Perturbations in Mortality," Contract No. ES-5-2101. Natl. Inst. Environ. Health Sci., Research Triangle Park, North Carolina, 1967.
70. R. E. Waller, *in* "Recent Advances in Community Medicine" (A. E. Bennett, ed.), Chapter 4, pp. 59–62. Churchill-Livingstone, Edinburgh and London, 1978.
71. R. E. Waller, *J. Royal Coll. Physicians, London* **5,** 362 (1971).
72. P. J. Lawther and R. E. Waller, *Environ. Health Perspect.* **22,** 71 (1978).
73. S. Chinn, C. du V. Florey, I. G. Baldwin, and M. Gorgol, *J. Epidemiol. Commun. Health* **35,** 174 (1981).
74. C. Fletcher, R. Peto, C. Tinker, and F. E. Speizer, "The Natural History of Chronic Bronchitis and Emphysema: An Eight Year Study of Early Chronic Obstructive Lung Disease in Working Men in London." Oxford Univ. Press, London and New York, 1976.
75. J. H. Angel, C. M. Fletcher, I. D. Hill *et al., Br. J. Dis. Chest* **59,** 66 (1965).
76. J. R. T. Colley and L. J. Brasser, "Chronic Respiratory Diseases in Children in Relation to Air Pollution." Regional Office for Europe of the World Health Organization, WHO, Copenhagen, 1980.
77. R. J. W. Melia, C. du V. Florey, and A. V. Swan, *J. Epidmiol. Commun. Health* **35,** 161 (1981).
78. R. J. W. Melia, C. du V. Florey, and S. Chinn, *J. Epidemiol. Commun. Health* **35,** 168 (1981).
79. J. R. Goldsmith and R. Toeplitz, *J. Epidemiol. Commun. Health* **37,** 165 (1983).
80. F. Boudik, J. Ulrich, F. Svec, J. Bobak, and J. R. Goldsmith, *in* "Ecology of Chronic Non-Specific Respiratory Diseases" (Z. Brzezinski, J. Kopczynski, and F. Sawicki, eds.), pp. 34–42. Panstwowy Zaklad Wydawnictw Lekarkskios, Warsaw, 1972.
81. A. Zapletal, J. Jech, T. Paul, and M. Samanek, *in* "Ecology of Chronic Non-Specific Respiratory Disease" (Z. Brzezinski, J. Kopczynski, and F. Sawicki, eds.), pp. 43–49. Panstwowy Zaklad Wydawnictw Lekarkskios, Warsaw, 1972.
82. F. Sawicki, *in* "Ecology of Chronic Non-Specific Respiratory Diseases" (Z. Brzezinski, J. Kopczynski, and F. Sawicki, eds.), pp. 3–13. Panstwowy Zaklad Wyndawnictw Lekarkskios, Warsaw, 1972.
83. PAARC (Group Cooperatif) (in French), Pollution Atmospherique et Affections

Respiratoires Chroniques ou a répétition. I. Méthodes et Sujets. II. Résultats et Discussion, *Bull. Eur. Physiopathol. Respir.* **18,** 87 (1982).
84. B. G. Ferris, Jr., F. E. Speizer, J. D. Spengler, D. Dockery, Y. M. M. Bishop, M. Wolfson, and C. Humble, *Am. Rev. Respir. Dis.* **120,** 767 (1979).
84a. B. G. Ferris, Jr., F. E. Speizer, Y. M. M. Bishop, J. D. Spengler, and J. H. Ware, *in* "Atmospheric Sulfur Deposition: Environmental Impact and Health Effects" (D. S. Shriner, C. R. Richmond, and S. E. Lindberg, eds.), Chapter 13. Ann Arbor Science (Butterworth), Woburn, Massachusetts, 1980.
85. D. W. Dockery and J. D. Spengler, *J. Air Pollut. Control Assoc* **31,** 153 (1981).
85a. D. W. Dockery and J. D. Spengler, *Atmos. Environ.* **15,** 22 (1981).
86. J. M. Samet, F. E. Speizer, Y. M. M. Bishop, J. D. Spengler, and B. G. Ferris, Jr., *J. Air Pollut. Control. Assoc.* **31,** 236 (1981).
87. J. H. Stebbings, K. E. Fogelman, K. E. McClain, and T. E. Townsend, *J. Air. Pollut. Control Assoc.* **26,** 545 (1976).
88. J. H. Stebbings and D. G. Fogelman, *Am. J. Epidemiol.* **110,** 27 (1979).
89. D. W. Dockery, J. H. Ware, B. G. Ferris, Jr., F. E. Speizer, N. R. Cook, and S. M. Herman, *J. Air Pollut. Control Assoc.* **32,** 937 (1982).
90. B. G. Ferris, Jr., F. E. Speizer, and J. H. Ware, *in* "Measurement of Risks" (G. G. Berg and H. D. Maillie, eds.), p. 211. Plenum, New York, 1981.
91. M. Saric, M. Fugas, and O. Hrustic, *Arch. Environ. Health* **36,** 101 (1982).
92. World Health Organization, "Environmental Health Criteria No. 7: Photochemical Oxidants." WHO, Geneva, 1979.
93. Committee on Medical and Biologic Effects of Environmental Pollutants, "Ozone and Other Photochemical Oxidants." Natl. Res. Counc.—Nat. Acad. Sci., Washington, D.C., 1977.
94. J. A. Golden, J. A. Nadel, and H. A. Boushey, *Am. Rev. Respir. Dis.* **118,** 287 (1978).
95. Y. Matsumura, *Am. Rev. Respir. Dis.* **102,** 438 (1970).
96. J. D. Davis, J. Gallo, E. P. C. Hu, R. C. Boucher, and P. A. Bromberg, *Am. Rev. Respir. Dis.* **121,** 231 (abstr.) (1980).
97. M. J. Holtzman, J. H. Cunningham, J. R. Sheller, G. B. Irsigler, J. A. Nadel, and H. A. Boushey, *Am. Rev. Respir. Dis.* **120,** 1059 (1979).
98. M. J. Holtzman, L. M. Fabri, P. M. O'Byrne, B. D. Gold, H. Aizawa, E. H. Walters, S. E. Alpert, and J. A. Nadel, *Am. Rev. Respir. Dis.* **126,** 686 (1983).
98a. H. A. Boushey and M. J. Holtzman, *Am. Rev. Respir. Dis.* **131,** 312–313 (1985).
99. H. E. Stokinger and L. D. Scheel, *Arch. Environ. Health* **4,** 327 (1962).
100. H. E. Stokinger and D. L. Coffin, *in* "Air Pollution" (A. C. Stern, ed.), Vol. II, 3rd ed., Chapter 6. Academic Press, New York, 1977.
101. S. M. Horvath, J. A. Gliner, and L. J. Folinsbee, *Am. Rev. Respir. Dis.* **123,** 496 (1981).
102. D. V. Bates, G. M. Bell, C. D. Burnham *et al., J. Appl. Physiol.* **32,** 176 (1972).
103. J. D. Hackney, W. S. Linn, R. D. Buckley, and J. J. Hislop, *Environ. Health Perspect.* **18,** 141 (1976).
104. W. S. Linn, D. A. Medway, U. T. Anzar, L. M. Valencia, C. E. Spier, F. S.-D. Tsao, D. A. Fischer, and J. D. Hackney, *Am. Rev. Respir. Dis.* **125,** 491 (1983).
105. P. A. Bromberg and M. J. Hazucha, *Am. Rev. Respir. Dis.* **125,** 489 (1983).
105a. E. L. Avol, W. S. Linn, D. A. Shamoo, L. M. Valencia, V. T. Anzar, T. G. Vengt, and J. D. Hackney. *Am. Rev. Respir. Dis.* **132,** 619–622 (1985).
106. C. E. Schoettlin and E. Landau, *Public Health Rep.* **76,** 545 (1961).
107. J. H. Stebbings and C. G. Hayes, *Environ. Res.* **11,** 89 (1976).

108. A. S. Whittemore and E. L. Korn, *Am. J. Public Health* **70,** 687 (1980).
109. R. L. Katz and R. Frezieres, *in* "Environmental Epidemiology: Community Studies of Environmental Health Problems" (J. R. Goldsmith, ed.), Chapter 10. CRC Press, Boca Raton, Florida, 1985.
109a. I. Goldstein and D. Hartel, *in* "Environmental Epidemiology: Community Studies of Environmental Health Problems" (J. R. Goldsmith, ed.), Chapter 9. CRC Press, Boca Raton, Florida, 1985.
110. J. Kagawa and T. Toyama, *Arch. Environ. Health* **30,** 117 (1975).
110a. M. Imai, K. Koshida, D. J. Kotchmar, and S. D. Lee, *J. Air Pollut. Control Assoc.* **35,** 103–108 (1985).
111. R. T. Zagraniski, B. P. Leaderer, and J. A. J. Stolwijk, *Environ. Res.* **19,** 306 (1979).
112. M. Lippmann, P. J. Lioy, G. Leikauf, K. B. Green, D. Baxter, M. Morandi, B. S. Pasternak, D. Fife, and F. E. Speizer, *Adv. Environ. Toxicol.* **5** (1983).
113. D. Reed, S. Glaser, and J. Kalder, *Am. J. Ind. Med.* **1,** 43 (1980).
114. R. Detels, S. N. Rokaw, A. H. Coulson, D. P. Tashkin, J. W. Sayre, and F. J. Massey, Jr., *Am. J. Epidemiol.* **109,** 33 (1979).
115. S. N. Rokaw, R. Detels, A. H. Coulson, J. W. Sayre, D. P. Tashkin, S. S. Alwright, and F. J. Massey, Jr., *Chest* **78,** 252 (1980).
116. H. Mencke, J. Casagrande, and B. E. Henderson, *Science* **183,** 210 (1974).
116a. U.S. Environmental Protection Agency "Draft Air Quality Criteria for Oxides of Nitrogen," EPA-450/5-82-002, Research Triangle Park, North Carolina (1982).
117. R. Ehrlich, *Bacteriol. Rev.* **30,** 604 (1966).
118. C. M. Shy, *Arch. Environ. Health* **27,** 124 (1973).
119. L. A. Wallace, *J. Air Pollut. Control Assoc.* **33,** 678 (1983).
120. J. R. Goldsmith and W. A. Aronow, *Environ. Res.* **10,** 236 (1975).
121. World Health Organization, "Environmental Health Criteria No. 13: Carbon Monoxide." WHO, Geneva, 1979.
121a. Workshop on Carbon Monoxide and Cardiovascular Disease, *Prev. Med.* **8,** 261 (1979).
122. C. Collier and J. R. Goldsmith, *Atmos. Environ.* **17,** 723 (1983).
122a. K. R. Mahaffey, J. L. Annest, J. Roberts, and R. S. Murphy, *N. Engl. J. Med.* **307,** 573 (1982).
122b. M. J. Duggan, *Environ. Health Perspect.* **50,** 371 (1983).
123. J. S. Lin-Fu, *N. Engl. J. Med.* **307,** 615 (1982).
124. Royal Commission on Environmental Pollution, "Ninth Report: Lead in the Environment." HMSO Cmnd 8852. H. M. Stationery Office, London, 1983.
125. Y. Yaffe, C. P. Flessel, J. J. Wesolowski, A. Del Rosario, G. H. Guirguis, V. Matias, J. W. Gramlich, W. R. Kelly, T. E. Degarmo, and G. C. Coleman, *Arch. Environ. Health* **38,** 237 (1983).
126. M. Rabinowitz, G. Wetherill, and J. Kopple, *J. Lab. Clin. Med.* **90,** 238 (1977).
127. I. S. Fachetti and F. Geiss, "Isotopic Lead Experiment: Status Report" (EUR 8352 EN). Commission of the European Communities, Brussels, 1982 [cited by P. C. Elwood, *Lancet* **1,** 869 (1983)].
128. I. Billick, *Environ. Health Perspect.* **42,** 73 (1982).
128a. V. N. Houck, Testimony before Subcommittee on Environment, Energy, and Natural Resources, U.S. House of Representatives, 14 April, 1982.
129. M. Rabinowitz and H. L. Needleman, *Lancet* **1,** 63 (1983).
130. A. C. Chamberlain, *Atmos. Environ.* **17,** 677 (1983).
131. J. R. Goldsmith, *Atmos. Environ.* **17,** 2365 (1983).
132. R. Russel-Jones, *Atmos. Environ.* **17,** 2367 (1983).
133. S. Piomelli, *Proc. Natl. Acad. Sci. U.S.A.* **79,** 3335 (1982).

134. J. C. Ruddock, *JAMA, J. Am. Med. Assoc.* **82,** 1682 (1924) (cited by H. L. Needleman, *in* "Lead Toxicity" (R. L. Singhal and J. A. Thomas, eds.), p. 17. Urban & Schwartzenburg, Baltimore, Maryland 1980).
135. B. K. Byers and E. Lord, *Am. J. Dis. Child.* **66,** 471 (1943) (cited by H. L. Needleman, *in* "Lead Toxicity" (R. L. Singhal and J. A. Thomas, eds.), p. 1–19. Urban & Schwartzenburg, Baltimore, Maryland 1980).
136. J. J. Chisholm and F. Kaplan, *J. Pediatr.* **73,** 942 (1968).
137. O. David, S. P. Hoffman, J. Verd *et al., J. Abnorm. Child Psychol.* **5,** 405 (1977).
138. M. Rutter, *Dev. Med Child Neurol.* **22,** Suppl. 1 (1980).
139. H. L. Needleman, C. Gunnoe, A. Leviton, R. Reed, H. Peresie, C. Maher, and P. Barrett, *N. Engl. J. Med.* **300,** 689 (1979).
140. J. M. Ratcliffe, *Brt. J. Prev. Soc. Med.* **31,** 258 (1977).
141. W. Yule, R. Landsdown, I. G. Millar, and C. Z. Urbanowitz, *Dev. Med. Child Neurol.* **23,** 567 (1981).
142. M. Smith, H. T. Delves, R. Landsdown, B. Clayton, and P. Graham, *Dev. Med. Child Neurol.* **25,** Suppl. 47 (1983).
143. G. Winnecke, K. G. Hrdina, and A. Brockhaus, *Int. Arch. Occup. Environ. Health* **51,** 169 (1983).
144. G. Winnecke, U. Kramer, A. Brockhaus, U. Ewers, G. Kujanek, H. Lechner, and W. Janke, *Int. Arch. Occup. Environ. Health* **51,** 231 (1983).
145. Subcommittee on Health and the Environment, Committee on Energy and Commerce, "Lead Poisoning in Children," Dec. 2, 1982 (Serial 97-184). U.S. House of Representatives, Washington, D.C., 1982.
146. Subcommittee on Environment, Energy and Natural Resources, "Public Health Aspects of Lead in Gasoline," 14 April 1982. U.S. House of Representatives, Washington, D.C., 1982.
147. British Medical Journal (Editorial), *Br. Med. J.* **284,** 529 (1982).
148. H. L. Needleman, P. C. Elwood, J. Gallagher, C. Toothill, and R. Russel-Jones (Corresp.), *Br. Med. J.* **284,** 1189 (1982).
149. Lancet (Editorial), *Lancet* **1,** 1337 (1982).
149a. Lancet (Editorial), *Lancet* **1,** 912 (1983).
150. R. Russel-Jones (Corresp.), *Lancet* **1,** 1042 (1983).
151. E. Marshal (news item), *Science* **222,** 906 (1983).
152. L. Friberg and V Vahter, *Lancet* **2,** 220 (1983).
153. R. A. Fisher, "Statistical Methods for Research Workers," 6th ed., p. 104. Oliver & Boyd London, 1936.
154. J. R. Goldsmith and S. Beeser, *Ann. Singapore Acad. Med., Suppl.* **3,** 294 (1984).
155. J. L. Annest, J. L. Pirkle, D. Makuc, J. W. Neese, D. D. Bayse, and M. G. Kovar, *N. Engl. J. Med.* **308,** 1372 (1983).
155a. E. Charney, B. Kessler, M. Farfel, and D. Jackson, *N. Engl. J. Med.* **309,** 1089 (1983).
155b. J. L. Pirkle, J. Schwartz, J. R. Landis, and W. R. Harlon, *Am. J. Epidermiol.* **121,** 246–258 (1985).
155c. T. Moreau, G. Orssaud, B. Juguet, and G. Busquet, *Rev. Epidem. et Sante Publ.* **30,** 395–397 (1982) (in French).
155d. G. Orssaud, J. R. Claude, T. Moreau, J. Lellouch, B. Juguet, and B. Fest, *Brit. Med. J.* **290,** 244 (1985).
156. Committee on Indoor Air Pollutants, National Research Council, "Indoor Pollutants." Natl. Acad. Press, Washington, D.C., 1981.
157. J. E. Yocom, *J. Air Pollut. Control. Assoc.* **32,** 500 (1982).
158. J. D. Spengler and K. Sexton, *Science* **221,** 9 (1983).

159. W. Aroem, E. Light, H. J. Roberts, J. D. Spengler, and K. Sexton (Corresp), *Science* **223,** 6 (1984).
160. J. D. Spengler, B. G. Ferris, Jr., and D. W. Dockery, *Environ. Sci. Technol.* **13,** 1279 (1979).
161. K. Sexton, R. Letz, and J. D. Spengler, *Environ. Res.* **32,** 151 (1983).
162. R. J. W. Melia, C. du V. Florey, D. G. Altman, and A. V. Swan, *Br. Med. J.* **2,** 149 (1977).
163. M. D. Keller, R. R. Lanese, R. I. Mitchell, and R. W. Cote, *Environ. Res.* **19,** 495 (1979).
164. F. E. Speizer, B. G. Ferris, Jr., Y. M. M. Bishop, and J. D. Spengler, *Am. Rev. Respir. Dis.* **121,** 3 (1980).
165. R. J. W. Melia, C. du V. Florey, and S. Chinn, *Int. J. Epidemiol.* **8,** 333 (1979).
166. B. D. Goldstein, R. J. W. Melia, S. Chinn, C. du V. Florey, D. Clark, and H. H. John, *Int. J Epidemiol.* **8,** 339 (1979).
167. C. du V. Florey, R. J. W. Melia, S. Chinn, B. D. Goldstein, A. G. F. Brooks, H. H. John, I. B. Craighead, and X. Webster, *Int. J. Epidemiol.* **8,** 346 (1979).
167a. J. H. Ware, D. W. Dockery, A. Spiro, III, F. E. Speizer, and B. G. Ferris, Jr., *Am. Rev. Respir. Dis.* **129,** 366 (1984).
168. H. R. Hosein and A. Bouhuys, *Br. Med. J.* **1,** 125 (1979).
168a. R. van der Lende, *et al., Bull. Eur. Physiopath. Respir.* **17,** 399–410 (1981).
168b. B. Remijn, P. Fischer, B. Brunekreef, *et al., Int. J. Epidemiol.* **14,** 215–220 (1985).
168c. P. Fischer, B. Remijn, B. Brunekreef, R. van der Lende, *et al., Int. J. Epidemiol.* **14,** 221–226 (1985).
168d. B. Brunekreef, P. Fischer, B. Remijn, R. van der Lende, *et al., Int. J. Epidemiol.* **14,** 227–230 (1985).
169. U.S. Department of Health, Education and Welfare, *in* "Smoking and Health," Report of the Surgeon General, DHEW Publ. No. (PHS) 79-50066, Chapter 11. U.S. Govt. Printing Office, Washington, D.C., 1979.
170. S. T. Weiss, I. B. Tager, M. Schenker, and F. E. Speizer, *Am. Rev. Respir. Dis.* **128,** 933 (1983).
171. J. D. Spengler, K. R. Stone, and F. W. Lilley, *J. Air Pollut. Control Assoc.* **28,** 776 (1978).
172. S. Harlap and A. M. Davies, *Lancet* **1,** 529 (1974).
173. J. R. T. Colley, W. W. Holland, and R. T. Corkill, *Lancet* **2,** 1031 (1974).
174. D. M. Fergusson, L. J. Horwood, F. T. Shannon, and B. Taylor, *J. Epidemiol. Commun. Health* **35,** 180 (1981).
175. C. Burchfiel, W. Howatt, J. Keller, W. Butler, I. Higgins, and M. Higgins, *Am. Rev. Respir. Dis.* **127,** 156 (abstr.) (1983).
176. A. Goren, J. R. Goldsmith, and R. Toeplitz, *Am. Rev. Respir. Dis.* **127,** 181 (abstr.) (1983).
176a. F. Kauffmann, J.-F. Tessier, and P. Oriol, *Am. J. Epidemiol.* **117,** 269 (1983).
177. N. A. Ashford, C. W. Ryan, and C. C. Caldart, *Science* **222,** 894 (1983).
178. National Research Council—National Academy of Sciences, "Vapor Phase Organic Pollutants from Hydrocarbons." Natl. Acad. Press, Washington, D.C., 1975.
178a. J. J. Wesolowski, *J. Environ. Health* **46,** 311 (1984).
179. D. Main and T. Hogan, *Scand. J. Work Environ. Health* **9,** 60 (abstr.) (1983).
180. P. E. Enterline, *J. Air Pollut. Control Assoc.* **33,** 318 (1983).
181. J. R. Goldsmith, *Am. J. Ind. Med.* **3,** 341 (1982).
181a. W. J. Nicholson, G. Perkel, and I. J. Selikoff, *Am. J. Ind. Med.* **3,** 259 (1982).

181b. H. Anderson, R. Lilis, S. Daum, A. Fischbein, and I. J. Selikoff, *in* "Dust and Disease" (R. Lemen and J. M. Dement, eds.), pp. 145–156. Pathotox Publisher, Park Forest South, Illinois, 1979.
182. J. Fraumeni, Jr., "Persons at High Risk of Cancer." Academic Press, New York, 1975.
183. J. R. Goldsmith, *J. Environ. Pathol. Toxicol.* **3,** 205 (1980).
183a. G. B. Weinberg, L. H. Kuller, and C. K. Redmond, *Am. J. Epidemiol.* **115,** 40–58 (1982).
184. Atmospheric Pollution and Lung Cancer, *Environ. Health Perspect.* **22,** 1–126 (1978).
185. L. Friberg and R. Cederlof, *Environ. Health Perspect.* **22,** 45–66 (1978).
186. R. Cederlof, R. Doll, B. Fowler, L. Friberg, N. Nelson, and V. Vouk, *Environ. Health Perspect.* **22,** 9 (1978).
187. E. C. Hammond, L. Garfinkel, citing H. Demopoulos, M. Eisenbud, J. Fraumeni, G. T. O'Connor, P. Greenwald, J. Higginson, and H. F. Kraybill, *Prev. Med* **9,** 206 (1980).
188. Proceedings of the Symposium on Biological Tests in the Evaluation of Mutagenicity and Carcinogenicity of Air Pollutants with Special Reference to Motor Exhausts and Coal Combustion Products, *Environ. Health Perspect.* **47,** 1–344 (1983).
189. National Research Council "Health Effects of Exposures to Diesel Exhaust." Natl. Acad. Press, Washington, D.C., 1981.
190. T. Sugimura and S. Takayama, *Environ. Health Perspect.* **47,** 171 (1983).
191. D. Anderson, *Environ. Health Perspect.* **47,** 239 (1983).
192. I. Nordenson, A. Swains, E. Dahlgren, and L. Beckman, *Scand. J. Work Environ. Health* **7,** 14 (1981) [cited by J. Rantanen, *Environ. Health Perspect.* **47,** 325 (1983)].
193. D. Magie, "Report on Chromosomal Aberrations in Peripheral Lymphocytes of College Students as a Response to Photochemical Air Pollution" (EPA 600/1-81-007). U.S. Environ. Prot. Agency, Washington, D.C., 1981.
193a. Board of Toxicology and Environmental Health Hazards, National Research Council, "Asbestiform Fibers: Non-occupational Health Risks," National Academy Press, Washington, D.C., 1984.
194. F. E. Speizer, *Environ. Health Perspect.* **47,** 33 (1983).
195. J. R. Goldsmith, *Scand. J. Work Environ. Health* **8,** Suppl. 1, 83 (1982).
196. "Reproductive Toxicology," *Am. J. Ind. Med., Spec. Issue* **4,** 1–396 (1983).
197. K. Hemminki, T. Partanen, H. Vaino, and E. Hemminki, *in* "Occupational Cancer and Carcinogenesis" (H. Vainio, M. Sorsa, and K. Hemminki, eds.). Hemisphere Publ. Co., Washington, D.C., 1979.
198. J. R. Goldsmith, "Health and Annoyance Impact of Odor Pollution," Final Report, EPA Contract 68-02-0083. U.S. Environ. Prot. Agency, Washington, D.C., 1973.
199. Board of Toxicology and Environmental Health Hazards, National Research Council, "Odor Detection and Effects." Natl. Acad. Press, Washington, D.C., 1979.

List of books relevant to this chapter published since the lists in Chapter 8, Volume II, 3d edition (1977) can be found in Chapter 1, Table IX, Volume VII.

Index